AutoCAD and Its Applications
ADVANCED

2012

by

Terence M. Shumaker
Faculty Emeritus
Former Chairperson
Drafting Technology
Autodesk Premier Training Center
Clackamas Community College
Oregon City, OR

David A. Madsen
President
Madsen Designs Inc.
Faculty Emeritus
Former Chairperson
Drafting Technology
Autodesk Premier Training Center
Clackamas Community College
Oregon City, OR
Director Emeritus
American Design Drafting Association

Jeffrey A. Laurich
Instructor, Mechanical Design
Technology
Fox Valley Technical College
Appleton, WI

J.C. Malitzke
Former Department Chair
Computer Integrated Technologies
Former Manager and Instructor
Authorized Autodesk Training Center
Moraine Valley Community College
Palos Hills, IL

Craig P. Black
Instructor, Mechanical Design
Technology
Former Manager
Autodesk Premier Training Center
Fox Valley Technical College
Appleton, WI

Adam M. Ferris
Director of Digital Technology
Wilson Architects
Boston, MA

Publisher
The Goodheart-Willcox Company, Inc.
Tinley Park, Illinois
www.g-w.com

Library of Congress Catalog Card Number 2011009866

ISBN 978-1-60525-563-7

1 2 3 4 5 6 7 8 9 – 12 – 16 15 14 13 12 11

The Goodheart-Willcox Company, Inc. Brand Disclaimer: Brand names, company names, and illustrations for products and services included in this text are provided for educational purposes only and do not represent or imply endorsement or recommendation by the author or the publisher.

The Goodheart-Willcox Company, Inc. Safety Notice: The reader is expressly advised to carefully read, understand, and apply all safety precautions and warnings described in this book or that might also be indicated in undertaking the activities and exercises described herein to minimize risk of personal injury or injury to others. Common sense and good judgment should also be exercised and applied to help avoid all potential hazards. The reader should always refer to the appropriate manufacturer's technical information, directions, and recommendations; then proceed with care to follow specific equipment operating instructions. The reader should understand these notices and cautions are not exhaustive.

The publisher makes no warranty or representation whatsoever, either expressed or implied, including but not limited to equipment, procedures, and applications described or referred to herein, their quality, performance, merchantability, or fitness for a particular purpose. The publisher assumes no responsibility for any changes, errors, or omissions in this book. The publisher specifically disclaims any liability whatsoever, including any direct, indirect, incidental, consequential, special, or exemplary damages resulting, in whole or in part, from the reader's use or reliance upon the information, instructions, procedures, warnings, cautions, applications, or other matter contained in this book. The publisher assumes no responsibility for the activities of the reader.

The Goodheart-Willcox Company, Inc. Internet Disclaimer: The Internet resources and listings in this Goodheart-Willcox Publisher product are provided solely as a convenience to you. These resources and listings were reviewed at the time of publication to provide you with accurate, safe, and appropriate information. Goodheart-Willcox Publisher has no control over the referenced Web sites and, due to the dynamic nature of the Internet, is not responsible or liable for the content, products, or performance of links to other Web sites or resources. Goodheart-Willcox Publisher makes no representation, either expressed or implied, regarding the content of these Web sites, and such references do not constitute an endorsement or recommendation of the information or content presented. It is your responsibility to take all protective measures to guard against inappropriate content, viruses, or other destructive elements.

Library of Congress Cataloging-in-Publication Data

Shumaker, Terence M.
    AutoCAD and its applications. Advanced 2012 / by
Terence M. Shumaker ... [et al.]. 19th ed.
        p. cm.
    Includes index.
    ISBN 978-1-60525-563-7
    1. Computer graphics. 2. AutoCAD. I. Title.

T385.S46123 2012
    620′.00420285536--dc22   2011009866

Introduction

AutoCAD and Its Applications—Advanced provides complete instruction in mastering three-dimensional design and modeling using AutoCAD. This text also provides complete instruction in customizing AutoCAD and introduces programming AutoCAD. These topics are covered in an easy-to-understand sequence and progress in a way that allows you to become comfortable with the commands as your knowledge builds from one chapter to the next. In addition, *AutoCAD and Its Applications—Advanced* offers:

- Examples and discussions of industrial practices and standards.
- Professional tips explaining how to effectively and efficiently use AutoCAD.
- Exercises to reinforce the chapter topics. These exercises should be completed where indicated in the text as they build on previously learned material.
- Review questions at the end of each chapter for testing knowledge of commands and key AutoCAD concepts.
- A large selection of modeling and customizing problems supplement each chapter. Problems are presented as 3D illustrations, actual plotted drawings, and engineering sketches.

Fonts Used in This Text

Different typefaces are used throughout each chapter to define terms and identify AutoCAD commands. Important terms appear in ***bold-italic face, serif*** type. AutoCAD menus, commands, variables, dialog box names, and toolbar button names are printed in **bold-face, sans serif** type. File names, folder names, and paths appear in the body of the text in Roman, sans serif type. Keyboard keys are shown inside of square brackets [ ] and appear in Roman, sans serif type. For example, [Enter] means to press the enter (return) key.

Other Text References

This text focuses on advanced AutoCAD applications. Basic AutoCAD applications are covered in *AutoCAD and Its Applications—Basics*, which is available from Goodheart-Willcox Publisher. *AutoCAD and Its Applications* texts are also available for previous releases of AutoCAD.

Introducing the AutoCAD Commands

There are several ways to select AutoCAD drawing and editing commands. Selecting commands from the ribbon is slightly different than entering them from the keyboard. When a command is introduced, the command-entry methods are illustrated in the margin next to the text reference.

The example in the margin next to this paragraph illustrates the various methods of initiating the **CONE** command to draw a solid cone primitive while the 3D environment is active. The 3D environment consists of a drawing file based on the acad3D.dwt template and the 3D Modeling workspace current, as described in Chapter 1. This book assumes the 3D environment is current for all procedures and discussions.

Flexibility in Design

Flexibility is the keyword when using *AutoCAD and Its Applications—Advanced*. This text is an excellent training aid for both individual and classroom instruction. It is also an invaluable resource for any professional using AutoCAD. *AutoCAD and Its Applications—Advanced* teaches you how to apply AutoCAD to common modeling and customizing tasks.

When working through the text, you will see a variety of notices. These include Professional Tips, Notes, and Cautions that help you develop your AutoCAD skills.

PROFESSIONAL TIP

These ideas and suggestions are aimed at increasing your productivity and enhancing your use of AutoCAD commands and techniques.

A note alerts you to important aspects of a command, function, or activity that is being discussed. These aspects should be kept in mind while you are working through the text.

CAUTION

A caution alerts you to potential problems if instructions or commands are incorrectly used or if an action can corrupt or alter files, folders, or storage media. If you are in doubt after reading a caution, always consult your instructor or supervisor.

AutoCAD and Its Applications—Advanced provides several ways for you to evaluate your performance. Included are:

- **Exercises.** The companion website contains exercises for each chapter. These exercises allow you to perform tasks that reinforce the material just presented. You can work through the exercises at your own pace. However, the exercises are intended to be completed when called out in the text.

- **Chapter reviews.** Each chapter includes review questions at the end of the chapter. Questions require you to give the proper definition, command, option, or response to perform a certain task. You may also be asked to explain a topic or list appropriate procedures. An electronic version of the chapter review for each chapter is available on the companion website.
- **Drawing problems.** There are a variety of drawing, design, and customizing problems at the ends of chapters. These are presented as real-world CAD drawings, 3D illustrations, and engineering sketches. The problems are designed to make you think, solve problems, use design techniques, research and use proper drawing standards, and correct errors in the drawings or engineering sketches. Graphics are used to represent the discipline to which a drawing problem applies.

 These problems address mechanical drafting and design applications, such as manufactured part designs.

 These problems address architectural and structural drafting and design applications, such as floor plans, furniture, and presentation drawings.

 These problems address piping drafting and design applications, such as tank drawings and pipe layout.

 These problems address a variety of general drafting, design, and customization applications. These problems should be attempted by everyone learning advanced AutoCAD techniques for the first time.

 Some problems presented in this text are given as engineering sketches. These sketches are intended to represent the kind of material from which a drafter is expected to work in a real-world situation. As such, engineering sketches often contain errors or slight inaccuracies and are most often not drawn according to proper drafting conventions and applicable standards. Additionally, other drawings may contain errors or inaccuracies. Errors in these problems are intentional to encourage you to apply appropriate techniques and standards in order to solve the problem. As in real-world applications, sketches should be considered preliminary layouts. Always question inaccuracies in sketches and designs and consult the applicable standards or other resources.

Companion Website

The companion website is located at www.g-wlearning.com/CAD. Select the entry for *AutoCAD and Its Applications—Advanced 2012* to access the material for this book. The companion website contains the exercises and chapter review questions for each chapter. Additionally, Chapters 27 through 30 are provided on the companion website, not in the printed book. The appendix material is also presented on the companion website. The icon shown in the margin here appears throughout the text to indicate a reference to the companion website.

As you work through each chapter, exercises on the companion website are referenced. The exercises are intended to be completed as the references are encountered in the text. The solid modeling tutorial in Appendix A should be completed after Chapter 15. The surface modeling tutorial in Appendix B should be completed after Chapter 11. The remaining appendix material is intended as reference material.

Also included on the companion website are the exercise activities. The exercise activities are intended to supplement the exercises on the companion website. These activities are referenced within the appropriate exercises and can be completed as additional practice.

About the Authors

Terence M. Shumaker is Faculty Emeritus, the former Chairperson of the Drafting Technology Department, and former Director of the Autodesk Premier Training Center at Clackamas Community College in Oregon City, OR. Terence taught at the community college level for over 28 years. He has professional experience in surveying, civil drafting, industrial piping, and technical illustration. He is the author of Goodheart-Willcox's *Process Pipe Drafting* and coauthor of the *AutoCAD and Its Applications* series (Releases 10 through 2011 editions) and *AutoCAD Essentials*.

David A. Madsen is the president of Madsen Designs Inc. (www.madsendesigns.com). David is Faculty Emeritus and the former Chairperson of Drafting Technology and the Autodesk Premier Training Center at Clackamas Community College in Oregon City, OR. David was an instructor and a department chairperson at Clackamas Community College for nearly 30 years. In addition to teaching at the community college level, David was a Drafting Technology instructor at Centennial High School in Gresham, OR. David is a former member of the American Design Drafting Association (ADDA) Board of Directors. He was honored with Director Emeritus status by the ADDA in 2005. David has extensive experience in mechanical drafting, architectural design and drafting, and building construction. He holds a Master of Education degree in Vocational Administration and a Bachelor of Science degree in Industrial Education. David is coauthor of the *AutoCAD and Its Applications* series (Releases 10 through 2011 editions), *Architectural Drafting Using AutoCAD, Geometric Dimensioning and Tolerancing*, and other drafting and design textbooks.

Jeffrey A. Laurich has been an instructor in Mechanical Design Technology at Fox Valley Technical College in Appleton, WI, since 1991. He has also taught business and industry professionals in the Autodesk Premier Training Center at FVTC. Jeff teaches drafting, AutoCAD, Design of Tooling, GD&T, and 3ds Max. He created a certificate program at FVTC entitled Computer Rendering and Animation that integrates the 3D capabilities of AutoCAD and 3ds Max. Jeff has professional experience in furniture design, surveying, and cartography and holds a degree in Natural Resources Technology. In his consulting business, Jeff uses 3ds Max to create renderings and animations for manufacturers and architects.

J.C. Malitzke is the former department chair of Computer Integrated Technologies at Moraine Valley Community College in Palos Hills, IL. J.C. was a professor of Mechanical Design and Drafting/CAD and taught for the Authorized Autodesk Training Center at Moraine Valley. J.C. has been teaching for 35 years and actively using and teaching Autodesk products for nearly 26 years. He is a founding member and past chair of the Autodesk Training Center Executive Committee and the Autodesk Leadership Council. J.C. has been the co-author and principal investigator on two National Science Foundation grants. He has won numerous awards, including: Educator of the Year by the Illinois Drafting Educators Association; the Instructor

Quality Award by Autodesk; Autodesk University Instructor Award; Professor of the Year, Co-Innovator of the Year, and Co-Master Teacher awards at Moraine Valley Community College; and the Illinois Outstanding Faculty Member of the Year awarded by the Illinois Community College Trustees Association. J.C. was also one of the recipients of the Top 10 Most Popular Autodesk University Classes Ever for his class Compelling 3D Features in AutoCAD. J.C. holds a Bachelor's degree in Education and a Master's degree in Industrial Technology from Illinois State University.

Craig P. Black is an instructor in the Mechanical Design Technology department and the former manager of the Autodesk Premier Training Center at Fox Valley Technical College in Appleton, WI. He has been teaching at Fox Valley Technical College since 1990. In 2009, Craig created the CAD Management certificate program at FVTC. Craig has served two terms on the Autodesk Training Center Executive Committee (now known as the Autodesk Leadership Council) and chaired the committee in 2001. He has been working with Autodesk software products for more than 25 years. He has presented on various topics at a number of Autodesk University annual training sessions and has been contracted to teach training sessions on Autodesk products across the United States. In addition to teaching, Craig also does AutoCAD customization and AutoLISP and DCL programming for area businesses and industries. Prior to his current position, Craig worked in the civil, architectural, electrical, and mechanical drafting and design disciplines.

Adam M. Ferris is the Director of Digital Technology at Wilson Architects in Boston, MA. Wilson Architects is a medium-size architecture firm specializing in the science and technology field. Adam is currently working on and defining new procedures on BIM efficiency and delivery methods with his clients. Adam is an award-winning Autodesk Certified Instructor with over 17 years of experience on training Autodesk products in the Boston area. He has trained over 4000 students in New England. Prior to joining Wilson Architects, Adam owned an Autodesk products/CAD consulting firm aiding businesses in efficient AutoCAD configurations and management, network management, and 3D visualization with Autodesk 3ds Max. Adam attends Autodesk University annually and has lectured there several times.

Acknowledgments

The authors and publisher would like to thank the following individuals and companies for their assistance and contributions.

Autodesk, Inc.
Bill Fane
CADENCE magazine
EPCM Services Ltd.
Fitzgerald, Hagan, & Hackathorn
Kunz Associates

ADDA Technical Publication

The content of this text is considered a fundamental component to the design drafting profession by the American Design Drafting Association. This publication covers topics and related material relevant to the delivery of the design drafting process. Although this publication is not conclusive, it should be considered a key reference tool in furthering the knowledge, abilities, and skills of a properly trained designer or drafter in the pursuit of a professional career.

Brief Contents

Three-Dimensional Design and Modeling

Model Visualization and Presentation

Customizing AutoCAD

Programming AutoCAD

Image Management and Drawing Distribution

Expanded Contents

Three-Dimensional Design and Modeling

Model Visualization and Presentation

Customizing AutoCAD

Programming AutoCAD

Image Management and Drawing Distribution

www.g-wlearning.com/CAD

Companion Website Contents

www.g-wlearning.com/CAD

Chapters 27 through 30

Chapter Exercises

Chapter Reviews

Chapter Drawing Files (DWG Files)
   Exercise Files
   Exercise Activity Files

Appendices
   Appendix A Solid Modeling Tutorial
   Appendix B Surface Modeling Tutorial
   Appendix C Common File Extensions
   Appendix D AutoCAD Command Aliases
   Appendix E Advanced Application Commands
   Appendix F Advanced Application System Variables
   Appendix G Basic AutoLISP Commands
   Appendix H Toolbar Flyouts and Marking Menus

Reference Materials
   Drafting Symbols
   Drawing Sheet Sizes, Settings, and Scale Parameters
   Standards and Related Documents
   Project and Drawing Problem Planning Sheet
   Standard Tables

Related Web Links

Chapter 1

Introduction to Three-Dimensional Modeling

Learning Objectives

After completing this chapter, you will be able to:

✓ Describe how to locate points in 3D space.
✓ Describe the right-hand rule of 3D visualization.
✓ Explain the function of the ribbon.
✓ Identify the functions of the viewport controls and the view cube.
✓ Display 3D objects from preset isometric viewpoints.
✓ Display 3D objects from any desired viewpoint.
✓ Set a visual style current.

Three-dimensional (3D) design and modeling is a powerful tool for use in design, visualization, testing, analysis, manufacturing, assembly, and marketing. Three-dimensional models also form the basis of computer animations, architectural walkthroughs, and virtual worlds used in the entertainment industry and for gaming platforms. Drafters who can design objects, buildings, and "worlds" in 3D are in demand for a wide variety of positions, both inside and outside of the traditional drafting and design disciplines.

The first 15 chapters of this book present a variety of techniques for drawing and designing in 3D. The skills you learn will provide you with the ability to construct any object in 3D and prepare you for entry into an exciting aspect of graphic communication.

To be effective in creating and using 3D objects, you must first have good 3D visualization skills. These skills include the ability to see an object in three dimensions and to visualize it rotating in space. Visualization skills can be obtained by using 3D techniques to construct objects and by trying to see two-dimensional sketches and drawings as 3D models. This chapter provides an introduction to several aspects of 3D drawing and visualization. Subsequent chapters expand on these aspects and provide a detailed examination of 3D drawing, editing, visualization, and display techniques.

Using Rectangular 3D Coordinates

In two-dimensional drawing, you see one plane defined by two dimensions. These dimensions are usually located on the X and Y axes and what you see is the XY plane. However, in 3D drawing, another coordinate axis—the Z axis—is added. This results in two additional planes—the XZ plane and the YZ plane. If you are looking at a standard AutoCAD screen after AutoCAD is launched using the acad.dwt template, the positive Z axis comes directly out of the screen toward you. AutoCAD can only draw lines in 3D if it knows the X, Y, and Z coordinate values of each point on the object. For 2D drawing, only two of the three coordinates (X and Y) are needed.

Compare the 2D and 3D coordinate systems shown in **Figure 1-1**. Notice that the positive values of Z in the 3D coordinate system come up from the XY plane. In a new drawing based on the acad.dwt template, consider the surface of your computer screen as the XY plane. Anything behind the screen is negative Z and anything in front of the screen is positive Z.

The object in **Figure 1-2A** is a 2D drawing showing the top view of an object. The XY coordinate values of the origin and each point are shown. Think of the object as being drawn directly on the surface of your computer screen. However, this is actually a 3D object. When displayed in a pictorial view, the Z coordinates can be seen. Notice in **Figure 1-2B** that the first two values of each coordinate match the X and Y values of the 2D view. Three-dimensional coordinates are always expressed as (X,Y,Z). The 3D object was drawn using positive Z coordinates. Therefore, the object

Figure 1-1.
A comparison of 2D and 3D coordinate systems.

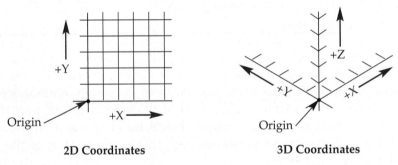

2D Coordinates   3D Coordinates

Figure 1-2.
A—The points making up a 2D object require only two coordinates. B—Each point of a 3D object must have an X, Y, and Z value. Notice that the first two coordinates (X and Y) are the same for each endpoint of a vertical line.

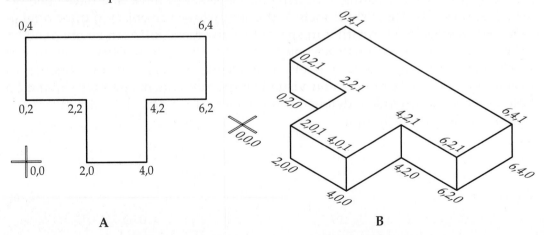

A   B

comes out of your computer screen when it is viewed from directly above. The object can also be drawn using negative Z coordinates. In this case, the object would extend behind, or into, the screen.

Study the nature of the rectangular 3D coordinate system. Be sure you understand Z values before you begin constructing 3D objects. It is especially important that you carefully visualize and plan your design when working with 3D constructions.

PROFESSIONAL TIP

All points in three-dimensional space can be drawn using one of three coordinate entry methods—rectangular, spherical, or cylindrical. This chapter uses the rectangular coordinate entry method. Complete discussions on the spherical and cylindrical coordinate entry methods are provided in Chapter 6.

Exercise 1-1

Complete the exercise on the companion website.
www.g-wlearning.com/CAD

Right-Hand Rule of 3D Drawing

In order to effectively draw in 3D, you must be able to visualize objects in 3D space. The *right-hand rule* is a simple method for visualizing the 3D coordinate system. It is a representation of the positive coordinate values in the three axis directions. The AutoCAD world coordinate system (WCS) and a user coordinate system (UCS) are based on this concept of visualization.

To use the right-hand rule, position the thumb, index finger, and middle finger of your right hand as shown in **Figure 1-3**. Imagine that your thumb is the X axis, your index finger is the Y axis, and your middle finger is the Z axis. Hold your hand in front of you so that your middle finger is pointing directly at you, as shown in **Figure 1-3**. This is the plan view of the XY plane. The positive X axis is pointing to the right and

Figure 1-3.
Positioning your hand to use the right-hand rule to understand the relationship of the X, Y, and Z axes.

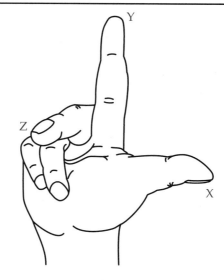

the positive Y axis is pointing up. The positive Z axis comes toward you and the origin of this system is the palm of your hand.

The concept behind the right-hand rule can be visualized even better if you are sitting at a computer and the AutoCAD drawing window is displayed. Make sure the current drawing is based on the acad.dwt template. If the UCS icon is not displayed in the lower-left corner of the screen, turn it on by using the **UCSICON** command. Now, orient your right hand as shown in **Figure 1-3** and position it next to the UCS (or WCS) icon. Your index finger and thumb should point in the same directions as the Y and X axes, respectively. Your middle finger will be pointing out of the screen directly at you, representing the Z axis. See **Figure 1-4**. Notice the illustration on the right in the figure. This is the shaded UCS icon. It is displayed when the visual style is not 2D Wireframe. Visual styles are introduced later in this chapter.

The right-hand rule can be used to eliminate confusion when changing the orientation of the UCS. For example, as you will learn in Chapter 6, a simple way to change the UCS is to rotate it. The UCS can rotate on any of the three axes, just like a wheel rotates on an axle. Therefore, if you want to visualize how to rotate about the X axis, keep your thumb stationary and turn your hand either toward or away from you. If you wish to rotate about the Y axis, keep your index finger stationary and turn your hand to the left or right. When rotating about the Z axis, you must keep your middle finger stationary and rotate your entire arm.

If your 3D visualization skills are weak or you are having trouble visualizing different orientations of the UCS, use the right-hand rule. It is a useful technique for improving your 3D visualization skills. Rotating the UCS around one or more of the axes becomes easier once you begin drawing 3D objects. A complete discussion of UCSs is provided in Chapter 6.

<table>
<tr><td>Ribbon</td></tr>
<tr><td>View
> Coordinates</td></tr>
<tr><td>Show UCS Icon</td></tr>
<tr><td>Type</td></tr>
<tr><td>UCSICON</td></tr>
</table>

UCSICON

Figure 1-4.
Using the right-hand rule to visualize the X, Y, and Z axis directions.

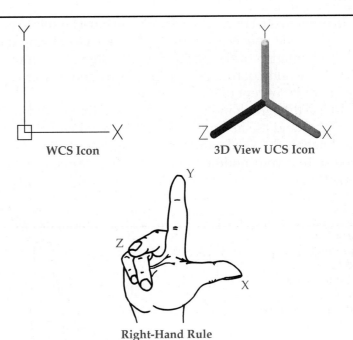

WCS Icon

3D View UCS Icon

Right-Hand Rule

Basic Overview of the Interface

AutoCAD provides different working environments tailored to either 2D drawing or 3D drawing or annotating a drawing. These environments are called *workspaces* and can be quickly restored. There are four default workspaces available in AutoCAD. The workspace for 2D development based on the traditional AutoCAD screen layout is called AutoCAD Classic. The Drafting & Annotation workspace is designed for drawing in 2D and annotating a drawing. It is similar to a streamlined version of the AutoCAD Classic layout with the ribbon displayed in place of toolbars and pull-down menus. The workspace for basic 3D modeling and editing is called 3D Basics. The full range of 3D modeling features is found in the 3D Modeling workspace.

Workspaces can be created, customized, and saved to allow a variety of graphical user interface configurations. The 3D Modeling workspace provides quick access to the full range of tools required to construct, edit, view, and visualize 3D models. It is the workspace used throughout this book. This section provides an overview of the 3D Modeling workspace and the layout of the ribbon and its panels.

 In order to use the default 3D "environment," you must start a new drawing file based on the acad3D.dwt template and set the 3D Modeling workspace current. All discussions in the remainder of this book assume that AutoCAD is in this default 3D environment.

Workspaces

A *workspace* is a drawing environment in which menus, toolbars, palettes, and ribbon panels are displayed for a specific task. A workspace stores not only which of these tools are visible, but also their on-screen locations. You can quickly change workspaces using the **Workspace Switching** tool on either the status bar or the **Quick Access** toolbar, the **WSCURRENT** command, or the **Workspaces** toolbar, as shown in **Figure 1-5**. By default, toolbars are only displayed in the AutoCAD Classic workspace. The default 3D Modeling workspace is shown in **Figure 1-6**.

Figure 1-5.
Switching workspaces.
A—Using the **Workspaces** toolbar (which is not displayed by default in the 3D Modeling workspace).
B—Using the **Workspace Switching** tool.

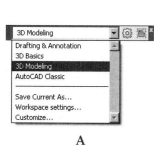

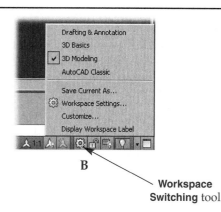

A **B**

Workspace Switching tool

Figure 1-6.

The 3D Modeling workspace with a drawing file based on the acad3D.dwt template.

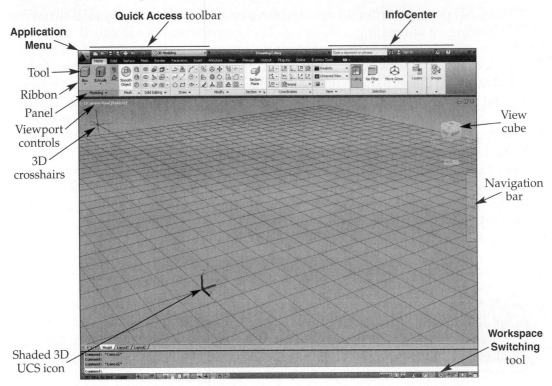

Application Menu

Quick Access toolbar

InfoCenter

Tool

Ribbon

Panel

Viewport controls

3D crosshairs

View cube

Navigation bar

Shaded 3D UCS icon

Workspace Switching tool

Ribbon Panels

Within the 3D Modeling workspace, the **Home** tab on the ribbon displays 11 panels. The tools in these panels provide all of the functions needed to design and view your 3D model. See **Figure 1-7**. The panel title appears at the bottom of the panel. The arrow button at the right-hand end of the tab names controls the display of the ribbon tabs, panels, and panel titles. Picking it cycles through the display of the full ribbon, tab and panel titles only, or tab titles only.

Many panels contain more tools than those displayed in the default view. These panels can be expanded by picking the flyout arrow to the right of the panel title. See **Figure 1-8A**. The flyout portion is displayed as long as the cursor remains over the panel. It retracts when the cursor moves off of the panel. To retain the flyout display, pick the pushpin icon in the lower-left corner of the expanded panel. See **Figure 1-8B**. The panel continues to be displayed until you pick the pushpin icon again to release it.

Figure 1-7.

The ribbon is composed of tabs and panels that provide access to 3D modeling and viewing commands without using toolbars and menus.

Tabs

Pick to minimize to just titles or tabs

Panels

Figure 1-8.
A—Picking the flyout arrow expands the panel. The flyout is displayed as long as the cursor remains over the panel. It retracts when the cursor moves off of the panel. B—Multiple panel flyouts can be displayed using the pushpin, but may overlap in the process. Pick the panel you wish to use to display its tools in full.

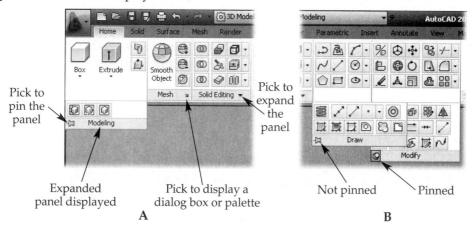

Multiple panel flyouts can be displayed using the pushpin, but they may overlap in the process. This may obscure menu tools in adjacent panels. Simply pick on the panel you wish to use and its tools are displayed in full.

In some cases, a dialog box or palette may be related to the tools in a panel. This is indicated by a diagonal arrow in the lower-right corner of the panel title. See **Figure 1-8A**. Picking on the arrow displays the related item.

You can display only those panels that you need. Right-click anywhere on the ribbon to display the shortcut menu. Select **Show Panels** to display a cascading menu that contains a list of the panels available for that tab. See **Figure 1-9**. The panels that are currently displayed in the ribbon have a check mark next to their name. Select any of the checked panels that you wish to remove from the current display. Unchecked panels can be displayed by selecting their name. Each tab has its own set of available panels.

Each panel contains command tools. These are discussed in detail in the appropriate chapters of this book.

Figure 1-9.
Panels in a tab can be displayed or hidden using the shortcut menu.

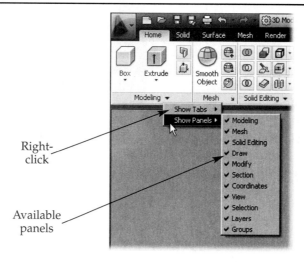

The Drafting & Annotation, 3D Basics, and 3D Modeling workspaces display the ribbon by default. In each workspace, the ribbon displays a set of panels specifically related to that workspace. As soon as you select a different workspace or re-select the current workspace, the panels associated with the workspace are displayed in the ribbon. The ribbon is not displayed by default in the AutoCAD Classic workspace. When the ribbon is displayed in this workspace, the tabs and panels available are the same as for the Drafting & Annotation workspace. The panels associated with a workspace can be customized, as discussed later in this book.

Displaying 3D Views

AutoCAD provides several methods of changing your viewpoint to produce different pictorial views. The default view in the 2D environment based on the acad.dwt template is a plan, or top, view of the XY plane. The default view in the 3D environment based on the acad3D.dwt template is a pictorial, or 3D, view. The *viewpoint* is the location in space from which the object is viewed. The methods for changing your viewpoint include preset isometric and orthographic viewpoints, the view cube, and camera lens settings. Camera settings are discussed in detail in Chapter 20.

Viewport Controls

The *viewport controls* in the upper-left corner of the drawing window provide a quick way to change the viewpoint, configure viewports, and manage the model display. See **Figure 1-10**. There are three viewport controls. When the cursor is moved

Figure 1-10.
Using the AutoCAD viewport controls. Picking on a control displays a flyout. A—The **Viewport Controls** flyout. B—The **View Controls** flyout. C—The **Visual Style Controls** flyout.

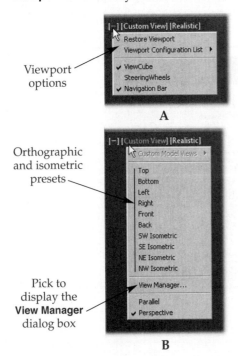

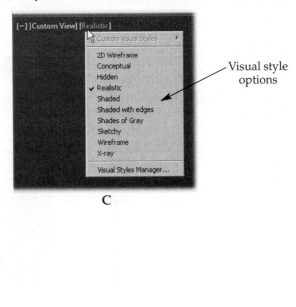

over each of these controls, the name of the control is displayed in the tooltip. The current setting for each control is shown in brackets. Picking on a control displays a flyout. A brief explanation of the viewport controls is provided here to introduce you to the options and tools available.

The options in the **Viewport Controls** flyout, **Figure 1-10A**, allow you to work with viewports, which are described in detail in Chapter 7. The default setting (-) indicates that the current viewport configuration is "minimized" to a single viewport. The **Viewport Controls** flyout also contains options to display AutoCAD navigation tools, including the view cube, steering wheels, and the navigation bar. The view cube is introduced in this chapter, and other navigation tools are introduced in Chapter 4.

The **View Controls** flyout, **Figure 1-10B**, enables you to change the viewpoint to one of the preset orthographic or isometric viewpoints. This is discussed in more detail in the next section. The default setting when you start a new drawing with the acad3D.dwt template is Custom View. The other options in the **View Controls** flyout can be used to display a perspective or parallel projection or to access the **View Manager** dialog box, as discussed later in this chapter.

The **Visual Style Controls** flyout, **Figure 1-10C**, allows you to quickly display the drawing in one of 10 different visual styles. The default setting when you start a new drawing with the acad3D.dwt template is Realistic. Use of visual styles is introduced in this chapter and discussed completely in Chapter 16.

Isometric and Orthographic Viewpoint Presets

A 2D isometric drawing is based on angles of 120° between the three axes. AutoCAD provides preset viewpoints that allow you to view a 3D object from one of four isometric locations. See **Figure 1-11**. Each of these viewpoints produces an isometric view of the object. In addition, AutoCAD has presets for the six standard orthographic views of an object. The isometric and orthographic viewpoint presets are based on the world coordinate system (WCS).

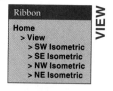

The four preset isometric views are southwest, southeast, northeast, and northwest. The six orthographic presets are top, bottom, left, right, front, and back. To switch your viewpoint to one of these presets, pick the view name in the **View Controls** flyout in the viewport controls, as previously discussed. Refer to **Figure 1-10B**. You can also select one of the presets in the **View** panel in the **Home** tab of the ribbon. See **Figure 1-12**. The presets are also available in the **Views** panel in the **View** tab.

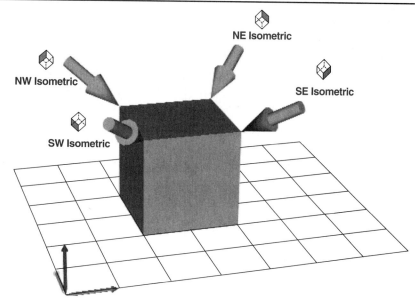

Figure 1-11.
There are four preset isometric viewpoints in AutoCAD. This illustration shows the direction from which the cube will be viewed for each of the presets. The grid represents the XY plane of the WCS.

Figure 1-12.
Selecting preset
views using the **View**
panel in the **Home**
tab of the ribbon.

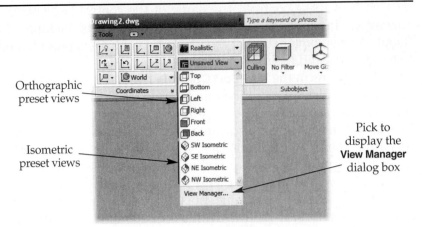

Orthographic
preset views

Isometric
preset views

Pick to
display the
View Manager
dialog box

You can also change the viewpoint to one of the orthographic or isometric presets by using the **VIEW** command to access the **View Manager** dialog box.

Once you select a view, the viewpoint in the current viewport is automatically changed to display an appropriate isometric or orthographic view. Since these presets are based on the WCS, selecting a preset produces the same view of the object regardless of the current user coordinate system (UCS).

A view that looks straight down on the current drawing plane is called a *plan view*. An important aspect of the orthographic presets is that selecting one not only changes the viewpoint, but, by default, it also changes the UCS to be plan to the orthographic view. All new objects are created on that UCS instead of the WCS (or previous UCS). Working with UCSs is explained in detail in Chapter 6. However, in order to change the UCS to the WCS, type UCS to access the **UCS** command and then type W for the **World** option.

When an isometric or other 3D view is displayed, you can easily switch to a plan view of the current UCS by typing the **PLAN** command and selecting the **Current** option. The **PLAN** command is discussed in more detail in Chapter 4.

Selecting an orthographic view of a model using one of the methods described previously produces a plan view, but it may not achieve the results you desire. Three-dimensional models can be displayed in AutoCAD using either parallel or perspective projection. Displaying a plan view in either projection is possible. However, a true plan view, as used in 2D orthographic projections, can only be created when the model is displayed as a parallel projection. You can quickly change the display from perspective to parallel, or vice versa, by picking either **Parallel** or **Perspective** in the **View Controls** flyout in the viewport controls. The current projection is checked in the flyout.

Exercise 1-2

Complete the exercise on the companion website.
www.g-wlearning.com/CAD

Introduction to the View Cube

You are not limited to the preset isometric viewpoints. In fact, you can view a 3D object from an unlimited number of viewpoints. The *view cube* allows you to display all of the preset isometric and orthographic views without making menu or ribbon selections. It also provides quick access to additional pictorial views and easy dynamic manipulation of all orthographic views. It allows you to dynamically rotate the view of the objects to create a new viewpoint.

The view cube is displayed by default in the upper-right corner of the drawing window. See **Figure 1-13**. It can be quickly turned on or off by selecting **ViewCube** from the **Viewport Controls** flyout in the viewport controls. Refer to **Figure 1-10A**. The **NAVVCUBE** command controls the display of the view cube.

As the cursor is moved over the view cube, individual faces, edges, and corners are highlighted. If you pick one of the named faces on the view cube, that orthographic plan view is displayed. However, the UCS is not changed. If you pick one of the corners, an isometric view is displayed that corresponds to one of the preset isometric views. If you pick an edge, an isometric view is displayed that looks at the edge you selected. If you pick on the cube and drag the mouse, the view changes dynamically.

If you get lost while changing the viewpoint, just pick the **Home** icon (the small house) on the view cube to display the view defined as the home view. By default, the southwest isometric view is the home view. You can test this by selecting the **Home** icon and looking at the compass at the base of the view cube. Note that the left face of the cube aligns with the west point on the compass. The front surface

Ribbon
View
> Windows
User Interface
Type
NAVVCUBE
CUBE

NAVVCUBE

Figure 1-13.
Using the view cube to change the viewpoint.

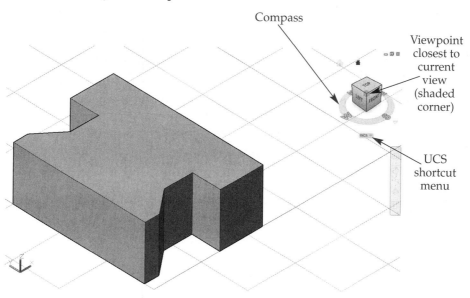

Compass

Viewpoint closest to current view (shaded corner)

UCS shortcut menu

aligns with the south point on the compass. Therefore, the top-front corner points to the southwest.

One edge, corner, or face of the view cube is always shaded or highlighted. Refer to **Figure 1-13**. This shading indicates the viewpoint on the cube that is closest to the current view.

The **NAVVCUBE** command has many options. This discussion is merely an introduction to the command. The command options and view cube features are covered in detail in Chapter 4.

PROFESSIONAL TIP

The **UNDO** command reverses the effects of the **NAVVCUBE** command.

Exercise 1-3

Complete the exercise on the companion website.
www.g-wlearning.com/CAD

Introduction to 3D Model Display Using Visual Styles

The *display* of a 3D model is how the model is presented. This does not refer to the viewing angle, but rather colors, edge display, and shading or rendering. An object can be shaded from any viewpoint. A model can be edited while still keeping the object shaded. This can make it easier to see how the model is developing without having to reshade the drawing. However, when editing a shaded object, it may also be more difficult to select features.

A 3D model can be displayed in a variety of visual styles. A *visual style* is a combination of settings that control the display of edges and shading in a viewport. There are 10 basic visual styles—2D wireframe, conceptual, hidden, realistic, shaded, shaded with edges, shades of gray, sketchy, wireframe, and X-ray. A *wireframe display* shows all lines on the object, including those representing back or internal features. A *hidden display* suppresses the display of lines that would normally be hidden.

A *shaded display* of the model can be created by setting the visual style to Conceptual, Realistic, or one of the shaded visual styles. The X-ray style presents the model in muted, translucent colors, and all hidden lines are displayed. This may be a good choice to use in the design of a model. The Realistic visual style is considered the most realistic *shaded* display. A more detailed shaded model, a *rendered display* of the model, can be created with the **RENDER** command. A rendering is the most realistic presentation.

Examples of the basic visual styles are shown in **Figure 1-14**. To change styles, select one from the **Visual Style Controls** flyout in the viewport controls. Refer to **Figure 1-10C**. You can also use the **VSCURRENT** command or the drop-down list in the **View** panel on the **Home** tab or the **Visual Styles** panel on the **View** tab in the ribbon. See **Figure 1-15**.

VSCURRENT

Ribbon
Home
> View
View
> Visual Styles
Type
VSCURRENT
VS

Figure 1-14.
The default AutoCAD visual styles.

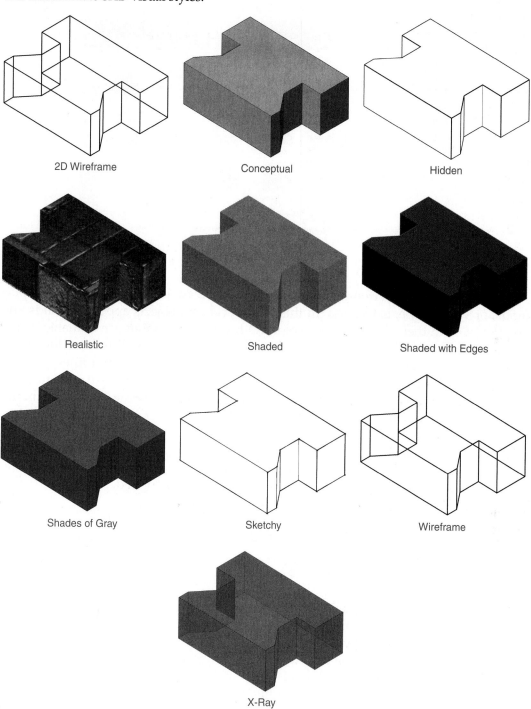

2D Wireframe

Conceptual

Hidden

Realistic

Shaded

Shaded with Edges

Shades of Gray

Sketchy

Wireframe

X-Ray

Figure 1-15.
Visual styles can be managed using the **Visual Style Controls** flyout in the viewport controls or one of the **Visual Styles** drop-down lists in the ribbon. Shown is the **Visual Styles** drop-down list in the **Visual Styles** panel in the **View** tab.

Pick a visual style

Pick to display the visual styles

Current visual style

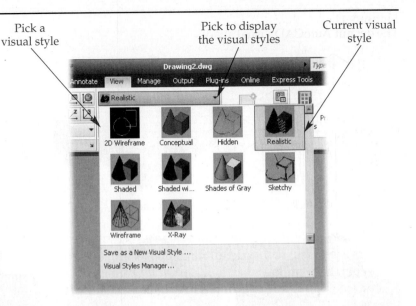

In the default 3D environment based on the acad3D.dwt template, the default display mode, or visual style, is Realistic. In this visual style, all objects appear as solids and are displayed in their assigned layer colors. Other display options are available. These options are discussed in detail in Chapter 16, but are given here as an introduction.

- **2D Wireframe.** Displays all lines of the model using assigned linetypes and lineweights. The 2D UCS icon and 2D grid are displayed, if turned on. If the **HIDE** command is used to display a hidden-line view, use the **REGEN** command to redisplay the wireframe view.
- **Wireframe.** Displays all lines of the model. The 3D grid and the 3D UCS icon are displayed, if turned on.
- **Hidden.** Displays all visible lines of the model from the current viewpoint and hides all lines not visible. Objects are not shaded or colored.
- **Sketchy.** Edges appear hand-sketched.
- **Shades of Gray.** Gray shades are shown with highlighted edges.
- **Conceptual.** The object is smoothed and shaded with transitional colors to help highlight details.
- **Realistic.** The model is shaded and smoothed using assigned layer colors and materials.
- **Shaded.** A smooth-shaded model is displayed, but edges are not shown.
- **Shaded with Edges.** Edges are displayed on the smooth-shaded model.
- **X-Ray.** The model appears transparent.

A variety of options can be used to change individual components of each of these visual styles. A complete discussion of visual styles is given in Chapter 16. Detailed discussions on rendering, materials, lights, and animations appear in Chapters 16 through 20.

When the visual style is 2D Wireframe, you can quickly view the model with hidden lines removed by typing HIDE. The **HIDE** command can be used at any time to remove hidden lines from a wireframe display. If **HIDE** is used when any other visual style is current, the Hidden visual style is set current.

Exercise 1-4

Complete the exercise on the companion website.
www.g-wlearning.com/CAD

Rendering a Model

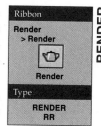

Ribbon

Render
> Render

Render

Type

RENDER
RR

RENDER

The **RENDER** command creates a realistic image of a model, **Figure 1-16**. However, rendering an image takes longer than shading an image. There are a variety of settings that you can change with the **RENDER** command that allow you to fine-tune renderings. These include lights, materials, backgrounds, fog, and preferences. Render settings are discussed in detail in Chapters 16 through 19.

When the command is initiated, the render window is displayed and the image is rendered. See **Figure 1-17**. The rendering that is produced is based on a variety of advanced render settings that are discussed in Chapters 16 through 19. The default render settings create an image using a single light source located behind the viewer. The light intensity is set to 1 and, if no materials are applied, the objects are rendered with a matte material that is the same color as the object display color.

Figure 1-16.
Rendering produces the most realistic display and can show shadows and materials.

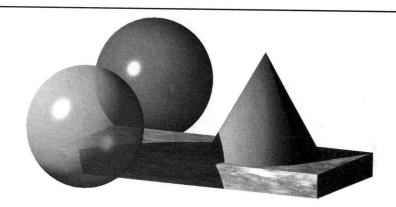

Figure 1-17.
The rendered model is displayed in the **Render** window.

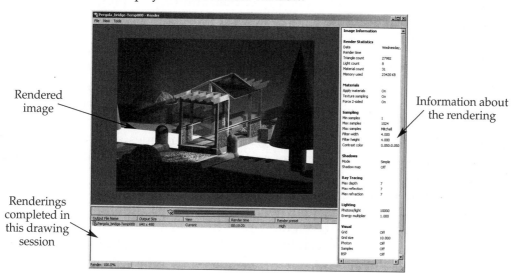

Rendered image

Renderings completed in this drawing session

Information about the rendering

NOTE If the image is rendered in the viewport, clean the screen using the **ZOOM**, **PAN**, **REGEN**, or **REDRAW** command. Setting the rendering destination as the viewport is discussed in Chapter 16.

Exercise 1-5

Complete the exercise on the companion website.
www.g-wlearning.com/CAD

3D Construction Techniques

Before constructing a 3D model, you should determine the purpose of your design. What will the model be used for—manufacturing, analysis, or presentation? This helps you determine which tools you should use to construct and display the model. Three-dimensional objects can be drawn as solids, meshes, or surfaces and displayed in wireframe, hidden-line removed, and shaded views.

A *wireframe object*, or model, is an object constructed of lines in 3D space. Wireframe models are hard to visualize because it is difficult to determine the angle of view and the nature of the surfaces represented by the lines. The **HIDE** command has no effect on a true wireframe model because there is nothing to hide. All lines are always visible because there are no surfaces or faces between the lines. True wireframe models have very limited applications.

Surface modeling represents solid objects by creating a skin in the shape of the object. However, there is nothing inside of the object. Think of a surface model as a balloon filled with air. A surface model looks more like the real object than a wireframe and can be used for rendering. Surface models are often constructed for applications such as civil engineering terrain modeling, automobile body design, sheet metal design and fabrication, and animation. Surface modeling techniques are discussed in Chapter 11.

Like surface modeling, *solid modeling* represents the shape of objects, but it also provides data related to the physical properties of the objects. Solid models can be analyzed to determine mass, volume, moments of inertia, and centroid. A solid model is not just a skin, it represents a solid object. Some third-party programs allow you to perform finite element analysis on the model. Solid model files can also be exported for use in stereolithography and rapid prototyping. These processes can render a plastic or polymer prototype for analysis and testing. This is discussed in Chapter 15. In addition, solid models can be rendered. Most 3D objects are created as solid models.

In AutoCAD, solid models can be created from primitives. *Primitives* are basic shapes used as the foundation to create complex shapes. Some of these basic shapes include boxes, cylinders, spheres, and cones. Detailed shapes and primitives can be created using 3D mesh primitives, mesh modeling techniques, and surface modeling. A 3D *mesh object*, which is a type of surface model, can have a free-flowing shape because the size of the mesh can be adjusted to achieve various levels of smoothness. Mesh objects can be converted to solids for use in model construction. See Chapter 3 for a detailed discussion on 3D mesh modeling. Solid primitives can also be modified to create a finished product. See **Figure 1-18**.

Surface and solid models can be exported from AutoCAD for use in animation and rendering software, such as Autodesk 3ds Max®. Rendered models can be used in

Figure 1-18.
A—These two
cylinders and
the box are solid
primitives. B—With
a couple of quick
modifications,
the large cylinder
becomes a shaft with
a machined keyway.

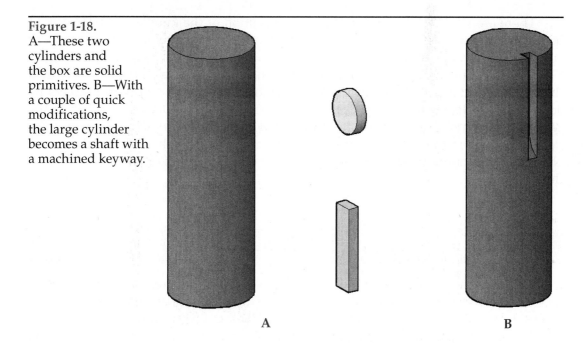

A

B

any number of presentation formats, including slide shows, black-and-white or color prints, and animations recorded to video files. Surface and solid models can also be used to create virtual worlds for entertainment and gaming applications.

3D Object Snaps

The construction and editing of a 3D model can be more efficient with the use of 3D object snaps. These work in the same manner as the standard 2D object snaps and can be set using the **3D Object Snap** tab of the **Drafting Settings** dialog box. If you use 3D object snaps, turn on only those options you need to construct the object. See **Figure 1-19**.

DSETTINGS

Figure 1-19.
The 3D object snaps
can be set in the
Drafting Settings
dialog box.

There are six 3D object snaps. With the exception of the knot snap, these should be familiar to you from your work in 2D. The knot 3D object snap option refers to a fit curve spline and the point at which one curve ends and the next curve begins.

The 3D autosnap markers that appear when drawing can be displayed in a different color than 2D autosnap markers. Set the color for 3D autosnap markers by accessing the **Display** tab in the **Options** dialog box. Pick the **Colors...** button in the **Window Elements** area to access the **Drawing Window Colors** dialog box. Select the desired context in the **Context:** list and then select a color for the 3d Autosnap marker element in the **Interface element:** list. Drawing window customization is discussed in detail in Chapter 21.

Sample 3D Construction

The example provided here illustrates how you can move from a layout sketch to an exact 2D drawing with parametric constraints, then quickly extrude the object into a 3D solid model. Techniques illustrated here are discussed throughout the book. The object shown is the plastic housing for a 24/15-pin cable connector. The final rendition of this object would include internal cutouts, rounded edges, and holes, but are not shown in this example.

Creating a 2D Sketch

A 2D layout can be drawn quickly using the Drafting & Annotation workspace. In doing so, you can create a quick sketch without close attention to exact dimensions or you can use direct distance entry and polar tracking for accuracy. If appropriate, constraints can then be used to apply accurate dimensions and geometric relationships, such as perpendicularity and parallelism, to the object. See **Figure 1-20A**. In this example, only half of the object is drawn, since it is symmetrical. Then the **MIRROR** command is used to complete the layout. See **Figure 1-20B**.

Converting to a Region

The object is still composed of just 2D lines. Therefore, it must be changed to an object that can be converted into a solid model. The **REGION** command is used for this purpose. After selecting this command, you are prompted to select objects. Be sure to

Figure 1-20.
A—Half of the 2D object is drawn using geometric constraints. B—The object is mirrored to create a full profile of the connector body. C—The object is now a "region" and is displayed in the southwest isometric viewpoint.

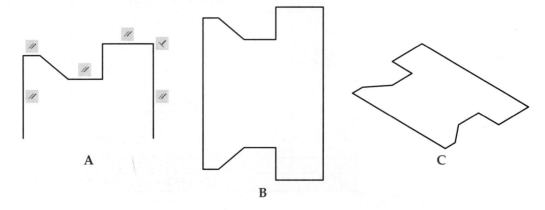

pick all of the lines of the object. After pressing [Enter], you are prompted that a loop is extracted, a region is created, and, if constraints were applied, that they are removed. The appearance of the object will not change.

Converting to a Solid

The object is now a "region" that can be quickly converted to a solid. But first change to the 3D Modeling workspace, then use the viewport controls or the view cube to display the object from the southwest isometric viewpoint. Working with models is more intuitive when a 3D display is used. See **Figure 1-20C**.

The region can be converted to a solid using either the **EXTRUDE** or **PRESSPULL** command. For example, pick the **Extrude** button in the **Modeling** panel on the **Home** tab of the ribbon and select the region. Using direct distance entry, enter a height of .55, as shown in **Figure 1-21**, and press [Enter].

Using a Visual Style

The object you have created is now a 3D solid. If the drawing was based on the acad3D.dwt template, the model is displayed in the Realistic visual style. If the drawing was based on the acad.dwt template, the model is displayed in the 2D Wireframe visual style.

As mentioned previously, there are 10 different visual styles from which to choose. Select an option from the **Visual Style Controls** flyout in the viewport controls or the **View** panel on the **Home** tab of the ribbon. The Conceptual visual style provides a quick display of the model using shaded tones of the object color, as shown in **Figure 1-22**.

The techniques used in this example are just a brief introduction to the creation of 3D models. Detailed descriptions of modeling, display, and editing techniques are included in the following chapters.

Figure 1-21.
The object is extruded into a solid and displayed as a wireframe.

Select the command

Enter the extrusion distance

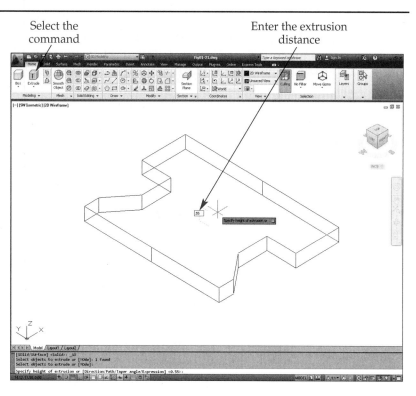

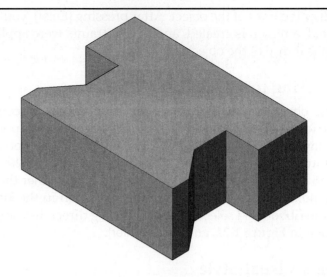

Figure 1-22.
The Conceptual visual style provides a quick display of the model using shaded tones of the object colors.

Guidelines for Working with 3D Drawings

Working in 3D, like working with 2D drawings, requires careful planning to efficiently produce the desired results. The following guidelines can be used when working in 3D.

Planning

- Determine the type of final drawing you need and the manner in which it will be displayed. Then, choose the method of 3D construction that best suits your needs—wireframe, surface, mesh, or solid.
- If appropriate for the project, use 2D constraints to create a 2D layout or sketch.
- For an object requiring only one pictorial view, it actually may be quicker to draw an object in 3D rather than in AutoCAD's isometric mode. AutoCAD's 3D solid modeling tools enable you to quickly create an accurate model, then display it in the required isometric format using preset views. The **VIEWBASE** command can then be used to create a 2D drawing of the model.
- It is best to use AutoCAD's 3D commands to construct objects and layouts that need to be viewed from different angles for design purposes.
- Construct only the details needed for the function of the drawing. This saves space and time and makes visualization much easier.
- Use 2D or 3D object snap modes in a pictorial view in conjunction with UCS icon manipulation to save having to create new UCSs.
- Keep in mind that when the grid is displayed, the pattern appears at the current elevation and parallel to the XY plane of the current UCS.
- Create layers having different colors for different drawing objects. Turn them on and off as needed or freeze those not being used.

Editing

- Use the **Properties** palette to change the color, layer, or linetype of 3D objects.
- Use grips to edit a solid-modeled object (see Chapter 12).
- Do as much editing as possible from a 3D viewpoint. It is quicker and the results are immediately seen.

Displaying

- Use the **HIDE** command and visual styles to help visualize complex drawings.
- To quickly change views, use the preset isometric views, view cube, and **PLAN** command.
- Create and save 3D views for quicker pictorial views. This avoids having to repeatedly use the view cube or other methods to change the viewing angle.
- Freeze unwanted layers before displaying objects in 3D and especially before using **HIDE**. AutoCAD regenerates layers that are turned off, which may cause an inaccurate hidden display to be created. Frozen layers are not regenerated.
- You may have to slightly move objects that touch or intersect if the display removes a line you need to see or plot. However, be sure to move the objects back to maintain accuracy in the model.

Chapter Review

1. What are the three coordinates needed to locate any point in 3D space?
2. In a 2D drawing, what is the value for the Z coordinate?
3. What purpose does the right-hand rule serve?
4. Which three fingers are used in the right-hand rule?
5. What is the definition of a *viewpoint*?
6. What is the function of the *ribbon* and its panels?
7. How do you turn the display of individual panels on or off in the ribbon?
8. How can you quickly change the display from perspective projection to parallel projection, or vice versa?
9. How many preset isometric viewpoints does AutoCAD have? List them.
10. How does changing the UCS impact using one of the preset isometric viewpoints?
11. List the six preset orthographic viewpoints.
12. When selecting a preset orthographic viewpoint, what happens to the UCS?
13. Which AutoCAD tool allows you to dynamically change your viewpoint using an on-screen cube icon?
14. Define *wireframe display*.
15. Define *hidden display*.
16. Define *surface model*.
17. Define *solid model*.
18. Define *primitive*.
19. Define *mesh object*.
20. What is a *visual style*?

Creating Primitives and Composites

Learning Objectives

After completing this chapter, you will be able to:

✓ Construct 3D solid primitives.
✓ Explain the dynamic feedback presented when constructing solid primitives.
✓ Create complex solids using the **UNION** command.
✓ Remove portions of a solid using the **SUBTRACT** command.
✓ Create a new solid from the interference volume between two solids.
✓ Create regions.

Overview of Solid Modeling

In Chapter 1, you were introduced to the three basic types of 3D models—wireframe objects, solid models, and surface models. A solid model is probably the most useful and, hence, most common type of 3D model. A solid model accurately and realistically represents the shape and form of a final object. In addition, a solid model contains data related to the object's volume, mass, and centroid.

Solid modeling is very flexible. A model can start with solid primitives, such as a box, cone, or cylinder, and a variety of editing functions can then be performed. Think of creating a solid model as working with modeling clay. Starting with a basic block of clay, you can add more clay, remove clay, cut holes, round edges, etc., until you have arrived at the final shape and form of the object.

Snaps can be used on solid objects. For example, using 2D object snaps, you can snap to the center of a solid sphere using the **Center** object snap. The **Endpoint** object snap can be used to select the corners of a box, apex of a cone, corners of a wedge, etc. Using 3D object snaps, you can snap to the vertex or midpoint of an edge or to the center of a face. Additionally, you can control the colors of the 2D and 3D autosnap markers in order to easily distinguish between the two when working with solids. Simply access the **Options** dialog box and pick the **Colors...** button in the **Display** tab to set the colors in the **Drawing Window Colors** dialog box. Drawing window customization is discussed in detail in Chapter 21.

Constructing Solid Primitives

As you learned in Chapter 1, a primitive is a basic building block. The eight *solid primitives* in AutoCAD are a box, cone, cylinder, polysolid, pyramid, sphere, torus, and wedge. These primitives can also be used as building blocks for complex solid models. This section provides detailed information on drawing all of the solid primitives. All of the 3D modeling primitive commands can be accessed using one of several methods. You can use the **Primitive** panel in the **Solid** tab of the ribbon, the **Modeling** panel in the **Home** tab of the ribbon, or the **Modeling** toolbar. You can also type the name of the 3D modeling primitive. Using the **Modeling** panel in the **Home** tab of the ribbon is shown in **Figure 2-1**.

The information required to construct a solid primitive depends on the type of primitive being drawn. For example, to draw a solid cylinder, you must provide a center point for the base, a radius or diameter of the base, and the height of the cylinder. A variety of command options are available when creating primitives, but each primitive is constructed using just a few basic dimensions. These are shown in **Figure 2-2**.

Certain familiar editing commands can be used on solid primitives. For example, you can fillet or chamfer the edges of a solid primitive. In addition, there are other

Figure 2-1.
The **Modeling** panel in the **Home** tab of the ribbon.

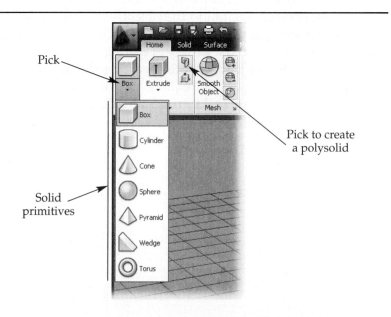

Figure 2-2.
An overview of AutoCAD's solid primitives and the dimensions required to draw them.

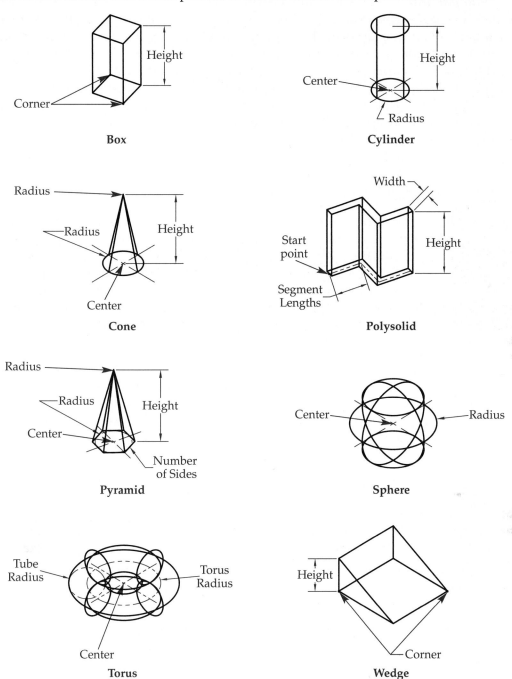

editing commands that are specifically for use on solids. You can also perform Boolean operations on solids. These operations allow you to add one solid to another, subtract one solid from another, or create a new solid based on how two solids overlap.

PROFESSIONAL TIP

Solid objects of a more free-form nature can be created by first constructing a mesh primitive, editing it, and then converting it to a solid. This process is covered in detail in Chapter 3.

Using Dynamic Input and Dynamic Feedback

Dynamic input allows you to construct models in a "heads up" fashion with minimal eye movement around the screen. When a command is initiated, the command prompts are then displayed in the dynamic input area, which is at the lower-right corner of the crosshairs. As the pointer is moved, the dynamic input area follows it. The dynamic input area displays values of the cursor location, dimensions, command prompts, and command options (in a drop-down list). Coordinates and dimensions are displayed in boxes called *input fields.* When command options are available, a drop-down list arrow appears. Press the down arrow key on the keyboard to display the list. You can use the pointer to select the option or press the down arrow key until a dot appears by the desired option and press [Enter].

The *autocomplete* feature of AutoCAD, if turned on, provides a list of commands and system variables at the dynamic input area as you type. As additional characters are typed, the list of selections changes. You can then use the pointer to select an item or press the down arrow key until the desired item is highlighted and press [Enter]. Options controlling the operation of the autocomplete feature can be accessed using the **AUTOCOMPLETE** command.

As an example, if you select the modeling command **BOX**, the first item that appears in the dynamic input area is the prompt to specify the first corner and a display of the X and Y coordinate values of the crosshairs. At this point, you can use the pointer to specify the first corner or type coordinate values. Type the X value and then a comma or the [Tab] key to move to the Y value input box. This locks the typed value and any movement of the pointer will not change it.

CAUTION

When using dynamic input to enter coordinate values from the keyboard, it is important that you avoid pressing [Enter] until you have completed the coordinate entry. When you press [Enter], all of the displayed coordinate values are accepted and the next command prompt appears.

In addition to entering coordinate values for sizes of solid primitives, you can provide direct distance dimensions. For example, the second prompt of the **BOX** command is for the second corner of the base. When you move the pointer, two dimensional input fields appear. Also, notice that a preview of the base is shown in the drawing area. This is the *dynamic feedback* that AutoCAD provides as you create a solid primitive. See **Figure 2-3A**. Once you enter or pick the first corner of the base, the dynamic input area changes to display X and Y coordinate boxes. In this case, the values entered are the X and Y coordinates of the opposite corner of the box base. If X, Y, *and* Z coordinates are entered, the point is the opposite corner of the box.

After establishing the location and size of the box base, the next prompt asks you to specify the height. Again, you can either enter a direct dimension value and press [Enter] or select the height with the pointer. See **Figure 2-3B**. AutoCAD provides dynamic feedback on the height of the box as the pointer is moved.

Figure 2-3.
A—Specifying the base of a box with dynamic input on. Notice the preview of the base.
B—Setting the height of a box with dynamic input on. Notice the preview of the height.

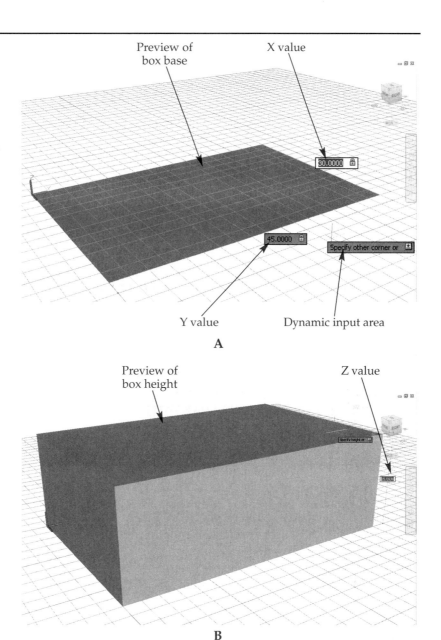

Preview of box base

X value

Y value

Dynamic input area

A

Preview of box height

Z value

B

The techniques previously described can be used with any form of dynamic input. The current input field is always highlighted. You can always enter a value and use the [Tab] key to lock the input and move to the next field.

Box

A *box* has six flat sides forming square corners. It can be constructed starting from an initial corner or the center. See **Figure 2-4**. A cube can be constructed, **Figure 2-4A**, as well as a box with unequal-length sides, **Figure 2-4B**.

When the command is initiated, you are prompted to select the first corner or enter the **Center** option. The first corner is one corner on the base of the box. The center is the geometric center of the box, as shown in **Figure 2-4B**. If you select the **Center** option, you are next prompted to select the center point.

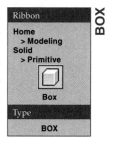

Ribbon

Home
> Modeling
Solid
> Primitive

Box

Type

BOX

BOX

Figure 2-4.
A—A box created
using the **Cube**
option. B—A box
created by selecting
the center point.

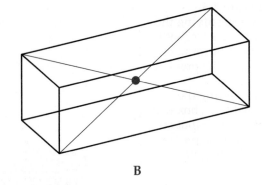

A   B

After selecting the first corner or center, you are prompted to select the other corner or enter the **Cube** or **Length** option. The "other" corner is the opposite corner of the box base if you enter an XY coordinate or the opposite corner of the box if you enter an XYZ coordinate. If the **Length** option is entered, you are first prompted for the length of one side. If dynamic input is on, you can also specify a rotation angle. After entering the length, you are prompted for the width of the box base. If the **Cube** option is selected, the length value is applied to all sides of the box.

Once the length and width of the base are established, you are prompted for the height, unless the **Cube** option was selected. Either enter the height or select the **2Point** option. This option allows you to pick two points on screen to set the height. The box is created.

PROFESSIONAL TIP

Using the UCS icon grips, you can select a surface that is not parallel to the current UCS on which to locate the object. Methods for working with user coordinate systems are discussed in detail in Chapter 6.

Exercise 2-1

Complete the exercise on the companion website.
www.g-wlearning.com/CAD

Cone

CONE

Ribbon

Home
> Modeling
Solid
> Primitive

Cone

Type

CONE

A *cone* has a circular or elliptical base with edges that converge at a single point. The cone may be *truncated* so the top is flat and the cone does not have an apex. See **Figure 2-5.** When the command is initiated, you are prompted for the center point of the cone base or to enter an option. If you pick the center, you must then set the radius of the base. To specify a diameter, enter the **Diameter** option after specifying the center.

The **3P**, **2P**, and **Ttr** options are used to define a circular base using either three points on the circle, two points on the circle, or two points of tangency on the circle and a radius. The **Elliptical** option is used to create an elliptical base.

If the **Elliptical** option is entered, you are prompted to pick both endpoints of one axis and then one endpoint of the other axis of an ellipse that defines the base. If the **Center** option is entered after the **Ellipse** option, you are asked to select the center of the ellipse and then pick an endpoint on each of the axes.

Figure 2-5.
A—A circular cone. B—A frustum cone. C—An elliptical cone.

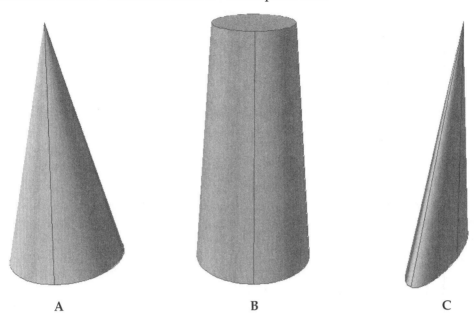

| A | B | C |

After the base is defined, you are asked to specify a height. You can enter a height or enter the **2Point**, **Axis endpoint**, or **Top radius** option. The **2Point** option is used to set the height by picking two points on screen. The distance between the points is the height. The height is always applied perpendicular to the base.

The **Axis endpoint** option allows you to orient the cone at any angle, regardless of the current UCS. For example, to place a tapered cutout in the end of a block, first create a construction line. Refer to **Figure 2-6**. Then, locate the cone base and give a coordinate location of the apex, or axis endpoint. You can then use editing commands to subtract the cone from the box to create the tapered hole. See Chapters 10 and 12 for details on model editing.

The **Top radius** option allows you to specify the radius of the top of the cone. If this option is not used, the radius is zero, which creates a pointed cone. Setting the radius to a value other than zero produces a *frustum cone*, or a cone where the top is truncated and does not come to a point.

Cylinder

A *cylinder* has a circular or elliptical base and edges that extend perpendicular to the base. See **Figure 2-7**. When the command is initiated, you are prompted for the center point of the cylinder base or to enter an option. If you pick the center, you must then set the radius of the base. To specify a diameter, enter the **Diameter** option after specifying the center.

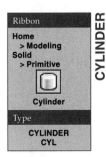

Figure 2-6.
A—Cones can be positioned relative to other objects using the **Axis endpoint** option.
B—The cone is subtracted from the box.

| A | B |

Figure 2-7.
A—A circular cylinder. B—An elliptical cylinder.

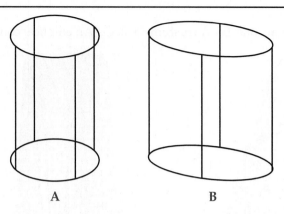

A   B

The **3P**, **2P**, and **Ttr** options are used to define a circular base using either three points on the circle, two points on the circle, or two points of tangency on the circle and a radius. The **Elliptical** option is used to create an elliptical base.

If the **Elliptical** option is entered, you are prompted to pick both endpoints of one axis and then one endpoint of the other axis of an ellipse defining the base. If the **Center** option is entered, you are asked to select the center of the ellipse and then pick an endpoint on each of the axes.

After the base is defined, you are asked to specify a height or to enter the **2Point** or **Axis endpoint** option. The **2Point** option is used to set the height by picking two points on screen. The distance between the points is the height. The **Axis endpoint** option allows you to orient the cylinder at any angle, regardless of the current UCS, just as with a cone.

The **Axis endpoint** option is useful for placing a cylinder inside of another object to create a hole. The cylinder can then be subtracted from the other object to create a hole. Refer to **Figure 2-8**. If the axis endpoint does not have the same X and Y coordinates as the center of the base, the cylinder is tilted from the XY plane.

If polar tracking is on when using the **Axis endpoint** option, you can rotate the cylinder axis 90° from the current UCS Z axis and then turn the cylinder to any preset polar increment. See **Figure 2-9A**. If the polar tracking vector is parallel to the Z axis of the current UCS, the tooltip displays a positive or negative Z value. See **Figure 2-9B**.

Polysolid

The *polysolid* primitive is simply a polyline constructed as a solid object by applying a width and height to the polyline. Many of the options used to create polylines are used with the **POLYSOLID** command. The principal difference is that a solid object is constructed using **POLYSOLID**.

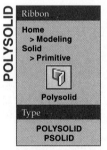

POLYSOLID

Ribbon

Home
 > Modeling
Solid
 > Primitive

Polysolid

Type

POLYSOLID
PSOLID

Figure 2-8.
A—A cylinder is drawn inside of another cylinder using the **Axis endpoint** option. B—The large cylinder has a hole after **SUBTRACT** is used to remove the small cylinder.

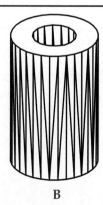

A   B

Figure 2-9.
A—If polar tracking is on, you can rotate the cylinder axis 90° from the current UCS Z axis and then move the cylinder to any angle in the XY plane. B—If the polar tracking vector is moved parallel to the current Z axis of the UCS, the tooltip displays a positive or negative Z dimension.

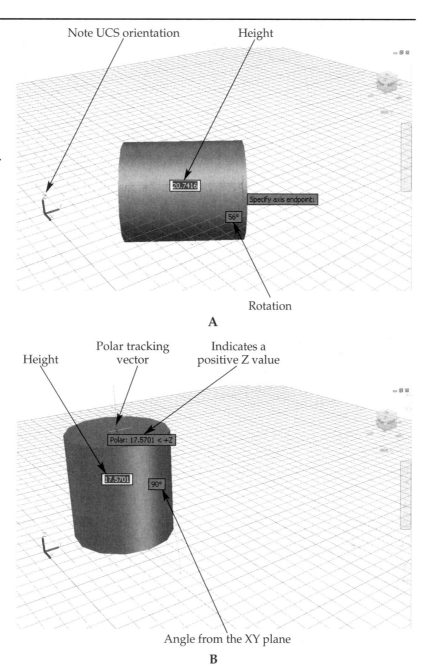

Note UCS orientation

Height

20.7416

Specify axis endpoint:

56°

Rotation

A

Height

Polar tracking vector

Indicates a positive Z value

Polar: 17.5701 < +Z

17.5701

90°

Angle from the XY plane

B

When the command is initiated, you are prompted to select the first point or enter an option. By default, the width of the polysolid is equally applied to each side of the line you draw. This is center justification. Using the **Justify** option, you can set the justification to center, left, or right. The justification applies to all segments created in the command session. See **Figure 2-10**. If you select the wrong justification option, you must exit the command and begin again.

The default width is .25 units and the default height is four units. These values can be changed using the **Height** and **Width** options of the command. The height value is saved in the **PSOLHEIGHT** system variable. The width value is saved in the **PSOLWIDTH** system variable. Using these system variables, the default width and height can be set outside of the command.

The **Object** option allows you to convert an existing 2D object into a polysolid. AutoCAD entities such as lines, circles, arcs, polylines, polygons, and rectangles can be converted. The 2D object cannot be self intersecting. Some objects, such as 3D polylines and revision clouds, cannot be converted.

Figure 2-10.
When you begin the **POLYSOLID** command, use the **Justify** option to select the alignment.

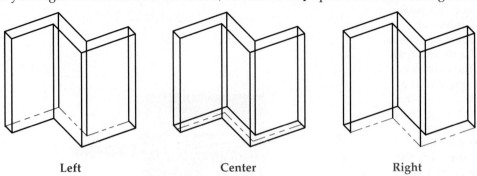

Left   Center   Right

Once you have set the first point on the polysolid, pick the endpoint of the first segment. Continue adding segments as needed. The **Undo** option allows you to remove the last segment drawn and continue from the previous point without exiting the command. To complete the command, press [Enter].

After the first point is set, you can enter the **Arc** option. The current segment will then be created as an arc instead of a straight line. See **Figure 2-11**. Arc segments will be created until you enter the **Line** option. The suboptions for the **Arc** option are:

- **Close.** If there are two or more segments, this option creates an arc segment between the active point and the first point of the polysolid. You can also close straight-line segments.
- **Direction.** Specifies the tangent direction for the start of the arc.
- **Line.** Returns the command to creating straight-line segments.
- **Second point.** Locates the second point of a two-point arc. This is not the endpoint of the segment.

PROFESSIONAL TIP

The **Object** option of the **POLYSOLID** command is a powerful tool for converting 2D objects to 3D solids. For example, you can create a single-line wall plan using a polyline and then quickly convert it to a 3D model.

Figure 2-11.
The **Arc** option of the **POLYSOLID** command is used to create curved segments.

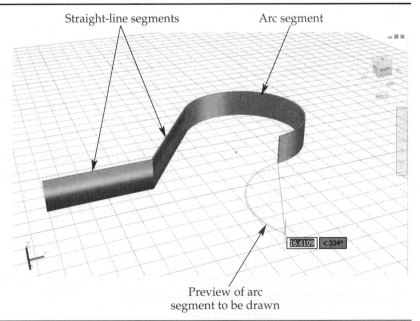

Straight-line segments

Arc segment

Preview of arc segment to be drawn

Pyramid

A *pyramid* has a base composed of straight-line segments and edges that converge at a single point. The pyramid base can be composed of three to 32 sides and is drawn much like a 2D polygon. A pyramid primitive may be drawn with a pointed apex or as a *frustum pyramid*, which has a truncated, or flat, apex. See **Figure 2-12**.

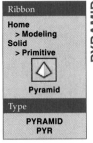

Once the command is initiated, you are prompted for the center of the base or to enter an option. To set the number of sides on the base, enter the **Sides** option. Then, enter the number of sides. You are returned to the first prompt.

The base of the pyramid can be drawn by either picking the center and the radius of a base circle or by picking the endpoints of one side. The default method is to pick the center. Simply specify the center and then set the radius. To pick the endpoints of one side, enter the **Edge** option. Then, pick the first endpoint of one side followed by the second endpoint. If dynamic input is on, you can also set a rotation angle for the pyramid.

If drawing the base from the center point, the polygon is circumscribed about the base circle by default. To inscribe the polygon on the base circle, enter the **Inscribed** option before setting the radius. To change back to a circumscribed polygon, enter the **Circumscribed** option before setting the radius.

After locating and sizing the base, you are prompted for the height. To create a frustum pyramid, enter the **Top radius** option. Then, set the radius of the top circle. The top will be either inscribed or circumscribed based on the base circle. You are then returned to the height prompt.

The height value can be set by entering a direct distance. You can also use the **2Point** option to set the height. With this option, pick two points on screen. The distance between the two points is the height value. The **Axis endpoint** option can also be used to specify the center of the top in the same manner as for a cone or cylinder.

PROFESSIONAL TIP

Six-sided frustum pyramids can be used for bolt heads and as "blanks" for creating nuts.

Figure 2-12.
A sampling of pyramids that can be constructed with the **PYRAMID** command.

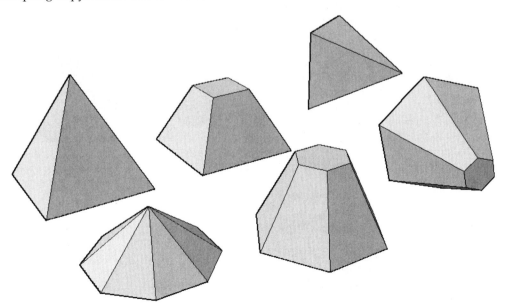

Sphere

SPHERE

Ribbon

Home
> Modeling
Solid
> Primitive

Sphere

Type

SPHERE

A *sphere* is a round, smooth object like a baseball or globe. Once the command is initiated, you are prompted for the center of the sphere or to enter an option. If you pick the center, you must then set the radius of the sphere. To specify a diameter, enter the **Diameter** option after specifying the center. The **3P**, **2P**, and **Ttr** options are used to define the sphere using either three points on the surface of the sphere, two points on the surface of the sphere, or two points of tangency on the surface of the sphere and a radius.

Spheres and other curved objects can be displayed in a number of different ways. The manner in which you choose to display these objects should be governed by the display requirements of your work. Notice in **Figure 2-13** the lines that define the shape of the spheres in a wireframe display. These lines are called *contour lines*, also known as *tessellation lines* or *isolines*. The **Visual Styles Manager** can be used to set the display of contour lines and silhouettes on spheres and other curved 3D surfaces for a given visual style. See **Figure 2-14**.

With the **Visual Styles Manager** displayed, select the 2D Wireframe image tile. The Contour lines setting in the **2D Wireframe options** area establishes the number of lines used to show the shape of curved objects. A similar setting appears in all of the other visual styles if their Show property in the **Edge Settings** area is set to Isolines. The default value is four, as shown in **Figure 2-13A**, but it can be set to a value from zero to 2047. **Figure 2-13B** displays spheres with 20 contour lines. It is best to use a lower number during construction and preliminary displays of the model and, if needed, higher settings for more realistic visualization. The contour lines setting is also available in the **Display** tab of the **Options** dialog box or by typing ISOLINES.

The Draw true silhouettes setting in the **2D Wireframe options** area of the **Visual Styles Manager** controls the display of silhouettes on 3D solid curved surfaces. The setting is either Yes or No. Notice the sphere silhouette in **Figures 2-13C** and **2-13D**. The Draw true silhouettes setting is stored in the **DISPSILH** system variable.

Figure 2-13.
Different displays of spheres established with visual style settings. A—The Draw true silhouettes setting is No and four contour lines are used. B—The Draw true silhouettes setting is No and 20 contour lines are used. C—The Draw true silhouettes setting is Yes and four contour lines are used. D—The Draw true silhouettes setting is Yes and the **HIDE** command is used with the 2D Wireframe visual style set current.

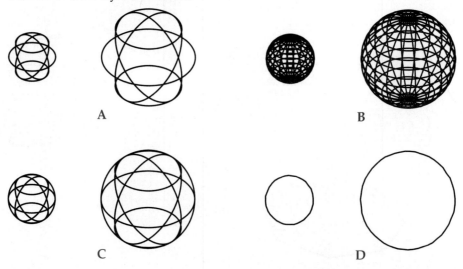

A   B

C   D

Figure 2-14.
The **2D Wireframe options** area of the **Visual Styles Manager** is used to control the display of contour lines and silhouettes on spheres and other curved 3D surfaces in a given visual style.

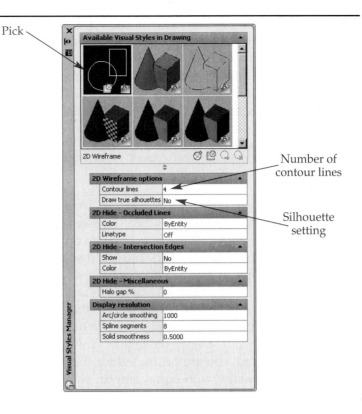

Pick

Number of contour lines

Silhouette setting

Torus

A basic *torus* is a cylinder bent into a circle, similar to a doughnut or inner tube. There are three types of tori. See **Figure 2-15.** A torus with a tube diameter that touches itself is called *self intersecting* and has no center hole. To create a self-intersecting torus, the tube radius must be greater than the torus radius. The third type of torus looks like a football. It is drawn by entering a negative torus radius and a positive tube radius of greater absolute value, i.e. –1 and 1.1.

Once the command is initiated, you are prompted for the center of the torus or to enter an option. If you pick the center, you must then set the radius of the torus. To specify a diameter, enter the **Diameter** option after specifying the center. This defines

Ribbon

Home
> Modeling
Solid
> Primitive

Torus

Type

TORUS
TOR

TORUS

Figure 2-15.
The three types of tori are shown as wireframes and with hidden lines removed.

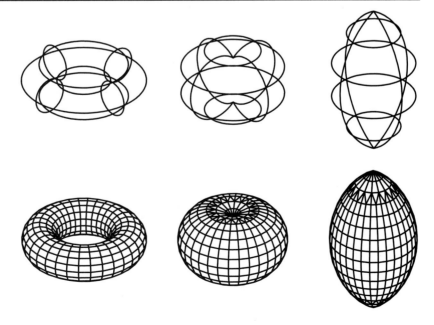

a base circle that is the centerline of the tube. The **3P**, **2P**, and **Ttr** options are used to define the base circle of the torus using either three points, two points, or two points of tangency and a radius.

Once the base circle of the torus is defined, you are prompted for the tube radius or to enter an option. The tube radius defines the cross-sectional circle of the tube. To specify a diameter of the cross-sectional circle, enter the **Diameter** option. You can also use the **2Point** option to pick two points on screen that define the diameter of the cross-sectional circle.

Wedge

WEDGE

Ribbon
Home
> Modeling
Solid
> Primitive

Wedge

Type
WEDGE
WE

A *wedge* has five sides, four of which are at right angles and the fifth at an angle other than 90°. See **Figure 2-16**. Once the command is initiated, you are prompted to select the first corner of the base or to enter an option. By default, a wedge is constructed by picking diagonal corners of the base and setting a height. To pick the center point, enter the **Center** option. The center point of a wedge is the middle of the angled surface. You must then pick a point to set the width and length before entering a height.

After specifying the first corner or the center, you can enter the length, width, and height instead of picking a second corner. When prompted for the second corner, enter the **Length** option and specify the length. You are then prompted for the width. After the width is entered, you are prompted for the height.

To create a wedge with equal length, width, and height, enter the **Cube** option when prompted for the second corner. Then, enter a length. The same value is automatically used for the width and height.

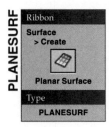

Exercise 2-2

Complete the exercise on the companion website.
www.g-wlearning.com/CAD

Constructing a Planar Surface

PLANESURF

Ribbon
Surface
> Create

Planar Surface

Type
PLANESURF

A *planar surface* primitive is an object consisting of a single plane and is created parallel to the current XY plane. The surface that is created has zero thickness and is composed of a mesh of lines. It is created with the **PLANESURF** command. The command prompts you to specify the first corner and then the second corner of a rectangle. Once drawn, the surface is displayed as a mesh with lines in the local directions, similar to the X and Y directions. See **Figure 2-17A**. These lines are called *isolines* and do not include the object's boundary. The **SURFU** and **SURFV** system variables determine how many isolines are created in the local U and V directions when the planar

Figure 2-16.
A—A wedge drawn by picking corners and specifying a height. B—A wedge drawn using the **Center** option. Notice the location of the center.

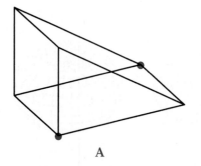

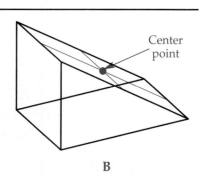

Center point

A

B

Figure 2-17.
A—A rectangular planar surface with four isolines in the U direction and eight isolines in the V direction. B—These two arcs and two lines form a closed area and lie on a single plane. C—The arcs and lines are converted into a planar surface. D—The planar surface is converted into a solid.

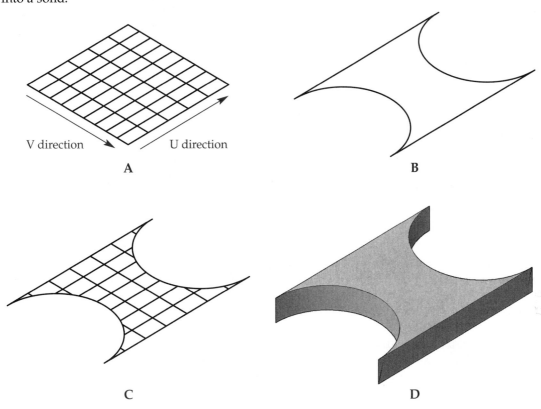

surface is drawn. The isoline values can be changed later using the **Properties** palette. The maximum number of isolines in either direction is 200.

The **Object** option of the **PLANESURF** command allows you to convert a 2D object into a planar surface. Any existing object or objects lying in a single plane and forming a closed area can be converted to a planar surface. The objects in **Figure 2-17B** are two arcs and two lines connected at their endpoints. The resulting planar surface is shown in **Figure 2-17C**.

Although a planar surface is not a solid, it can be converted into a solid in a single step. For example, the object in **Figure 2-17C** is converted into a solid using the **THICKEN** command. See **Figure 2-17D**. The object that started as two arcs and two lines is now a solid model and can be manipulated and edited like any other solid. This capability allows you to create intricate planar shapes and quickly convert them to a solid for use in advanced modeling applications. Model editing procedures are discussed in detail in Chapters 10, 12, and 13.

Creating Composite Solids

A *composite solid* is a solid model constructed of two or more solids, often primitives. Solids can be subtracted from each other, joined to form a new solid, or overlapped to create an intersection or interference. The commands used to create composite solids are found in the **Solid Editing** panel of the **Home** tab or the **Boolean** panel of the **Solid** tab in the ribbon. See **Figure 2-18**.

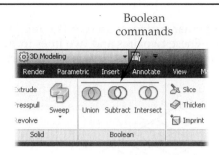

Figure 2-18.
Selecting a Boolean command in the **Boolean** panel on the **Solid** tab of the ribbon. These commands are also located in the **Solid Editing** panel on the **Home** tab.

Introduction to Booleans

There are three operations that form the basis of constructing many complex solid models. Joining two or more solids is called a *union* operation. Subtracting one solid from another is called a *subtraction* operation. Forming a solid based on the volume of overlapping solids is called an *intersection* operation. Unions, subtractions, and intersections as a group are called *Boolean operations*. George Boole (1815–1864) was an English mathematician who developed a system of mathematical logic where all variables have the value of either one or zero. Boole's two-value logic, or *binary algebra*, is the basis for the mathematical calculations used by computers and, with respect to modeling, for those required in the construction of composite solids.

 Boolean operations used to create composite solids can also be used on meshes that have been converted to solids. See Chapter 3 for a complete discussion of meshes.

Joining Two or More Solid Objects

The **UNION** command is used to combine solid objects, **Figure 2-19**. The solids do not need to touch or intersect to form a union. Therefore, accurately locate the primitives when drawing them. After selecting the objects to join, press [Enter] and the action is completed.

In the examples shown in **Figure 2-19B**, notice that lines, or edges, are shown at the new intersection points of the joined objects. This is an indication that the features are one object, not separate objects.

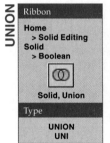

UNION

Ribbon
Home
> Solid Editing
Solid
> Boolean
Solid, Union
Type
UNION
UNI

Figure 2-19.
A—The solid primitives shown here have areas of intersection and overlap. B—Composite solids after using the **UNION** command. Notice the lines displayed where the previous objects intersected.

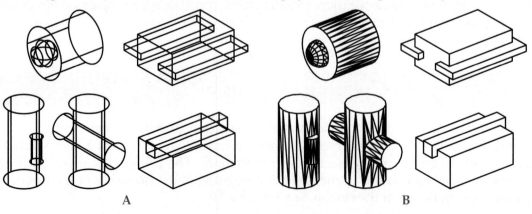

A   B

Subtracting Solids

The **SUBTRACT** command allows you to remove the volume of one or more solids from another solid. Several examples are shown in **Figure 2-20**. The first object selected in the subtraction operation is the object *from* which volume is to be subtracted. The next object is the object to be subtracted from the first. The completed object will be a new solid. If the result is the opposite of what you intended, you may have selected the objects in the wrong order. Just undo the operation and try again.

Creating New Solids from the Intersection of Solids

When solid objects intersect, the overlap forms a common volume, a space that both objects share. This shared space is called an *intersection*. An intersection (common volume) can be made into a composite solid using the **INTERSECT** command. **Figure 2-21** shows several examples. A solid is formed from the common volume. The original objects are removed.

The **INTERSECT** command is also useful in 2D drawing. For example, if you need to create a complex shape that must later be used for inquiry calculations or hatching, draw the main object first. Then, draw all intersecting or overlapping objects. Next, create regions of the shapes. Finally, use **INTERSECT** to create the final shape. The resulting shape is a region and has solid properties. Regions are discussed later in this chapter.

SUBTRACT

Ribbon
Home > Solid Editing **Solid** > Boolean
Solid, Subtract

Type
SUBTRACT **SU**

INTERSECT

Ribbon
Home > Solid Editing **Solid** > Boolean
Solid, Intersect

Type
INTERSECT **IN**

Exercise 2-3

Complete the exercise on the companion website.
www.g-wlearning.com/CAD

Creating New Solids Using the Interfere Command

When you use the **SUBTRACT**, **UNION**, and **INTERSECT** commands, the original solids are deleted. They are replaced by the new composite solid. The **INTERFERE** command does not do this. A new solid is created from the interference (common volume) as if the **INTERSECT** command were used, but the original objects are retained and the new solid can be either deleted or retained.

Once the command is initiated, you are prompted to select the first set of solids or to enter an option. The **Settings** option opens the **Interference Settings** dialog box, which is used to change the visual style and color of the interference solid and the

INTERFERE

Ribbon
Home > Solid Editing **Solid** > Solid Editing
Interfere

Type
INTERFERE **INF**

Figure 2-20.
A—The solid primitives shown here have areas of intersection and overlap.
B—Composite solids after using the **SUBTRACT** command.

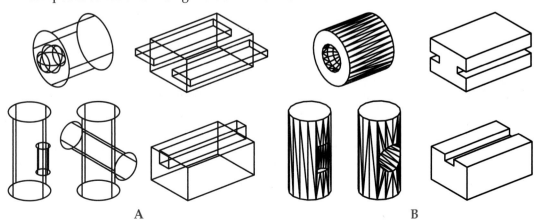

A   B

Figure 2-21.
A—The solid
primitives shown
here have areas
of intersection
and overlap.
B—Composite
solids after using
the **INTERSECT**
command.

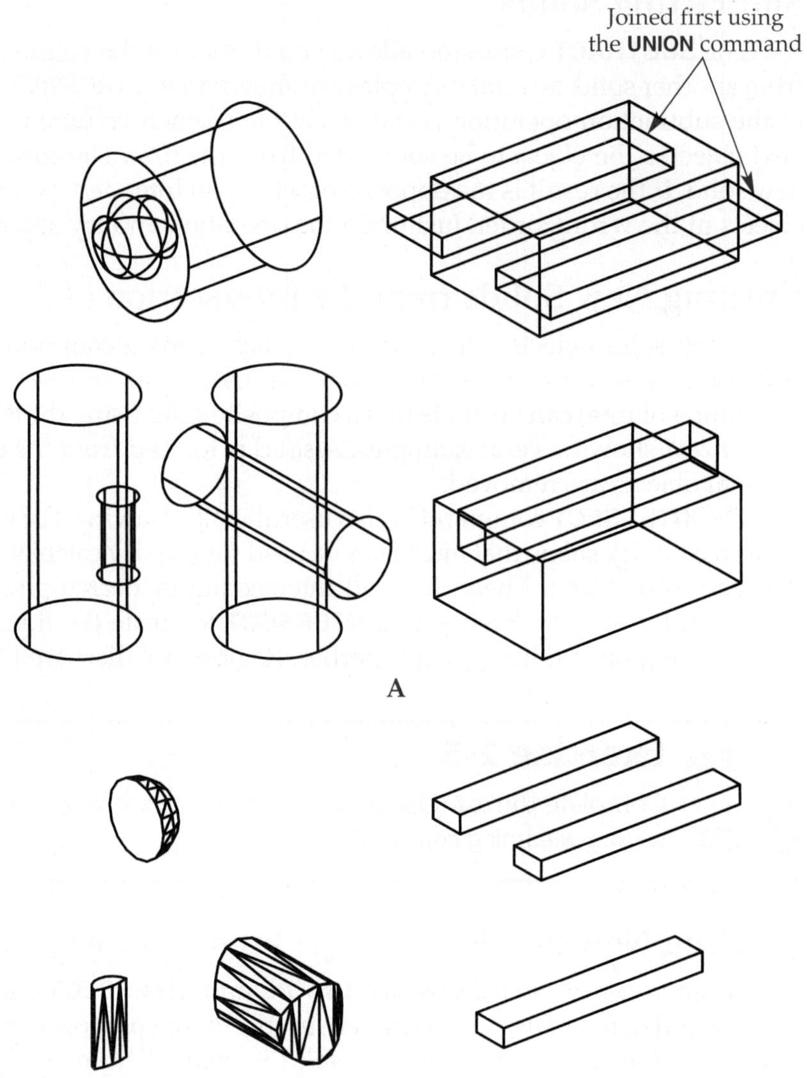

Joined first using
the **UNION** command

A

B

visual style of the viewport. The **Nested selection** option allows you to check the inter-ference of separate solid objects within a nested block. A *nested block* is one that is composed of other blocks. When any needed options are set, select the first set of solids and press [Enter].

You are prompted to select the second set of solids or to enter an option. Entering the **Check first set** option tells AutoCAD to check the objects in the first set for interfer-ence. There is no second set when this option is used. Otherwise, select the second set of solids and press [Enter].

AutoCAD zooms in on the highlighted interference solid and displays the **Interference Checking** dialog box. See **Figure 2-22.** The visual style is set to a wire-frame display by default and the interference solid is shaded in a color, which is red by default.

In the **Interfering objects** area of the **Interference Checking** dialog box, the number of objects selected in the first and second sets is displayed. The number of interfering pairs found in the selected objects is also displayed.

The buttons in the **Highlight** area of the dialog box are used to highlight the previous or next interference object. If the **Zoom to pair** check box is checked, AutoCAD zooms to the interference objects when the **Previous** or **Next** button is selected.

Figure 2-22.
The **Interference Checking** dialog box is used to check for interference between solids. To retain the interference solid, uncheck the **Delete interference objects created on Close** check box.

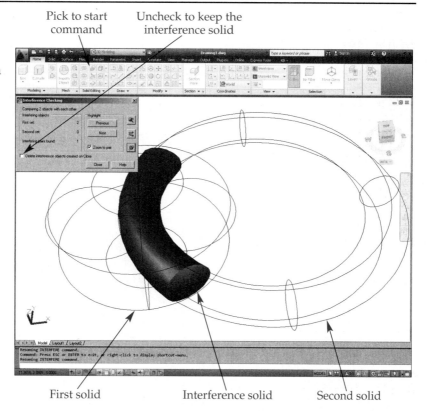

Pick to start command

Uncheck to keep the interference solid

First solid   Interference solid   Second solid

To the right of the **Highlight** area are three navigation buttons—**Zoom Realtime**, **Pan Realtime**, and **3D Orbit**. Selecting one of these navigation options temporarily hides the dialog box and activates the selected command. This allows you to navigate in the viewport. When the navigation mode is exited, the dialog box is redisplayed.

By default, the **Delete interference objects created on Close** check box is checked. With this setting, AutoCAD deletes the object(s) created by interference. In order to retain the new solid(s), uncheck this box.

An example of interference checking and the result is shown in **Figure 2-23**. Notice that the original solids are intact, but new lines indicate the new solid. The new solid is retained as a separate object because the **Delete interference objects created on Close** check box was unchecked. The new solid can be moved, copied, and manipulated just like any other object. **Figure 2-23C** shows the new object after it has been moved and a conceptual display generated.

When the **INTERFERE** command is used, AutoCAD compares the first set of solids to the second set. Any solids that are selected for both the first and second sets are automatically included as part of the first selection set and eliminated from the second. If you do not select a second set of objects or the **Check first set** option is used, AutoCAD calculates the interference between the objects in the first selection set.

Exercise 2-4

Complete the exercise on the companion website.
www.g-wlearning.com/CAD

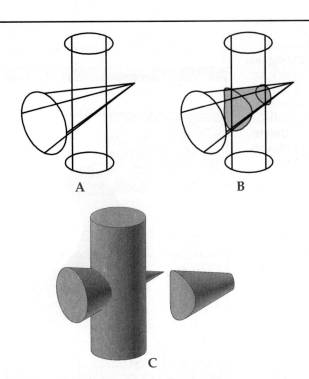

Figure 2-23.
A—Two solids form an area of intersection. B—After using **INTERFERE**, a new solid is defined (shown here in color) and the original solids remain. C—The new solid can be moved or copied.

A

B

C

Creating a Helix

A *helix* is a spline in the form of a spiral and can be created as a 2D or 3D object. See Figure 2-24. It is not a solid object. However, it can be used as the path or framework for creating solid objects such as springs and spiral staircases.

When the command is initiated, you are prompted for the center of the helix base. After picking the center, you are prompted to enter the radius of the base. If you want to specify the diameter, enter the **Diameter** option. After the base is defined, you are prompted for the radius of the top. You can use the **Diameter** option to enter a diameter. The top and bottom can be different sizes. Entering different sizes creates a tapered helix, if the helix is 3D. A 2D helix should have different sizes for the top and bottom.

After the top and bottom sizes are set, you are prompted to set the height or enter an option. To specify the number of turns in the helix, enter the **Turns** option. Then, enter the number of turns. The maximum is 500 and you can enter values less than one, but greater than zero.

By default, the helix turns in a counterclockwise manner. To change the direction in which the helix turns, enter the **Twist** option. Then, enter CW for clockwise or CCW for counterclockwise.

Figure 2-24.
Three types of helices. From left to right, equal top and bottom diameters, unequal top and bottom diameters, and unequal top and bottom diameters with the height set to zero.

The height of the helix can be set in one of three ways. First, you can enter a direct distance. To do this, type the height value or pick with the mouse to set the height. To create a 2D helix, enter a height of zero.

You can also set the height for one turn of the helix using the **Turn height** option. In this case, the total height is the number of turns multiplied by the turn height. If you provide a value for the turn height and then specify the helix height, the number of turns is automatically calculated and the helix is drawn. Conversely, if you provide values for both the turn height and number of turns, the helix height is automatically calculated.

Finally, you can pick a location for the axis endpoint using the **Axis endpoint** option. This is the same option available with a cone, cylinder, or pyramid.

As an example, a solid model of a spring can be created by constructing a helix and a circle and then using the **SWEEP** command to sweep the circle along the helix path. See **Figure 2-25**. The **SWEEP** command is discussed in detail in Chapter 9.

First, determine the diameter of the spring wire and then draw a circle using that value. For this example, you will create two springs, each with a wire diameter of .125 units. Therefore, draw two circles of that diameter, **Figure 2-26**. Their locations are not important. Next, determine the diameter of the spring and draw a corresponding helix. For this example, draw a helix anywhere on screen with a bottom diameter of one unit and a top diameter of one unit. Set the number of turns to eight and specify a height of two units. Draw another helix with the same settings, except make the top diameter .5 units.

Initiate the **SWEEP** command. You are first prompted to select the objects to sweep; pick one circle and press [Enter]. Next, you are prompted to select the sweep path. Select one of the helices. The first sweep, or spring, is completed. Repeat the procedure for the other circle and helix. The drawing is now composed of the two original, single-line helices and the two new swept solids. The circles are consumed by the **SWEEP** command.

Figure 2-25.
A helix can be used as a path to create a spring.

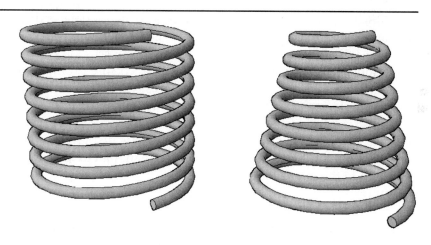

Figure 2-26.
To create a spring, first draw a circle the same diameter as the spring wire. Then, draw the helix and sweep the circle along the helix. Shown here are the two helices used to create the springs in Figure 2-25.

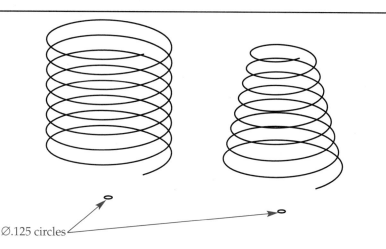

Ø.125 circles

Exercise 2-5

Complete the exercise on the companion website.
www.g-wlearning.com/CAD

Working with Regions

A *region* is a closed, two-dimensional solid. It is a solid model without thickness (Z value). A region can be analyzed for its mass properties. Therefore, regions are useful for 2D applications where area and boundary calculations must be quickly obtained from a drawing.

Boolean operations can be performed on regions. When regions are unioned, subtracted, or intersected, a *composite region* is created. A composite region is also called a *region model*.

A region can be quickly given a thickness, or *extruded*, to create a 3D solid object. This means that you can convert a 2D shape into a 3D solid model in just a few steps. An application is drawing a 2D section view, converting it into a region, and extruding the region into a 3D solid model. Extruding is covered in Chapter 8.

Constructing a 2D Region Model

The following example creates, as a region, the plan view of a base for a support bracket. In Chapter 8, you will learn how to extrude the region into a solid. First, start a new drawing. Next, create the profile geometry in **Figure 2-27** using the **RECTANGLE** and **CIRCLE** commands. These commands create 2D objects that can be converted into regions. The **PLINE** and **LINE** commands can also be used to create closed 2D objects.

The **REGION** command allows you to convert closed, two-dimensional objects into regions. When the command is initiated, you are prompted to select objects. Select the rectangle and four circles and then press [Enter]. The rectangle and each circle are now separate regions and the original objects are deleted. You may need to switch to a wireframe visual style in order to see the circles. You can individually pick the regions. If you pick a circle, notice that a grip is displayed in the center, but not at the four quadrants. This is because the object is not a circle anymore. However, you can still snap to the quadrants.

In order to create the proper solid, the circular regions must be subtracted from the rectangular region. Using the **SUBTRACT** command, select the rectangle as the object to be subtracted *from* and then all of the circles as the objects to subtract. Now, if you select the rectangle or any of the circles, you can see that a single region has been created from the five separate regions. If you set the Conceptual or Realistic visual style current, you can see that the circles are now holes in the region. See **Figure 2-28**.

REGION

Ribbon
Home > Draw
Region
Type
REGION REG

Figure 2-27.
These 2D shapes can be made into a region. The region can then be made into a 3D solid.

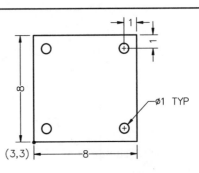

Figure 2-28.
Once the circular regions are subtracted from the rectangular region, they appear as holes. This is clear when a shaded visual style is set current.

Using the Boundary Command to Create a Region

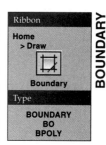

The **BOUNDARY** command is often used to create a polyline for hatching or an inquiry. In addition, this command can be used to create a region. When the command is initiated, the **Boundary Creation** dialog box is displayed. See **Figure 2-29**.

To create a region, select **Region** from the **Object type:** drop-down list in the **Boundary retention** area of the dialog box. Also, you can refine the boundary selection method by turning island detection on or off. When the **Island detection** check box above the **Boundary retention** area is checked, island detection is on. This is the default creation method. When an internal point is selected in the object, AutoCAD creates separate regions from any islands that reside within the object. Unchecking the **Island detection** check box turns island detection off. In this case, when an internal point is selected in the object, AutoCAD ignores islands that reside within the object when creating the region.

Next, select the **Pick Points** button. The dialog box is closed and you are prompted to select an internal point. Pick a point inside of the object that you wish to convert to a region. Press [Enter] when you are finished and the region is created. You can always check to see if an object is a polyline or region by selecting the object, right-clicking, and selecting **Properties...** from the shortcut menu.

Calculating the Area of a Region

A region is not a polyline. It is an enclosed area called a *loop*. Certain properties of the region, such as area, are stored as a value of the region. The **MEASUREGEOM** command can be used to determine the length of all sides and the area of the loop. This can be a useful advantage of a region.

For example, suppose a parking lot is being repaved. You need to calculate the surface area of the parking lot to determine the amount of material needed. This total surface area excludes the space taken up by planting dividers, sidewalks, and lampposts because you will not be paving under these items. If the parking lot and all

Figure 2-29.
Regions can be created using the **Boundary Creation** dialog box.

Pick to select a point inside the boundary

Turn island detection on and off

Select the type of object to be created

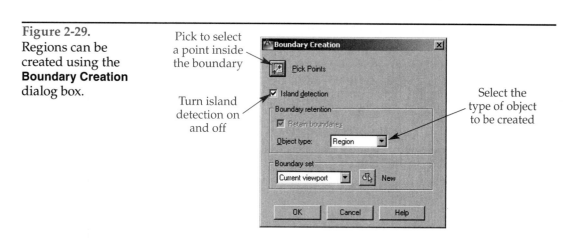

objects inside of it are drawn as a region model, the **MEASUREGEOM** command can give you this figure in one step using the **Object** option. If a polyline is used to draw the parking lot, all internal features must be subtracted each time the **MEASUREGEOM** command is used.

PROFESSIONAL TIP

Regions can prove valuable when working with many items:
- Roof areas excluding chimneys, vents, and fans.
- Bodies of water, such as lakes, excluding islands.
- Lawns and areas of grass excluding flower beds, trees, and shrubs.
- Landscaping areas excluding lawns, sidewalks, and parking lots.
- Concrete surfaces, such as sidewalks, excluding openings for landscaping, drains, and utility covers.

You can find many other applications for regions that can help in your daily tasks.

Exercise 2-6

Complete the exercise on the companion website.
www.g-wlearning.com/CAD

Chapter Review

Answer the following questions. Write your answers on a separate sheet of paper or complete the electronic chapter review on the companion website.
www.g-wlearning.com/CAD

1. What is a *solid primitive*?
2. How is a solid cube created?
3. How is an elliptical cylinder created?
4. Where is the center of a wedge located?
5. What is a *frustum pyramid*?
6. What is a *polysolid*?
7. Name at least four AutoCAD 2D entities that can be converted to a polysolid.
8. What type of entity does the **HELIX** command create and how can it be converted into a solid model?
9. What is a *composite solid*?
10. Which type of mathematical calculations are used in the construction of solid models?
11. How are two or more solids combined to make a composite solid?
12. What is the function of the **INTERSECT** command?
13. How does the **INTERFERE** command differ from **INTERSECT** and **UNION**?
14. What is a *region*?
15. How can a 2D section view be converted to a 3D solid model?

Drawing Problems

Draw the objects in the following problems using the appropriate solid primitive commands and Boolean operations. If appropriate, apply geometric constraints to the base 2D drawing. Do not add dimensions to the models. Save the drawings as **P2**-*(problem number). Display and plot the problems as indicated by your instructor.*

1.

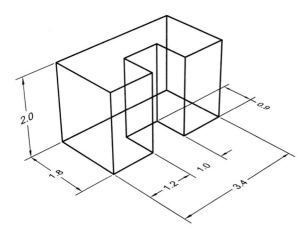

2.

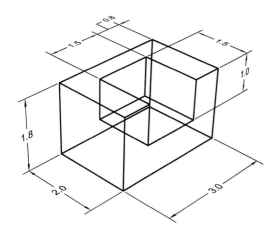

3.

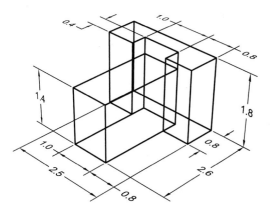

4.

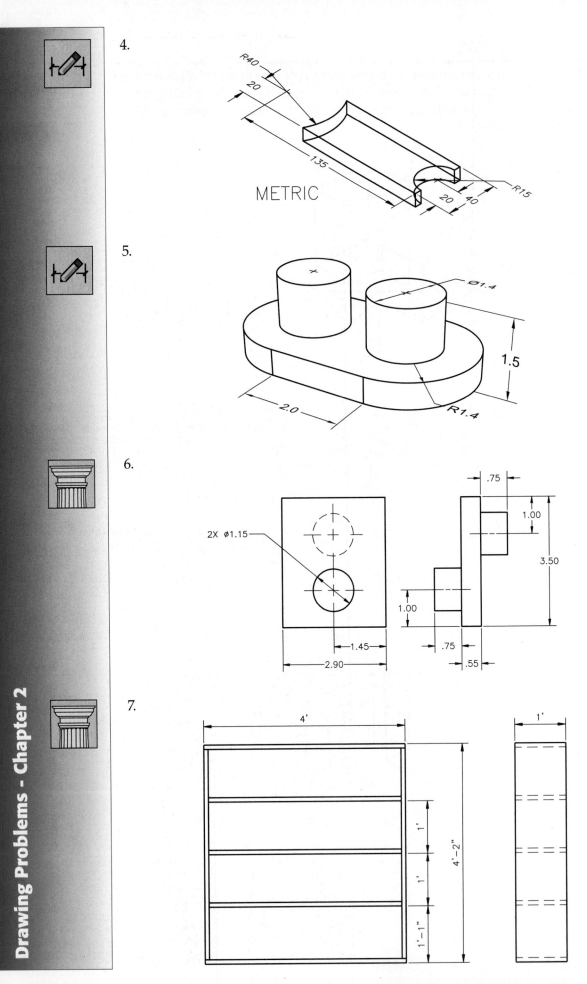

5.

6.

7.

8.

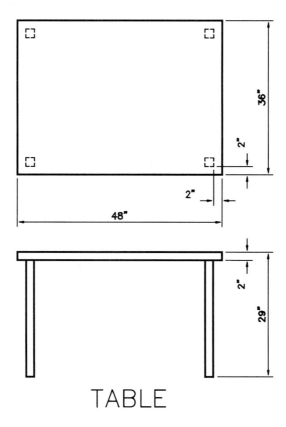

TABLE

9.

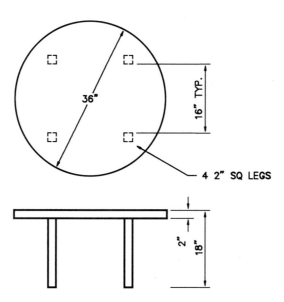

10.

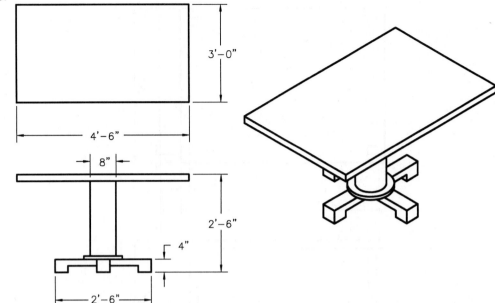

11. Draw the seat for a kitchen chair shown below. In later chapters, you will complete this chair. The seat is 1" thick.

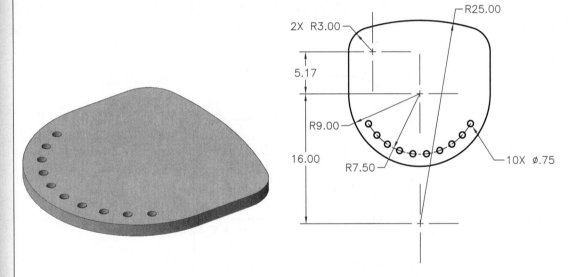

Mesh Modeling

Chapter 3

Learning Objectives

After completing this chapter, you will be able to:

- ✓ Explain tessellation divisions and values.
- ✓ Create mesh primitives.
- ✓ Create a smoothed mesh object.
- ✓ Create a refined mesh object.
- ✓ Generate a mesh by converting a solid or surface.
- ✓ Generate a solid or surface by converting a mesh.
- ✓ Execute editing techniques on mesh objects.
- ✓ Create a split face on a mesh.
- ✓ Produce an extruded mesh face.
- ✓ Apply a crease to mesh subobjects.
- ✓ Create and close mesh object gaps.
- ✓ Create a new mesh face by collapsing a mesh face or edge.
- ✓ Merge mesh faces to form a single mesh face.
- ✓ Construct a new mesh face by spinning a triangular mesh face.
- ✓ Construct mesh forms.

Overview of Mesh Modeling

Mesh primitives and mesh forms can be used to create freeform designs. See **Figure 3-1**. The tools for creating and editing meshes extend the capability of AutoCAD's 3D modeling tools. There are two key workflows that the designer considers:

- The creation of 3D models, which can be solids, surfaces, or meshes.
- Editing the 3D models to create unique shapes.

Mesh models can be created as mesh primitives, mesh forms, or freeform mesh shapes. A *mesh model* consists of vertices, edges, and faces. You can modify or refine a mesh by adding smoothness, creases, extrusions, and splits. You can also distort a mesh to create unique freeform shapes.

A mesh model is a type of surface model. *Subdivision surfaces* is another term for mesh models. Mesh models do *not* have volume or mass. Rather, mesh models only define the shape of the design.

Figure 3-1.
Constructing an ergonomic mouse as a mesh model. A—The basic mesh primitive.
B—Using editing tools, the mesh is reformed. C—The completed mesh model.

Figure 3-1.
Constructing an ergonomic mouse as a mesh model. A—The basic mesh primitive.
B—Using editing tools, the mesh is reformed. C—The completed mesh model.

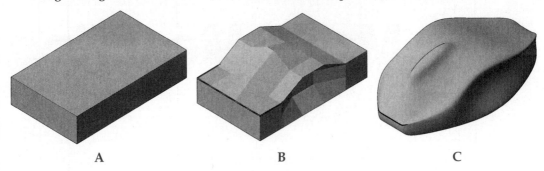

A   B   C

Mesh objects can be created using one of these methods:

- Construct mesh primitives (**MESH** command).
- Convert an existing solid or surface into a mesh object (**MESHSMOOTH** command).
- Construct mesh forms that are ruled, revolved, tabulated, or edge-defined objects (**RULESURF**, **REVSURF**, **TABSURF**, and **EDGESURF** commands).
- Convert legacy surface objects into mesh objects using commands such as **3DFACE**, **3DMESH**, and **PFACE**.

The tools used to create and modify meshes are found on the **Mesh** tab in the ribbon. The commands used to create mesh primitives are similar to those used to create solid primitives, which are discussed in Chapter 2. However, mesh primitives have face mesh objects that are divided into smaller faces. These divisions are based on tessellation division values (smoothness), as discussed in this chapter.

Tessellation Division Values

Tessellation divisions are the basic foundation for the smoothness of a mesh object. Tessellation divisions on a mesh object consist of planar shapes that fit together to form the surface. See **Figure 3-2**. These divisions display the edges of a mesh face that can then be edited.

When creating mesh primitives, set the mesh tessellation divisions *before* creating a mesh primitive shape. Setting the proper values for the mesh tessellation divisions ensures the model has enough faces, edges, and vertices for editing. The default tessellation divisions are listed in the **Mesh Primitive Options** dialog box, which is discussed in the next section.

You can change the default smoothness in the **Mesh Primitive Options** dialog box or, if the primitive is being drawn using the command line, by entering the **Settings** option of the **MESH** command. **Figure 3-3** shows an example of a box mesh primitive created

Figure 3-2.
Tessellation divisions are key to mesh modeling. They define the smoothness of the mesh model.

One tessellation division

Tessellation lines

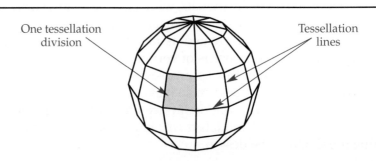

Figure 3-3.
This box mesh primitive is drawn with the default settings.
A—Displayed with the 2D Wireframe visual style current.
B—After the **HIDE** command is used.

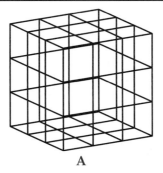
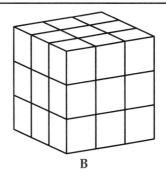

A   B

using the default tessellation divisions. The default settings create a box with no smoothness, length divisions of three, width divisions of three, and height divisions of three.

Drawing Mesh Primitives

A primitive is a basic building block. Just as there are solid primitives in AutoCAD, there are mesh primitives. The seven *mesh primitives* are the mesh box, mesh cone, mesh cylinder, mesh pyramid, mesh sphere, mesh wedge, and mesh torus. See **Figure 3-4.** These primitives can be used as the starting point for creating complex freeform mesh models.

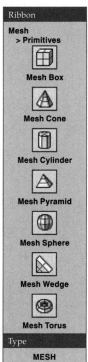

Figure 3-4.
An overview of AutoCAD's mesh primitives and the dimensions required to draw them.

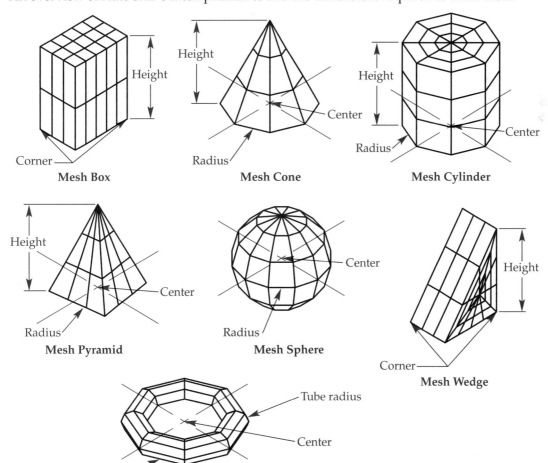

The **Mesh Primitive Options** dialog box is used to set the number of tessellation subdivisions for each mesh primitive object created. See **Figure 3-5**. This dialog box is displayed by picking the dialog box launcher button at the lower-right corner of the **Primitives** panel in the **Mesh** tab of the ribbon or by using the **MESHPRIMITIVEOPTIONS** command. Set the number of tessellation subdivisions in the **Mesh Primitive Options** dialog box *before* creating a mesh primitive. There is no way to change the number of subdivisions after the primitive is created.

The tessellation subdivisions for each primitive are based on the dimensions required to create the primitive. For example, a box has length, width, and height subdivisions. On the other hand, a mesh cylinder has axis, height, and base subdivisions. To set the subdivisions, select the primitive in the tree on the left-hand side of the **Mesh Primitive Options** dialog box. The subdivision properties are then displayed below the tree. Enter the number of subdivisions as required and then close the dialog box. All new primitives of that type will have this number of subdivisions until the setting is changed in the dialog box. Existing primitives are *not* affected.

The following is an example of how to create a mesh box primitive. The mesh box is the mesh primitive used for most base shapes. Refer to **Figure 3-4** for the information required to draw mesh primitives.
1. Open the **Mesh Primitive Options** dialog box.
2. Select the box primitive in the tree.
3. Enter 3 in each of the Length, Width, and Height property boxes to set the subdivisions.
4. Close the dialog box.
5. Select the command for drawing a mesh box.
6. Specify the first corner of the base of the mesh box.
7. Specify the opposite corner of the base of the mesh box.
8. Specify the height of the mesh box.

PROFESSIONAL TIP

Creating mesh primitives is similar to creating solid primitives. Creating solid primitives is discussed in Chapter 2.

Figure 3-5.
The **Mesh Primitive Options** dialog box.

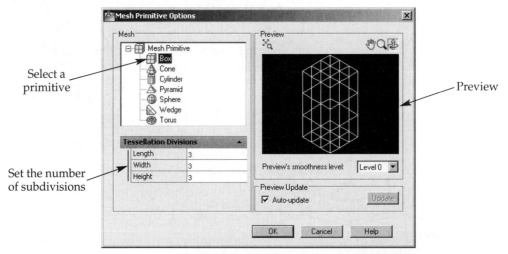

AutoCAD and Its Applications—Advanced

Exercise 3-1

Complete the exercise on the companion website.
www.g-wlearning.com/CAD

Converting Between Mesh and Surface or Solid Objects

AutoCAD offers the flexibility of converting between solid, surface, and mesh objects. This allows you to select the type of modeling that offers the best tools for the task at hand, then convert the model into a form appropriate for the end result. The next sections discuss converting between solid, surface, and mesh objects and the settings that control the conversion.

PROFESSIONAL

TIP

The **DELOBJ** system variable plays an important role when working with meshes, solids, and surfaces. Set the **DELOBJ** system variable to 0 to retain the geometry used to create the mesh, solid, or surface. When the variable is set to 3 (the default value), the geometry used to create the mesh, solid, or surface is deleted.

Mesh Tessellation Options

The **Mesh Tessellation Options** dialog box contains many mesh options, **Figure 3-6**. This dialog box is displayed by picking the dialog box launcher button at the lower-right corner of the **Mesh** panel in the **Mesh** tab of the ribbon or by using the **MESHOPTIONS** command.

When converting to mesh objects, the resulting mesh is one of three different mesh types. The **FACETERMESHTYPE** system variable controls which type of mesh is created. Selecting Smooth Mesh Optimized in the **Mesh type:** drop-down list converts objects to the

Figure 3-6.
The **Mesh Tessellation Options** dialog box.

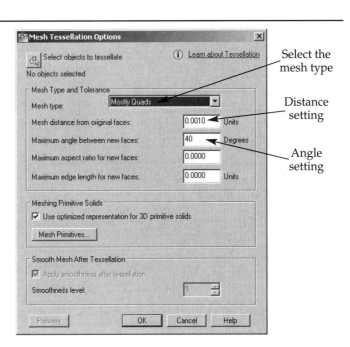

optimized mesh type (**FACETERMESHTYPE** = 0). This is the default and recommended setting. Selecting Mostly Quads in the **Mesh type:** drop-down list creates faces that are mostly quadrilateral (**FACETERMESHTYPE** = 1). Selecting Triangle in the **Mesh type:** drop-down list creates faces that are mostly triangular (**FACETERMESHTYPE** = 2).

The **Mesh distance from original faces** setting is the maximum deviation of the mesh faces (**FACETERDEVSURFACE**). Simply put, this setting determines how closely a converted mesh shape matches the original solid or surface shape. No smoothness values are set for this system variable.

The **Maximum angle between new faces** setting is the maximum angle of a surface normal of two adjoining faces (**FACETERDEVNORMAL** system variable). The higher the value, the more faces created in very curved areas and the less faces created in flat areas. Increasing the value is good for objects that have curved areas, such as fillets, rounds, holes, or other tightly curved areas. No smoothness values are set for this system variable.

The **Maximum aspect ratio for new faces** setting is the upper limit for the ratio of height to width for new faces (**FACETERGRIDRATIO** system variable). By adjusting this value, long faces can be avoided, such as those that would be created from a cylindrical object during the conversion process. A low value will create a cleaner look in the formed faces. The default setting is 0, which specifies that no limitation is applied to the aspect ratio. A setting of 1 specifies that the height must be equal to the width. A setting greater than 1 specifies that the height may exceed the width. A setting between 0 and 1 specifies that the width may exceed the height. No smoothness values are set for this system variable.

The **Maximum edge length for new faces** setting is the maximum length any edge can be (**FACETERMAXEDGELENGTH** system variable). The default setting is 0, which allows the size of the mesh to be determined by the size of the 3D model. A setting of 1 or higher results in a reduced number of faces and less accuracy compared to the original model. No smoothness values are set for this system variable.

> **NOTE**
> Converting swept solids and surfaces, regions, closed polylines, 3D face objects, and legacy polygon and polyface mesh objects may produce unexpected results. If this happens, undo the operation and try making setting adjustments in the **Mesh Tessellation Options** dialog box for better results.

Converting from a Solid or Surface to a Mesh

MESHSMOOTH

Ribbon
Home
> Mesh
Mesh
> Mesh

Smooth Object

Type
MESHSMOOTH
SMOOTH
CONVTOMESH

The **MESHSMOOTH** command is used to convert a solid or surface object into a mesh object. See **Figure 3-7**. The command is easy to use. First, enter the command. Then, select the solids or surfaces to be converted. Finally, press [Enter] and the objects are converted into a mesh. If you select an object that is not a primitive, you may receive a message indicating that the command works best on primitives. If you receive this message, choose to create the mesh.

PROFESSIONAL TIP

With a good understanding of solid modeling, create a solid model first. Then, convert it to a mesh object using the **MESHSMOOTH** command and edit the mesh to create a freeform design.

Figure 3-7.
Converting an existing solid or surface into a mesh object. A—The existing object is a surface (left) or solid (right). B—After converting the object into a mesh using the **MESHSMOOTH** command.

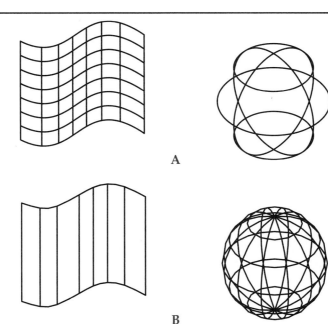

A

B

Exercise 3-2

Complete the exercise on the companion website.
www.g-wlearning.com/CAD

Converting From a Mesh to a Solid or Surface

Mesh objects can be converted into solids or surfaces using the **CONVTOSOLID** and **CONVTOSURFACE** commands. The faces on the resulting solid or surface can be smoothed or faceted and optimized or not. This is controlled by the **SMOOTHMESHCONVERT** system variable. Select one of the four possible settings *before* converting a mesh to a solid or surface. The settings are described as follows:

- **Smoothed and optimized.** Coplanar faces are merged into a single face. The overall shape of some faces can change. Edges of faces that are not coplanar are rounded. (**SMOOTHMESHCONVERT** = 0)
- **Smoothed and not optimized.** Each original mesh face is retained in the converted object. Edges of faces that are not coplanar are rounded. (**SMOOTHMESHCONVERT** = 1)
- **Faceted and optimized.** Coplanar faces are merged into a single, flat face. The overall shape of some faces can change. Edges of faces that are not coplanar are creased or angular. (**SMOOTHMESHCONVERT** = 2)
- **Faceted and not optimized.** Each original mesh face is converted to a flat face. Edges of faces that are not coplanar are creased or angular. (**SMOOTHMESHCONVERT** = 3)

To convert a mesh to a solid, first set the smoothing option, as described above. Then, select the **CONVTOSOLID** command. Next, select the mesh objects to convert and press [Enter]. The objects are converted from mesh objects to solid objects based on the selected smoothing option.

To convert a mesh to a surface, first set the smoothing option, as described above. Then, select the **CONVTOSURFACE** command. Next, select the mesh objects to convert and press [Enter]. The objects are converted from mesh objects to surface objects based on the selected smoothing option.

The examples shown in **Figure 3-8A** are simple mesh objects. In **Figure 3-8B**, the mesh objects have been converted into solid objects with the **Faceted, Optimized** button selected in the **Convert Mesh** panel on the **Mesh** tab in the ribbon. In **Figure 3-8C**, the mesh objects have been converted into surface objects with the **Smooth, Optimized** button selected.

PROFESSIONAL TIP

There are some mesh shapes that cannot be converted to a 3D solid. If using grips to edit the mesh, gaps or holes between the faces may be created. Smooth the mesh object to close the gaps or holes. Also, during the mesh editing process, mesh faces may be created that intersect each other and cannot be converted to a 3D solid. Converting this type of a mesh into a 3D solid will result in the following error message:

Mesh not converted because it is not closed or it self-intersects. Object cannot be converted.

In some cases, there may be a mesh shape that cannot be converted to a solid object, but can be converted to a surface.

Figure 3-8.
A—Three basic mesh objects.
B—The mesh objects converted into faceted solids.
C—The mesh objects converted into smoothed surfaces.

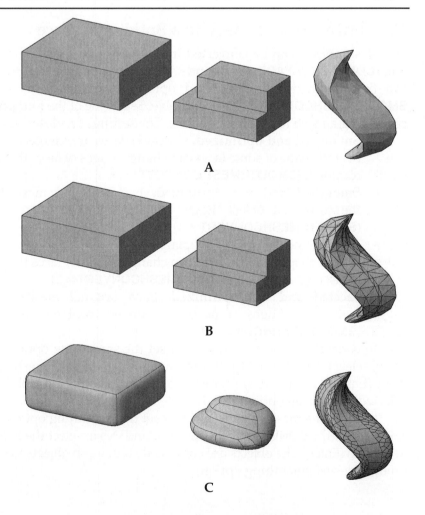

A

B

C

AutoCAD and Its Applications—Advanced

Exercise 3-3

Complete the exercise on the companion website.
www.g-wlearning.com/CAD

Smoothing and Refining a Mesh Object

The roundness of a mesh object is increased by increasing the smoothness level. The smoothness can be set before creating the mesh, as described earlier in this chapter. The mesh object can also be refined or have its smoothness increased after it is created. This is described in the following sections. When smoothing or refining a mesh object, the number of tessellation subdivisions is either increased or decreased. The lowest smoothness level is 0 and the highest smoothness level is 4. The default smoothness level is 0.

PROFESSIONAL TIP

When creating primitives, begin with the least amount of faces possible. You can always refine the mesh model to create more faces.

Adjusting the Smoothness of a Mesh

When a mesh is smoothed, it changes form to more closely represent a rounded shape. The **MESHSMOOTHMORE** command is used to increase the level of smoothness of a mesh object. The maximum level of smoothness attained with this command is controlled by the **SMOOTHMESHMAXLEV** system variable. The default setting is 4. For example, a mesh object with a level of smoothness of 0 is considered to have no smoothness. If the value of the **SMOOTHMESHMAXLEV** system variable is 4, you can increase the smoothness of the mesh object up to level 4.

Once the **MESHSMOOTHMORE** command is selected, pick the mesh objects for which to increase the smoothness. Then, press [Enter] and the level is increased by one. See **Figure 3-9**. If the selection set includes objects that are not meshes, you have the opportunity to either filter out the non-mesh objects or convert them to meshes.

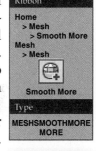

Ribbon
Home
 > Mesh
  > Smooth More
Mesh
 > Mesh

Smooth More

Type
MESHSMOOTHMORE
MORE

MESHSMOOTHMORE

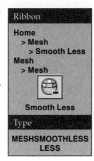

Ribbon
Home
 > Mesh
  > Smooth Less
Mesh
 > Mesh

Smooth Less

Type
MESHSMOOTHLESS
LESS

MESHSMOOTHLESS

Figure 3-9.
Using the **MESHSMOOTHMORE** command to increase the level of smoothness from level 0 to level 4. From left to right, smoothness levels 0, 1, 2, 3, 4.

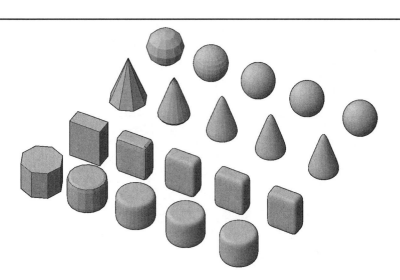

When a mesh is desmoothed, it changes form to more closely represent a boxed shape. The **MESHSMOOTHLESS** command is used to decrease the level of smoothness of a mesh object. Once the command is selected, pick the mesh objects for which to decrease the smoothness. Then, press [Enter] and the level is decreased by one. If the selection set includes objects that are not meshes, you have the opportunity to either filter out the non-mesh objects or convert them to meshes.

You can change the **SMOOTHMESHMAXLEV** system variable setting to create a higher level of mesh model smoothness. The setting can range from 1 to 255. However, a recommended range of smoothness levels is 1–5. Using a level of 5 as the upper limit should be sufficient enough to create a very smooth model. Using a higher level may generate too many faces (a dense mesh) and affect system performance. In this case, you may receive an error message stating that the operation cannot be completed because your system does not have enough physical memory.

PROFESSIONAL TIP

The **MESHSMOOTHMORE** and **MESHSMOOTHLESS** commands change the smoothing one level at a time. You can use the **Properties** palette to change the smoothness to any level in one step.

Exercise 3-4

Complete the exercise on the companion website.
www.g-wlearning.com/CAD

Refining a Mesh

MESHREFINE

Ribbon
Home
> Mesh
Mesh
> Mesh

Mesh Refine

Type
MESHREFINE
REFINE

Refining a mesh increases the number of subdivisions in a mesh object. Refining adds detail to the mesh to make it more realistic with smooth, flowing lines. This also gives a greater selection for editing the mesh faces, vertices, or edges. The **MESHREFINE** command is used to refine a mesh. When the command is used, the number of subdivisions is quadrupled. See **Figure 3-10**. The smoothness level of the original object must be level 1 or higher.

The entire mesh can be refined (all faces at once) or a selected face can be refined. When you refine a mesh *model*, the increased amount of face subdivisions becomes a new smoothness level of 0. However, if you refine an individual *face*, the level of smoothness is not reset.

Figure 3-10.
Refining a mesh.
A—The original mesh with a smoothness level of 1.
B—The refined mesh, which now has more faces and a smoothness level of 0.

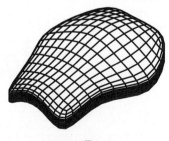

A   B

Be careful not to create too dense of a mesh. Creating a mesh that is too dense may result in subobjects (the mesh shapes) that are very small. This may make it difficult to select subobjects and edit the mesh.

Exercise 3-5

Complete the exercise on the companion website.
www.g-wlearning.com/CAD

Editing Meshes

As discussed earlier, the second key workflow of mesh modeling is the ability to edit the mesh. The tools for editing a mesh are found in the **Mesh Edit** and **Selection** panels on the **Mesh** tab of the ribbon. See **Figure 3-11.** The face, edge, and vertex subobjects can be edited to change the shape of the mesh. These subobjects can be moved, rotated, or scaled. A face on the mesh can be split or extruded.

Subobject filters are used to assist in the selection of a mesh face, edge, or vertex *before* it is moved, rotated, or scaled. This is especially true for a very dense mesh. First, right-click in the drawing window and select **Subobject Selection Filter** to display the cascading menu, **Figure 3-12.** Then, select the filter you wish to use. You can also use the filter in the **Selection** panel of the **Mesh** tab in the ribbon.

Figure 3-11.
The tools for editing a mesh are found in the **Mesh Edit** and **Selection** panels of the **Mesh** tab in the ribbon.

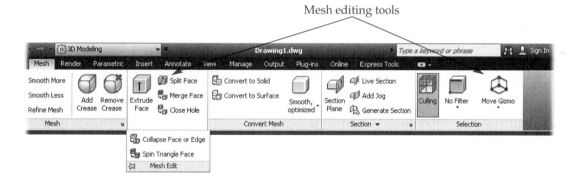

Figure 3-12.
Selecting a subobject filter in the shortcut menu.

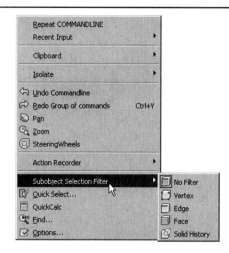

Subobject editing is discussed in detail in Chapter 12. The same procedures discussed in that chapter for solids can be applied to mesh models.

 Use the [Ctrl] key and left mouse button to select the subobjects or use the **SUBOBJSELECTIONMODE** system variable for subobject filtering. These settings apply:
- 0 = off
- 1 = vertices
- 2 = edges
- 3 = faces
- 4 = solid history subobjects

Gizmos

A *gizmo*, also called a grip tool, appears when a subobject is selected. This tool is used to specify how the transformation (movement, rotation, or scaling) is applied. A visual style other than 2D Wireframe must be current in order for the gizmo to appear. Using gizmos (grip tools) is discussed in detail in Chapter 12.

The move gizmo allows movement of a subobject along the X, Y, or Z axis or on the XY, XZ, or YZ plane. The rotate gizmo allows rotation about the X, Y, or Z axis. The scale gizmo allows scaling along the X, Y, or Z axis or XY, XZ, or YZ plane.

You can switch between the three gizmos by picking the button in the drop-down list in the **Selection** panel on the **Mesh** tab of the ribbon. You can also cycle between the gizmos by pressing [Enter].

For example, to use the move gizmo, pick any axis to move along that axis. See **Figure 3-13**. To move along a plane, pick the rectangular area at the intersection of the axes. To use the rotate gizmo, pick the circle with the center about which you wish to rotate the selection. To use the scale gizmo, pick an axis to nonuniformly scale along that axis. Pick the triangular area at the intersection of the axes to uniformly scale the selection. Pick the area between the inner and outer triangular areas to nonuniformly scale along that plane.

When you make a subobject selection on a mesh model, the ribbon displays *context-sensitive panels* based on the selection. For example, if a face subobject is

Figure 3-13.
Editing a mesh using one of the three gizmos is an important part of mesh modeling. A—Move gizmo. B—Rotate gizmo. C—Scale gizmo.

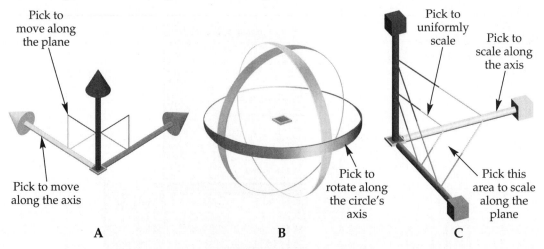

selected, the ribbon displays the **Crease** and **Edit Face** panels, **Figure 3-14**. Context-sensitive panels allow improved editing efficiency by displaying tools commonly used for the selected subobject. These panels are indicated by the green bar behind the panel name. Once the selection is canceled, the context-sensitive panels are no longer displayed.

Extrude a Face

You can select a face subobject on a mesh and extrude it. Extruding a mesh face adds new features to the mesh. This creates new faces that can be edited. The **MESHEXTRUDE** command is used to extrude mesh faces. The command sequence to extrude a mesh face is similar to extruding to create a solid shape. This command is covered in detail in Chapter 8.

To extrude a face, enter the command. The command automatically turns on the face subobject filter. This is indicated by the **Face** button on the **Selection** panel in the **Mesh** tab on the ribbon. Next, pick the face subobject(s) to extrude. Finally, enter an extrusion height. In **Figure 3-15**, a camera is being developed as a mesh model. The **MESHEXTRUDE** command is used to extrude an individual face as the first step in creating the lens tube.

<table>
<tr><td>Ribbon</td></tr>
<tr><td>Mesh
> Mesh Edit</td></tr>
<tr><td>Extrude Face</td></tr>
<tr><td>Type</td></tr>
<tr><td>MESHEXTRUDE</td></tr>
</table>

MESHEXTRUDE

PROFESSIONAL TIP

Extrude a face using the **MESHEXTRUDE** command instead of moving a face. This will give greater editing control over an individual face.

Figure 3-14.
A context-sensitive panel is only displayed when certain objects or subobjects are selected. In this case, the **Crease** and **Edit Face** panels are displayed when a face subobject is selected.

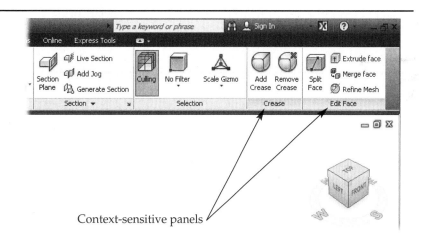

Context-sensitive panels

Figure 3-15.
Extruding a mesh face. A—Select the face to extrude. B—The process adds faces to the model.

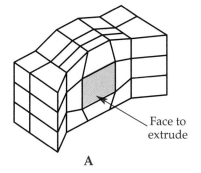

Face to extrude

A

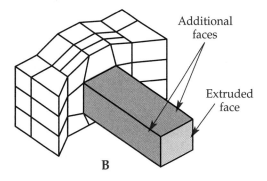

Additional faces

Extruded face

B

Split a Face

Splitting a mesh face is used to increase the number of faces on the model without refining a mesh. Splitting one face creates two faces. This is easier than refining the mesh. Think of these new split faces as subdivisions of an existing face.

The **MESHSPLIT** command is used to split a face. Once the command is entered, select the face to split. Then, specify the starting point and the ending point of a line defining the split. You can specify any two points on the mesh face. **Figure 3-16** shows a mesh model with three faces that have been split.

Once the faces are split, mesh editing options, such as extruding, moving, rotating, or scaling, can be used on the new faces. In **Figure 3-17**, the split faces from **Figure 3-16** have been extruded, and the model has been smoothed to level 3.

MESHSPLIT

Ribbon

Mesh
> Mesh Edit

Split Face

Type

MESHSPLIT
SPLIT

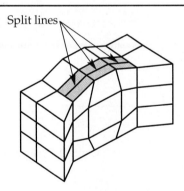

Exercise 3-6

Complete the exercise on the companion website.
www.g-wlearning.com/CAD

Applying a Crease to a Mesh Model

A *crease* is a sharpening of a mesh subobject, much like a crease in folded paper. A crease sharpens or squares off an edge or flattens a face. This prevents the subobject from being smoothed. Creases can be applied to faces, edges, and vertices. A smoothness level of 1 or higher must be assigned to the mesh object for the creases to have an effect, but they can be applied at any smoothness level. Once a crease is applied, any existing smoothing is removed from the subobject. If the mesh smoothness is increased, any creased subobjects are not smoothed. Also, creasing an edge *before* an object is smoothed will limit the mesh editing capabilities.

MESHCREASE

Ribbon

Mesh
> Mesh

Add Crease

Type

MESHCREASE
CREASE

Figure 3-16.
Three faces have been split on this mesh model. Each face has been split into two faces.

Split lines

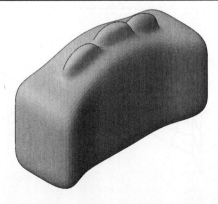

Figure 3-17.
Smoothing is applied to this model. Notice how its edges are rounded.

AutoCAD and Its Applications—Advanced

An example of where a crease might be applied is the bottom of a computer mouse. As the design of the computer mouse begins, the base of the mesh model is rounded when the model is smoothed. See **Figure 3-18A**. By creasing all bottom edges of the mouse, the bottom is squared off. See **Figure 3-18B**. A flat bottom is a requirement of the design intent because the mouse must sit flat on a desk.

In another example, the trigger pads on the game controller shown in **Figure 3-19** have been creased to create a flat surface. When the conceptual design is finished, it can be converted to a solid. This will give the designer a unique shape to which buttons and joysticks can be added. This unique shape would be difficult, if not impossible, to create from scratch as a solid model.

Once the command is entered, select the subobjects to crease. You do not need to press the [Ctrl] key to select subobjects, but filters should be used to make selection easier. Once the subobjects are selected, press [Enter]. This prompt appears:

Specify crease value or [Always] <Always>:

Figure 3-18.
Adding creases to a mesh. A—Before the creases are applied, the bottom of the mouse is rounded. This is obvious if the object is rotated (right). B—After the creases are added, the bottom is flat.

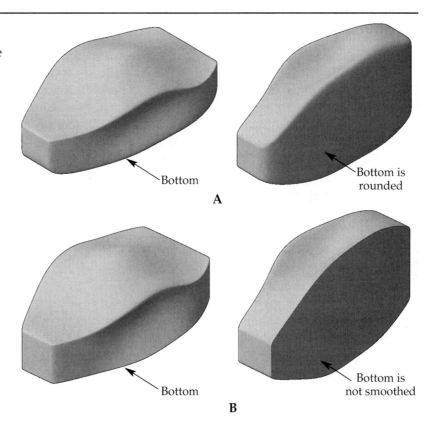

Bottom

Bottom is rounded

A

Bottom

Bottom is not smoothed

B

Figure 3-19.
In this example, creases have been added to the mesh to create flat areas on the game pad for buttons.

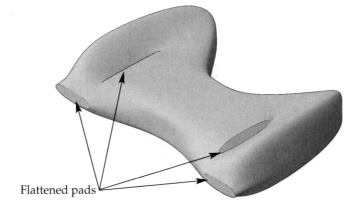

Flattened pads

If you enter a crease value, this is the highest smoothing level for which the crease is retained. If the smoothness level is set higher than the crease value, the creased subobject is smoothed. The **Always** option forces the crease to be retained for all smoothness levels. In most cases, this is the recommended option.

Exercise 3-7

Complete the exercise on the companion website.
www.g-wlearning.com/CAD

Removing a Crease

MESHUNCREASE

Ribbon

Mesh
> Mesh

Remove Crease

Type

MESHUNCREASE
UNCREASE

Removing a crease is a simple process. The **MESHUNCREASE** command is used to do this. Enter the command and then select the subobjects from which a crease is to be removed. Then, press [Enter]. The crease is removed and smoothing is applied as appropriate.

Creating a Mesh Gap

As the designer creates meshes, editing the mesh faces is the next step toward final design. There will be times when the designer needs to modify a mesh by deleting faces. The techniques of erasing or deleting a mesh face aid in the design. To delete a mesh face, use the **ERASE** command or press the [Delete] key after selecting the mesh face. By erasing or deleting the mesh face, a gap in the mesh occurs. **Figure 3-20** shows a game pad controller with *multiple* top faces removed. Use the [Ctrl] key or the face subobject filter to select the top faces and then press the [Delete] key.

PROFESSIONAL TIP

If a mesh face is erased or removed, the mesh object is not considered a *watertight* mesh object. A mesh that is not watertight cannot be converted to a solid object. However, it can be converted to a surface object.

Close Gaps in Mesh Objects

MESHCAP

Ribbon

Mesh
> Mesh Edit

Close Hole

Type

MESHCAP

After a mesh face has been erased or deleted, the designer has the ability to close the mesh gap. The **MESHCAP** command is used to close the gap, **Figure 3-21**. This is done by creating a new face between selected, continuous edges. Once the command

Figure 3-20.
Deleting faces creates a gap on the mesh model.

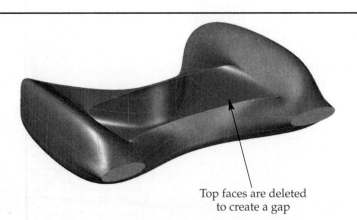

Top faces are deleted
to create a gap

Figure 3-21.
A gap on a mesh model can be closed with the **MESHCAP** command.

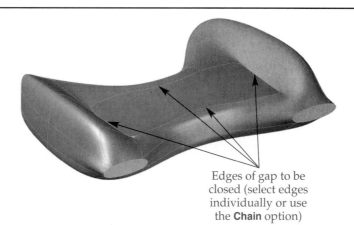

Edges of gap to be closed (select edges individually or use the **Chain** option)

is entered, select the edges of the surrounding mesh object faces. The selected edges do not need to form a closed loop. If a closed loop is not formed, AutoCAD automatically closes the loop. Then, press [Enter] to complete the command and close the mesh gap.

To select multiple edges at once, use the **Chain** option of the **MESHCAP** command. A quick way to access this option is to right-click and select **Chain** from the shortcut menu after entering the **MESHCAP** command. The **Chain** option allows you to select multiple continuous edges with a single pick.

PROFESSIONAL TIP

When using the **MESHCAP** command, the selected mesh face edges should be on the same plane whenever possible. Also, the **MESHCAP** command cannot be used in cases where the edges are shared by adjacent faces forming closed and bounded geometry, such as the geometry for a hole feature.

Exercise 3-8

Complete the exercise on the companion website.
www.g-wlearning.com/CAD

Collapse a Mesh Face or Edge

After a mesh is created, the surrounding mesh faces may be converged to create a different and unique mesh face shape. This is called collapsing a mesh. The designer can collapse surrounding mesh faces at the *center* of a selected mesh edge or face. New mesh faces are created. This helps the designer create new mesh face shapes for further editing.

The **MESHCOLLAPSE** command is used to collapse a mesh. Once the command is selected, pick the face or edge to collapse. The command is immediately applied and then ends.

Figure 3-22 shows an in-progress conceptual design for a computer mouse. The side must be collapsed to a point and then moved inward. Enter the command and select the middle edge. Be sure to use filters to help select the edge. Once the edge is collapsed, the new vertex can be selected for subobject editing and moved inward.

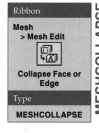

Ribbon

Mesh
> Mesh Edit

Collapse Face or Edge

Type

MESHCOLLAPSE

MESHCOLLAPSE

Figure 3-22.
Collapsing a mesh. A—The edge to be selected. B—The mesh is collapsed and the new vertex is edited.

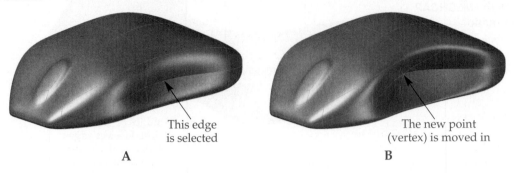

This edge
is selected

The new point
(vertex) is moved in

A

B

Exercise 3-9

Complete the exercise on the companion website.
www.g-wlearning.com/CAD

Merge Mesh Faces

MESHMERGE

Ribbon

Mesh
> Mesh Edit

Merge Face

Type

MESHMERGE

The designer can merge adjacent mesh faces into a new, single mesh face. The **MESHMERGE** command is used to do this. Two or more adjacent faces can be selected to merge to create a different or unique face. For best results, the faces should be on the same plane. Do *not* try to merge faces that are not adjacent or are on corners.

Figure 3-23 shows a tape dispenser. The three faces on the front need to be merged to create a new mesh shape. Enter the command and set the face subobject filter. Then, select the three adjoining faces to merge. Finally, press [Enter] to apply the command.

Exercise 3-10

Complete the exercise on the companion website.
www.g-wlearning.com/CAD

Figure 3-23.
Merging faces. A—The faces to be merged. B—The resulting model.

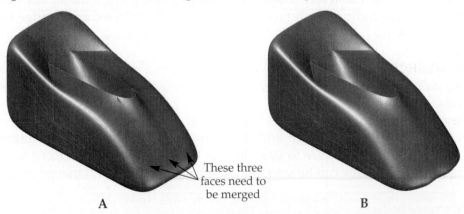

These three
faces need to
be merged

A

B

Spin Mesh Faces

Mesh faces have different and unique shapes when created. Additional unique shapes can be created by spinning a triangular mesh face. Spinning a mesh face spins the adjoining edge of *two* triangle mesh faces. Modifying a mesh face by spinning rotates the newly shared edge. The **MESHSPIN** command is used to spin a face.

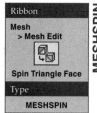

Ribbon
Mesh
> Mesh Edit
Spin Triangle Face
Type
MESHSPIN

MESHSPIN

PROFESSIONAL TIP

Use the **Vertex** option of the **MESHSPLIT** command to create triangular mesh faces.

Figure 3-24 shows a pocket camera. The camera's top button has been split to create two triangular faces. Before the button top can be extruded and smoothed, the top adjoining faces must be spun.

Select the command, then pick the two triangular faces. Press [Enter] to complete the command. Notice how the diagonal line of the triangular faces is now in a different orientation. You can then extrude one of the faces up or down to create a new button. Smooth the camera to create its final design shape.

Drawing Mesh Forms

You are not limited to the mesh primitives discussed earlier. You can also create *mesh forms* based on arcs, lines, and polylines. Mesh forms can be created by four basic methods:

- Using the **RULESURF** command to add a mesh between two objects such as arcs, lines, or polylines. This is called a *ruled-surface mesh*.
- Revolving an open shape with the **REVSURF** command. This is called a *revolved-surface mesh*.

Figure 3-24.
Spinning a face. A—The face has been split, but the orientation is wrong. B—The face is spun and the orientation is correct.

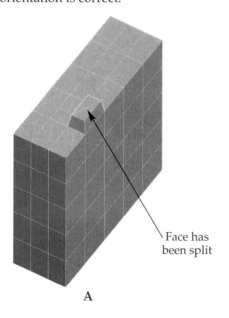

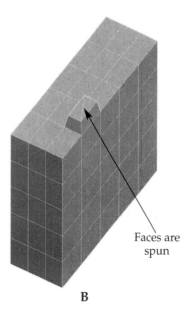

Face has been split

Faces are spun

A

B

- Using the **TABSURF** command to extrude an open shape along a directional path. This is called a *tabulated-surface mesh*.
- Using the **EDGESURF** command to create a mesh based on four adjoining edges, such as lines, arcs, polylines, or splines. This is called an *edged-surface mesh*.

These four commands can create either meshes or legacy surfaces. The **MESHTYPE** system variable controls which type of object is created. The default setting of 1 results in the commands creating mesh objects. A setting of 0 results in the commands creating legacy surfaces (polygon or polyface objects).

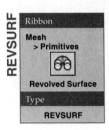

It is strongly recommended the **MESHTYPE** system variable be set to 1 so the **RULESURF**, **REVSURF**, **TABSURF**, and **EDGESURF** commands will create mesh objects. Legacy surfaces are not widely used.

Ruled-Surface Mesh

RULESURF

Ribbon

Mesh
> Primitives

Ruled Surface

Type

RULESURF

A mesh can be constructed between two objects using the **RULESURF** command, **Figure 3-25**. This is called a *ruled-surface mesh*. The two objects can be points, lines, arcs, circles, polylines, splines, or enclosed objects. A ruled-surface mesh can be created between a point and any of the objects listed. However, AutoCAD cannot generate a ruled-surface mesh between an open object and a closed object, such as an arc and a circle.

The number of tessellation subdivisions that compose the ruled-surface mesh is determined by the **SURFTAB1** system variable. The greater the number of subdivisions, the smoother the surface appears.

When using **RULESURF** to create a mesh between two objects, it is important to select both objects near the same end. If you pick near opposite ends of each object, the resulting mesh may not be what you want. The mesh may be twisted or reversed, somewhat like a butterfly shape. Refer to **Figure 3-25**.

Revolved-Surface Mesh

REVSURF

Ribbon

Mesh
> Primitives

Revolved Surface

Type

REVSURF

With the **REVSURF** command, you can draw a profile and then rotate that profile around an axis to create a symmetrical object. This is called a *revolved-surface mesh*. The profile, or path curve, can be drawn using lines, arcs, circles, ellipses, elliptical arcs, polylines, or donuts. The rotation axis can be a line or an open polyline. Notice the initial layout path curves shown in **Figure 3-26** and the resulting meshes. The **REVSURF** command is powerful because it can create a symmetrical surface from any profile.

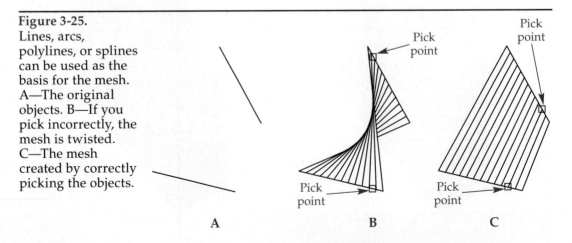

Figure 3-25.
Lines, arcs, polylines, or splines can be used as the basis for the mesh. A—The original objects. B—If you pick incorrectly, the mesh is twisted. C—The mesh created by correctly picking the objects.

Pick point

Pick point

Pick point

Pick point

A   B   C

Figure 3-26.
Creating revolved-surface mesh objects.

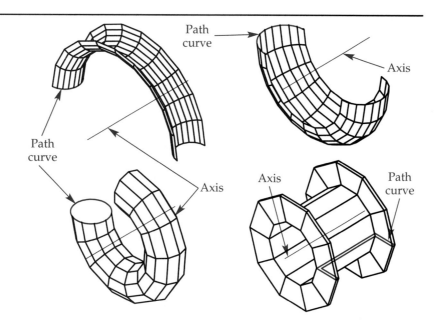

After selecting the path curve and the axis, you are prompted to specify the start angle. This allows you to specify an offset angle at which to start the surface revolution. This is useful when the profile is not revolved through 360°. The next prompt requests the included angle. This lets you create the object through 360° of rotation or just a portion of that, as shown in **Figure 3-26**.

The **SURFTAB1** and **SURFTAB2** system variables control the number of tessellation subdivisions on a revolved-surface mesh. The **SURFTAB1** value determines the number of divisions in the direction of rotation around the axis. The **SURFTAB2** value divides the path curve into the specified number of subdivisions of equal size.

Tabulated-Surface Mesh

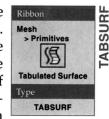

A *tabulated-surface mesh* is similar to a ruled-surface mesh. However, the shape is based on only one entity, **Figure 3-27**. This entity is called the path curve. Lines, arcs, circles, ellipses, 2D polylines, and 3D polylines can all be used as the path curve. A line called the direction vector is also required. This line indicates the direction and length of the tabulated-surface mesh. AutoCAD finds the endpoint of the direction vector closest to your pick point. It sets the direction toward the *opposite* end of the vector line. The mesh follows the direction and length of the direction vector. **Figure 3-27** shows the difference the location of the pick point makes when selecting the vector. The **SURFTAB1** system variable controls the number of "steps" that are constructed.

Edged-Surface Mesh

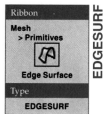

The **EDGESURF** command allows you to construct a 3D mesh between four edges, **Figure 3-28**. The edges can be lines, polylines, splines, or arcs. The endpoints of the objects must precisely meet. However, a closed polyline cannot be used. The four objects can be selected in any order. The resulting mesh is smooth and is called an *edged-surface mesh*.

The number of tessellation subdivisions is determined by the system variables **SURFTAB1** and **SURFTAB2**. The default value for each of these variables is 6. Higher values increase the smoothness of the mesh. The current values are always displayed on the command line when one of the mesh surfacing commands is executed.

Figure 3-27.
Creating a tabulated-surface mesh object. Lines, arcs, polylines, or splines can be used as the shape and vector. A—The objects on which the mesh will be based. B—The completed tabulated-surface mesh.

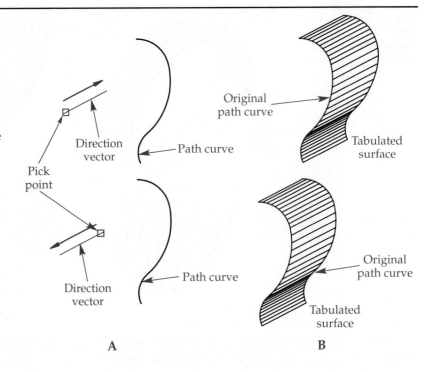

A

B

Figure 3-28.
Creating an edged-surface mesh object. Four lines, arcs, polylines, or splines can be used as the four edges.

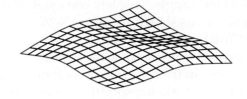

Exercise 3-11

Complete the exercise on the companion website.
www.g-wlearning.com/CAD

Chapter Review

Answer the following questions. Write your answers on a separate sheet of paper or complete the electronic chapter review on the companion website.
www.g-wlearning.com/CAD

1. Of what does a mesh model consist?
2. What is another term for *mesh models*?
3. What are *tessellation divisions*?
4. When creating a mesh primitive, when should mesh tessellation divisions be set?
5. For what is the **Mesh Primitive Options** dialog box used?
6. How is a mesh box created?
7. How is a mesh sphere created?
8. How is a mesh torus created?

9. List the four commands used to create mesh forms.
10. What is the purpose of the **DELOBJ** system variable?
11. Which command converts a mesh object to a surface object?
12. Which command converts a mesh object to a solid object?
13. Describe the smoothness of a mesh object.
14. Which command is used to convert an existing solid or surface to a mesh object?
15. Name the system variable that controls the maximum level of smoothness attained with the **MESHSMOOTHMORE** command.
16. List two ways to decrease the smoothness of a mesh.
17. What happens to the mesh when you refine it?
18. How many types of subobjects does a mesh have? List them.
19. Which keyboard key is used to select subobjects for editing?
20. What is a *context-sensitive panel*?
21. Name the three operations that can be performed with a gizmo.
22. How do you cycle through the three different gizmos?
23. Which command is used to extrude a mesh face?
24. Briefly describe the process for extruding a mesh face.
25. What is the process for splitting a mesh face?
26. Why would you crease a mesh model?
27. Which command is used to remove a crease?
28. Explain why you would erase or delete a mesh face during the design process.
29. Which command is used to close gaps in a mesh object?
30. What is the purpose of collapsing a mesh face or edge?

Drawing Problems

1. In this problem, you will create an ergonomic computer mouse as a mesh shape. Change visual styles as needed throughout your work.
   A. Set the tessellation divisions for the box primitive to length = 5, width = 3, and height = 3.
   B. Create a 5 × 3 × 1.04 mesh box primitive.
   C. Move the middle faces inward .25 units on both sides of the mesh box, as shown in A.
   D. Move edges on the mesh model to form the mouse shape shown in B.
   E. Smooth the model.
   F. Continue editing faces, edges, and vertices to create a mouse shape similar to the one shown in C.
   G. Crease the model to create a flat bottom.
   H. Save the model as P03_01.

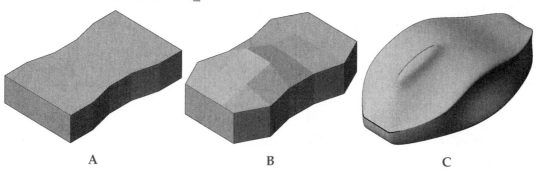

A   B   C

2. In this problem, you will create a wireless phone charger cradle as a mesh shape. Change visual styles as needed throughout your work.
   A. Set the tessellation divisions for the box primitive to length = 5, width = 3, and height = 2.
   B. Create a 3 × 2 × 1.5 mesh box primitive.
   C. Move the middle-back edges up 1 unit and the front edges down .75 units, as shown in A.
   D. Crease the middle face on the top of the model, then move it down 1 unit.
   E. Crease all of the faces on the bottom.
   F. Smooth the mesh to a level 3 mesh.
   G. On the top of the model, nonuniformly scale the middle face (which is now round) to create an elliptical shape. Then, move the face toward the back of the model.
   H. Move the middle three edges in the front of the model .5 units toward the back of the model, as shown in B.
   I. Save the model as P03_02.

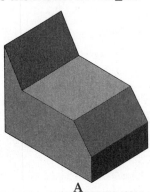
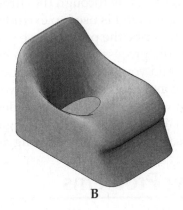

A

B

3. In this problem, you will create a game pad controller as a mesh shape. Change visual styles as needed throughout your work.
   A. Set the tessellation divisions for the box primitive to length = 5, width = 3, and height = 2.
   B. Create a 6 × 4 × 1 mesh box primitive.
   C. Move the middle faces on the back side (long side) inward 1.25 units.
   D. Move the top edges on the right and left side up 1 unit, as shown in A.
   E. Move the front middle face inward .5 units.
   F. Crease the front two corner faces. The top corner of these faces is formed by the edges moved up earlier.
   G. Smooth the model to level 3.
   H. Move the two creased areas outward .25 units.
   I. Split the top face to the left of the indention into three unique faces, as shown in B.
   J. Extrude the three split faces upward .375 units.
   K. Save the model as P03_03.

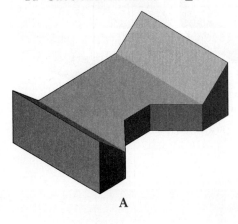
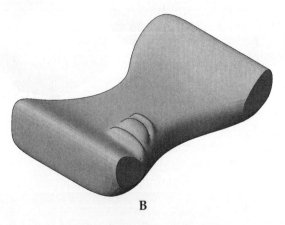

A

B

4. In this problem, you will create a starship for a gaming application as a mesh shape. Change visual styles as needed throughout your work.
   A. Set the tessellation divisions for the sphere primitive to axis = 12 and height = 6.
   B. Create a Ø4 mesh sphere primitive.
   C. Move the vertex on the top of the sphere up 4 units.
   D. Move the vertex on the right and left sides outward 2 units, as shown in A.
   E. Move the vertex on the bottom of the sphere into the sphere by 1 unit.
   F. Smooth the model to level 2.
   G. On the bottom of the ship, as indicated in A, crease the four middle faces, then move them into the sphere by 1 unit.
   H. Extrude the four middle faces a distance of .75 with a taper of 30°. This forms the landing gear.
   I. Select two faces on the front of the ship and crease them to create windows, as shown in B.
   J. Increase the smoothness to level 4.
   K. Save the model as P03_04.

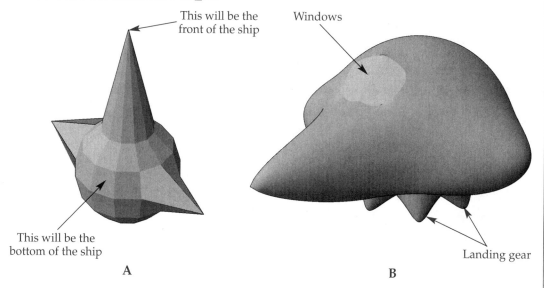

This will be the front of the ship

Windows

This will be the bottom of the ship

Landing gear

A

B

5. In this problem, you will create a home theater chair as a mesh shape. Change visual styles as needed throughout your work.
   A. Set the tessellation divisions for the box primitive to length = 5, width = 3, and height = 2.
   B. Create a 3 × 2 × 1.5 mesh box primitive.
   C. Move the top-back edges (short side) up 1 unit and the top-front edges down .75 units.
   D. Crease the one face in the middle of the top, then move it down 1 unit, as shown in A.
   E. Crease all faces on the bottom.
   F. Smooth the model to level 3.
   G. Nonuniformly scale the middle face on the top, which is now circular, to create an elliptical shape. Then, move the face .5 units toward the chairback.
   H. Move the three top edges on the front .5 units toward the chairback.
   I. Move the middle face in the front top down 1 unit, as shown in B.
   J. Rotate the top three back edges 45° to create a sloped chairback.
   K. Move the single edge in the middle of the back inward .5 units, as shown in C.
   L. Increase the smoothness to level 4.
   M. Save the model as P03_05.

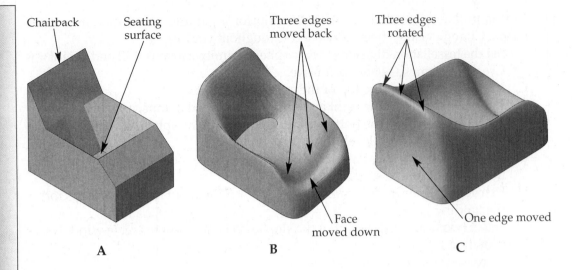

Chairback   Seating surface   Three edges moved back   Three edges rotated

Face moved down   One edge moved

A   B   C

6. In this problem, you will create a conceptual design for a pocket camera as a mesh shape. Change visual styles as needed throughout your work.
   A. Set the tessellation divisions for the box primitive to length = 5, width = 3, and height = 5.
   B. Create a $3 \times 1 \times 3$ mesh box primitive.
   C. Extrude the top face second from the edge to create a button. Extrude to a height of .25 with a taper of 10°, as shown in A. Crease the top of the extruded face.
   D. Nonuniformly scale the middle face on the front by a factor of two. Then, crease the face.
   E. Move the five edges on the left side of the front outward .25 units, as shown in B.
   F. Move the three edges on the right-hand side of the top inward .375 units.
   G. Move the three edges on the right-hand side of the bottom inward .375 units.
   H. Crease all faces on the bottom.
   I. Smooth the model to level 4.
   J. Extrude the middle face on the front (which was earlier creased) inward .125 units, as shown in C.
   K. Save the model as P03_06.

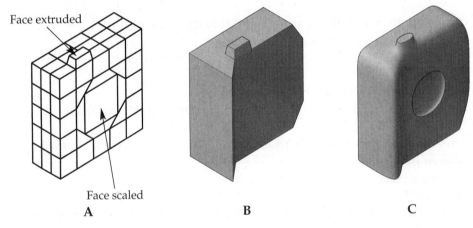

Face extruded

Face scaled

A   B   C

Viewing and Displaying Three-Dimensional Models

Learning Objectives

After completing this chapter, you will be able to:

✓ Use the viewport controls to display views and control the display of view-navigation tools.
✓ Use the navigation bar to perform a variety of display manipulation functions.
✓ Use the view cube to dynamically rotate the view of the model in 3D space.
✓ Use the view cube to display orthographic plan views of all sides on the model.
✓ Use steering wheels to display a 3D model from any angle.
✓ Hide, isolate, and unisolate objects.

AutoCAD provides several tools with which you can display and present 3D models in pictorial and orthographic views:

- Preset isometric viewpoints, discussed in Chapter 1.
- Dynamic model display using the view cube. This on-screen tool provides access to preset and dynamic display options.
- Complete 3D model display using steering wheels.
- The **3DORBIT**, **3DFORBIT**, and **3DCORBIT** commands. These commands provide dynamic display and continuous orbiting functions for demonstrations and presentations.

Once a viewpoint has been selected, you can enhance the display by applying visual styles. The **View** panel in the **Home** tab of the ribbon provides a variety of ways to display a model, including wireframe representation, hidden line removal, and simple rendering. An introduction to visual styles is provided in Chapter 1 and additional details are provided in Chapter 16.

A more advanced rendering can be created with the **RENDER** command. It produces the most realistic image with highlights, shading, and materials, if applied. **Figure 4-1** shows a 3D model of a cast iron plumbing cleanout after using **HIDE**, setting the Conceptual visual style current, and using **RENDER**. Notice the difference in the three displays.

Figure 4-1.
A—Hidden display (hidden lines removed). B—The Conceptual visual style set current.
C—Rendered with lights and materials.

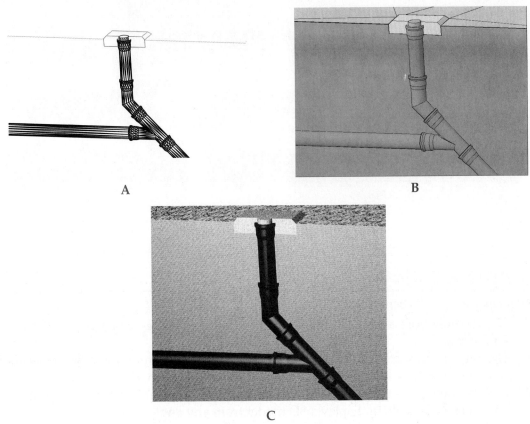

A

B

C

PROFESSIONAL TIP

The viewport controls, displayed in the upper-left corner of the AutoCAD drawing window, are discussed in Chapter 1. These controls provide quick access to preset isometric and orthographic views and settings for the view cube, navigation bar, and steering wheels. Use the viewport controls whenever possible to increase your drawing efficiency.

Keep in mind that undocked palettes may hide the viewport controls. If this happens, either dock the palette or move it out of the way.

Using the Navigation Bar to Display Models

By default, the navigation bar appears below the view cube on the right side of the screen. See Figure 4-2. It allows you to quickly use the navigation tools described in this chapter, in addition to the **ZOOM** and **PAN** commands and three orbit options. The **NAVBAR** command is used to turn on or off the display of the navigation bar. The navigation bar can be controlled using the viewport controls by selecting **Navigation Bar** in the **Viewport Controls** flyout. You can also select **Navigation Bar** in the drop-down menu displayed by picking the **User Interface** button on the **Windows** panel in the **View** tab of the ribbon. The appearance and location of the navigation bar can be customized, as discussed later in this chapter.

Figure 4-2.
The navigation
bar appears below
the view cube
by default. This
bar allows you
to quickly use
navigation tools for
drawing-display
purposes.

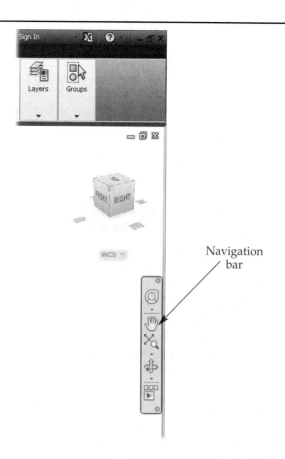

Navigation
bar

Dynamically Displaying Models with the View Cube

The *view cube* navigation tool allows you to quickly change the current view of the model to a preset pictorial or orthographic view or any number of dynamically user-defined 3D views. The tool is displayed, by default, in the upper-right corner of the drawing area. If the view cube is not displayed, pick **View Cube** in the **Viewport Controls** flyout in the viewport controls. You can also pick **View Cube** in the drop-down menu displayed by picking the **User Interface** button on the **Windows** panel in the **View** tab of the ribbon.

The **View Cube Settings** dialog box offers many settings for expanding and enhancing the view cube. This dialog box is discussed in detail later in this chapter.

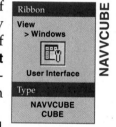

PROFESSIONAL
TIP

The location of the view cube can be quickly changed by picking the customize button on the navigation bar. Then, select **Docking positions** in the shortcut menu and select the location of your choice in the cascading menu.

Understanding the View Cube

The view cube tool is composed of a cube labeled with the names of all six orthographic faces. See **Figure 4-3**. A compass rests at the base of the cube and is labeled with the four compass points (N, S, E, and W). The compass can be used to change the view. It also provides a visual cue to the orientation of the model in relation to the current user coordinate system (UCS). The *user coordinate system* describes the

Figure 4-3.
The cube in the view cube tool is labeled with the names of all six orthographic faces. The view cube also contains a **Home** icon, compass, and UCS shortcut menu.

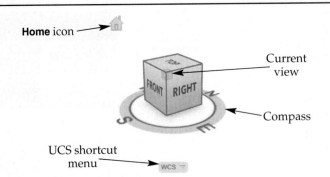

Home icon

Current view

Compass

UCS shortcut menu

orientation of the X, Y, and Z axes. A complete discussion of user coordinate systems is given in Chapter 6.

Below the cube and compass is a button labeled WCS, which displays a shortcut menu. The label on this button displays the name of the current user coordinate system, which is, by default, the world coordinate system (WCS). This shortcut menu gives you the ability to switch from one user coordinate system to another. A user coordinate system (UCS) is any coordinate system that is not the world coordinate system. The benefit of creating new user coordinate systems is the increase in efficiency and productivity when working on 3D models.

When the cursor is not over the view cube, the tool is displayed in its dimmed or *inactive state*. When the cursor is moved onto the view cube, the tool is displayed in its *active state* and a little house appears to the upper-left of the cube. This house is the **Home** icon. Picking this icon always restores the same view, called the *home view*. Later in this chapter, you will learn how to change the home view.

Notice that a corner, edge, or face on the cube is highlighted. In the default view of a drawing based on the acad3D.dwt template, the corner between the top, right, and front faces is shaded. The shaded part of the cube represents the standard view that is closest to the current view. Move the cursor over the cube and notice that edges, corners, and faces are highlighted as you move the pointer. If you pick a highlighted edge, corner, or face, a view is displayed that looks directly at the selected feature. By selecting standard views on the view cube, you can quickly move between pictorial and orthographic views.

PROFESSIONAL TIP

Picking the top face of the view cube displays a plan view of the current UCS XY plane. This is often quicker than using the **PLAN** command, which is discussed later in this chapter.

Dynamic Displays

The easiest way to change the current view using the view cube is to pick and drag the cube. Simply move the cursor over the cube, press and hold the left mouse button, and drag the mouse to change the view. The view cube and model view dynamically change with the mouse movement. When you have found the view you want, release the mouse button. This is similar to the **Orbit** tool in steering wheels, which are discussed later in the chapter.

When dragging the view cube, you are not restricted to the current XY plane, as you are when using the compass. The compass is discussed later in this chapter.

Pictorial Displays

The view cube provides immediate access to 20 different preset pictorial views. When one of the corners of the cube is selected, a standard isometric pictorial view is displayed. See **Figure 4-4A**. Eight corners can be selected. Picking one of the four corners on the top face of the cube restores one of the preset isometric views. You can select a preset isometric view using the viewport controls or the **Home** or **View** tab of the ribbon. However, using the view cube may be a quicker method.

Selecting one of the edges of the cube sets the view perpendicular to that edge. See **Figure 4-4B**. There are 12 edges that can be selected. These standard views are not available on the ribbon.

PROFESSIONAL TIP

The quickest method of dynamically rotating a 3D model is achieved by using the mouse wheel button. Simply press and hold the [Shift] key while pressing down and holding the mouse wheel. Now move the mouse in any direction and the view rotates accordingly. This is a transparent function that executes the **3DORBIT** command and can be used at any time. Since it is transparent, it can even be used while you are in the middle of a command. This is an excellent technique to use because it does not require selecting another tool or executing a command.

Projection

The view displayed in the drawing window can be in one of two projections. The *projection* refers to how lines are applied to the viewing plane. In a pictorial view, lines in a *perspective projection* appear to converge as they recede into the background. The points at which the lines converge are called *vanishing points*. In 2D drafting, it is common to represent an object in pictorial as a one- or two-point perspective, especially in architectural drafting. In a *parallel projection*, lines remain parallel as they recede. This is how an orthographic or axonometric (isometric, dimetric, or trimetric) view is created.

Figure 4-4.
A—When one of the corners of the cube is selected, a standard isometric view is displayed. B—Selecting one of the edges of the cube produces the same rotation in the XY plane as an isometric view, but a zero elevation view in the Z plane.

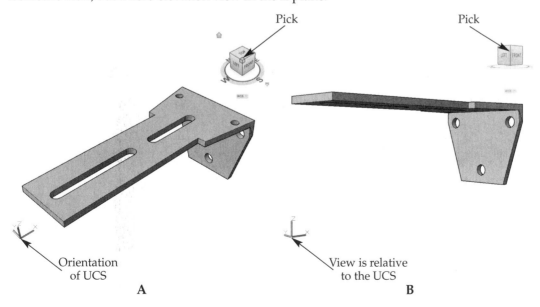

Orientation of UCS

View is relative to the UCS

A   B

To quickly change the projection using the viewport controls, pick **Parallel** or **Perspective** from the **View Controls** flyout. To change the projection using the view cube, right-click on the view cube to display the shortcut menu. Three display options are given:

- **Parallel.** Displays the model as a parallel projection. This creates an orthographic or axonometric view.
- **Perspective.** Displays the model in the more realistic, perspective projection. Lines recede into the background toward invisible vanishing points.
- **Perspective with Ortho Faces.** Displays the model in perspective projection when a pictorial view is displayed and parallel projection when an orthographic view is displayed. The parallel projection is only set current if a face on the view cube is selected to display the orthographic view. It is not set current if the **PLAN** command is used.

Figure 4-5 illustrates the difference between parallel and perspective projection.

Exercise 4-1

Complete the exercise on the companion website.
www.g-wlearning.com/CAD

Orthographic Displays

The view cube faces are labeled with orthographic view names, such as Top, Front, Left, and so on. Picking on a view cube face produces an orthographic display of that face. Keep in mind, if the current projection is perspective, the view will not be a true orthographic view. See **Figure 4-6.** If you plan to work in perspective projection, but also want to view proper orthographic faces, turn on **Perspective with Ortho Faces** in the view cube shortcut menu. Another way to achieve a proper orthographic view is to turn on parallel projection.

When a face is selected on the view cube, the cube rotates to orthographically display the named face. The view rotates accordingly. In addition, notice that a series of triangles point to the four sides of the cube (when the cursor is over the tool). See **Figure 4-7A.** Picking one of these triangles displays the orthographic view corresponding to the face to which the triangle is pointing, **Figure 4-7B.** This is a quick and efficient method to precisely rotate the display between orthographic views.

When an orthographic view is displayed, two *roll arrows* appear on the view cube. Picking either of these arrows rotates the current view 90° in the selected direction and about an axis perpendicular to the view. See **Figure 4-7C.** Using the triangles and roll arrows on the view cube provides the greatest flexibility in manipulating the model between orthographic views.

Figure 4-5.
In a parallel projection, parallel lines remain parallel. In a perspective projection, parallel lines converge to a vanishing point. Notice the three receding lines on the boxes. If these lines are extended, they will intersect.

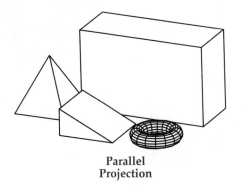

Parallel
Projection

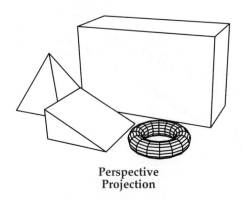

Perspective
Projection

Figure 4-6.
A—To properly view orthographic faces using the view cube, turn on **Perspective with Ortho Faces** in the shortcut menu. B—When an orthographic view is set current with perspective projection on, the view is not a true orthographic view. Notice how you can see the receding surfaces.

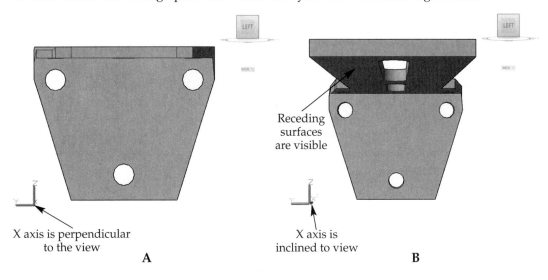

Receding surfaces are visible

X axis is perpendicular to the view

A

X axis is inclined to view

B

Figure 4-7.
A—The selected orthographic face is surrounded by triangles. Pick one of the triangles to display that orthographic face. B—The orthographic face corresponding to the picked triangle is displayed. Picking a roll arrow rotates the current view 90° in the selected direction. C—The rotated view.

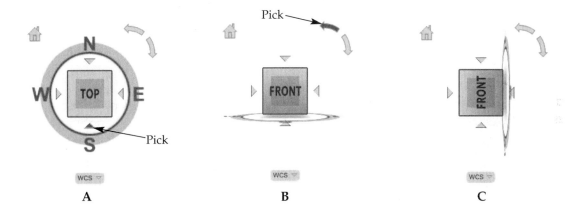

Pick

Pick

A　　　　　**B**　　　　　**C**

Setting Views with the Compass

The compass allows you to dynamically rotate the model in the XY plane. To do so, pick and hold on one of the four labels (N, S, E, or W). Then, drag the mouse to rotate the view. The view pivots about the Z axis of the current UCS. Try this a few times and notice that you can completely rotate the model by continuously moving the cursor off of the screen. For example, pick the letter W and move the pointer either right or left (or up or down). Notice as you continue to move the mouse in one direction, the model continues to rotate in that direction.

You can use the compass to display the model in a view plan to the right, left, front, or back face of the cube. When the pointer is moved to one of the four compass directions, the letter is highlighted, **Figure 4-8A**. Simply single pick on the compass label that is next to the face you wish to view. See **Figure 4-8B**.

Figure 4-8.
A—When the pointer is moved over one of the four compass directions, the letter is highlighted. B—If you pick the letter, the orthographic view from that compass direction is displayed.

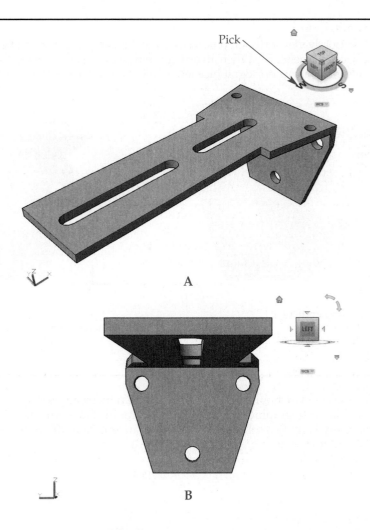

A

B

Home View

The default home view is the southwest isometric view of the WCS. Picking the **Home** icon in the view cube always displays the view defined as the home view, regardless of the current UCS. You can easily set the home view to any display you wish. First, using any navigation method, display the model as required. Next, right-click on the view cube and pick **Set Current View as Home** from the shortcut menu. Now, when you pick the **Home** icon in the view cube, this view is set current. Remember, the **Home** icon does not appear until the cursor is over the view cube.

PROFESSIONAL TIP

Should you become disoriented after repeated use of the view cube, it is far more efficient to pick the **Home** icon than it is to use **UNDO** or try to select an appropriate location on the view cube.

UCS Settings

The UCS shortcut menu in the view cube lists the WCS and the names of all named UCSs in the current drawing. If there are no named UCSs, the listing is **WCS** and **New UCS**. See **Figure 4-9A**. To create a new UCS, pick **New UCS** from the shortcut menu. Next, use the appropriate **UCS** command options to create the new UCS. See Chapter 6 for complete coverage of the **UCS** command.

As new UCSs are created and saved, their names are added to the UCS shortcut menu. See **Figure 4-9B**. Now, if you wish to work on the model using a specific UCS, simply select it from the list. The UCS is then restored.

Exercise 4-2

Complete the exercise on the companion website.
www.g-wlearning.com/CAD

View Cube Settings Dialog Box

The appearance and function of the view cube can be changed using the **View Cube Settings** dialog box, **Figure 4-10**. This dialog box is displayed by selecting **View Cube Settings...** from the view cube shortcut menu or by using the **Settings** option of the **NAVVCUBE** command. The next sections discuss the options found in the dialog box.

NAVVCUBE

Display

Options in the **Display** area of the **View Cube Settings** dialog box control the appearance of the view cube tool. The thumbnail dynamically previews any changes made to the display options. There are four options in this area of the dialog box.

Figure 4-9.
The UCS shortcut menu below the view cube lists all named UCSs in the current drawing. A—The menu entries are **WCS** and **New UCS**. In this example, there are no saved UCSs in the drawing. B—The names of new UCSs are added to the UCS shortcut menu. The current UCS is indicated with a check mark.

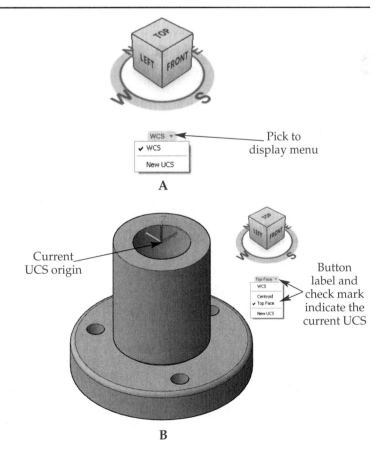

Figure 4-10.
The **View Cube**
Settings dialog box.

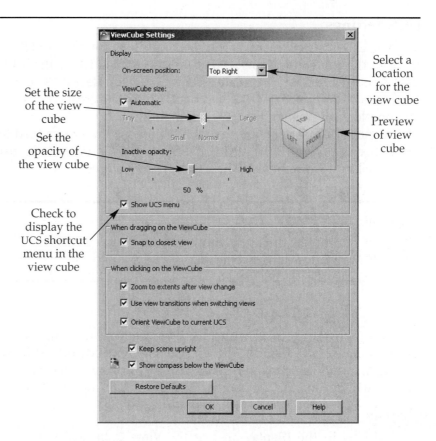

Set the size
of the view
cube

Set the
opacity of
the view cube

Check to
display the
UCS shortcut
menu in the
view cube

Select a
location
for the
view cube

Preview
of view
cube

On-Screen Position. The view cube can be placed in one of four locations in the drawing area: top-right, bottom-right, top-left, or bottom-left corner. By default, it is located in the top-right corner of the screen. To change the location, select it in the **On-screen position:** drop-down list. The **NAVVCUBELOCATION** system variable controls this setting.

View Cube Size. The view cube can be displayed in one of four sizes. Use the **View Cube Size:** slider to set the size to tiny, small, normal, or large. The slider is unavailable if the **Automatic** check box is checked, in which case AutoCAD sets the size based on the available screen area. The **NAVVCUBESIZE** system variable controls this setting.

Inactive Opacity. When the view cube is inactive, it is displayed in a semitransparent state. Remember, the view cube is in the inactive state whenever the pointer is not over it. When the view cube is in the active state, it is displayed at 100% opacity. Use the **Inactive opacity:** slider to set the level of opacity (transparency). The value can be from 0% to 100%; the default value is 50%. The percentage is displayed below the slider. A value of zero results in the view cube being hidden until the cursor is moved over it. If opacity is set to 100%, there is no difference between the inactive and active states. The **NAVVCUBEOPACITY** system variable controls this setting.

Show UCS Menu. By default, the UCS shortcut menu is displayed in the view cube. This menu is displayed by picking the button below the cube. If you wish to remove the UCS menu from the view cube display, uncheck the **Show UCS menu** check box.

When Dragging on the View Cube

By default, when you drag the view cube, the view "snaps" to the closest standard view that can be displayed by the view cube. This is because the **Snap to closest view** check box is checked by default. Uncheck this check box if you want the view to freely rotate without snapping to a preset standard view as you drag the view cube.

When Clicking on the View Cube

Options in the **When clicking on the View Cube** area control how the final view is displayed and how the labels on the view cube can be related to the UCS. There are three options in this area.

Zoom to Extents After View Change. If the **Zoom to extents after view change** check box is checked, the model is zoomed to the extents of the drawing whenever the view cube is used to change the view. Uncheck this if you want to use the view cube to change the display without fitting the model to the current viewport.

Use View Transitions When Switching Views. The default transition from one view to the next is a smooth rotation of the view. Although this transition may look nice, if you are working on a large model it may require more time and computer resources than you are willing to use. Therefore, if you are switching views a lot using the view cube, it may be more efficient to uncheck the **Use view transitions when switching views** check box. When this is unchecked, a view change just cuts to the new view without a smooth transition. This option does not affect the view when dragging the view cube or its compass.

Orient View Cube to Current UCS. As you have seen, the view cube is aligned to the current UCS by default. In other words, the top face of the cube is always perpendicular to the Z axis of the UCS. However, this can be turned off by unchecking the **Orient View Cube to current UCS** check box. The **NAVVCUBEORIENT** system variable controls this setting. When unchecked, the faces of the view cube are not reoriented when the UCS is changed. It may be easier to visualize view changes if the view cube faces are oriented to the current UCS.

When the **Orient View Cube to current UCS** check box is unchecked, WCS is displayed above the UCS shortcut menu in the view cube (unless the WCS is current). See **Figure 4-11**. This is a reminder that the labels on the view cube relate to the WCS and not to the current UCS.

Additional Options

Three additional options are available near the bottom of the **View Cube Settings** dialog box. These options are located below the **When clicking on the View Cube** area.

Keep Scene Upright. If the **Keep scene upright** check box is checked, the view of the model cannot be turned upside down. When unchecked, you may accidentally rotate the view so it is upside down. This can be confusing, so it is best to leave this box checked.

Show Compass below the View Cube. The compass is displayed by default in the view cube. However, if you do not find the compass useful, you can turn it off. To hide the compass, uncheck the **Show compass below the View Cube** check box.

Restore Defaults. After making several changes in the **View Cube Settings** dialog box, you may get confused about the effect of different option settings on the model display. In this case, it is best to pick the **Restore Defaults** button to return all settings to their original values. Then change one setting at a time and test it to be sure the view cube functions as you intended.

Figure 4-11.
When the **Orient View Cube to current UCS** option is off in the **View Cube Settings** dialog box, the UCS shortcut menu has WCS displayed above it.

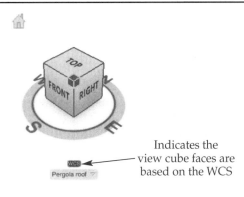

Indicates the view cube faces are based on the WCS

Exercise 4-3

Complete the exercise on the companion website.
www.g-wlearning.com/CAD

Creating a Continuous 3D Orbit

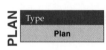

3DCORBIT

Ribbon

View
> Navigate

Continuous Orbit

Type

3DCORBIT

The **3DCORBIT** command provides the ability to create a continuous orbit of a model. By moving your pointing device, you can set the model in motion in any direction and at any speed, depending on the power of your computer. An impressive display can be achieved using this command. The command is located in the **Orbit** drop-down list in the **Navigate** panel of the **View** tab on the ribbon. See **Figure 4-12A**. Additionally, it can be selected on the navigation bar.

Once the command is initiated, the continuous orbit cursor is displayed. See **Figure 4-12B**. Press and hold the pick button and move the pointer in the direction that you want the model to rotate and at the desired speed of rotation. Release the button when the pointer is moving at the appropriate speed. The model will continue to rotate until you pick the left mouse button, press [Enter] or [Esc], or right-click and pick **Exit** or another option in the shortcut menu. At any time while the model is orbiting, you can left-click and adjust the rotation angle and speed by repeating the process for starting a continuous orbit.

Plan Command Options

PLAN

Type

Plan

The **PLAN** command, introduced in Chapter 1, allows you to create a plan view of any user coordinate system (UCS) or the world coordinate system (WCS). The **PLAN** command automatically performs a **ZOOM Extents**. This fills the drawing window with the plan view. The command options are:

- **Current UCS.** This creates a view of the object that is plan to the current UCS.
- **UCS.** This displays a view plan to a named UCS. The preset UCSs are not considered named UCSs.
- **World.** This creates a view of the object that is plan to the WCS. If the WCS is the current UCS, this option and the **Current UCS** option produce the same results.

Figure 4-12.
A—Selecting the **3DCORBIT** command. B—This is the continuous orbit cursor in the **3DCORBIT** command (or **Continuous** option of the **3DORBIT** command). Pick and hold the left mouse button. Then, move the cursor in the direction in which you want the view to rotate and release the mouse button.

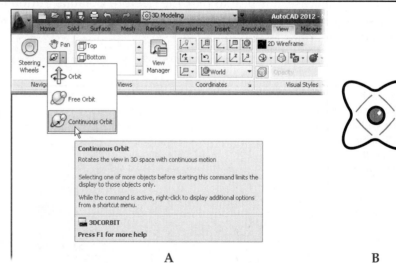

A   B

This command may have limited usefulness when working with 3D models. The dynamic capabilities of the view cube are much more intuitive and may be quicker to use. However, you may find instances where the **PLAN** command is easier to use, such as when the view cube is not currently displayed.

Displaying Models with Steering Wheels

Steering wheels, or *wheels*, are dynamic menus that provide quick access to view-navigation tools. A steering wheel follows the cursor as it is moved around the drawing. Each wheel is divided into wedges and each wedge contains a tool. See **Figure 4-13**. The **NAVSWHEEL** command is used to display a steering wheel, but a quicker method is to pick **Steering Wheels** from the **Viewport Controls** flyout in the viewport controls. The steering wheel drop-down menu on the navigation bar can also be used. All of the steering wheel options discussed in this chapter can be selected in the drop-down menu on the navigation bar.

AutoCAD provides two basic types of wheels: View Object and Tour Building. The options contained in these two wheels are combined to form the Full Navigation wheel. Each of these three types can be used in a full-wheel display or a minimized format. This section discusses the Full Navigation wheel and the two basic wheels in their full and mini formats. There is also a 2D Wheel that is displayed in paper (layout) space or when the **NAVSWHEELMODE** system variable is set to 3. The 2D Wheel provides basic view-navigation tools for 2D drafting.

Ribbon

NAVSWHEEL

View
> Navigate

Steering Wheels

Type

NAVSWHEEL
WHEEL

Toolbar

Navigation Bar

Steering Wheels

Using a Steering Wheel

Once a steering wheel is displayed, move the cursor around the screen. Notice as you move the cursor, the wheel follows it. When you stop the cursor, the wheel stops. If you move the cursor anywhere inside of the wheel, the wheel remains stationary. Note also that as you move the cursor inside of the wheel, a wedge (tool) is highlighted. If you pause the cursor over a tool, a tooltip is displayed that describes the tool.

To use a specific tool, simply pick and hold on the highlighted wedge, then move the cursor as needed to change the view of the model. Once you release the pick button, the tool ends and the wheel is redisplayed. Some options display a *center point* about

Figure 4-13.
A—Full Navigation wheel. B—View Object wheel. C—Tour Building wheel.

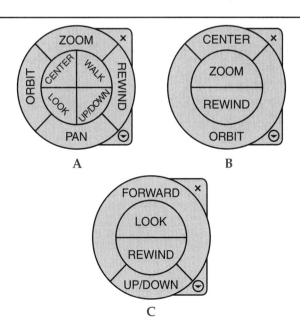

which the display will move. The **Center** tool is used to set the center point. These features are all discussed in the next sections.

To change between wheels, right-click to display the shortcut menu. Then, select **Full Navigation Wheel** from the menu to display that wheel. Or, select **Basic Wheels** to display a cascading menu. Select either **View Object Wheel** or **Tour Building Wheel** from the cascading menu to display that wheel. Refer to **Figure 4-14**.

Exercise 4-4

Complete the exercise on the companion website.
www.g-wlearning.com/CAD

Full Navigation Wheel

The Full Navigation wheel contains all of the tools available in the View Object and Tour Building wheels. Refer to **Figure 4-13A**. These tools are discussed in the following sections. The Full Navigation wheel also contains the **Pan** and **Walk** tools, which are not available in either of the other two wheels. These tools are discussed next. The tools shared with the View Object and Tour Building tools are discussed in the sections corresponding to those tools.

In addition, a number of settings are available to change the appearance of the steering wheel and the manner in which some of the tools function. Refer to the Steering Wheel Settings section later in this chapter for a complete discussion of these settings.

Pan

The **Pan** tool allows you to move the model in the direction that you drag the cursor. This tool functions exactly the same as the AutoCAD **RTPAN** command. When you pick and hold on the tool, the cursor changes to four arrows with the label Pan Tool below the cursor.

Figure 4-14.
The shortcut menu allows you to switch between steering wheels.

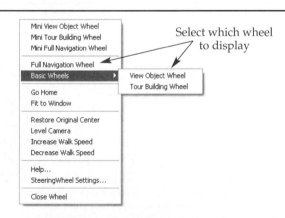

If you use the **Pan** tool with the perspective projection current, it may appear that the model is slowly rotating about a point. This is not the case. What you are seeing is merely the effect of the vanishing points. As you pan, the relationship between the viewpoint and the vanishing points changes. You can quickly test this by closing the wheel, right-clicking on the view cube, and picking **Parallel** from the shortcut menu. Now display the wheel again and use the **Pan** tool. Notice the difference. The model pans without appearing to rotate.

Walk

The **Walk** tool is used to simulate walking toward, through, or away from the model. When you pick and hold on the tool, the center circle icon is displayed at the bottom-center of the drawing area. See **Figure 4-15**. The cursor changes to an arrow pointing away from the center of the circle as you move it off of the circle. The arrow indicates the direction in which the view will move as you move the mouse. This gives the illusion of walking in that direction in relation to the model.

If you hold down the [Shift] key while clicking the **Walk** tool, the **Up/Down** slider is displayed. This allows you to change the screen Y axis orientation of the view. This equates to elevating the camera view relative to the object. Releasing the [Shift] key returns you to the standard walk mode. The up and down arrow keys can also be used to change the "height" of the view. The speed of walking can be increased with the plus key (+).

The **Up/Down** slider is also used with the **Up/Down** tool on the Tour Building wheel. It is explained in the Tour Building Wheel section.

View Object Wheel

If you think of your model as a building, the tools in the View Object wheel are used to view the outside of the building. This wheel contains four navigation tools: **Center**, **Zoom**, **Rewind**, and **Orbit**. Refer to **Figure 4-13B**.

Center

The **Center** tool is used to set the center point for the current view. Many tools, such as **Zoom** and **Orbit**, are applied in relation to the center point. Pick and hold the **Center** tool, then move the cursor to a point on the model and release. The display immediately changes to center the model on that point. See **Figure 4-16**. The selected point must be on an object, but it does not need to be on a solid. Notice that the center point icon resembles a globe with three orbital axes. These axes relate to the three axes of the model shown on the UCS icon.

Zoom

The **Zoom** tool is used to dynamically zoom the view in and out, just as with the AutoCAD **RTZOOM** command. The tool uses the center point set with the **Center** tool.

Figure 4-15.
The **Walk** tool displays the center circle icon. As you move the cursor around the icon, one of the arrows shown here is displayed to indicate the direction in which the view is being moved.

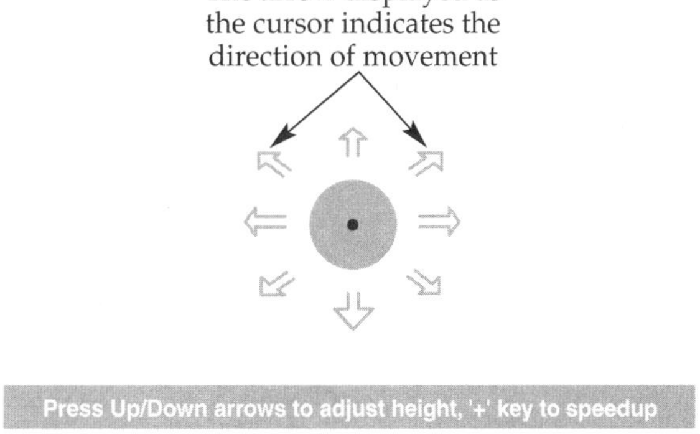

The arrow displayed as the cursor indicates the direction of movement

Press Up/Down arrows to adjust height, '+' key to speedup

Figure 4-16.
The **Center** tool allows you to select a new center point for the current view. This becomes the point about which many steering wheel tools operate.

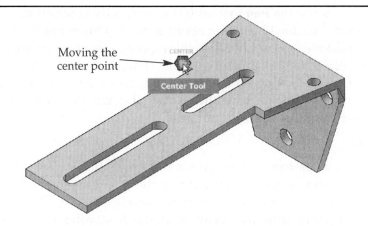

Moving the center point

Center Tool

When you pick and hold the **Zoom** tool in the wheel, the center point is displayed at its current location. The cursor changes to a magnifying glass with the label **Zoom Tool** displayed below it. See **Figure 4-17**. There are three different ways to use the **Zoom** tool, as discussed in this section.

When the **Zoom** tool is accessed in the Full Navigation wheel, the center point is relocated to the position of the steering wheel. If you wish to zoom on the existing center point when using the Full Navigation wheel, first press the [Ctrl] key and then access the **Zoom** tool. This prevents the tool from relocating the center point. You can also move the center point using the **Center** tool, then switch to the View Object wheel and access the **Zoom** tool from that wheel. In either case, the zoom is relative to the location of the center point.

Once you close the steering wheel, the center point is reset. The next time a steering wheel is displayed, the center point will be in the middle of the current view.

Pick and Drag. Pick and drag the cursor to dynamically zoom. This is similar to performing a realtime zoom. As the cursor is moved up or to the right, the viewpoint moves closer (zooms in). Move the pointer to the left or down and the viewpoint moves farther away (zooms out). The zooming is based on the current center point. When you have achieved the appropriate zoom location, release the pointer button.

Single Click. If you select the **Zoom** tool with a single click (using the View Object wheel), the view of the model zooms in by an incremental percentage. Each time you single click on the tool, the view is zoomed by 25%. The zoom is in relation to the center point.

Figure 4-17.
The **Zoom** tool operates in relation to the center point. If the tool is selected from the Full Navigation wheel, the center point is automatically relocated to the cursor location.

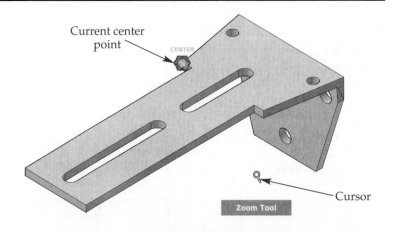

Current center point

CENTER

Zoom Tool

Cursor

In order for this function to work when the **Zoom** tool is accessed from the Full Navigation wheel, you must check the **Enable single click incremental zoom** check box in the **Steering Wheels Settings** dialog box. This dialog box is discussed in detail later in the chapter.

Shift and Click. If you press and hold the [Shift] key and then pick the **Zoom** tool (using the View Object wheel), the view is zoomed out by 25%. As with the single-click method, the **Enable single click incremental zoom** check box must be checked in order to use this with the Full Navigation wheel.

Rewind

The **Rewind** tool allows you to step back through previous views. A single pick on this tool displays the previous view. If you pick and hold the tool, a "slide show" of previous views is displayed as thumbnail images. See **Figure 4-18**. The most recent view is displayed on the right-hand side. The oldest view is displayed on the left-hand side. The slide representing the current view is highlighted with an orange frame and a set of brackets. While holding the pick button, move the cursor to the left. Notice that the set of brackets moves with the cursor. As a slide is highlighted, the corresponding view is restored in the viewport. Release the pick button when you find the view you want and it is set current.

The navigation history is maintained in the drawing file and is different for each open drawing. However, it is not saved when a drawing is closed. The **Steering Wheels Settings** dialog box allows you to control when thumbnail images are created and saved in the navigation history. This dialog box is discussed later in the chapter.

Orbit

The **Orbit** tool allows you to completely rotate your point of view around the model in any direction. The view pivots about the center point set with the **Center** tool. When using the Full Navigation wheel, the center point can be quickly set by pressing and holding the [Ctrl] key, then picking and holding the **Orbit** tool. Next, drag the center point to the desired pivot point on the model and release. Now you can use the **Orbit** tool.

To use the **Orbit** tool, pick and hold on the tool in the steering wheel. The current center point is displayed with the label Pivot. Also, the cursor changes to a point surrounded by two circular arrows. See **Figure 4-19**. Move the cursor around the screen and the view of the model pivots about the center point. If this is not the result you wanted, just reset the pivot point.

Figure 4-18.
A single pick on the **Rewind** tool displays the previous view. A "slide show" of previous views is displayed as thumbnail images if you press and hold the pick button.

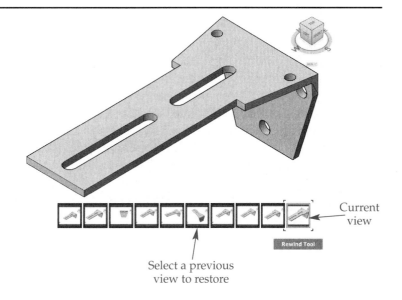

Current view

Rewind Tool

Select a previous view to restore

Figure 4-19.
The **Orbit** tool
allows you to rotate
your point of view
completely around
the model in any
direction. This is the
cursor displayed for
the tool.

Tour Building Wheel

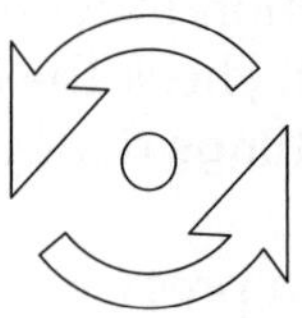

Where the tools in the View Object wheel are used to view the outside of the "building," the tools in the Tour Building wheel are used to move around inside of the "building." This wheel contains four navigation tools. The **Forward, Look**, and **Up/ Down** tools are discussed here. The **Rewind** tool is addressed in the View Object Wheel section, discussed earlier.

Forward

The principal tool in the Tour Building wheel is the **Forward** tool. It is similar to the **Walk** tool in the Full Navigation wheel. However, it only allows forward movement from the current viewpoint. This tool requires a center point to be set on the model from within the tool. The existing center point cannot be used.

First, move the cursor and steering wheel to the point on the model that will be the target (center point). Next, pick and hold the **Forward** tool. The pick point becomes the center point and a drag distance indicator is displayed. See **Figure 4-20**. This indicator shows the starting viewpoint, the center point, and the surface of the model that you selected. Hold the mouse button down while moving the pointer up. The orange location slider moves to show the current viewpoint relative to the center point. As you move closer to the model, the green center point icon increases in size, which also provides a visual cue to the zoom level.

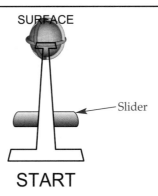

Figure 4-20.
The drag distance indicator is displayed when using the **Forward** tool. This indicator shows the start point of the view and the selected surface of the model. The slider indicates the current view position relative to the starting point.

SURFACE

Slider

START

Look

The **Look** tool is used to rotate the view about the center of the view. When the tool is activated, the cursor appears as a half circle with arrows. See **Figure 4-21.** As you move the cursor down, the model moves up in the view as if you are actually tilting your head down to see the top of the model. Similarly, as you "look" away from the model to the right or left, the model appears to move away from your line of sight. If this does not seem intuitive to you, it is easy to change. Open the **Steering Wheels Settings** dialog box and check the **Invert vertical axis for Look tool** check box. Now when using the **Look** tool, moving the cursor in one direction also moves the object in the same direction.

The distance between you and the model remains the same and the orientation of the model does not change. Therefore, you would not want to use this tool if you wanted to see another side of the model.

Up/Down

As the name indicates, the **Up/Down** tool moves the view up or down along the Y axis of the screen, regardless of the orientation of the current UCS. Pick and hold on the tool and the vertical distance indicator appears. See **Figure 4-22.** Two marks on this indicator show the upper and lower limits within which the view can be moved. The orange slider shows the position of the view as you move the cursor. When the tool is first activated, the view is at the top position. The **Up/Down** tool has limited value. The **Pan** tool is far more versatile.

Figure 4-21.
The **Look** tool cursor appears as a half circle with arrows.

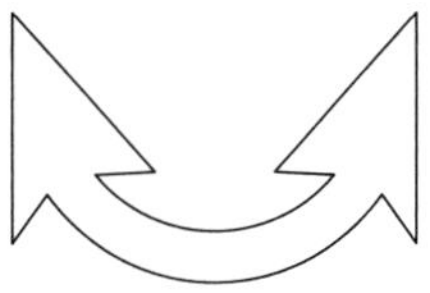

Figure 4-22.
The indicator displayed when using the **Up/Down** tool shows the upper and lower limits within which the view can be moved. The top position is the location of the view when the tool is selected.

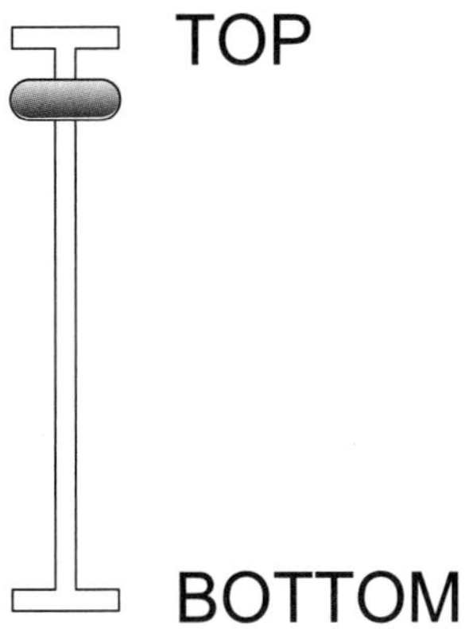

TOP

BOTTOM

Use steering wheel tools such as **Forward, Look, Orbit,** and **Walk** to manipulate a view and fully explore the model. The **Rewind** tool can then be used to replay all of the previous views saved in the navigation history. This process reveals model views that can be saved as named views for later use in model construction or for shots created with show motion (discussed in Chapter 5). Remember, views created using steering wheels are not saved with the drawing file. Therefore, it may be a time-saver to create and save named views if there is a possibility they will be needed later. Named views are created using the **New View/Shot Properties** dialog box, which is accessed by picking **View Manager...** from the **View Controls** flyout in the viewport controls and then picking the **New...** button. Using the **New View/Shot Properties** dialog box is discussed in detail in Chapter 5.

Exercise 4-5

Complete the exercise on the companion website.
www.g-wlearning.com/CAD

Mini Wheels

The three wheels discussed previously were presented in their *full wheel* formats. As you gain familiarity with the use and function of each wheel and its tools, you may wish to begin using the abbreviated formats. The abbreviated formats are called *mini wheels*. See **Figure 4-23**. The mini wheels can be selected in the steering wheel shortcut menu. When you select a mini wheel, it replaces the cursor pointer.

As you move the mouse, the mini wheel follows. Slowly move the mouse in a small circle and notice that each wedge of the wheel is highlighted. The name of the currently highlighted tool appears below the mini wheel. When the tool you need is highlighted, simply click and hold to activate the tool. All tools in the mini wheels function the same as those in the full-size wheels. The only difference is the appearance of the wheels.

Steering Wheel Shortcut Menu

A quick method for switching between the different wheel formats is to use the steering wheel shortcut menu. Pick the menu arrow at the lower-right corner of a full wheel to display the menu. See **Figure 4-24**. Additionally, you can right-click when any full or mini wheel is displayed to access the menu. To select a wheel or change wheel formats, simply select the appropriate entry in the menu. Note that a check mark is *not* placed by the current wheel.

Figure 4-23.
In addition to full-size steering wheels, mini wheels can be used. A—Mini Full Navigation wheel. B—Mini View Object wheel. C—Mini Tour Building wheel.

Current tool

Name of tool

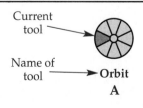

Orbit
A

Pan
B

Walk
C

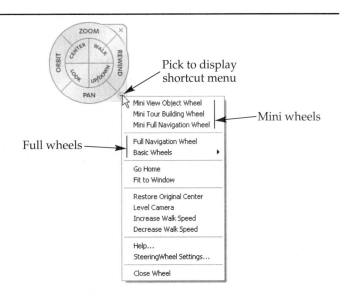

Figure 4-24.
Select the down arrow to display the steering wheel shortcut menu. Both full-size wheels and mini wheels can be displayed using this menu.

Pick to display shortcut menu

Mini wheels

Full wheels

In addition to selecting a wheel, the shortcut menu provides options for viewing the model. These additional options are described as follows:

- **Go Home.** Returns the display to the home view. This is the same as picking the **Home** icon in the view cube.
- **Fit to Window.** Resizes the current view to fit all objects in the drawing inside of the window. This is essentially a zoom extents operation.
- **Restore Original Center.** Restores the original center point of the drawing using the current drawing extents. This does not change the current zoom factor. Therefore, if you are zoomed close into an object and pick this option, the object may disappear from view by moving off of the screen.
- **Level Camera.** The camera (your viewpoint) is rotated to be level with the XY ground plane.
- **Increase Walk Speed.** The speed used by the **Walk** tool is increased by 100%.
- **Decrease Walk Speed.** The speed used by the **Walk** tool is decreased by 50%.
- **Help.** Displays the online documentation (help file) for steering wheels.
- **Steering Wheel Settings.** Displays the **Steering Wheels Settings** dialog box, which is discussed in the next section.
- **Close Wheel.** Closes the steering wheel. This is the same as pressing [Esc] to close the wheel.

Exercise 4-6

Complete the exercise on the companion website.
www.g-wlearning.com/CAD

Steering Wheel Settings

The **Steering Wheels Settings** dialog box provides options for wheel appearance. See **Figure 4-25.** It also contains settings for the display and operation of several tools. It is displayed by picking **Steering Wheels Settings** in the steering wheel shortcut menu.

Changing Wheel Appearance

The two areas at the top of the **Steering Wheels Settings** dialog box allow you to change the size and opacity of all wheels. The settings in the **Big Wheels** area are for the full-size wheels. The settings in the **Mini Wheels** area are for the mini wheels. The

Figure 4-25.
The **Steering Wheels Settings** dialog box provides options for wheel appearance and the display and operation of several tools.

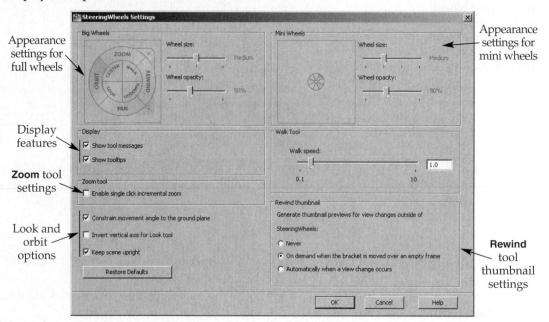

Appearance settings for full wheels

Display features

Zoom tool settings

Look and orbit options

Appearance settings for mini wheels

Rewind tool thumbnail settings

Wheel size: slider in each area is used to display the wheels in small, normal, or large size. See **Figure 4-26.** The mini wheel has a fourth, extra large size. These sliders set the **NAVSWHEELSIZEBIG** and **NAVSWHEELSIZEMINI** system variables.

The **Wheel opacity:** slider in each area controls the transparency of the wheels. These sliders can be set to a value from 25% to 90% opacity. The sliders control the **NAVSWHEELOPACITYBIG** and the **NAVSWHEELOPACITYMINI** system variables. The appearance of the full wheel in three different opacity settings is shown in **Figure 4-27.**

Display Features

The **Display** area of the **Steering Wheels Settings** dialog box controls two features of the wheel display. These settings determine whether messages and tooltips are displayed.

When a tool is selected, its name is displayed below the cursor. In addition, some tools have features or restrictions that can be indicated in a tool message. To see these messages, the **Show tool messages** check box must be checked. Otherwise, the messages are not displayed, but the restrictions remain in effect.

A tooltip is a short message that appears below the wheel when the cursor is hovered over the wheel. When the **Show tooltips** check box is checked, you can hold the cursor stationary over a tool for approximately three seconds and the tooltip will appear. Then, as you move the cursor over tools in the wheel, the appropriate tooltip is immediately displayed.

Figure 4-26.
The size of a steering wheel can be set in the **Steering Wheels Settings** dialog box or by using a system variable.

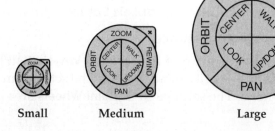

Small   Medium   Large

Figure 4-27.
The opacity of a steering wheel can be changed. A—Opacity of 25%. B—Opacity of 50%.
C—Opacity of 90%.

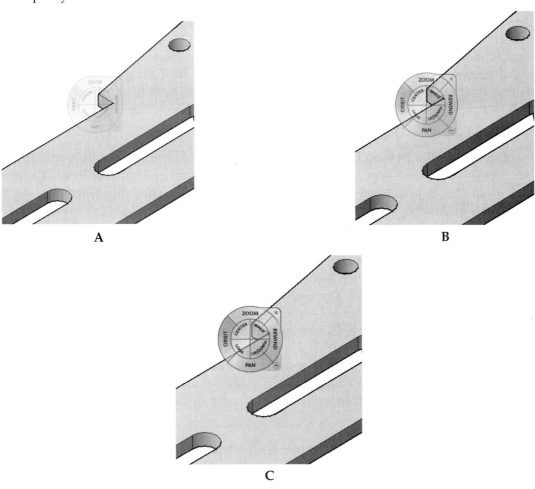

A

B

C

Walk Tool

By default, when the **Walk** tool is used, you move parallel to the ground plane. This is because the **Constrain movement angle to ground plane** check box is checked. Test this by selecting the **Walk** tool and then move the cursor toward the top of the screen. You appear to be "walking" over the top of the model. Now, open the **Steering Wheels Settings** dialog box and uncheck the **Constrain movement angle to ground plane** check box. This allows you to "fly" in the direction the cursor is moved when using the **Walk** tool. Exit the dialog box and again select the **Walk** tool. Move the cursor toward the top of the screen. This time it appears that you are flying directly toward or into the model.

The speed at which you walk through or around the model is controlled by the **Walk speed:** slider. The value can also be changed by typing in the text box at the right-hand end of the slider. The greater the value, the faster you will move as the cursor is moved away from the center circle icon.

Zoom Tool

As discussed earlier, a single click on the **Zoom** tool zooms in on the current view by a factor of 25%. This is controlled by the **Enable single click incremental zoom** check box. When this check box is not checked, a single click on the tool has no effect. Some users find the single-click zoom confusing.

Look and Orbit Tool Options

By default, when the **Look** tool is used and the cursor is moved downward, the view of the model moves up, just as if you were moving your eyes down. If you check the **Invert vertical axis for Look tool** option, this movement is reversed. In this case, the model moves in the same direction as the cursor.

If the **Keep scene upright** check box is not checked, it is possible to turn the model upside down while using the **Orbit** tool. This may not be desirable because it can be disorienting. To prevent this, be sure to leave the option checked. Uncheck this option only when you want to use **Orbit** in a "free-floating" mode.

Rewind Thumbnail Options

Options in the **Rewind thumbnail** section of the **Steering Wheels Settings** dialog box control when and how thumbnail images are generated for use with the **Rewind** tool when view changes are made without using a wheel. There are three options in this area. Only one option can be on.

When the **Never** radio button is on, thumbnail images are never generated for view changes made outside of a wheel. Thumbnail images are created for view changes made with a wheel, which is true for all three options.

When the **On demand when the bracket is moved over an empty frame** radio button is on, thumbnail images are not automatically generated for view changes made outside of a wheel. The frames for these views display a double arrow icon when the **Rewind** tool in a wheel is used. However, as the brackets are moved over these frames, thumbnail images are generated. This is the default setting. See **Figure 4-28**.

When the **Automatically when a view change occurs** option is selected, a thumbnail image is generated any time a view change is made outside of a wheel. When the **Rewind** tool in a wheel is used, the frames for these views automatically display thumbnail images.

Restore Defaults

Picking the **Restore Defaults** button in the **Steering Wheels Settings** dialog box returns all of the settings in the dialog box to their default values. Select this when at any time you are not sure how the settings are affecting the appearance and function of the steering wheel tools. Then, make changes one at a time as needed.

Exercise 4-7

Complete the exercise on the companion website.
www.g-wlearning.com/CAD

Figure 4-28.
A—By default, when the **Rewind** tool is selected, frames representing view changes made outside of a steering wheel display double-arrow icons. B—As the brackets are moved over the blank frames, thumbnail images are generated of the views created outside of a wheel.

View change made outside of a wheel

Thumbnail is generated

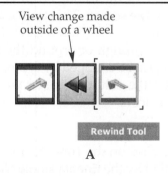

Rewind Tool

A

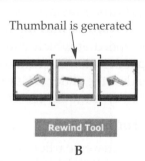

Rewind Tool

B

The navigation bar is a flexible tool that can be customized or positioned to suit the needs of your drawing project. The three principal navigation tools are for the view cube, steering wheels, and show motion (discussed in Chapter 5). The view cube is displayed on the navigation bar when the normal display of the view cube is turned off.

The navigation bar can be placed at any location around the edge of the screen. In addition, the bar can be linked to the location of the view cube. The navigation bar can be quickly turned on or off by picking **Navigation Bar** in the **Viewport Controls** flyout in the viewport controls.

Repositioning the Navigation Bar

The navigation bar can be moved around the screen by picking the customize button at the lower-right corner of the bar, and then selecting **Docking positions** to display the cascading menu. See **Figure 4-29**. The default setting is for the bar to be linked to the location of the view cube. This means if the position of the view cube is changed, the navigation bar will follow. Test this by first confirming that **Link to View Cube** is checked in the menu. Next, move the view cube to one of the other locations, such as top left, using the **View Cube Settings** dialog box. Notice that the view cube and navigation bar move to the new location. Picking the **Undo** button on the **Quick Access** toolbar can quickly restore the previous location.

You can also move the navigation bar to a position of your choosing without moving the view cube. Simply uncheck the **Link to View Cube** option. Notice that a band appears at the top of the bar. When the pointer is moved into this band and you click, it becomes a move cursor. See **Figure 4-30**. Drag the bar to any edge location on the screen. The bar will dock at the closest screen edge. You cannot place it in a floating position in the middle of the screen.

Figure 4-29.
The navigation bar can be moved around the screen by selecting the customize button at the lower-right corner of the bar, and then selecting **Docking positions** in the menu.

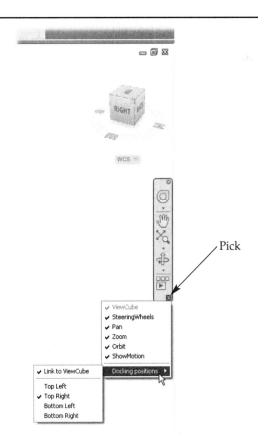

Pick

Figure 4-30.
Click and hold to
move the navigation
bar to any position
along the edge of the
screen.

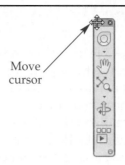

Move
cursor

Customizing the Navigation Bar

The tools that are displayed in the navigation bar can be changed by selecting the customize button at the lower-right corner of the bar. By default, all of the tools are checked. If you wish to hide a tool, just select it to uncheck it in the list. All of the tools in the navigation bar can be removed or redisplayed in this manner. If the view cube is displayed on the screen, it will be grayed out in the list. If the view cube is currently not displayed, an option for it appears in the menu. A button for the view cube also appears on the navigation bar. See **Figure 4-31**. The view cube can then be turned on using the navigation bar.

Hiding and Isolating Objects

Complex 3D models require the manipulation of viewpoints and even freezing layers in order to work on details. Instead of freezing layers, the **HIDEOBJECTS** and **ISOLATEOBJECTS** commands can be used to selectively view any aspect of the model for ease of construction and editing.

The **HIDEOBJECTS** command suppresses selected objects from view. This allows you to work on objects that are obscured by the selected objects. The **ISOLATEOBJECTS** command suppresses the display of all objects except those objects that are selected. In other words, it hides unselected objects.

Once the work is completed, simply use the **UNISOLATEOBJECTS** command to restore the view of all hidden objects. This command displays all objects hidden with either the **HIDEOBJECTS** or **ISOLATEOBJECTS** command.

The commands can also be selected from the **Isolate** flyout of the shortcut menu displayed by right-clicking in the drawing window. See **Figure 4-32**. In this menu, the

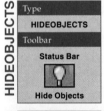

HIDEOBJECTS

Type
HIDEOBJECTS
Toolbar
Status Bar
Hide Objects

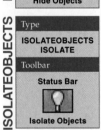

ISOLATEOBJECTS

Type
ISOLATEOBJECTS
ISOLATE
Toolbar
Status Bar
Isolate Objects

UNISOLATEOBJECTS

Type
UNISOLATEOBJECTS
UNISOLATE
UNHIDE
Toolbar
Status Bar
Unisolate Objects

Figure 4-31.
If the view cube
is not currently
displayed, a button
for displaying it
appears on the
navigation bar.

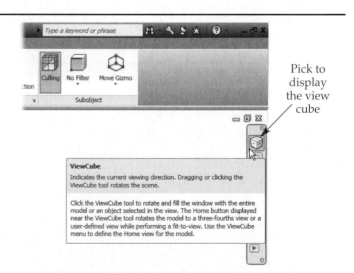

Pick to
display
the view
cube

Figure 4-32.
Using the shortcut
menu to hide and
show objects.

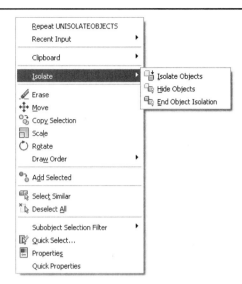

End Object Isolation selection issues the **UNISOLATEOBJECTS** command. A similar menu is displayed when the **Isolate Objects** button is picked on the status bar.

Consider an example of how to use these commands. The roof panel in **Figure 4-33A** needs to be hidden from view to construct framing details. Simply select **Hide Objects** from the **Isolate** flyout in the shortcut menu and then select the object(s) to hide. **Figure 4-33B** shows one roof panel hidden from view.

Conversely, suppose the floor framing joists need to be exposed for work. For example, you may need to design or determine hardware for the joist connections. In this case, select **Isolate Objects** from the **Isolate** flyout in the shortcut menu and then select the object(s) you want to view. After pressing [Enter], all of the unselected objects in the model are removed from the view, thus isolating the selected objects. The result is shown in **Figure 4-33C**.

PROFESSIONAL TIP

If you are confused about what parts of the model you have hidden or isolated, just pick **End Object Isolation** from the **Isolate** flyout of the shortcut menu to restore the entire model.

Figure 4-33.
A—The roof panel is in the way and needs to be hidden. B—The roof panel is hidden.
C—Only the floor joists were selected. The unselected objects are hidden.

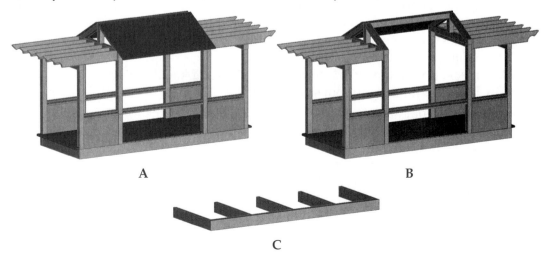

Chapter Review

1. How do you select a standard isometric preset view using the view cube?
2. How is a standard orthographic view displayed using the view cube?
3. What is the difference between *parallel projection* and *perspective projection*?
4. What happens when one of the four view cube compass letters is picked?
5. Which command generates a continuous 3D orbit?
6. Which command can be used to produce a view that is parallel to the XY plane of the current UCS?
7. What is a *steering wheel*?
8. Briefly describe how to use a steering wheel.
9. The principal tool in the Tour Building wheel is the **Forward** tool. What is the purpose of this tool?
10. What are the three principal navigation tools found in the navigation bar?

Drawing Problems

1. Open one of your 3D drawings from Chapter 2 or 3. Do the following.
   A. Use the view cube to create a pictorial view of the drawing.
   B. Set the Conceptual visual style current.
   C. Using a steering wheel, display different views of the model.
   D. Use the **Rewind** tool in the steering wheel to restore a previous view.
   E. Save the drawing as P04_01.

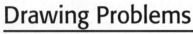

2. Open one of your 3D drawings from Chapter 2 or 3. Do the following.
   A. Toggle the projection from parallel to perspective and turn off the view cube compass.
   B. Pick an edge on the view cube. Set the parallel projection current.
   C. Display three additional views based on view cube edges. Alternate between parallel and perspective projection.
   D. Put the model into a continuous orbit.
   E. Save the drawing as P04_02.

3. Open the drawing Pergola_bridge.dwg provided on the companion website at www.g-wlearning.com/CAD. Do the following.
   A. Set the Conceptual visual style current.
   B. Right-click in the drawing area and select **Isolate>Hide Objects**. Hide the trees and the spherical structures.
   C. Right-click in the drawing area and select **Isolate>Isolate Objects**. Select only the roof components of the pergola. Use the view cube to rotate the view of the model for ease of selection if necessary.
   D. Unisolate all of the objects to redisplay the complete drawing.
   E. Close the drawing without saving.

4. Open one of your 3D drawings from Chapter 2 or 3. Do the following.
   A. Use a steering wheel to change the view of the model. Do this at least four times.
   B. Use the **Rewind** tool to restore a previous view.
   C. Repeat this for each of the previous views recorded with the steering wheel.
   D. Save the drawing as P04_04.

Using Show Motion to View a Model

Learning Objectives

After completing this chapter, you will be able to:

✓ Explain the show motion tool.
✓ Create still shots of 3D models.
✓ Create walk shots of 3D models.
✓ Create cinematic shots of 3D models.
✓ Replay single shots and a sequence of shots.
✓ Change the properties of a shot.

AutoCAD's *show motion* tool is a powerful function that allows you to create named views, animated shots, and basic walkthroughs. It can quickly display a variety of named shots. It is also used to create basic animated presentations and displays. This capability is especially useful for animating 3D models that do not require the complexity and detail of fully textured, rendered animations and walkthroughs. Advanced rendering and walkthroughs are presented later in this book.

A *view* is a single-frame display of a model or drawing from any viewpoint. A *shot* is the manner in which the model is put in motion and the way the camera *moves* to that view. Therefore, a single, named view can be modified to create several different shots using camera motion and movement techniques. Using show motion, you can create shots in one of three different formats:

- Still.
- Walk.
- Cinematic.

A *still shot* is exactly the same as a named view, but show motion allows you to add transition effects to display it. A *walk shot* requires that you use the **Walk** tool to define a camera motion path to create an animated shot. A *cinematic shot* is a single view to which you can add camera motion and movement effects to display the view. These shots are all created using the **New View/Shot Properties** dialog box. This is the same dialog box used to create named views.

Understanding Show Motion

Show motion is simply a means for creating, manipulating, and displaying named views. The **NAVSMOTION** command is used for show motion. The process involves the creation of shots using the **ShowMotion** toolbar and the **New View/Shot Properties** dialog box. Once a shot is created, you can give it properties that allow it to be displayed in many different ways. If a saved shot does not display in the manner you desire, it is easily modified.

A *view category* is a heading under which different views are filed. It is not necessary to create view categories, especially if you will be making just a few views. On the other hand, if you are working on a complex model and need to create a number of views with a variety of cinematic and motion characteristics, it may be wise to create view categories.

The process of using show motion to create, modify, and display shots begins by first using the **ShowMotion** toolbar. All of your work with shots will be performed using the **New View/Shot Properties** dialog box. Options in this dialog box change based on the type of shot that is selected. These options are discussed in the sections that follow relating to each kind of shot.

PROFESSIONAL TIP

As you first start working with show motion, you may want to create view categories named Still, Cinematic, and Walk. As you create shots, file each under its appropriate category name. This will assist you in seeing how the different shot-creation techniques work. Later, you can apply the view categories to specific components of a complex model. For example, a subassembly of a 3D model may have a view category that contains all three types of shots for use in different types of modeling work or presentations. Creating and using view categories is discussed in detail at the end of this chapter.

Show Motion Toolbar

NAVSMOTION

Type
> NAVSMOTION
> MOTION

Toolbar
> Navigation Bar
> ShowMotion

The **ShowMotion** toolbar is displayed at the bottom of the screen when the **NAVSMOTION** command is entered. See **Figure 5-1**. The quickest way to display the **ShowMotion** toolbar is to pick the **ShowMotion** button on the navigation bar. The **ShowMotion** toolbar is displayed at the bottom-center of the screen and provides controls for creating and manipulating views:

- **Unpin ShowMotion/Pin ShowMotion**
- **Play All**
- **Stop**
- **Turn on Looping**
- **New Shot...**
- **Close ShowMotion**

When the toolbar is pinned, it remains displayed if you execute other commands, minimize the drawing, change ribbon panels, or switch to another software application. The **Unpin ShowMotion** button is used to unpin the toolbar. The **Pin ShowMotion** button is then displayed in its place. If you unpin the toolbar, you must execute the **NAVSMOTION** command each time you wish to use show motion.

Figure 5-1.
The **ShowMotion** toolbar is displayed at the bottom of the screen and provides controls for creating and manipulating shots.

ShowMotion toolbar

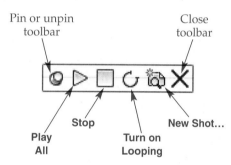

Pin or unpin toolbar

Close toolbar

Play All

Stop

Turn on Looping

New Shot...

Press the **Play All** button to play all of the shots displayed as thumbnails above the toolbar. The playback of these shots will loop (repeat) if the **Turn on Looping** button is selected. If looping is turned on, a shot, category, or all categories are displayed in a loop whenever the **Play All** button is selected. The **Turn on Looping** button is a toggle. The button image changes to indicate whether or not looping is turned on.

Picking the **New Shot...** button opens the **New View/Shot Properties** dialog box. This dialog box is discussed in the next section. Picking the **Close ShowMotion** button closes the **ShowMotion** toolbar.

New View/Shot Properties Dialog Box

All shot creation takes place inside of the **New View/Shot Properties** dialog box. It is accessed by using the **ShowMotion** toolbar as previously described or by using the **NEWSHOT** command. The dialog box can also be displayed from within the **View Manager** dialog box by picking the **New...** button. The **New View/Shot Properties** dialog box is shown in **Figure 5-2**.

You must supply a shot name and a shot type. A view category is not required, but can help organize shots. This feature is discussed later in the chapter. The dialog

Figure 5-2.
The **New View/Shot Properties** dialog box is used to create the three types of shots: still, walk, and cinematic.

Enter a name

Select a category

Select a view type

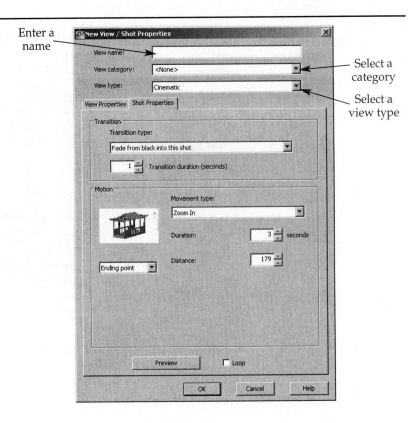

box provides two tabs containing options that allow you to define the overall view and then specify the types of movement and motion desired in the shot. These tabs are discussed in the next sections.

View Properties Tab

The **View Properties** tab of the **New View/Shot Properties** dialog box is composed of three areas in which you can specify the overall presentation of the model view. See **Figure 5-3**. The overall presentation includes the view boundary, visual settings, and background.

Figure 5-3.
The **View Properties** tab of the **New View/Shot Properties** dialog box.

Boundary setting

Visual settings

Background settings

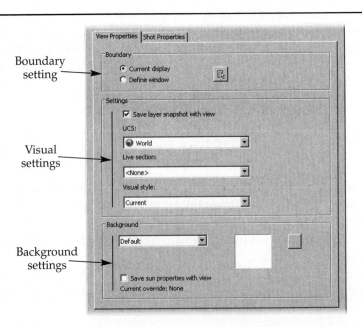

Boundary Settings

The setting in the **Boundary** area of the **View Properties** tab determines what is displayed in the shot. The **Current display** button is on by default. This means the shot will be composed of what is currently shown in the drawing area.

You can adjust the view by picking the **Define window** radio button. This temporarily closes the dialog box so you can draw a rectangular window to define the view. After picking the second corner, you can adjust the view by picking the first and second corners of the window again. Press [Enter] to accept the window and return to the dialog box.

Visual Settings

The **Settings** area of the **View Properties** tab contains options that apply to the overall display of the model in the shot. When the **Save layer snapshot with view** check box is checked, all of the current layer visibility settings are saved with the new shot. This is checked by default.

Any UCS currently defined in the drawing can be selected for use with the new shot. Use the **UCS:** drop-down list to select the UCS. When the shot is restored, that UCS is restored, too. If you select <None> in the drop-down list, there is no UCS associated with the shot.

Live sectioning is a tool that allows you to view the internal features of 3D solids that are cut by a section plane object. This feature is covered in detail in Chapter 15. When a section plane object is created, it is given a name. Therefore, if a model contains one or more section plane objects, their names appear in the **Live section:** drop-down list. If a section plane object is selected in this list, the new shot shows the live sectioning for that plane.

The **Visual style:** drop-down list contains all visual styles in the drawing plus the options of Current and <None>. Selecting a visual style from the list results in that style being set current when the shot is played. Selecting Current or <None> will cause the shot to be displayed in the visual style currently displayed on the screen. A detailed discussion of visual styles is provided in Chapter 16.

A shot created with live sectioning on will be displayed in that manner even though live sectioning may be currently turned off in the drawing. However, subsequent displays of the model that were created with live sectioning off are shown with it on. Keep this in mind as you develop shots for show motion.

Background Settings

The **Background** area of the **View Properties** tab provides options for changing the background of the new shot. The drop-down list in this area allows you to select a solid, gradient, image, or sun and sky background. You can also choose to retain the default background. If you pick Solid, Gradient, or Image, the **Background** dialog box appears. See **Figure 5-4**. If you select Sun & Sky, the **Adjust Sun & Sky Background** dialog box is displayed. See **Figure 5-5**. These dialog boxes are used to set the background. They are discussed in detail in Chapter 18.

The **Save sun properties with view** setting is used to apply the sunlight data to the view. If you are displaying an architectural model using sunlight, you will likely want this check box checked for show motion. Sunlight and geographic location are discussed in detail in Chapter 18.

Shot Properties Tab

Settings in the **Shot Properties** tab of the **New View/Shot Properties** dialog box provide options for controlling the transition and motion of shots. There are numerous

Figure 5-4.
The **Background** dialog box is used to add a background to the shot. Here, a gradient background is being created.

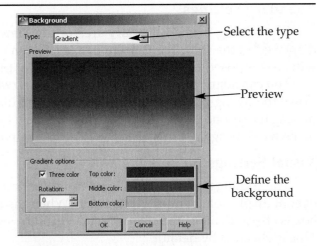

Select the type

Preview

Define the background

Figure 5-5.
When the background is set to **Sun & Sky**, the **Adjust Sun & Sky Background** dialog box is used to change settings for the background.

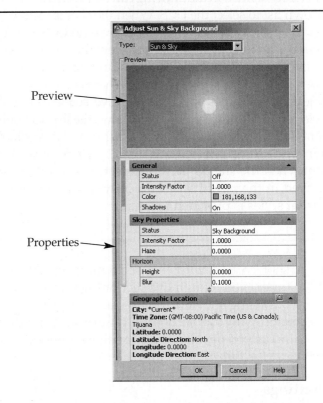

Preview

Properties

movement and motion options in this tab. The specific options available are based on the type of shot selected: still, walk, or cinematic. In addition, the cinematic shot contains a variety of motions that can be applied to the shot and each type of motion contains a number of variables. The options in the **Shot Properties** tab are discussed later in this chapter as they apply to different shots, movements, and motions.

Creating a Still Shot

A still shot is the same as a named view, but with a transition. Open the **New View/ Shot Properties** dialog box and enter a name in the **View name:** text box. Next, pick in the **View category:** text box and type a category name or select an existing category using the drop-down list. Remember, it is not necessary to create or select a category at this time. Now, select Still from the **View type:** drop-down list.

The Still view type is also used to create named views. A *named view* is a single-frame display of a model from a given viewpoint. Named views can be used for any purpose in AutoCAD, not just for show motion. For example, it is useful to create named views for use in rendering. To create a named view, enter a view name in the **New View/Shot Properties** dialog box and select Still for the view type. Select the **View Properties** tab and pick the **Current display** radio button to accept the model display as is. You can also pick the **Define window** radio button to return to the model and refine the display. Pick **OK** to complete the named view creation. The new view name will appear under the Model Views branch in the **View Manager** dialog box when using the **VIEW** command. The named view can be displayed at any time, and can be used for any show motion view. Named views are saved with the drawing and can be displayed by accessing the drop-down list in the **Views** panel on the **View** tab of the ribbon.

In the **Shot Properties** tab, select a transition from the **Transition type:** drop-down list. You can select from one of three transitions:

- **Fade from black into this shot.** The screen begins totally black and fades into the current background color.
- **Fade from white into this shot.** The screen begins totally white and fades into the current background color.
- **Cut to shot.** The view is immediately displayed without a transition and the shot movements are applied.

The fade transitions will not function unless hardware acceleration is enabled. This feature is normally activated by default upon software installation. However, you can confirm this by hovering over the **Performance Tuner** tool in the status bar. See **Figure 5-6**. Hover over the tool to display the tooltip, which should read **Hardware Acceleration on**. If hardware acceleration is not on, pick on the tool to display the shortcut menu and select **Hardware Acceleration** to place a check mark next to it. Refer to **Figure 5-6B**.

You can also verify that hardware acceleration is turned on by selecting **Performance Tuner...** from the shortcut menu. This accesses the **Adaptive Degradation**

Figure 5-6.
Hardware acceleration must be activated for the show motion fade transitions to function. A—Hover over the **Performance Tuner** tool to confirm that hardware acceleration is turned on. B—The shortcut menu displayed by picking on the **Performance Tuner** tool.

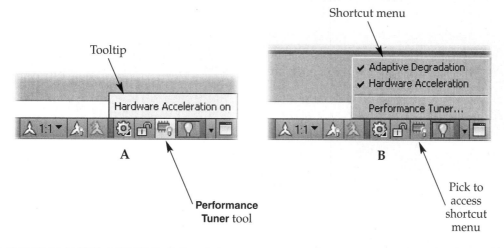

and Performance Tuning dialog box. In this dialog box, pick the **Manual Tune** button to display the **Manual Performance Tuning** dialog box, which is used to enable hardware acceleration and control the performance of various graphics effects. See **Figure 5-7**.

After hardware acceleration has been enabled, display the **New View/Shot Properties** dialog box, enter a name, and select a transition. Next, in the **Transition duration (seconds)** text box, enter a length of time over which the transition will occur. If you want a fade transition to be complete, be sure to enter a value in the **Duration:** text box that is equal to or greater than the transition duration. Pick the **Preview** button to view the shot, then edit the transition and motion values as needed.

When finished, pick the **OK** button to close the **New View/Shot Properties** dialog box. The thumbnail image for the new shot is displayed above the **ShowMotion** toolbar. See **Figure 5-8A**. The large thumbnail image represents the view category. Since there was no view category selected for the view, the name <None> is displayed. The small

Figure 5-7.
The **Manual Performance Tuning** dialog box.

Hardware acceleration turned on

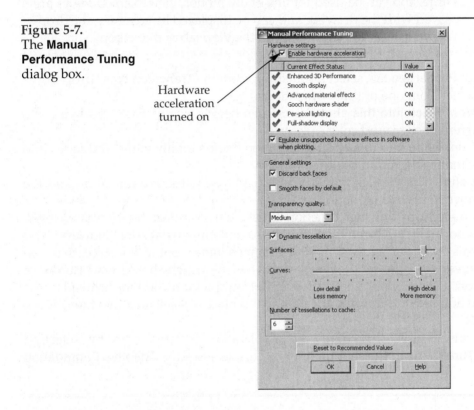

Figure 5-8.
A—The large thumbnail image on the bottom is for the category. The small thumbnail image is for the shot just created. Its name may be truncated. B—Move your cursor into the shot thumbnail image and it converts to a large image. The category thumbnail image is reduced to a small image.

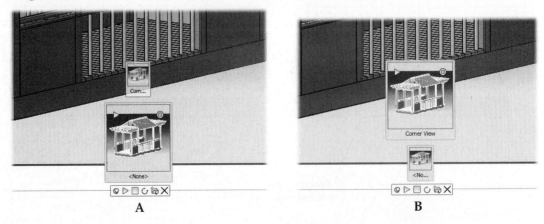

A

B

thumbnail image represents the shot just created. Its name may be truncated. Move the cursor into the shot thumbnail image and a large image is displayed. The name of the shot should appear in its entirety. The category thumbnail image is reduced to a small image. See **Figure 5-8B**.

If the **Loop** check box at the bottom of the **Shot Properties** tab is checked, the shot will continuously loop through the transition during playback. This is similar to picking the **Turn on Looping** button in the **ShowMotion** toolbar. All three shot types have this option.

Exercise 5-1

Complete the exercise on the companion website.
www.g-wlearning.com/CAD

Creating a Walk Shot

To create a walk shot, open the **New View/Shot Properties** dialog box, enter a name, and select a category, if needed. Next, select Recorded Walk from the **View type:** drop-down list. In the **View Properties** tab, set up the view as described for a still shot.

In the **Shot Properties** tab, set up the transition as described for a still shot. The only option in the **Motion** area of the tab is the **Start recording** button. See **Figure 5-9**. The **Duration:** text box is grayed out because the value is based on how long you record the walk. The camera drop-down list below the preview image is also grayed

Figure 5-9.
Creating a walk shot.

Transition settings

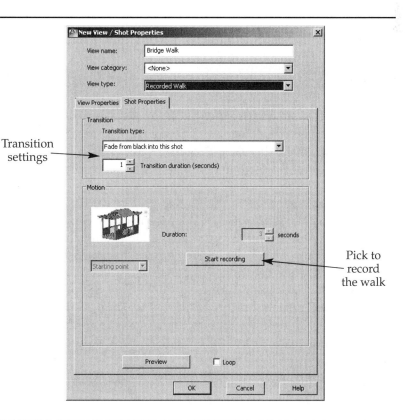

Pick to record the walk

out because the walk begins at the current display and ends at the point where you terminate it.

This type of shot uses the same **Walk** tool found in the steering wheels, discussed in Chapter 4. After picking the **Start recording** button, the **New View/Shot Properties** dialog box is hidden and the **Walk** tool message is displayed. Pick and drag to activate the **Walk** tool. As soon as you pick, the center circle icon is displayed. If needed, you can hold down the [Shift] key to move the view up or down. When the [Shift] key is released (with the mouse button still held down), you can resume walking. The mouse button must be depressed the entire time to record all movements. As soon as you release the button, the recording ends and the dialog box is redisplayed.

Finally, preview the shot. If you need to re-record it, pick the **Start recording** button and begin again. You cannot add to the shot; you must start over. Pick the **OK** button to save the shot and a new thumbnail is displayed in the **ShowMotion** toolbar.

 The walk shot capability of show motion is limited in its ability to produce a true "walkthrough." The path of your "walk" is inexact and it may take several times to create the effect you need. Should you wish to create a genuine walkthrough of an architectural or structural model, it is better to use tools such as **3DWALK** (walk-through), **3DFLY** (flyby), or **ANIPATH** (motion path animation). These powerful commands are used to create professional walk-throughs and animations and are covered in Chapter 20.

 Exercise 5-2

Complete the exercise on the companion website.
www.g-wlearning.com/CAD

Creating a Cinematic Shot

To create a cinematic shot, open the **New View/Shot Properties** dialog box, enter a name, and select a category, if needed. Next, select **Cinematic** from the **View type:** drop-down list. In the **View Properties** tab, set up the view as described for a still shot. In the **Shot Properties** tab, pick a transition type and duration as described for a still shot.

The specific options in the **Motion** area of the **Shot Properties** tab are discussed in the next sections, **Figure 5-10**. Pick a type of movement in the **Movement type:** drop-down list. The movement options available will change depending on the type of movement you select. Refer to the chart in **Figure 5-11** as a quick reference for move-ment options when creating a cinematic shot. Finally, pick the current position of the camera from the camera drop-down list below the preview image.

Pick the **Preview** button to view the shot. Make any changes required before picking the **OK** button to save the shot. Once the **OK** button is picked, the new view is displayed as a small thumbnail image above the large category thumbnail image. See **Figure 5-12**.

Figure 5-10.
Creating a cinematic
shot.

Transition
settings

Camera
drop-down
list

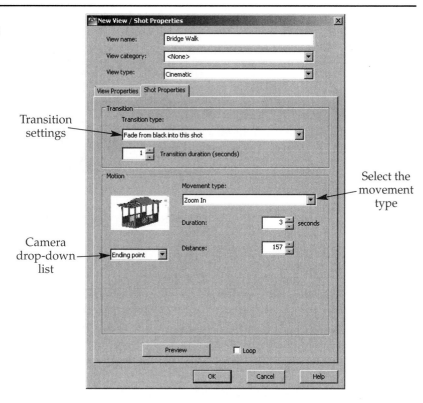

Select the
movement
type

Figure 5-11.
This chart shows the different movement options based on the movement type selected in the **Movement type:** drop-down list in the **New View/Shot Properties** dialog box.

	Duration	Distance	Look at Camera Point	Distance Up	Distance Back	Distance Down	Distance Forward
Zoom in	X	X					
Zoom out	X	X					
Track left	X	X	X				
Track right	X	X	X				
Crane up	X		X	X	X		
Crane down	X		X			X	X
Look	X						
Orbit	X						

Figure 5-12.
The thumbnail image for the new cinematic shot is displayed above the category thumbnail image.

Cinematic Basics

The motion and movement options available for a cinematic shot allow you to create a final display that appears to move into position as if the camera is traveling in a path toward the object. The **Motion** area of the **Shot Properties** tab contains a variety of options for creating an array of cinematic shots. Refer to Figure 5-10. The options can be confusing unless you understand some basics about essential components of a cinematic shot. The most important aspect is the preview of the current view. All of the motion actions revolve around this view. This image tile is the current position of the camera and is also referred to as the *key position* of a shot. This is the position that is displayed when you pick the **Go** button on a thumbnail image above the **ShowMotion** toolbar.

The two elements of a cinematic shot are motion and movement. *Motion* relates to the behavior of the object and how it appears to be in motion during the cinematic shot. In addition, it refers to the position of the model at a specified point in the animation. *Movement* in a cinematic shot is the manner in which the camera moves in relation to the object.

PROFESSIONAL TIP

If your goal is to create a series of shots that blend together, it is a good idea to first develop a storyboard of the entire sequence. This could be as simple as a few notes indicating how you want the shots to move or even a few sketches noting the required movements and motion values. Planning your shots will save time when you begin creating them in AutoCAD.

Camera Drop-Down List

The camera drop-down list is located below the preview image in the **Motion** area of the **Shot Properties** tab. The preview image represents the position of the camera based on the option selected in the camera drop-down list. The following three options are available in the drop-down list, Figure 5-13.

- **Ending point.** The view in the preview image is the display that will be shown at the end of the cinematic shot. All movement options take the shot to this point.
- **Starting point.** The view in the preview image is the display that will be shown at the start of the cinematic shot. All movement options begin at this point.
- **Half-way point.** The view in the preview image is the display that will be shown at the half-way point of the cinematic shot.

Figure 5-13.
Selecting what
the key position
represents.

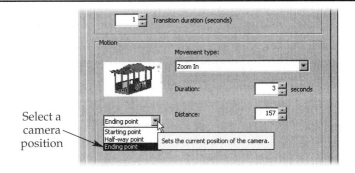

Select a
camera
position

These options, in part, determine how the cinematic shot is created. If Starting point is selected, the cinematic shot begins with the current view. During the animation, the model may move off of the screen. If this happens, you may want to select Ending point so the model appears to move into position and stay there.

PROFESSIONAL TIP

The **Preview** button in the **New View/Shot Properties** dialog box is the best way to test any changes you make to the motion options. Each time you make a change, pick the button to see if the effect is what you want. Previewing each change is far more efficient than making several changes before you examine the results.

Movement Type

The **Movement type:** drop-down list in the **Shot Properties** tab of the **New View/Shot Properties** dialog box is used to select the motion for the cinematic shot. There are eight types of camera movement that can be used with a cinematic shot:
- Zoom in.
- Zoom out.
- Track left.
- Track right.
- Crane up.
- Crane down.
- Look.
- Orbit.

The options available in the **Motion** area of the tab are based on which movement type is selected. The movement types and their options are discussed in the next sections.

Zoom In

When Zoom in is selected in the **Movement type:** drop-down list, the camera appears to zoom into the model in the shot. This movement type has two options, **Figure 5-14.** The value in the **Duration:** text box is the length of time over which the animation is recorded. The value in the **Distance:** text box is the distance the camera

Figure 5-14.
The settings for the
Zoom in movement
type.

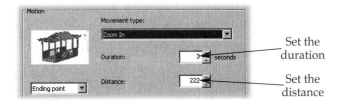

Set the
duration

Set the
distance

travels during the animation. The camera zooms in to cover the distance in the specified duration of time. When Zoom in is selected, the camera drop-down list is automatically set to Ending point. Keep in mind that with this option the current position of the camera represents the final display after the cinematic shot is complete.

Zoom Out

When Zoom out is selected in the **Movement type:** drop-down list, the camera appears to zoom away from the model in the shot. This movement type has **Duration:** and **Distance:** text options, as described for the Zoom in movement type.

When Zoom out is selected, the camera drop-down list is automatically set to Starting point. With this setting, the camera will zoom out the specified distance and the final image in the shot will be smaller than the preview image. If this is not the effect you want, it may be better to pick Ending point for the current position of the camera. Then, the model will appear to move from behind your view to stop at the current view.

Track Left

When Track left is selected in the **Movement type:** drop-down list, the camera will move from right to left. This results in the view moving from left to right. This movement occurs over the specified distance and duration. The Track left movement type has **Duration:** and **Distance:** text boxes, as described previously, and the **Always look at camera pivot point** check box, **Figure 5-15**.

If the **Always look at camera pivot point** check box is checked, the center of the view remains stationary. As a result, the view in the shot appears to rotate about this point instead of sliding across the screen. The best way to visualize this motion is to use the **Preview** button with the option checked and then unchecked. This option is available for all "track" and "crane" movement types.

When Track left is selected, the camera drop-down list is automatically set to Half-way point. This means the preview image is the middle point of the animation. It will be displayed at the midpoint of the **Duration:** value.

PROFESSIONAL TIP

If the **Distance:** value is large, the screen may be blank for a few moments until the camera moves enough to bring the model into view. If this is not what you want, decrease the **Distance:** value or check the **Always look at camera pivot point** check box.

Track Right

When Track right is selected in the **Movement type:** drop-down list, the camera will move from left to right. This results in the view moving from right to left. This movement type has **Distance:** and **Duration:** text boxes and the **Always look at camera**

Figure 5-15.
The settings for the Track left movement type.

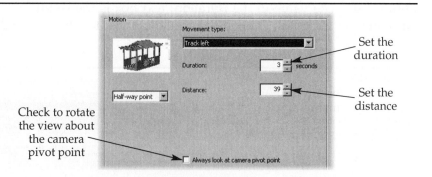

Set the duration

Set the distance

Check to rotate the view about the camera pivot point

pivot point check box. These options function in the same manner as described for the Track left movement type. When Track right is selected, the camera drop-down list is automatically set to Half-way point.

Crane Up

Where the "track" movement types move the view left and right, the "crane" movement types move the view up and down. When Crane up is selected in the **Movement type:** drop-down list, the camera will move from bottom to top and then backward. This results in the view moving from top to bottom and zooming out. This movement type has the **Distance:** and **Duration:** text boxes and the **Always look at camera pivot point** check box described previously. The **Always look at camera pivot point** check box is checked by default.

This movement type has three additional options, **Figure 5-16**. The value in the **Distance Up:** text box is the distance the camera is moved upward. The value in the **Distance Back:** text box is the distance the camera is moved backward. The backward movement is typically short compared to the upward movement.

You can add more interest to the motion in the shot by shifting the view left or right. First, enable the shift option by checking the check box below the **Distance Back:** setting. Refer to **Figure 5-16**. Then, select either Shift left or Shift right from the drop-down list. Finally, enter a distance in the text box next to the drop-down list. The camera will be shifted left or right for the distance specified in this text box, resulting in the view shifting in the opposite direction. This shifting option is available for both "crane" motion types.

When Crane up is selected, the camera drop-down list is automatically set to Starting point. This means the preview image shows the beginning of the shot before the movement is applied. If the **Always look at camera pivot point** check box is not checked, and depending on the movement settings, the model may move off of the screen in the shot.

Crane Down

When Crane down is selected in the **Movement type:** drop-down list, the camera will move from top to bottom and then forward. This results in the view moving from bottom to top and zooming in. This movement type has the **Distance:** and **Duration:** text boxes and the **Always look at camera pivot point** check box described previously. The **Always look at camera pivot point** check box is checked by default. This movement type also has the left/right shifting option described in the previous section.

Instead of **Distance Up:** and **Distance Back:** settings, this movement type has **Distance Down:** and **Distance Forward:** settings. The value in the **Distance Down:** text box is the distance the camera cranes down in the shot. The value in the **Distance Forward:** text box is the distance the camera is moved forward in the shot.

When Crane down is selected, the camera drop-down list is automatically set to Ending point. This means the preview image shows the end of the shot before the movement is applied. If the **Always look at camera pivot point** check box is not checked, and depending on the movement settings, the model may start off of the screen in the shot.

Figure 5-16.
The settings for the Crane up movement type.

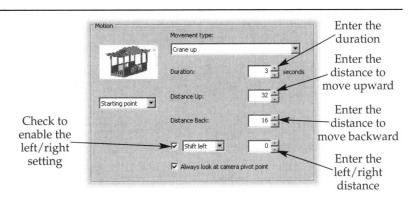

Check to enable the left/right setting

Enter the duration

Enter the distance to move upward

Enter the distance to move backward

Enter the left/right distance

All distance values related to the Crane up and Crane down movement types represent how far away from the view the camera must begin before the cinematic shot is started. For example, a large **Distance Back:** value means that the camera may have to begin at a point beyond or even around the view in order to travel the distance back to display the current view (when Ending point is selected in the camera drop-down list). It is always good practice to make a single change to the movements, then preview the shot. This is especially true for the Crane up and Crane down movement types.

Look

When Look is selected in the **Movement type:** drop-down list, the camera pans based on the values for left/right and up/down to display the view, **Figure 5-17.** For example, if the movement is set to look up 45° and Ending point is selected in the camera drop-down list, then the camera begins at a 45° angle below the view and looks up to display it.

The value in the **Duration:** text box is the length of time over which the animation is recorded. This option is the same as described for the other movement types.

The first drop-down list below the **Duration:** text box is used to specify either left or right movement. Select Degrees left in the drop-down list to have the camera move from right to left, resulting in the view moving from left to right. Select Degrees right to have the camera move from left to right and the view right to left. Next, specify the angular value for this movement in the degrees text box to the right of the drop-down list. For example, suppose you select Degrees left and enter an angle of 15°. In this case, the camera will start 15° to the *right* of the model and rotate to the left in the shot.

The second drop-down list below the **Duration:** text box is used to specify either up or down movement. Select Degrees up in the drop-down list to have the camera move from bottom to top, resulting in the view moving from top to bottom. Select Degrees down to have the camera and view move in the opposite direction. Specify the angular value for this movement in the degrees text box to the right of the drop-down list.

When Look is selected in the **Movement type:** drop-down list, the camera drop-down list is automatically set to Starting point. This means the shot will start with the current view and then apply the movement settings.

Orbit

When Orbit is selected in the **Movement type:** drop-down list, the camera rotates in place based on the values set for left/right and up/down movements. The Orbit movement type has the same options as the Look movement type, as described in the previous section.

Figure 5-17.
The settings for the Look movement type.

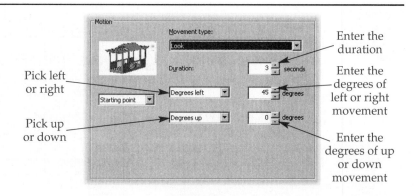

AutoCAD and Its Applications—Advanced

When Orbit is selected in the **Movement type:** drop-down list, the camera drop-down list is automatically set to Starting point. This means the shot will start with the current view and then apply the movement settings. Unlike the Look movement type, the view center remains stationary in the shot. As a result, you do not need to worry about the model moving off of the screen.

Exercise 5-3

Complete the exercise on the companion website.
www.g-wlearning.com/CAD

Displaying or Replaying a Shot

As each shot is created, its thumbnail image is placed above the **ShowMotion** toolbar. The shot name is displayed below the thumbnail image. By default, the category thumbnail image is large and each shot thumbnail image is small. If the cursor is moved over one of the shot thumbnail images, all of the shot thumbnail images are enlarged and the category thumbnail image is reduced in size.

Each thumbnail image is composed of an image of the view in the shot, the shot name, and viewing controls. See **Figure 5-18**. The viewing controls are only displayed when the cursor is moved over the thumbnail image. The three viewing controls are:

- **Play.** Plays the shot. If looping is enabled, the shot repeats. Otherwise, the shot is played once. When this button is picked, it changes to **Pause**. You can also play the shot by picking anywhere on its thumbnail image. The **VIEWPLAY** command can be used to replay a shot.

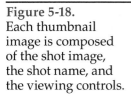

- **Pause.** Pauses the shot. When this button is picked, it changes to **Play**. Picking anywhere inside the image or on the **Play** button restarts playing of the shot.
- **Go.** Displays the key position of the view without playing the shot. The **VIEWGO** command also restores a named view.

You can play all of the shots in sequence by picking the **Play** button in the view category thumbnail image or by picking anywhere inside of the view category thumbnail image. Additionally, you can move to the key position of the first shot in a view category by picking the **Go** button on the view category thumbnail image.

Figure 5-18.
Each thumbnail image is composed of the shot image, the shot name, and the viewing controls.

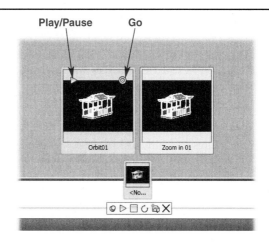

Exercise 5-4

Complete the exercise on the companion website.
www.g-wlearning.com/CAD

Thumbnail Shortcut Menu

Right-clicking on a shot or view category thumbnail image displays the shortcut menu shown in **Figure 5-19**. This menu provides quick access for modifying and manipulating shots and their thumbnail images.

Picking the **New View/Shot...** entry displays the **New View/Shot Properties** dialog box. Picking the **Properties...** entry displays the **View/Shot Properties** dialog box. This dialog box is the same as the **New View/Shot Properties** dialog box. However, all of the settings of the shot are displayed and can be changed. This option is only available to change the properties of a shot, not a category.

To rename a shot, pick the **Rename** entry in the shortcut menu. The name is highlighted below the thumbnail image. Type the new name and press [Enter]. To delete a shot, select **Delete** from the shortcut menu. There is no warning; the shot is simply deleted. The **UNDO** command can reverse this action.

If there is more than one shot in a category, you can rearrange the order. Right-click on a thumbnail image and pick either **Move left** or **Move right** in the shortcut menu. This moves the shot one step in the selected direction.

When a change is made to the model, it is not automatically reflected in the thumbnail images. If you do not update thumbnail images, they will remain in their original format, regardless of how many changes are made to the model. Selecting **Update the thumbnail for** in the shortcut menu displays a cascading menu with options for updating the thumbnail image:

- **This view.** Updates only the thumbnail image for the shot on which you right-clicked.
- **This category.** Updates all thumbnail images in the category. This option is only enabled when you right-click on a category thumbnail image.
- **All.** This option updates all thumbnail images in all categories and shots.

Creating and Using View Categories

A *view category* is a grouping that can be created in order to separate different types of shots. A view category can also contain shots arranged in a sequence that appear connected as they are played together. This is an optional feature, but an efficient method of separating different types of shots or grouping shots to use for a specific purpose. When a new category is created, it is represented by a category thumbnail image displayed above the **ShowMotion** toolbar.

Figure 5-19.
Right-click on a shot or view category thumbnail image in the **ShowMotion** toolbar to display this shortcut menu.

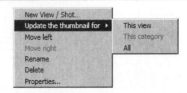

To create a new view category, simply pick in the **View category:** drop-down list text box in the **New View/Shot Properties** dialog box. Then, enter a name. See **Figure 5-20.** The name is added to the drop-down list. If you wish to organize shots by categories, be sure to select the view category from the drop-down list before picking the **OK** button to exit the dialog box and create the shot.

You must create a shot to create a view category. If you delete the only shot in a view category, you also delete the category. The view category will no longer be available in the **New View/Shot Properties** dialog box.

View Category Basics

After a new view category is created, it is represented by a thumbnail image above the **ShowMotion** toolbar. The category name is shown below the thumbnail image. As you create more shots within a category, the thumbnail images for the new shots are placed to the right of existing shots above the category thumbnail image. The thumbnail image for the first shot in a view category is displayed as the view category thumbnail image.

An entire view category and all of the views included in it can be quickly deleted. Simply right-click on the view category thumbnail image and pick **Delete** in the shortcut menu. An alert box appears asking you to confirm the deletion, **Figure 5-21.** The shots in a deleted view category cannot be recovered, so be sure there are no shots in the category you wish to save before deleting. However, you can use the **UNDO** command to reverse the deletion.

PROFESSIONAL TIP

Remember, you can change the order of the shots in a view category using the shortcut menu. Picking **Move left** or **Move right** moves the shot one step. Continue moving shots until the order is appropriate.

Playing and Looping Shots in a View Category

A view category is a useful tool for grouping shots to be played together in a sequence. To play the shots in a view category, move the cursor over the view category thumbnail image above the **ShowMotion** toolbar. Then, pick the **Play** button in

---

Figure 5-20.
To create a new view category, type its name in the **View category:** drop-down list text box.

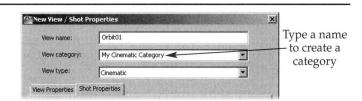

Type a name to create a category

Figure 5-21.
To remove a view category, right-click on its thumbnail image in the **ShowMotion** toolbar and pick **Delete** in the shortcut menu. This alert box appears to confirm the deletion.

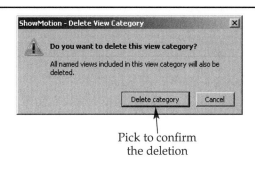

Pick to confirm the deletion

Type

**SEQUENCEPLAY
SPLAY**

the thumbnail image. All shots in the category will be played. You can also use the **SEQUENCEPLAY** command to play the shots in a view category.

If you want the shots in a category to run on a continuous loop, pick the **Turn on Looping** button in the **ShowMotion** toolbar. Then, when the **Play** button is picked, the shots in the view category will display until the **Pause** button is picked either on the **ShowMotion** toolbar or in the view category thumbnail image. Pressing the [Esc] key also stops playback. To return to single-play mode, pick the **Turn off Looping** button on the **ShowMotion** toolbar. This button replaces the **Turn on Looping** button.

Changing a Shot's View Category and Properties

If you put a shot in the wrong view category, it is simple to move the shot to a different category. Right-click on the shot thumbnail image and select **Properties...** in the shortcut menu. This displays the **New View/Shot Properties** dialog box. Select the proper category name in the **View category:** drop-down list and pick the **OK** button. The shot thumbnail image will move into position above its new view category thumbnail image.

To change shot properties, right-click on the shot thumbnail image and pick **Properties...** in the shortcut menu. The **New View/Shot Properties** dialog box is displayed. All properties of the shot can be changed. Change the settings as needed. Always preview the shot before you pick the **OK** button to save it.

PROFESSIONAL TIP

Try this procedure for creating a series of shots in a view category that are to be played in sequence.

1. Play each shot to determine where it should be located in the sequence.
2. Move shots left or right in the category as needed to create the proper sequence.
3. Play the category to determine if the shots properly transition from one to the next.
4. If a subsequent shot does not begin where the previous shot ended, right-click on that shot, pick **Properties...**, and make the necessary adjustments in the **View/Shot Properties** dialog box.
5. Play the category again.
6. Repeat the editing process with each shot until the category plays smoothly.

Exercise 5-5

Complete the exercise on the companion website.
www.g-wlearning.com/CAD

Chapter Review

Answer the following questions. Write your answers on a separate sheet of paper or complete the electronic chapter review on the companion website. www.g-wlearning.com/CAD

1. For what is the *show motion tool* used?
2. Define *view* as it relates to the show motion tool.
3. Define *shot* as it relates to the show motion tool.
4. List the formats in which a shot can be created.
5. List the six buttons on the **ShowMotion** toolbar.
6. What is *live sectioning* and how can it be included in a shot?
7. Which type of shot is the same as a named view, but with a transition?
8. Which type of shot requires you to navigate through the view as you record the motion?
9. What is the *key position* of a cinematic shot?
10. Define *motion* and *movement* as they relate to a cinematic shot.
11. What is the purpose of the camera drop-down list in the **Shot Properties** tab of the **New View/Shot Properties** or **View/Shot Properties** dialog box?
12. List the eight types of camera movement for a cinematic shot.
13. Briefly describe two ways to play a single shot.
14. What is a *view category*?
15. How is an entire category of shots replayed?

Drawing Problems

1. Open one of your 3D drawings from Chapter 2 or 3. Do the following.
   A. Display a pictorial view of the drawing.
   B. Create a still shot. Use settings of your choice. Create a view category and place the shot in it.
   C. Display a different view of the drawing and create another still shot. Place the shot in the view category you created.
   D. Display a third view of the drawing and create a third still shot. Place the shot in the view category you created.
   E. Play each shot. If necessary, rearrange the shots. Then, play all shots in the category.
   F. Save the drawing as P05_01.

2. Open one of your 3D drawings from Chapter 2 or 3. Do the following.
   A. Display the objects in the Conceptual or Realistic visual style.
   B. Toggle the projection from parallel to perspective.
   C. Create two still shots and four cinematic shots of the model. Use a different motion type with each of the cinematic shots.
   D. Create two new categories and place one still and two cinematic shots in each category.
   E. Edit the shots in each category to create a smooth motion sequence.
   F. Save the drawing as P05_02.

3. Open one of your 3D drawings from Chapter 2 that was created with solid primitives. Do the following.
   A. Display a pictorial view.
   B. Create a walk shot. Place it in a view category named Walk Shots.
   C. Play the shot. How does this compare to a cinematic shot as far as ease of creation?
   D. Save the drawing as P05_03.

Drawing Problems - Chapter 5

4. Open drawing P05_03 and do the following.
   A. Create a new cinematic shot using the Orbit motion type. Place it in a view category named Cinematic Shots.
   B. Create another cinematic shot using the Track left or Track right movement type. Check the **Always look at camera pivot** check box. Place the view in the Cinematic Shots category.
   C. Create a third cinematic shot using the Crane up or Crane down movement type. Make sure the **Always look at camera pivot** check box is checked. Place the view in the Cinematic Shots category.
   D. Play the Cinematic Shots category. Edit the shots as needed to create a smooth display.
   E. Create three still shots. Place them in a view category named Still Shots. Try creating three different gradient backgrounds to simulate dawn, noon, and dusk.
   F. Play the Still Shots category. Edit the shots as needed to create a smooth display.
   G. Save the drawing as P05_04.

Drawing Problems - Chapter 5

Understanding Three-Dimensional Coordinates and User Coordinate Systems

Learning Objectives

After completing this chapter, you will be able to:

✓ Describe rectangular, spherical, and cylindrical methods of coordinate entry.
✓ Draw 3D polylines.
✓ Describe the function of the world and user coordinate systems.
✓ Move the user coordinate system to any surface.
✓ Rotate the user coordinate system to any angle.
✓ Use the UCS icon grips to move and rotate the UCS.
✓ Use a dynamic UCS.
✓ Save and manage user coordinate systems.
✓ Restore and use named user coordinate systems.
✓ Control UCS icon visibility in viewports.

As you learned in Chapter 1, any point in space can be located using X, Y, and Z coordinates. This type of coordinate entry is called *rectangular coordinate entry*. Rectangular coordinates are most commonly used for coordinate entry. However, there are actually three ways in which to locate a point in space. The other two methods of coordinate entry are spherical coordinate entry and cylindrical coordinate entry. These two coordinate entry methods are discussed in the following sections. In addition, this chapter introduces working with user coordinate systems (UCSs).

Introduction to Spherical Coordinates

Locating a point in 3D space with *spherical coordinates* is similar to locating a point on Earth using longitudinal and latitudinal values, with the center of Earth representing the origin. Lines of longitude connect the North and South Poles and provide an east-west measurement on Earth's surface. Lines of latitude horizontally extend around Earth and provide a north-south measurement. The origin (Earth's center) can be that of the default world coordinate system (WCS) or the current user coordinate system (UCS). See **Figure 6-1A**.

When entering spherical coordinates, the longitude measurement is expressed as the angle *in* the XY plane and the latitude measurement is expressed as the angle *from*

Figure 6-1.
A—Lines of longitude, representing the highlighted latitudinal segments in the illustration, run from north to south. Lines of latitude, representing the highlighted longitudinal segments, run from east to west. B—Spherical coordinates require a distance, an angle in the XY plane, and an angle from the XY plane.

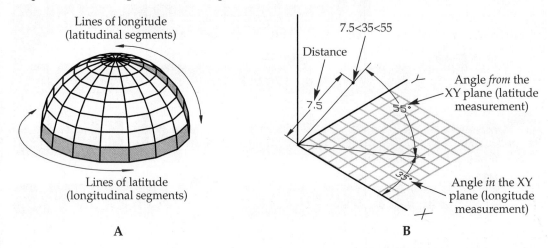

A

B

the XY plane. See **Figure 6-1B**. A distance from the origin is also provided. The coordinates represent a measurement from the equator toward either the North or South Pole on Earth's surface. The following spherical coordinate entry is shown in **Figure 6-1B**.

7.5<35<55

This coordinate represents an ***absolute*** spherical coordinate, which is measured from the origin of the current UCS. Spherical coordinates can also be entered as ***relative*** coordinates. For example, a point drawn with the relative spherical coordinate @2<35<45 is located two units from the last point, at an angle of 35° *in* the XY plane, and at a 45° angle *from* the XY plane.

If dynamic input is turned on, the "second" or "next" coordinate entry is automatically a *relative* entry (by default). The @ symbol is not entered. To enter *absolute* coordinates with dynamic input turned on, enter an asterisk (*) before the first coordinate.

PROFESSIONAL TIP

Spherical coordinates are useful for locating features on a spherical surface. For example, they can be used to specify the location of a hole drilled into a sphere or a feature located from a specific point on a sphere. If you are working on such a spherical object, you might consider locating a UCS at the center of the sphere, then creating several different user coordinate systems rotated at different angles on the surface of the sphere. Any time a location is required, spherical coordinates can be used. Working with UCSs is introduced later in this chapter.

Using Spherical Coordinates

Spherical coordinates are well-suited for locating points on the surface of a sphere. In this section, you will draw a solid sphere and then locate a second solid sphere with its center on the surface of the first sphere.

To draw the first sphere, select the **SPHERE** command. Specify the center point as 7,5 and the radius as 1.5 units. Display a southeast isometric pictorial view of the sphere. Alternately, you can use the view cube to create a different pictorial view. Also, set the Wireframe visual style current and switch to a parallel projection. Your drawing should look similar to **Figure 6-2A**.

Since you know the radius of the sphere, but the center of the sphere is not at the origin of the current UCS (the WCS), a relative spherical coordinate will be used to draw the second sphere. The sphere you drew is a solid and, as such, you can snap to its center using object snap. Set the center running object snap and then enter the **SPHERE** command again to draw the second sphere:

> Specify center point or [3P/2P/Ttr]: **FROM**↵
> Base point: (*use the* **Center** *object snap to select the center of the existing sphere*)
> <Offset>: **@1.5<30<60**↵ (*1.5 is the radius of the first sphere*)
> Specify radius or [Diameter]: **.4**↵

The objects should now appear as shown in **Figure 6-2B**. The center of the new sphere is located on the surface of the original sphere. This is clear after setting the Conceptual visual style current, **Figure 6-2C**. If you want the surfaces of the spheres to be tangent, add the radius value of each sphere (1.5 + .4) and enter this value when prompted for the offset from the center of the first sphere:

> <Offset>: **@1.9<30<60**↵

Notice in **Figure 6-2B** that the polar axes of the two spheres are parallel. This is because both objects were drawn using the same UCS, which can be misleading unless you understand how objects are constructed based on the current UCS. Test this by locating a cone on the surface of the large sphere, just below the small sphere. First, display a 3D wireframe view of the objects. Then, select the **CONE** command and continue as follows:

> Specify center point of base or [3P/2P/Ttr/Elliptical]: **FROM**↵
> Base point: **CEN**↵
> of (*pick the large sphere*)
> <Offset>: **@1.5<30<30**↵
> Specify base radius or [Diameter]: **.25**↵
> Specify height or [2Point/Axis endpoint/Top radius]: **1**↵

The result of this construction with the Conceptual visual style set current is shown in **Figure 6-3**. Notice how the axis of the cone is parallel to the polar axis of the sphere.

Figure 6-2.
A—A three-unit diameter sphere shown from the southeast isometric viewpoint.
B—A .8-unit diameter sphere is drawn with its center located on the surface of the original sphere. Also, lines have been drawn between the poles of the spheres. Notice how the polar axes are parallel. C—The objects after the Conceptual visual style is set current.

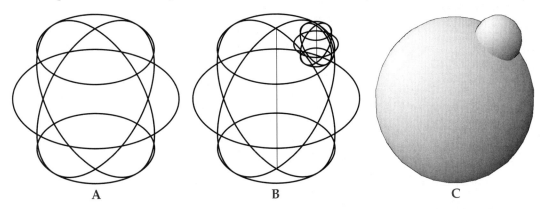

| A | B | C |

Figure 6-3.
The axis lines of objects drawn in the same user coordinate system are parallel. Notice that the cone does not project from the center of the large sphere.

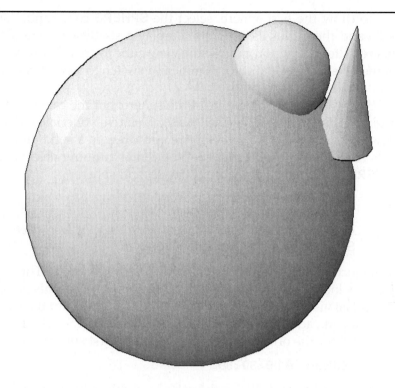

To draw the cone so that its axis projects from the center of the sphere, you will need to change the UCS. This is discussed later in the chapter.

Introduction to Cylindrical Coordinates

Locating a point in space with *cylindrical coordinates* is similar to locating a point on an imaginary cylinder. Cylindrical coordinates have three values. The first value represents the horizontal distance from the origin, which can be thought of as the radius of a cylinder. The second value represents the angle in the XY plane, or the rotation of the cylinder. The third value represents a vertical dimension measured up from the polar coordinate in the XY plane, or the height of the cylinder. See **Figure 6-4.** The absolute cylindrical coordinate shown in the figure is:

7.5<35,6

Figure 6-4.
Cylindrical coordinates require a horizontal distance from the origin, an angle in the XY plane, and a Z dimension.

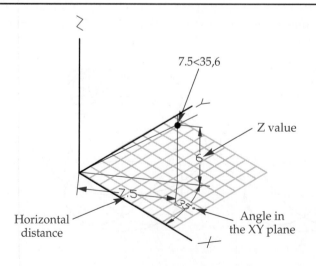

AutoCAD and Its Applications—Advanced

Like spherical coordinates, cylindrical coordinates can also be entered as relative coordinates. For example, a point drawn with the relative cylindrical coordinate @1.5<30,4 is located 1.5 units from the last point, at an angle of 30° in the XY plane of the previous point, and at a distance of four units up from the XY plane of the previous point.

 Turn off dynamic input before entering cylindrical coordinates. Dynamic input does not correctly interpret the entry when cylindrical coordinates are typed.

Using Cylindrical Coordinates

Cylindrical coordinates work well for attaching new objects to a cylindrical shape. An example of this is specifying coordinates for a pipe that must be attached to another pipe, tank, or vessel. In **Figure 6-5**, a pipe must be attached to a 12′ diameter tank at a 30° angle from horizontal and 2′-6″ above the floor. In order to properly draw the pipe as a cylinder, you will have to change the UCS, which you will learn how to do later in this chapter. An attachment point for the pipe can be drawn using the **POINT** command and cylindrical coordinates. First, set the drawing units to architectural. This can be done by selecting **Drawing Utilities>Units** from the **Application Menu** and specifying the length units as Architectural in the **Drawing Units** dialog box. Next, set the **PDMODE** system variable to 3. Enter the **POINT** command and continue as follows:

```
Current point modes: PDMODE=3 PDSIZE=0.0000
Specify a point: FROM↵
Base point: CEN↵
of (pick the base of the cylinder)
<Offset>: @6'<30,2'6"↵ (The radius of the tank is 6'.)
```

The point can now be used as the center of the pipe (cylinder), **Figure 6-5B**. However, if you draw the pipe now, it will be parallel to the tank (large cylinder). By changing the UCS, as shown in **Figure 6-5C**, the pipe can be correctly drawn. Working with the UCS is introduced later in this chapter.

Figure 6-5.
A—A plan view of a tank shows the angle of the pipe attachment. B—A 3D view from the southeast quadrant shows the pipe attachment point located with cylindrical coordinates. C—By creating a new UCS, the pipe can be drawn as a cylinder and correctly located without editing.

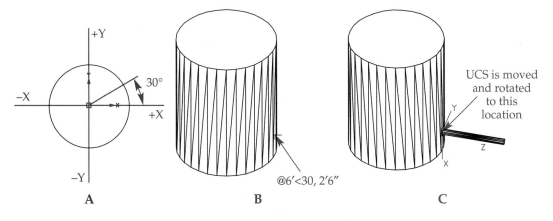

Exercise 6-1

Complete the exercise on the companion website.
www.g-wlearning.com/CAD

3D Polylines

A polyline drawn with the **PLINE** command is a 2D object. All segments of the polyline must be drawn parallel to the XY plane of the current UCS. A *3D polyline*, on the other hand, can be drawn in 3D space. The Z coordinate value can vary from point to point in the polyline.

The **3DPOLY** command is used to draw 3D polylines. Any form of coordinate entry is valid for drawing 3D polylines. If polar tracking is on when using the **3DPOLY** command, you can pick points in the Z direction if the polar tracking alignment path is parallel to the Z axis.

The **Close** option can be used to draw the final segment and create a closed shape. There must be at least two segments in the polyline to use the **Close** option. The **Undo** option removes the last segment without canceling the command.

The **PEDIT** command can be used to edit 3D polylines. The **PEDIT Spline Curve** option is used to turn the 3D polyline into a B-spline curve based on the vertices of the polyline. A regular 3D polyline and the same polyline turned into a B-spline curve are shown in **Figure 6-6**.

Exercise 6-2

Complete the exercise on the companion website.
www.g-wlearning.com/CAD

Introduction to Working with User Coordinate Systems

All points in a drawing or on an object are defined with XYZ coordinate values (rectangular coordinates) measured from the 0,0,0 origin. Since this system of coordinates is fixed and universal, AutoCAD refers to it as the *world coordinate system (WCS)*. A *user coordinate system (UCS)*, on the other hand, can be defined with its origin at any location and with its three axes in any orientation desired, while remaining at 90° to each other. The **UCS** command is used to change the origin, position, and rotation of the coordinate system to match the surfaces and features of an

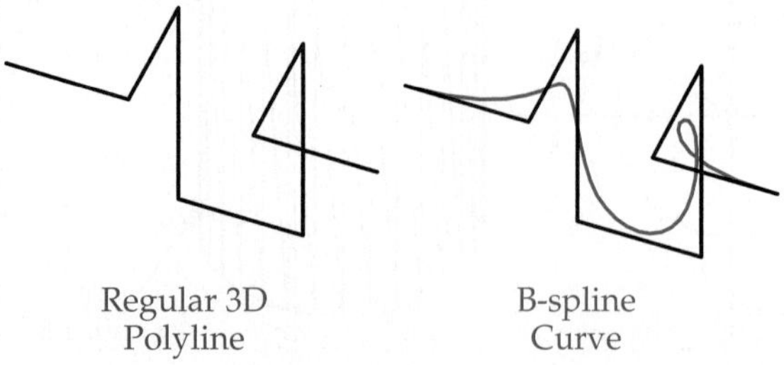

Figure 6-6.
A regular 3D polyline and the B-spline curve version after using the **PEDIT** command.

Regular 3D
Polyline

B-spline
Curve

object under construction. When set up to do so, the UCS icon reflects the changes in the orientation of the UCS and placement of the origin.

The **UCS** command is used to create and manage UCSs. This command and its options can be accessed on the **Coordinates** panel of the **Home** and **View** tabs on the ribbon or by typing the command. Three selections in the ribbon provide access to all UCS options. These selections are introduced here and discussed in detail later in this chapter.

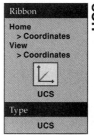

Ribbon

Home
> Coordinates
View
> Coordinates

UCS

Type

UCS

- **UCS.** Picking this button executes the **UCS** command. This command allows you to create a named UCS and provides access to all of the command options. The **UCS** command can also be selected by picking on the UCS drop-down list below the view cube. All of the command options are covered later in this chapter.
- **Named UCS Combo Control.** This drop-down list contains the six orthographic UCS options, which are covered later in this chapter. Any saved UCSs will be listed here, too.
- **UCS, Named UCS.** Picking this button displays the **UCS** dialog box. The three tabs in the dialog box contain a variety of UCS and UCS icon options and settings. These options and settings are described as you progress through the chapter.

Earlier in this chapter, you used spherical coordinates to locate a small sphere on the surface of a larger sphere. You also drew a cone with the center of its base on the surface of the large sphere. However, the axis of the cone, which is a line from the center of the base to the tip of the cone, is not pointing to the center of the sphere. Refer to **Figure 6-3**. This is because the Z axes of the large sphere and cone are parallel to the world coordinate system (WCS) Z axis. The WCS is the default coordinate system of AutoCAD.

In order for the axis of the cone to project from the sphere's center point, the UCS must be changed using the **UCS** command. Working with different UCSs is discussed in the next section. However, the following is a quick overview and describes how to draw a cone with its axis projecting from the center of the sphere.

First, draw a three-unit diameter sphere with its center at 7,5. Display the drawing from the southeast isometric preset. To help see how the UCS is changing, make sure the UCS icon is displayed at the origin of the current UCS. The UCS icon drop-down list is found in the **Coordinates** panel of the **Home** and **View** tabs of the ribbon and provides access to three options of the **UCSICON** command. See **Figure 6-7**. Pick the **Show UCS Icon at Origin** button to ensure the icon is displayed at the origin. Also, set the X-ray visual style current, and set the opacity to 25%. In the **Visual Styles** panel of the **View** tab, use the **Opacity** slider to adjust the value. You can also type a numerical value to the right of the slider.

Now, the sphere is drawn and the UCS icon is displayed at the origin of the current UCS (or at the lower-left corner of the screen, depending on the zoom level). However, the WCS is still the current user coordinate system. You are ready to start changing the UCS to meet your needs.

Figure 6-7.
The UCS icon drop-down list provides access to the **Origin, Off**, and **On** options of the **UCSICON** command.

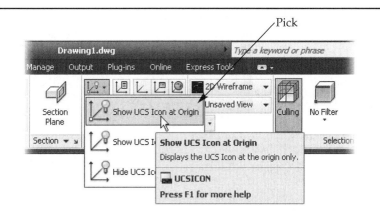

UCS

Ribbon

Home
> Coordinates
View
> Coordinates

Origin

Type

UCS

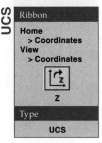

UCS

Ribbon

Home
> Coordinates
View
> Coordinates

Z

Type

UCS

First, turn ortho on to ensure that the current orientation of the UCS remains the same when its origin is changed. Next, move the UCS origin to the center of the sphere using the **Origin** option of the **UCS** command and the center object snap. Notice that the UCS icon is now displayed at the center of the sphere, **Figure 6-8**. Also, if the grid is displayed, the red X and green Y axes of the grid intersect at the center of the sphere.

Study **Figure 6-9** and continue as follows. Keep in mind that the point you are locating—the center of the cone on the sphere's surface—is 30° from the X axis and 30° from the XY plane. For ease of visualization, zoom in so the sphere fills the screen.

In the **Coordinates** panel on the ribbon, pick the **Z** button in the axis rotation flyout. Enter 30 for the rotation angle. Watch the position of the UCS icon change when you press [Enter]. See **Figure 6-9B**.

Figure 6-8.
The UCS origin is moved to the center of the sphere.

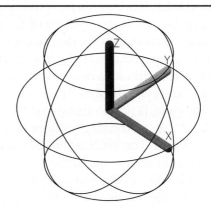

Figure 6-9.
A—The world coordinate system. B—The new UCS is rotated 30° in the XY plane about the Z axis. C—A line rotated up 30° from the XY plane represents the axis of the cone. D—The UCS is rotated 60° about the Y axis. The centerline of the cone coincides with the Z axis of this UCS.

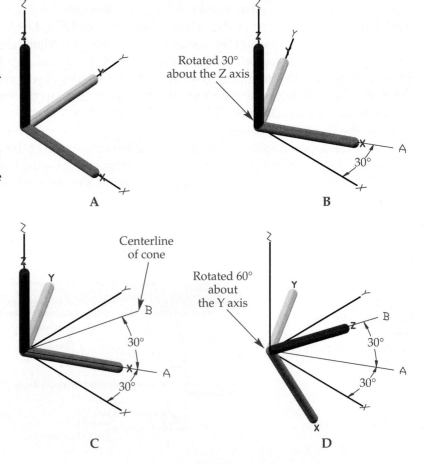

Ribbon

Home
> Coordinates
View
> Coordinates

Type

UCS

UCS

In the **Coordinates** panel on the ribbon, pick the **Y** button in the axis rotation flyout. Enter 60° for the rotation angle. Watch the position of the UCS icon change when you press [Enter]. See **Figure 6-9D**.

If the view cube is set to be oriented to the current UCS, then it rotates when the UCS is changed. Additionally, the grid rotates to match the new UCS. If the grid is not on, turn it on by pressing [Ctrl]+[G]. Remember, the grid is displayed on the XY plane of the current UCS.

The new UCS can be used to construct a cone with its axis projecting from the center of the sphere. **Figure 6-10A** shows the new UCS located at the center of the sphere. With the UCS rotated, rectangular coordinates can be used to draw the cone. Enter the **CONE** command and specify the center as 0,0,1.5 (the radius of the sphere is 1.5 units). Enter a radius of .25 and a height of 1. The completed cone is shown in **Figure 6-10B**. You can see that the axis projects from the center of the sphere. **Figure 6-10C** shows the objects after setting the Conceptual visual style current.

This same basic procedure can be used in the tank and pipe example presented earlier in this chapter. To correctly locate the pipe (cylinder), first move the UCS origin to the point previously located on the surface of the cylinder. Next, rotate the UCS 30° about the Z axis. Then, rotate the UCS 90° about the Y axis. The Z axis of this new UCS aligns with the long axis of the pipe. Finally, use rectangular coordinates to draw the cylinder with its center at the point drawn in **Figure 6-5B**.

Once you have changed to a new UCS, you can quickly return to the WCS by picking WCS in the view cube drop-down list or by using the **World** option of the **UCS** command. The WCS provides a common "starting place" for creating new UCSs.

Exercise 6-3

Complete the exercise on the companion website.
www.g-wlearning.com/CAD

Working with User Coordinate Systems

Once you understand a few of the basic options of user coordinate systems, creating 3D models becomes an easy and quick process. The following sections show how to display the UCS icon, change the UCS in order to work on different surfaces of a model, and name and save a UCS. As you saw in the previous section, working with UCSs is easy.

Figure 6-10.
A—A new UCS is created with the Z axis projecting from the center of the sphere.
B—A cone is drawn using the new UCS. The axis of the cone projects from the center of the sphere. C—The objects after the Conceptual visual style is set current.

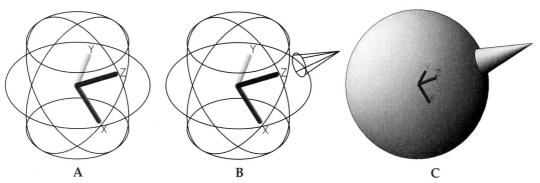

A   B   C

Displaying the UCS Icon

The symbol that identifies the orientation of the coordinate system is called the *UCS icon*. When AutoCAD is first launched based on the acad3D.dwt template, the UCS icon is located at the WCS origin in the middle of the viewport. The display of this symbol is controlled by the **UCSICON** command. Turn the UCS icon on and off using the **Hide UCS Icon** and **Show UCS Icon** buttons on the **Coordinates** panel in the ribbon. Refer to **Figure 6-7**.

If your drawing does not require viewports and altered coordinate systems, you may want to turn the icon off. The icon disappears until you turn it on again. If you redisplay the UCS icon and it does not appear at the origin, simply pick the **Show UCS Icon at Origin** button on the **Coordinates** panel in the ribbon.

You can also turn the icon on or off and set the icon to display at the UCS origin point using the options in the **Settings** tab of the **UCS** dialog box. See **Figure 6-11**. This dialog box is displayed by picking the **UCS, Named UCS...** button on the **Coordinates** panel in the ribbon or typing the **UCSMAN** command.

Notice that the **Allow Selecting UCS icon** option is checked by default in the **UCS Icon settings** area of the **Settings** tab. When this option is checked, it enables the display of the UCS icon grips. The UCS icon grips provide for dynamic and intuitive moving and rotating of the UCS icon. The **UCSSELECTMODE** system variable controls the **Allow Selecting UCS icon** setting and is set to 1 (on) by default.

PROFESSIONAL TIP

It is recommended that you have the UCS icon turned on at all times when working in 3D drawings. It provides a quick indication of the current UCS.

Changing the Coordinate System

To construct a three-dimensional object, you must draw shapes at many different angles. Different planes are needed to draw features on angled surfaces. To construct these features, it is easiest to rotate the UCS to match any surface on an object. The following example illustrates this process.

The object in **Figure 6-12** has a cylinder on the angled surface. The **EXTRUDE** command, which is discussed in Chapter 8, is used to create the base of the object. The cylinder is then drawn on the angled feature. In Chapter 14, you will learn how to dimension the object as shown in **Figure 6-12**.

Figure 6-11.
Setting UCS and UCS icon options in the **UCS** dialog box.

UCS icon options

UCS options

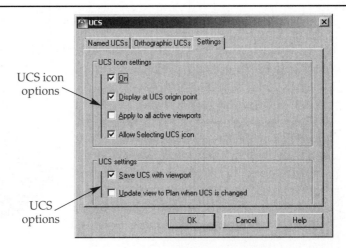

Figure 6-12.
This object can
be constructed
by changing the
orientation of the
coordinate system.

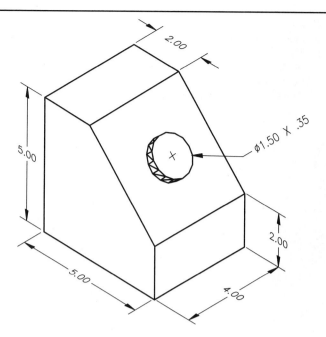

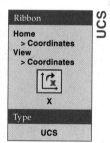

The first step in creating this model is to draw the side view of the base as a wireframe. You could determine the X, Y, and Z coordinates of each point on the side view and enter the coordinates. However, a lot of typing can be saved if all points share a Z value of 0. By rotating the UCS, you can draw the side view entering only X and Y coordinates. Start a new drawing and display the southeast isometric view. If the UCS icon is off, turn it on and display it at the origin.

Now, rotate the UCS 90° around the X axis. The quickest way to do this is to click directly on the UCS icon. Notice that as the pointer is moved over the UCS icon, the color of the icon changes to a light yellow. See **Figure 6-13A**. When you click on the icon, it displays grips. A square *origin grip* is displayed at the origin, and circular *axis grips* are displayed at the axis endpoints. See **Figure 6-13B**.

Once the UCS icon grips are displayed, move the pointer over the Y axis grip. The axis shortcut menu is displayed, **Figure 6-13C**. Select the **Rotate Around X Axis** option. The default value is 90. This is the desired rotation angle, so press [Enter]. The new UCS is now oriented parallel to the side of the new object. The UCS icon is displayed at the origin of the UCS. If needed, pan the view so the UCS icon is near the center of the view.

Next, use the **PLINE** command to draw the outline of the side view. Refer to the coordinates shown in **Figure 6-14**. When entering coordinates, you may want to turn off dynamic input. As an alternative, type a pound sign (#) before each coordinate with dynamic input turned on. This temporarily overrides the dynamic input of relative coordinates, which is controlled by the **DYNPICOORDS** system variable. The **PLINE** command is used instead of the **LINE** command for this model because a closed polyline can be extruded into a solid. Be sure to use the **Close** option to draw the final segment. A wireframe of one side of the object is created. Notice the orientation of the UCS icon.

Now, the **EXTRUDE** command is used to create the base as a solid. This command is covered in detail in Chapter 8. Make sure the same UCS used to create the wireframe side is current. Then, select the **EXTRUDE** command by picking the **Extrude** button on the **Modeling** panel of the **Home** tab in the ribbon. When prompted to select objects, pick the polyline and then press [Enter]. Next, move the mouse so the preview extends to the right and enter an extrusion height of 4. If dynamic input is off, you must enter –4.

The base is created as a solid. See **Figure 6-15**. You may want to switch to a parallel projection, as shown in the figure.

Figure 6-13.
Using the UCS icon to rotate the UCS. A—When the pointer is moved over the UCS icon, the color of the icon changes to light yellow. B—Selecting the UCS icon displays a grip at the origin and grips at all three axis endpoints. C—The shortcut menu displayed when the pointer is moved over the Y axis grip to highlight the grip. To rotate the UCS 90° about the X axis, select the **Rotate Around X Axis** option.

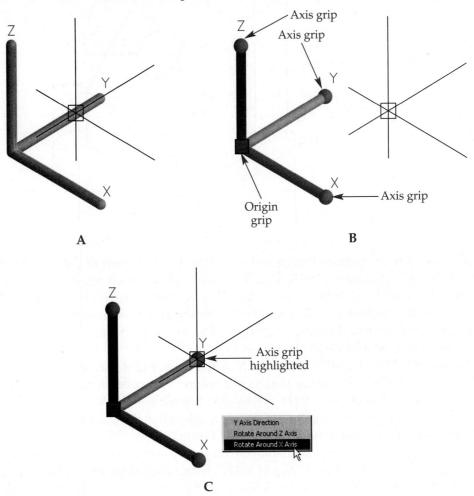

Figure 6-14.
A wireframe of one side of the base is created. Notice the orientation of the UCS.

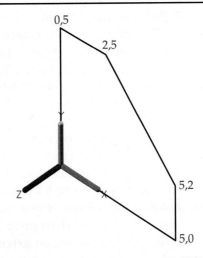

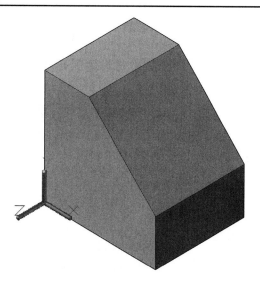

Figure 6-15.
The wireframe is extruded to create the base as a solid.

Saving a Named UCS

Once you have created a new UCS that may be used again, it is best to save it for future use. For example, you just created a UCS used to draw the wireframe of one side of the object. To save this UCS, use the UCS icon shortcut menu. Move the pointer over the UCS icon and right-click. See **Figure 6-16.** Select **Named UCS>Save** in the shortcut menu, type a name at the prompt, and press [Enter]. The UCS can also be saved by using the **Named>Save** option of the **UCS** command or the **UCS** dialog box.

If using the dialog box, right-click on the entry Unnamed and pick **Rename** in the shortcut menu. See **Figure 6-17.** You can also pick once or double-click on the highlighted name. Then, type the new name in place of Unnamed and press [Enter]. A name can have up to 255 characters. Numbers, letters, spaces, dollar signs ($), hyphens (–), and underscores (_) are valid. Use this method to save a new UCS or to rename an existing one. Now, the coordinate system is saved and can be easily recalled for future use.

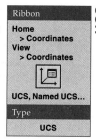

Ribbon	UCS
Home > Coordinates View > Coordinates UCS, Named UCS...	
Type	
UCS	

Figure 6-16.
Saving a new UCS. A—Move the pointer over the UCS icon and right-click to display the UCS icon shortcut menu. Pick **Named UCS>Save**.

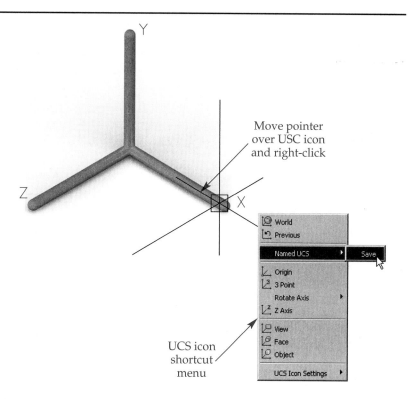

Figure 6-17.
Select **Rename** in the UCS dialog box to enter a name and save the Unnamed UCS.

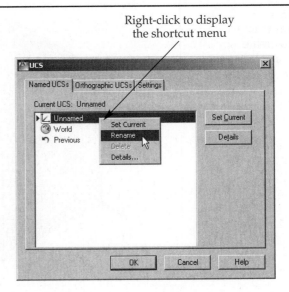

Right-click to display the shortcut menu

If Unnamed does not appear in the **UCS** dialog box, AutoCAD is having a problem. Exit the dialog box and enter the **UCS** command on the command line. Then, use the **Named>Save** option to save the unnamed UCS. The saved UCS will then appear in the **UCS** dialog box.

PROFESSIONAL TIP

Most drawings can be created by rotating the UCS as needed without saving it. If the drawing is complex with several planes, each containing a large amount of detail, you may wish to save a UCS for each detailed face. Then, restore the proper UCS as needed. For example, when working with architectural drawings, you may wish to establish a different UCS for each floor plan and elevation view and for roofs and walls that require detail work.

Working with the UCS Icon

The UCS icon can be manipulated much like any AutoCAD entity. As you have seen, when it is selected, grips are displayed, and the UCS can be rotated with ease. You can also use the dynamic moving capability of the UCS icon to relocate the UCS to any model face. This allows you to create new objects or features on existing faces with minimal effort. New user coordinate systems you create can be saved as needed for future use.

For example, to draw the cylinder on the angled face of the object shown in **Figure 6-12**, use the UCS icon grips to establish a UCS on the angled face. First, pick the UCS icon and move the pointer to the origin grip to highlight it. When the shortcut menu appears, select the **Move and Align** option. See **Figure 6-18A**. Now move the UCS icon onto the angled face of the object. Notice how the icon immediately rotates

Figure 6-18.
Using the UCS icon to establish a UCS on the angled face of the object. A—Moving the pointer over the origin grip displays the UCS icon shortcut menu. Select the **Move and Align** option. B—Use the **Endpoint** object snap to relocate the UCS on the angled face.

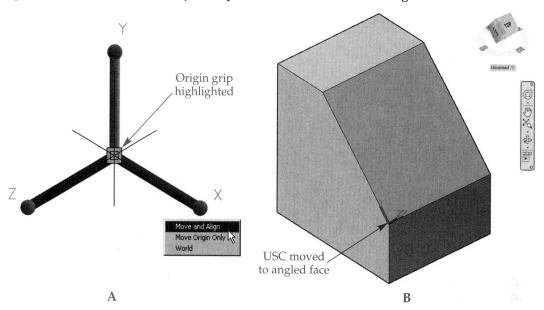

A   B

to match the orientation of the highlighted face it is resting on. Test this by moving the icon to each visible face. As you move the pointer over the object faces, note that hidden faces are not highlighted. Therefore, you cannot work on those faces. If you wish to work on a hidden face, you must first change the viewpoint to make that face visible. You can use the view cube or press and hold the [Shift] key while pressing the mouse wheel button to dynamically change the viewpoint without interrupting the current command.

Now return the UCS icon to the angled face, being careful to ensure that the icon is oriented in the manner shown in Figure 6-18B. You may have to first move the icon to the top face to achieve the proper UCS orientation, and then move to the angled face. Use the **Endpoint** object snap to select the lower-left corner of the angled face as the new origin.

Next, select the **CYLINDER** command. You are prompted to select a center point of the base. To locate a 1.5″ diameter cylinder in the center of the angled face, use the following procedure.

1. At the "specify center point of base" prompt, press [Shift] and right-click in the drawing area and pick **3D Osnap** from the shortcut menu, then pick **Center of face** from the cascading menu. See Figure 6-19A.
2. Move the pointer over the angled face. Notice the crosshairs are aligned with the angled face. A circular 3D snap pick point appears at the center of the face. Select this point.
3. Specify the 1.5 unit diameter for the base. See Figure 6-19B.
4. Specify a cylinder height of .35 units. See Figure 6-19C.
5. The cylinder is properly located on the angled face. See Figure 6-19D.

If you do not wish to retain the location of the current UCS after drawing a feature, change it as needed using the UCS icon. If necessary, you can also quickly restore the previous UCS. To do so, simply right-click on the UCS icon and pick **Previous** from the UCS icon shortcut menu. Refer to Figure 6-16. Do this after drawing the cylinder so that the UCS is oriented parallel to the side of the object.

Figure 6-19.
Moving the UCS allows you to draw a cylinder on the angled face of the object. A—To set the center point of the base, use the **Center of face** 3D object snap and select the center of the face. B—Set the radius or diameter of the base. C—Set the height of the cylinder. D—The completed cylinder.

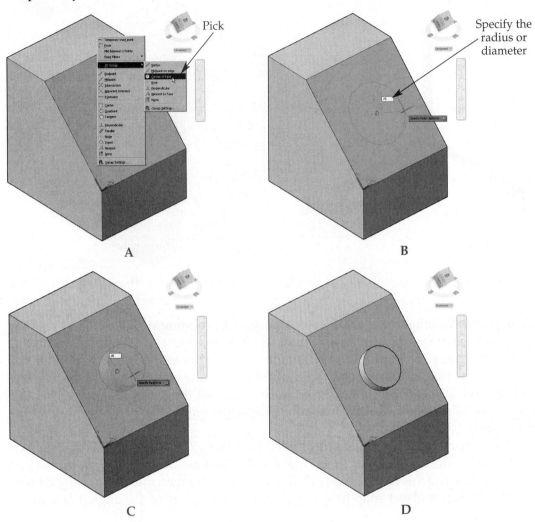

A

B

C

D

PROFESSIONAL
TIP

Selecting 3D object snaps can be automated for solid model construction. Simply right-click on the **3D Object Snap** button on the status bar and select the snap mode you wish to use. To set multiple snap modes, pick the **Settings...** option, and then select the appropriate modes in the **Drafting Settings** dialog box. Be sure to turn on the **3D Object Snap** button before using a modeling command.

Exercise 6-4

Complete the exercise on the companion website.
www.g-wlearning.com/CAD

Using a Dynamic UCS

A powerful tool for 3D modeling is the *dynamic UCS function*. A dynamic UCS is a UCS temporarily located on any existing face of a 3D model. The function is activated by picking the **Allow/Disallow Dynamic UCS** button on the status bar, pressing the [Ctrl]+[D] key combination, or setting the **UCSDETECT** system variable to 1. When the pointer is moved over a model surface, the XY plane of the UCS is aligned with that surface. This is especially useful when adding primitives or shapes to model surfaces. In addition, dynamic UCSs are useful when inserting blocks and xrefs, locating text, editing 3D geometry, editing with grips, and area calculations.

A limitation of using a dynamic UCS is that it operates as a temporary override to the current UCS. Once the operation is complete, the UCS remains in its previous location. A more flexible option is to use the full range of capabilities inherent in the UCS icon as previously discussed.

To see how to work with a dynamic UCS, recreate the cylinder on the angled face of the model completed in the previous discussion. First, delete the existing cylinder by picking on the cylinder and pressing the [Delete] key. Make sure that the UCS is oriented parallel to the side of the object. Refer to **Figure 6-15**. Next, select the **CYLINDER** command. Make sure the dynamic UCS function is on. Then, move the pointer over one of the surfaces of the object. Notice that the 3D crosshairs change when they are moved over a new surface. The red (X) and green (Y) crosshairs are flat on the face. For ease of visualizing the 3D crosshairs as they are moved across different surfaces, right-click on the **Allow/Disallow Dynamic UCS** button on the status bar and select **Display crosshair labels** to turn on the XYZ labels on the crosshairs.

The **CYLINDER** command is currently prompting to select a center point of the base. If you pick a point, this sets the center of the cylinder base and temporarily relocates the UCS so its XY plane lies on the selected face. Once the point is selected and the dynamic UCS created, the UCS moves to the temporary UCS. When the command is ended, the UCS reverts to its previous orientation. Using the **CYLINDER** command options and the steps presented earlier, draw a 1.5″ diameter cylinder in the center of the angled face.

When using the dynamic UCS function, experiment with the behavior of the crosshairs as they are moved over different surfaces. The orientation of the crosshairs is related to the edge of the face that they are moved over. Can you determine the pattern by which the crosshairs are turned? The X axis of the crosshairs is always aligned with the edge that is crossed.

If you want to temporarily turn off the dynamic UCS function while working in a command, press and hold the [Shift]+[Z] key combination while moving the pointer over a face. As soon as you release the keys, the dynamic UCS function is reinstated.

Type
UCSDETECT [Ctrl]+[D]

Toolbar
Status Bar
Allow/Disallow Dynamic UCS

UCSDETECT

Additional Ways to Change the UCS

As you have seen, the UCS can be moved to any location and rotated to any angle desired using the UCS icon grips. Location and alignment are controlled using the origin grip, and exact rotation of the three different UCS axes is controlled using the axis grips. The same UCS options are available when using the **UCS** command, but the UCS icon provides a quick alternative to entering the command or making ribbon or menu selections. The following sections discuss the UCS options available when using the UCS icon grips or the **UCS** command.

Moving and Aligning the UCS

The UCS icon origin grip is used to quickly move and align the UCS. Once the UCS icon is selected to display grips, you can pick on the origin grip and drag the icon dynamically. You can drag the UCS icon to a model surface and orient the UCS parallel to that surface. If you are using dynamic input, you can enter a distance and angle defining a new location. You can also move the pointer over the grip to display the shortcut menu, as previously discussed. Pick **Move Origin Only** from the shortcut menu to relocate the UCS in its current orientation to a different origin point. Pick **Move and Align** to relocate the UCS to a new location and align it with an object face. Use object snaps for accuracy.

When the UCS icon is moved along a curved surface, it will always remain normal to that surface. In other words, the Z axis of the UCS will project from the center point of the radius curve. This applies to an arc, circle, sphere, or surface curve. See **Figure 6-20**. In the example shown, the UCS is moved to a point on the outer surface of the model using the **Move and Align** option. Notice the direction of the Z axis in **Figure 6-20C**. If dynamic input is on, a tooltip is displayed when using the **Move and Align** option. You can press the [Ctrl] key to cycle through the options. The options are displayed on the command line to indicate the current mode.

Figure 6-20.
Using the UCS icon to move the UCS to a curved surface. In this example, the UCS icon origin grip is used to access the **Move and Align** option. A—The current UCS is located at the center point of the left curve of the model. This is apparent in the top view. B—A top view of the model. C—A new UCS is established on the outer surface by moving the UCS icon to the surface and picking. Note that the Z axis is normal to the surface.

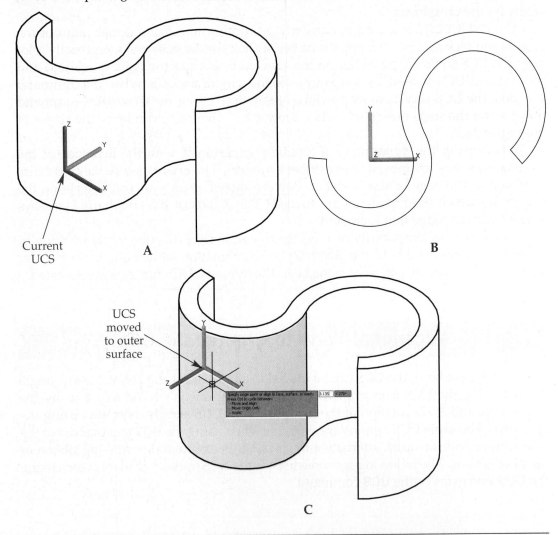

Current UCS

A

B

UCS moved to outer surface

C

AutoCAD and Its Applications—Advanced

Dynamically moving and aligning the UCS to any object surface can also be achieved by using the **Object** option of the **UCS** command. This option is also available in the UCS icon shortcut menu and is discussed later in this chapter.

Rotating the UCS on an Axis

The axis grips of the UCS icon enable you to rotate the icon in any direction, and to cycle between rotation options. For example, pick the UCS icon to display the grips, then move the pointer over the Z axis grip (but do not select it). See **Figure 6-21A**. The first item in the axis shortcut menu is **Z Axis Direction**. Picking this option is the same as picking the Z axis grip. Select this option and notice that it enables you to select the direction of the Z axis. Use object snaps to assist, or turn ortho on to restrain selections to 90°. In addition, if dynamic input is on, you can press the [Ctrl] key to cycle through the three options in the axis shortcut menu. See **Figure 6-21B**.

The other two options in the axis shortcut menu allow the Z axis to be moved in relation to one of the other two axes. Test each of these options to see how they function. The **Rotate Around X Axis** option limits rotation of the Z axis to around the X axis, and the **Rotate Around Y Axis** option limits rotation to around the Y axis.

Move the pointer to each of the other two UCS icon axis grips and note that a similar shortcut menu is displayed. Each axis of the UCS can be rotated in the same manner using the appropriate axis grip.

Additional UCS Options

Additional options for changing the UCS are discussed in the following sections. Some of these options can be selected from the UCS icon shortcut menu displayed by right-clicking on the UCS icon, shown in **Figure 6-16**, or by accessing the **Coordinates** panel in the **Home** or **View** tab on the ribbon. All of the options discussed are available when entering the UCS command on the command line.

Keep in mind that many of the procedures discussed in the following sections can be accomplished using the dynamic moving and rotating functions of the UCS icon grips.

Figure 6-21.
Using the UCS icon axis rotation options. A—The Z axis options appear after moving the pointer over the Z axis grip. Pick the Z Axis Direction option to set the direction of the Z axis. B—The new UCS after rotating the Z axis. Note that the other Z axis options can be accessed by pressing the [Ctrl] key.

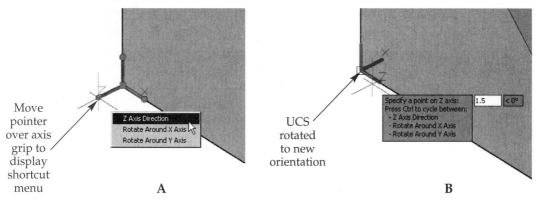

Selecting Three Points to Create a New UCS

UCS

Ribbon
Home
> Coordinates
View
> Coordinates

3 Point

Type
UCS

The **3 Point** option of the **UCS** command can be used to change the UCS to any flat surface. This option requires that you first locate a new origin, then a point on the positive X axis, and finally a point on the XY plane that has a positive Y value. See **Figure 6-22**. Use object snaps to select points that are not on the current XY plane. After you pick the third point—the point on the XY plane—the UCS icon changes its orientation to align with the plane defined by the three points. The **3 Point** option is available in the UCS icon shortcut menu.

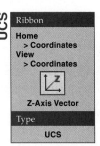

PROFESSIONAL TIP

When typing the **UCS** command, enter 3 at the Specify origin of UCS or [Face/NAmed/OBject/Previous/View/World/X/Y/Z/ZAxis] <World>: prompt. Notice that the option is not listed in the prompt.

Selecting a New Z Axis

UCS

Ribbon
Home
> Coordinates
View
> Coordinates

Z-Axis Vector

Type
UCS

The **ZAxis** option of the UCS command allows you to select the origin point and a point on the positive Z axis. Once the new Z axis is defined, AutoCAD sets the new X and Y axes. The **ZAxis** option is available in the UCS icon shortcut menu.

You will now add a cylinder to the lower face of the base created earlier. The cylinder extends into the base. Earlier, you moved the UCS to the angled face of the object, drew the cylinder on the angled face, and then restored the UCS to its previous position on the side face. If this is the current orientation, the Z axis does not project perpendicular to the lower face. Therefore, a new UCS must be created on the lower-right face. Change the UCS after entering the **ZAxis** option as follows.

1. Pick the origin of the new UCS. See **Figure 6-23A**. You may have to use an object snap to select the origin.
2. Pick a point on the positive portion of the new Z axis.
3. The new UCS is established and it can be saved if necessary.

Figure 6-22.
A—A new UCS can be established by picking three points. P1 is the origin, P2 is on the positive X axis, and P3 is on the XY plane and has a positive Y value. B—The new UCS is created.

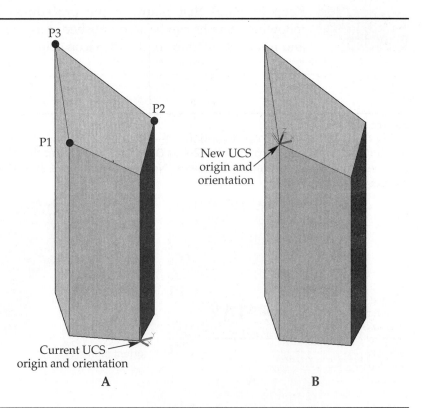

A   B

Figure 6-23.
A—Using the **ZAxis** option to establish a new UCS. B—The new UCS is used to create a cylinder, which is then subtracted from the base to create a hole.

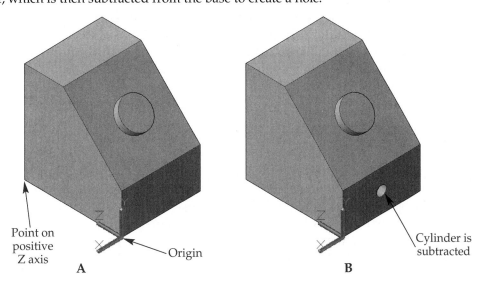

Point on
positive
Z axis

Origin

A

Cylinder is
subtracted

B

Now, use 3D object snaps, auto-tracking, or 2D object snaps to draw a ∅.5" cylinder centered on the lower face and extending 3" into the base. Then, subtract the cylinder from the base part to create the hole, as shown in **Figure 6-23B**.

Setting the UCS to an Existing Object

The **Object** option of the **UCS** command can be used to define a new UCS on an object. This option cannot be used with 3D polylines or xlines. There are also certain rules that control the orientation of the UCS. For example, if you select a circle, the center point becomes the origin of the new UCS. The pick point on the circle determines the direction of the X axis. The Y axis is relative to X and the UCS Z axis may or may not be the same as the Z axis of the selected object. The **Object** option is available in the UCS icon shortcut menu.

Look at **Figure 6-24A**. The circle is rotated an unknown number of degrees from the XY plane of the WCS. However, you need to create a UCS in which the circle is lying on the XY plane. Select the **Object** option of the **UCS** command and then pick

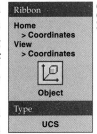

Ribbon

Home
> Coordinates
View
> Coordinates

Object

Type

UCS

UCS

Figure 6-24.
A—This circle is rotated off of the WCS XY plane by an unknown number of degrees. It will be used to establish a new UCS. B—The circle is on the XY plane of the new UCS. However, the X and Y axes do not align with the circle's quadrants. C—The **ZAxis** option of the **UCS** command is used to align the UCS with the quadrants of the circle.

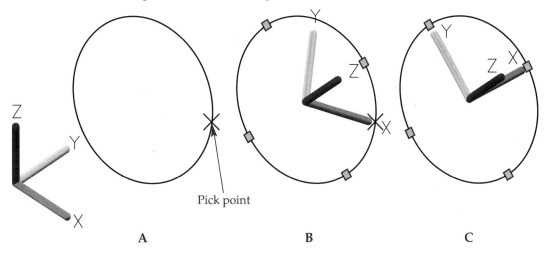

Pick point

A　　　　　　　　**B**　　　　　　　　**C**

the circle. The UCS icon may look like the one shown in **Figure 6-24B**. Notice how the X and Y axes are not aligned with the quadrants of the circle, as indicated by the grip locations. This may not be what you expected. The X axis orientation is determined by the pick point on the circle. Notice how the X axis is pointing at the pick point.

To rotate the UCS in the current plane so the X and Y axes of the UCS are aligned with the quadrants of the circle, use the **ZAxis** option of the **UCS** command. Select the center of the circle as the origin and then enter the absolute coordinate 0,0,1. This uses the current Z axis location, which also forces the X and Y axes to align with the object. Refer to **Figure 6-24C**. This method may not work with all objects.

Setting the UCS to the Face of a 3D Solid

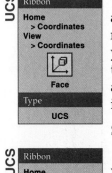

The **Face** option of the **UCS** command allows you to orient the UCS to any face on a 3D solid, surface, or mesh object. The **Face** option is available in the UCS icon shortcut menu. Select the option and then pick a face on the model. After you have selected a face, you have the options of moving the UCS to the adjacent face or flipping the UCS 180° on the X axis, Y axis, or both axes. Use the **Next**, **Xflip**, or **Yflip** option to move or rotate the UCS as needed. Once you achieve the UCS orientation you want, press [Enter] to accept. Notice in **Figure 6-25** how many different UCS orientations can be selected for a single face.

Setting the UCS Perpendicular to the Current View

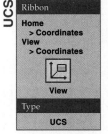

You may need to add notes or labels to a 3D drawing that are plan to the current view, such as the note shown in **Figure 6-26**. The **View** option of the **UCS** command makes this easy to do. The **View** option is available in the UCS icon shortcut menu. Immediately after selecting the **View** option, the UCS rotates to a position so the new XY plane is perpendicular to the current line of sight (parallel to the screen). Now, anything added to the drawing is plan to the current view. The **View** option works on the current viewport only; other viewports are unaffected.

Applying the Current UCS to a Viewport

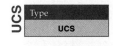

The **Apply** option of the **UCS** command allows you to apply the UCS in the current viewport to any or all model space or paper space viewports. Using the **Apply** option, you can have a different UCS displayed in every viewport or you can apply one UCS to all viewports. With the viewport that contains the UCS to apply active, enter the **Apply** option. This option is only available on the command line. However, the option does not appear in the command prompt. Enter either A or APPLY to select the option. Then, pick a viewport to which the current UCS will be applied and press [Enter]. To apply the current UCS to all viewports, enter the **All** option. See Chapter 7 for a complete discussion of model space viewports.

Figure 6-25.
Several different UCSs can be selected from a single pick point using the **Face** option of the **UCS** command. Given the pick point, five of the eight possibilities are shown here.

Figure 6-26.
The **View** option of the **UCS** command allows you to place text plan to the current view.

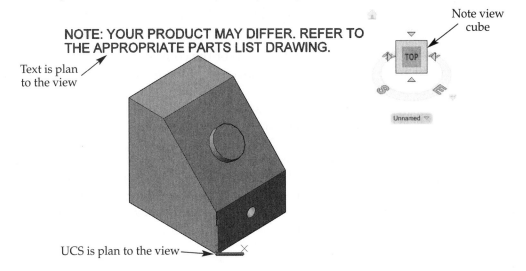

NOTE: YOUR PRODUCT MAY DIFFER. REFER TO
THE APPROPRIATE PARTS LIST DRAWING.

Text is plan
to the view

Note view
cube

UCS is plan to the view

Preset UCS Orientations

AutoCAD has six preset orthographic UCSs that match the six standard orthographic views. With the current UCS as the top view (plan), all other views are arranged as shown in **Figure 6-27**. These orientations can be selected by using the **Named UCS Combo Control** drop-down list in the **Coordinates** panel of the **Home** or **View** tab in the ribbon, entering the command on the command line, or using the **Orthographic UCSs**

Figure 6-27.
The standard orthographic UCSs coincide with the six basic orthographic views.

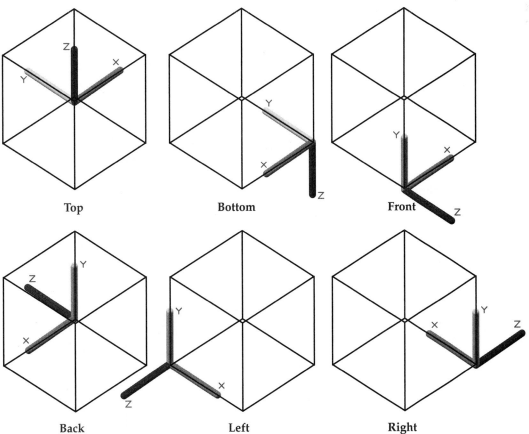

Top   Bottom   Front

Back   Left   Right

tab of the **UCS** dialog box. When using the command line, type the name of the UCS (FRONT, BACK, RIGHT, etc.) at the first prompt:

Specify origin of UCS or [Face/NAmed/OBject/Previous/View/World/X/Y/Z/ZAxis]
<World>: **FRONT** *or* **FR.**↵

Note that there is no orthographic option listed for the command.

The **Relative to:** drop-down list at the bottom of the **Orthographic UCSs** tab of the **UCS** dialog box specifies whether each orthographic UCS is relative to a named UCS or absolute to the WCS. For example, suppose you have a saved UCS named Front Corner that is rotated 30° about the Y axis of the WCS. If you set current the top UCS relative to the WCS, the new UCS is perpendicular to the WCS, **Figure 6-28A**. However, if the top UCS is set current relative to the named UCS Front Corner, the new UCS is also rotated from the WCS, **Figure 6-28B**.

The Z value, or depth, of a preset UCS can be changed in the **Orthographic UCSs** tab of the **UCS** dialog box. First, right-click on the name of the UCS you wish to change. Then, pick **Depth** from the shortcut menu, **Figure 6-29A**. This displays the **Orthographic**

Figure 6-28.
The **Relative to:** drop-down list entry in the **Orthographic UCSs** tab of the **UCS** dialog box determines whether the orthographic UCS is based on a named UCS or the WCS. The UCS icon here represents the named UCS. A—Relative to the WCS. B—Relative to the named UCS.

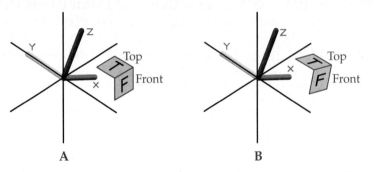

A   B

Figure 6-29.
A—The Z value, or depth, of a preset UCS can be changed by right-clicking on its name and selecting **Depth**. B—Enter a new depth value or pick the **Select new origin** button to pick a new location on screen.

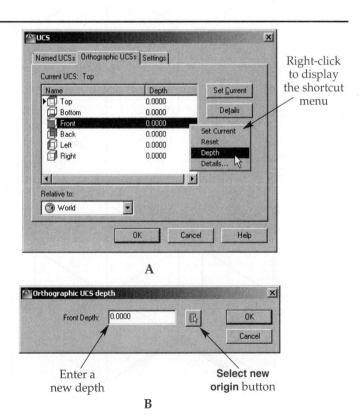

Right-click to display the shortcut menu

A

Enter a new depth

Select new origin button

B

UCS depth dialog box. See **Figure 6-29B**. You can either enter a new depth value or specify the new location on screen by picking the **Select new origin** button. Once the new depth has been selected, it is reflected in the preset UCS list.

CAUTION

Changing the **Relative to:** setting affects *all* preset UCSs and *all* preset viewpoints! Therefore, leave this set to World unless absolutely necessary to change it.

Exercise 6-5

Complete the exercise on the companion website.
www.g-wlearning.com/CAD

Managing User Coordinate Systems and Displays

You can create, name, and use as many user coordinate systems as needed to construct your model or drawing. As you saw earlier, AutoCAD allows you to name (save) coordinate systems for future use. User coordinate systems can be created, renamed, set current, and deleted using the **Named UCSs** tab of the **UCS** dialog box, **Figure 6-30**.

The **Named UCSs** tab contains the **Current UCS:** list box. This list box contains the names of all saved coordinate systems plus World. If other coordinate systems have been used in the current drawing session, Previous appears in the list. Unnamed appears if the current coordinate system has not been named (saved). The current UCS is indicated by a small triangle to the left of its name in the list and by the label at the top of the list. To make any of the listed coordinate systems active, highlight the name and pick the **Set Current** button.

A list of coordinate and axis values of the highlighted UCS can be displayed by picking the **Details** button. This displays the **UCS Details** dialog box shown in **Figure 6-31**.

If you right-click on the name of a UCS in the list in the **Named UCSs** tab, a shortcut menu is displayed. Using this menu, you can rename the UCS. You can also set the UCS current or delete it using the shortcut menu. The Unnamed UCS cannot be deleted, nor can World be deleted.

Figure 6-30.
The **UCS** dialog box allows you to rename, list, delete, and set current an existing UCS.

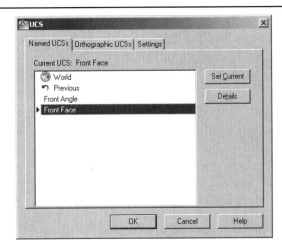

Figure 6-31.
The **UCS Details**
dialog box displays
the coordinate
values of the
selected UCS.

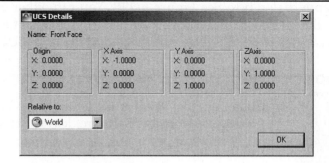

PROFESSIONAL TIP

You can also manage UCSs on the command line using the **UCS** command. In addition, using the UCS icon, you can select a UCS from the UCS icon shortcut menu to make it current.

Setting an Automatic Plan Display

After changing the UCS, a plan view is often needed to give you a better feel for the XYZ directions. While you should try to draw in a pictorial view when possible as you construct a 3D object, some constructions may be much easier in a plan view. AutoCAD can be set to automatically make your view of the drawing plan to the current UCS. This is especially useful if you will be changing the UCS often, but want to work in a plan view.

The **UCSFOLLOW** system variable is used to automatically display a plan view of the current UCS. When it is set to 1, a plan view is automatically created in the current viewport when the UCS is changed. Viewports are discussed in Chapter 7. The default setting of **UCSFOLLOW** is 0 (off). After setting the variable to 1, a plan view will be automatically generated the next time the UCS is changed. The **UCSFOLLOW** variable generates the plan view only after the UCS is changed, not immediately after the variable is changed. However, if you select a different viewport, the previous viewport is set plan to the UCS if **UCSFOLLOW** has been set to 1 in that viewport. The **UCSFOLLOW** variable can be individually set for each viewport.

PROFESSIONAL TIP

To get the plan view displayed without changing the UCS, use the **PLAN** command, which is discussed in Chapter 4.

AutoCAD and Its Applications—Advanced

UCS Settings and Variables

As discussed in the previous section, the **UCSFOLLOW** system variable allows you to change how an object is displayed in relation to the UCS. There are also system variables that display a variety of information about the current UCS. These variables include:

- **UCSAXISANG.** (stored value) The default rotation angle for the **X**, **Y**, or **Z** option of the **UCS** command.
- **UCSBASE.** The name of the UCS used to define the origin and orientation of the orthographic UCS settings. It can be any named UCS.
- **UCSDETECT.** (on or off) Turns the dynamic UCS function on and off. The **Allow/Disallow Dynamic UCS** button on the status bar controls this variable, as does the [Ctrl]+[D] key combination.
- **UCSNAME.** (read only) Displays the name of the current UCS.
- **UCSORG.** (read only) Displays the XYZ origin value of the current UCS.
- **UCSORTHO.** (on or off) If set to 1 (on), the related orthographic UCS setting is automatically restored when an orthographic view is restored. If turned off, the current UCS is retained when an orthographic view is restored. Depending on your modeling preferences, you may wish to set this variable to 0.
- **UCSSELECTMODE.** (on or off) Enables the selection and manipulation of the UCS icon with grips. The default setting is 1 (on).
- **UCSVIEW.** (on or off) If this variable is set to 1 (on), the current UCS is saved with the view when a view is saved. Otherwise, the UCS is not saved with the view.
- **UCSVP.** Controls which UCS is displayed in viewports. The default value is 1, which means that the UCS configuration in the viewport is independent from all other UCS configurations. If the setting is 0, the UCS configuration in the current viewport is displayed. Each viewport can be set to either 0 or 1.
- **UCSXDIR.** (read only) Displays the XYZ value of the X axis direction of the current UCS.
- **UCSYDIR.** (read only) Displays the XYZ value of the Y axis direction of the current UCS.

UCS options and variables can also be managed in the **Settings** tab of the **UCS** dialog box. Refer to **Figure 6-11**. The options in the **UCS settings** area are:

- **Save UCS with viewport.** If checked, the current UCS settings are saved with the viewport and the **UCSVP** system variable is set to 1. This variable can be set for each viewport in the drawing. Viewports in which this setting is turned off, or unchecked, will always display the UCS settings of the current active viewport.
- **Update view to Plan when UCS is changed.** This setting controls the **UCSFOLLOW** system variable. When checked, the variable is set to 1. When unchecked, the variable is set to 0.

Chapter Review

Answer the following questions. Write your answers on a separate sheet of paper or complete the electronic chapter review on the companion website.
www.g-wlearning.com/CAD

1. Explain *spherical coordinate entry.*
2. Explain *cylindrical coordinate entry.*
3. A new point is to be drawn 4.5″ from the last point. It is to be located at a 63° angle in the XY plane, and at a 35° angle from the XY plane. Write the proper spherical coordinate notation.
4. Write the proper cylindrical coordinate notation for locating a point 4.5″ in the horizontal direction from the origin, 3.6″ along the Z axis, and at a 63° angle in the XY plane.
5. Name the command that is used to draw 3D polylines.
6. Why is the command in question 5 needed?
7. Which command option is used to change a 3D polyline into a B-spline curve?
8. What is the *WCS*?
9. What is a *user coordinate system (UCS)*?
10. What effect does the **Show UCS Icon at Origin** option have on the UCS icon display?
11. Describe how to rotate the UCS so that the Z axis is tilted 30° toward the WCS X axis.
12. How do you return to the WCS from any UCS?
13. Which command controls the display of the user coordinate system icon?
14. What system variable controls the display of grips on the UCS icon? In which dialog box can it be set?
15. Briefly describe how to access the **Move and Align** option using the UCS icon grips.
16. How can you return to a previous UCS configuration using the UCS icon?
17. What is a *dynamic UCS* and how is one activated?
18. What is the function of the **3 Point** option of the **UCS** command?
19. How do you automatically create a display that is plan to a new UCS?
20. What is the function of the **Object** option of the **UCS** command?
21. What is the function of the **Apply** option of the **UCS** command?
22. In which dialog box is the **Orthographic UCSs** tab located?
23. Which command displays the **UCS** dialog box?
24. What appears in the **Named UCSs** tab of the **UCS** dialog box if the current UCS has not been saved?

Drawing Problems

For Problems 1–4, draw each object using solid primitives and Boolean commands to create composite solids. Measure the objects directly to obtain the necessary dimensions. If appropriate, apply geometric constraints to the base 2D drawing. Plot the drawings at a 3:1 scale using display methods specified by your instructor. Save the drawings as P06_*(problem number).*

1.

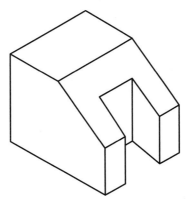

2.

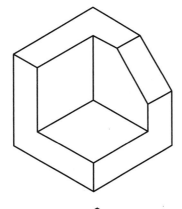

3.

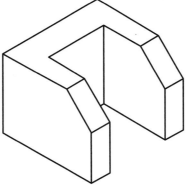

4.

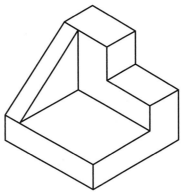

5. Create the mounting bracket shown. Save the file as P06_05.

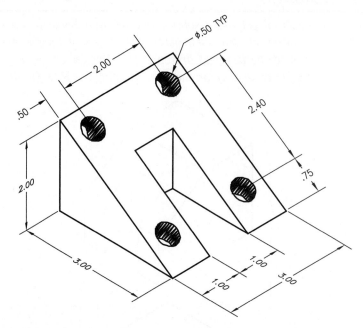

6. Create the computer speaker as shown below. The large-radius, arched surface is created by drawing a three-point arc. The second point of the arc passes through the point located by the .26 and 2.30 dimensions. Save the file as P06_06.

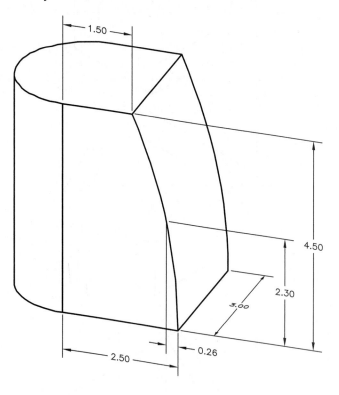

For Problems 7–9, draw each object using solid primitives and Boolean commands to create composite solids. Use the dimensions provided. Save the drawings as P06_(problem number).

7.

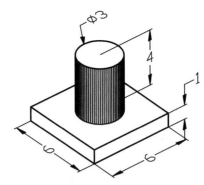

Pedestal #1

8.

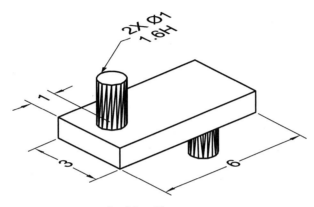

Locking Plate

9.

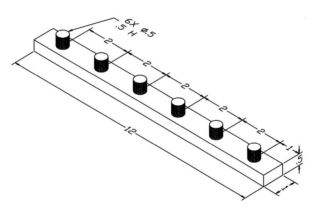

Pin Bar

10. Draw the ∅8″ pedestal shown. It is .5″ thick. The four feet are centered on a ∅7″ circle and are .5″ high. Save the drawing as P06_10.

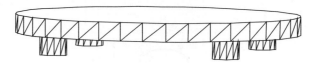

Pedestal #2

11. Four legs (cones), each 3″ high with a ∅1″ base, support this ∅10″ globe. Each leg tilts at an angle of 15° from vertical. The base is ∅12″ and .5″ thick. The bottom surface of the base is 8″ below the center of the globe. Save the drawing as P06_11.

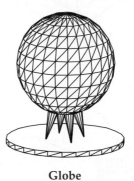

Globe

12. The table legs (A) are 2″ square and 17″ tall. They are 2″ in from each edge. The tabletop (B) is 24″ × 36″ × 1″. Save the drawing as P06_12.

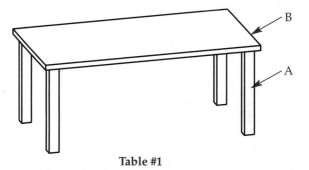

Table #1

13. The table legs (A) for the large table are ∅2″ and 17″ tall. The tabletop (B) is 24″ × 36″ × 1″. The table legs (C) for the small table are ∅2″ and 11″ tall. The tabletop (D) is 24″ × 14″ × 1″. All legs are 1″ in from the edges of the table. Save the drawing as P06_13.

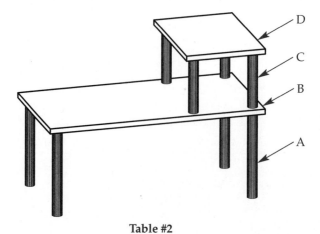

Table #2

14. The spherical objects (A) are Ø4". Object B is 6" long and Ø1.5". Save the drawing as P06_14.

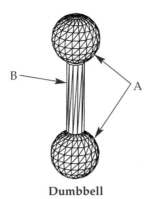

Dumbbell

15. Create the model of the globe using the dimensions shown. Save the file as P06_15.

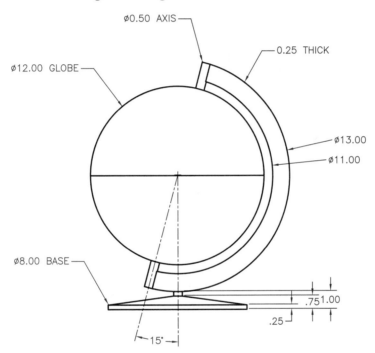

16. Object A is a Ø8" cylinder that is 1" tall. Object B is a Ø5" cylinder that is 7" tall. Object C is a Ø2" cylinder that is 6" tall. Object D is a .5" × 8" × .125" box, and there are four pieces. The top surface of each piece is flush with the top surface of Object C. Object E is a Ø18" cone that is 12" tall. Create a smaller cone and hollow out Object E. Save the drawing as P06_16.

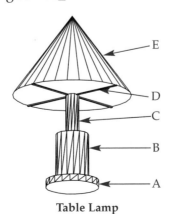

Table Lamp

17. Objects A and B are brick walls that are 5' high. The walls are two bricks thick. Research the dimensions of standard brick and draw accordingly. Wall B is 7' long and Wall A is 5' long. Lamps are placed at each end of the walls. Object C is Ø2" and 8" tall. The center is offset from the end of the wall by a distance equal to the width of one brick. Object D is Ø10". Save the drawing as P06_17.

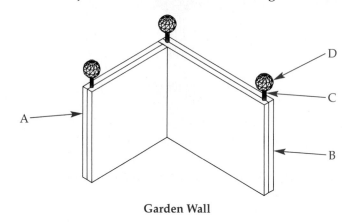

Garden Wall

18. Object A is Ø18" and 1" tall. Object B is Ø1.5" and 6' tall. Object C is Ø6" and .5" tall. Object D is a Ø10" sphere. Object E is a U-shaped bracket to support the shade (Object F). There are two items; draw them an appropriate size. Object F has a Ø22" base and is 12" tall. Save the drawing as P06_18.

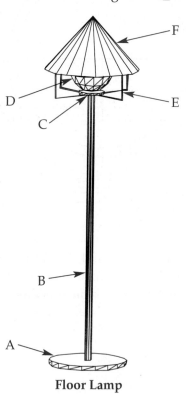

Floor Lamp

19. This is a concept sketch of a desk organizer. Create a solid model using the dimensions given. Create and save new UCSs as needed. Inside dimensions of compartments can vary, but the thickness between compartments should be consistent. Do not add dimensions to the drawing. Plot your drawing on a B-size sheet of paper in a visual style specified by your instructor. Save the drawing as P06_19.

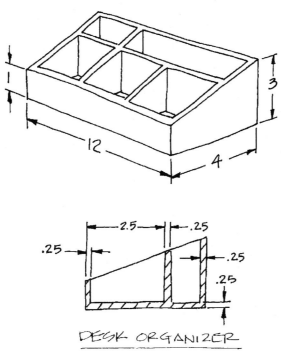

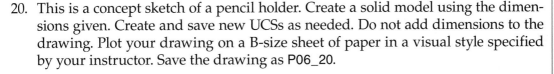

DESK ORGANIZER

20. This is a concept sketch of a pencil holder. Create a solid model using the dimensions given. Create and save new UCSs as needed. Do not add dimensions to the drawing. Plot your drawing on a B-size sheet of paper in a visual style specified by your instructor. Save the drawing as P06_20.

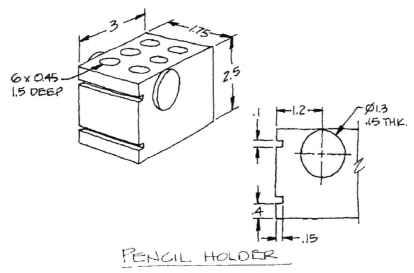

PENCIL HOLDER

21. This is an engineering sketch of a window blind mounting bracket. Create a solid model using the dimensions given. Create and save new UCSs as needed. Do not add dimensions to the drawing. Create two plots, each of a different view, on B-size paper in the visual styles specified by your instructor. Save the drawing as P06_21.

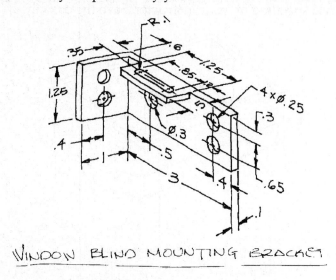

WINDOW BLIND MOUNTING BRACKET

Chapter

Using Model Space Viewports

Learning Objectives

After completing this chapter, you will be able to:

✓ Describe the function of model space viewports.
✓ Create and save viewport configurations.
✓ Alter the current viewport configuration.
✓ Use multiple viewports to construct a drawing.

A variety of views can be displayed in a drawing at one time using model space viewports. This is especially useful when constructing 3D models. Using the **VPORTS** command, you can divide the drawing area into two or more smaller areas. These areas are called *viewports*. Each viewport can be configured to display a different 2D or 3D view of the model.

The *active viewport* is the viewport in which a command will be applied. Any viewport can be made active, but only one can be active at a time. As objects are added or edited, the results are shown in all viewports. A variety of viewport configurations can be saved and recalled as needed. This chapter discusses the use of viewports and shows how they can be used for 3D constructions.

Understanding Viewports

The AutoCAD drawing area can be divided into a maximum of 64 viewports. However, this is impractical due to the small size of each viewport. Usually, the maximum number of viewports practical to display at one time is four. The number of viewports you need depends on the model you are drawing. Each viewport can show a different view of an object. This makes it easier to construct 3D objects.

 The **MAXACTVP** (maximum active viewports) system variable sets the number of viewports that can be used at one time. The initial value is 64, which is the highest setting.

There are two types of viewports used in AutoCAD. The type of viewport created depends on whether it is defined in model space or paper space. *Model space* is the space, or mode, where the model or drawing is constructed. *Paper space*, or layout space, is the space where a drawing is laid out to be plotted. Viewports created in model space are called *tiled viewports*. Viewports created in paper space are called *floating viewports*.

Model space is active by default when you start AutoCAD. Model space viewports are created with the **VPORTS** command. Model space viewport configurations are for display purposes only and cannot be plotted. If you plot from model space, the content of the active viewport is plotted. Model space viewports are described as *tiled viewports*. Tiled viewports are not AutoCAD objects. They are referred to as *tiled* because the edges of each viewport are placed side to side, as with floor tile, and they cannot overlap.

Floating (paper space) viewports are used to lay out the views of a drawing before plotting. They are described as *floating* because they can be moved around and overlapped. Paper space viewports are objects and they can be edited. These viewports can be thought of as "windows" cut into a sheet of paper to "see into" model space. You can then display different scaled drawings (views) in these windows. For example, architectural details or sections and details of complex mechanical parts may be displayed in paper space viewports at different scales. Detailed discussions of paper space viewports are provided in *AutoCAD and Its Applications—Basics*.

The **VPORTS** command can be used to create viewports in a paper space layout. The process is very similar to that used to create model space viewports, which is discussed next. You can also use the **MVIEW** command to create paper space viewports.

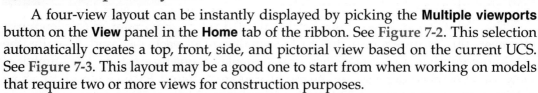

Creating Viewports

Creating model space viewports allows you to work with multiple views of the same model. To work on a different view, simply pick with your pointing device in the viewport in which you wish to work. The picked viewport becomes active. Using viewports is a good way to construct 3D models because all views are updated as you draw. However, viewports are also useful when creating 2D drawings.

The project on which you are working determines the number of viewports needed. Keep in mind that the more viewports you display on your screen, the smaller the view in each viewport. Small viewports may not be useful to you. Four different viewport configurations are shown in **Figure 7-1**. As you can see, when 16 viewports are displayed, the viewports are very small. Normally, two to four viewports are used.

Quick Viewport Layout

A four-view layout can be instantly displayed by picking the **Multiple viewports** button on the **View** panel in the **Home** tab of the ribbon. See **Figure 7-2**. This selection automatically creates a top, front, side, and pictorial view based on the current UCS. See **Figure 7-3**. This layout may be a good one to start from when working on models that require two or more views for construction purposes.

The screen display can be quickly returned to a single view by picking **Single viewport** on the **View** panel in the **Home** tab of the ribbon. Keep in mind, the resulting view will be the viewport that is current, which is the one surrounded by a highlighted frame.

Figure 7-1.
A—Two vertical viewports. B—Two horizontal viewports. C—Three viewports, with the largest viewport positioned at the right. D—Sixteen viewports.

Viewport controls

Crosshairs and navigation bar appear in the active viewport

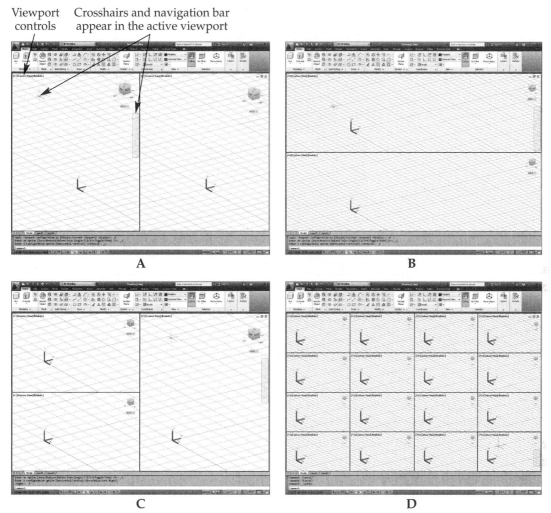

A

B

C

D

Figure 7-2.
A four-viewport layout can be instantly displayed by picking the **Multiple viewports** button in the **View** panel on the **Home** tab of the ribbon.

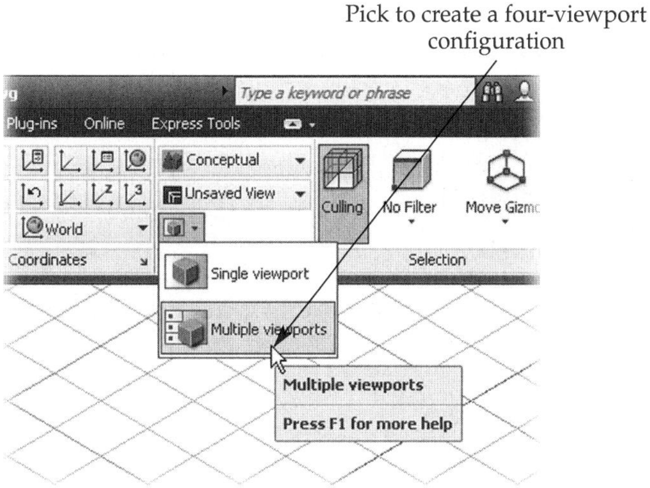

Pick to create a four-viewport configuration

Figure 7-3.
Picking the **Multiple viewports** button in the **View** panel on the **Home** tab of the ribbon automatically creates a top, front, side, and pictorial view based on the current UCS.

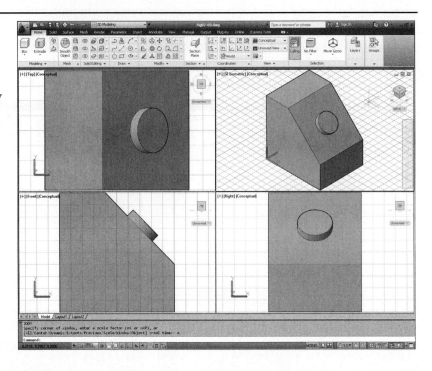

Viewport Configurations

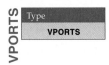

A layout of viewport configurations can be quickly created by using the **Viewport Controls** flyout in the viewport controls located in the upper-left corner of the drawing window. See **Figure 7-4**. The viewport controls are located in every model space viewport, and may provide the quickest access to all viewport configurations and settings. Selecting **Viewport Configuration List** in the **Viewport Controls** flyout displays a menu with 12 different preset viewport configurations. Selecting **Configure...** in this menu displays the **Viewports** dialog box, **Figure 7-5**. This dialog box, which is also accessed with the **VPORTS** command, is used to save and manage viewport configurations. Another way to access the preset viewport configurations is to use the **Viewport Configurations List** drop-down list in the **Viewports** panel on the **View** tab of the ribbon. See **Figure 7-6**.

Figure 7-4.
Preset viewport configurations can be accessed in the **Viewport Controls** flyout located in the viewport controls.

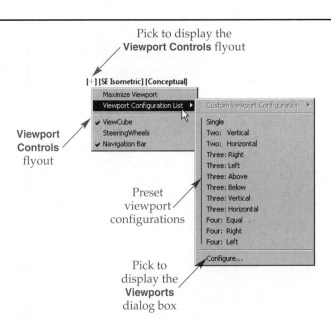

Figure 7-5.
Viewport configurations are created using the **New Viewports** tab of the **Viewports** dialog box.

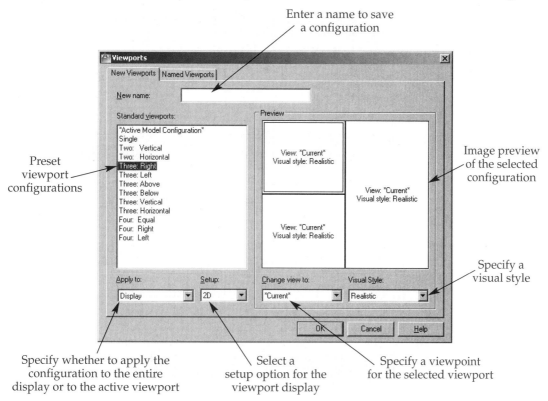

Enter a name to save a configuration

Preset viewport configurations

Image preview of the selected configuration

Specify a visual style

Specify whether to apply the configuration to the entire display or to the active viewport

Select a setup option for the viewport display

Specify a viewpoint for the selected viewport

Figure 7-6.
The **Viewport Configurations List** drop-down list in the **Viewports** panel on the **View** tab of the ribbon offers a quick way to recall a preset viewport configuration.

As shown in **Figure 7-5**, there are two tabs in the **Viewports** dialog box. The preset viewport configurations are available in the **New Viewports** tab. The preset viewport configurations include six different options for three-viewport configurations. See **Figure 7-7**. When you pick the name of a configuration in the **Standard viewports:** list, the viewport arrangement is displayed in the **Preview** area. After you have made a selection, you can save the configuration by entering a name in the **New name:** text box and then picking **OK** to close the dialog box. When the **Viewports** dialog box closes, the configuration is displayed on screen.

Figure 7-7.
Twelve preset tiled viewport configurations are provided in the **Viewports** dialog box.

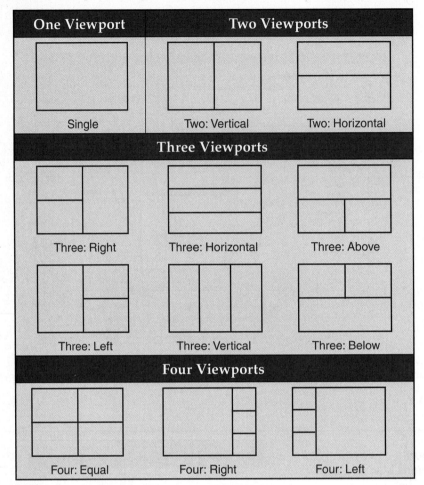

PROFESSIONAL
TIP

Notice in **Figure 7-1** that the UCS icon and view cube are displayed in all viewports. This is an easy way to tell that several separate screens are displayed, rather than different views of the drawing. Additionally, the navigation bar is only displayed in the current viewport.

Making a Viewport Active

After a viewport configuration has been created, a thick line surrounds the active viewport. When the screen cursor is moved inside of the active viewport, it appears as crosshairs. When moved into an inactive viewport, the standard Windows cursor appears.

Any viewport can be made active by moving the cursor into the desired viewport and pressing the pick button. You can also press the [Ctrl]+[R] key combination to switch viewports, or use the **CVPORT** (current viewport) system variable. Only one viewport can be active at a time.

Command: **CVPORT**↵
Enter new value for CVPORT <*current*>: **3**↵

The current value given is the ID number of the active viewport. The ID number is automatically assigned by AutoCAD, starting with 2. To change viewports with the **CVPORT** system variable, simply enter a different ID number. This technique may be used in custom programming for AutoCAD. Using the **CVPORT** system variable is also a good way to determine the ID number of a viewport. The number 1 is not a valid viewport ID number.

PROFESSIONAL TIP

Each viewport can have its own view, viewpoint, UCS, zoom scale, limits, grid spacing, and snap setting. Specify the drawing aids in all viewports before saving the configuration. When a viewport is restored, all settings are restored as well.

Managing Defined Viewports

If you are working with several different viewport configurations, it is easy to restore, rename, or delete existing viewports. You can do so using the **Viewports** dialog box. To access a list of named viewports, open the dialog box and select the **Named Viewports** tab. See **Figure 7-8**. To display a viewport configuration, highlight its name in the **Named viewports:** list and then pick the **OK** button.

Assume you have saved the current viewport configuration. Now, you want to work in a specific viewport, but do not need other viewports displayed on screen. First, pick the viewport you wish to work in to make it active. Next, select **Viewport Configuration List** in the **Viewport Controls** flyout in the viewport controls and select **Single**. You can also use the **Viewport Configurations List** drop-down list in the **Viewports** panel on the **View** tab of the ribbon. The active viewport is displayed as the only viewport. To restore the original viewport configuration, select **Viewport Configuration List** in the **Viewport Controls** flyout in the viewport controls, and then select **Custom Viewport Configuration**. This displays a list of named viewport configurations. Select the desired name and that configuration is displayed. Alternatively, you can display the **Viewports** dialog box, pick the **Named Viewports** tab, and then select the name of the saved viewport configuration. The **Preview** area displays the selected viewport configuration. Pick the **OK** button to exit.

Ribbon
View > Viewports
Named
Type
VPORTS

VPORTS

Figure 7-8.
The **Named Viewports** tab of the **Viewports** dialog box lists all named viewports and displays the selected configuration in the **Preview** area.

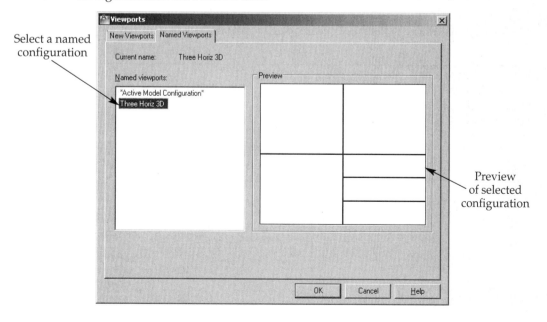

Viewport configurations can also be renamed and deleted using the **Named Viewports** tab of the **Viewports** dialog box. To rename a viewport configuration, right-click on the name and pick **Rename** from the shortcut menu. You can also single-click on a highlighted name. When the name becomes highlighted text, type the new name and press [Enter]. To delete a viewport configuration, right-click on the name and pick **Delete** from the shortcut menu. You can also press the [Delete] key to delete the highlighted viewport configuration. Press **OK** to exit the dialog box.

PROFESSIONAL TIP

The **-VPORTS** command can be used to manage viewports on the command line. This may be required for some LISP programs where the dialog box cannot be used. AutoLISP is covered in Chapters 27 and 28, which are provided on the companion website. www.g-wlearning.com/CAD

Using the Viewports Panel

The **Viewports** panel in the **View** tab of the ribbon is shown in **Figure 7-9**. It is used with both model space and paper space viewports. Picking the **Named** button on the panel displays the **Viewports** dialog box. The **Join** button is used to combine two viewports into a single viewport, as described in the next section. Expanding the **Viewports** panel displays the **Viewport Configurations List** drop-down list, which contains the 12 preset viewport configurations. Selecting a configuration in the list sets it current. The **Rectangular** and **Clip** buttons apply to paper space viewports in a layout. A complete discussion of paper space viewports is given in *AutoCAD and Its Applications—Basics*.

Joining Two Viewports

You can join two adjacent viewports in an existing configuration to form a single viewport. This process is often quicker than creating an entirely new configuration. However, the two viewports must form a rectangle when joined, **Figure 7-10**.

When you enter the **Join** option, AutoCAD first prompts you for the *dominant viewport*. All aspects of the dominant viewport are used in the new (joined) viewport. These aspects include the limits, grid, UCS, and snap settings.

Select dominant viewport <current viewport>: *(select the viewport by picking in it or press* [Enter] *to set the current viewport as the dominant viewport)*
Select viewport to join: *(select the other viewport)*

Ribbon

View
> Viewports

Join Viewports

Type

-VPORTS

Figure 7-9.
The **Viewports** panel in the **View** tab of the ribbon.

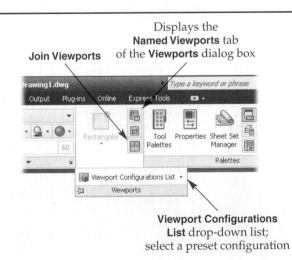

Join Viewports

Displays the **Named Viewports** tab of the **Viewports** dialog box

Viewport Configurations List drop-down list; select a preset configuration

Figure 7-10.
Two viewports can be joined if they will form a rectangle. If the two viewports will not form a rectangle, they cannot be joined.

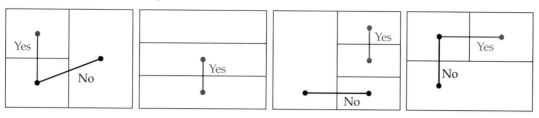

The two selected viewports are joined into a single viewport. If you select two viewports that do not form a rectangle, AutoCAD returns the message:

The selected viewports do not form a rectangle.

PROFESSIONAL TIP

Create only the number of viewports and viewport configurations needed to construct your drawing. Using too many viewports reduces the size of the image in each viewport and may confuse you. Also, it helps to zoom each view so that the objects fill the viewport.

Exercise 7-1

Complete the exercise on the companion website.
www.g-wlearning.com/CAD

Applying Viewports to Existing Configurations and Displaying Different Views

You have total control over what is displayed in model space viewports. In addition to displaying various viewport configurations, you can divide an existing viewport into additional viewports or assign a different viewpoint to each viewport. The options for these functions are provided in the **New Viewports** tab of the **Viewports** dialog box. These options are located along the bottom of the **Viewports** dialog box, as shown in **Figure 7-5**:
- **Apply to.**
- **Setup.**
- **Change view to.**
- **Visual style.**

Apply To

When a preset viewport configuration is selected from the **Standard viewports:** list, it can be applied to either the entire display or the current viewport. The previous examples have shown how to create viewports that replace the entire display. Applying a configuration to the active viewport rather than the entire display can be useful when you need to display additional viewports.

For example, first create a configuration of three viewports using the Three: Right configuration option. Then, with the right (large) viewport active, open the **Viewports** dialog box again. Notice that the drop-down list under **Apply to:** is grayed out. Now, pick one of the standard configurations. This enables the **Apply to:** drop-down list.

The default option is Display, which means the selected viewport configuration will replace the current display. Pick the drop-down list arrow to reveal the second option, Current Viewport. Pick this option and then pick the **OK** button. Notice that the selected viewport configuration has been applied to only the active (right) viewport. See **Figure 7-11**.

Setup

Viewports can be set up to display views in 2D or 3D. The 2D and 3D options are provided in the **Setup:** drop-down list. Displaying different views while working on a drawing allows you to see the results of your work on each view, since changes are reflected in each viewport as you draw. The selected viewport **Setup:** option controls the types of views available in the **Change view to:** drop-down list.

Change View To

The views that can be displayed in a selected viewport are listed in the **Change view to:** drop-down list. If the **Setup:** drop-down list is set to 2D, the views available to be displayed are limited to the current view and any named views. If 3D is active, the options include all of the standard orthographic and isometric views along with named views. When an orthographic or isometric view is selected for a viewport, the resulting orientation is shown in the **Preview** area. To assign a different viewpoint to a viewport, simply pick within a viewport in the **Preview** area to make it active and then pick a viewpoint from the **Change view to:** drop-down list. Important: note that if you set a viewport to one of the orthographic preset views, the UCS is also (by default) changed in that viewport to the corresponding preset.

Visual Style

A visual style can be specified for a viewport. Pick within a viewport in the **Preview** area to make it active and then select a visual style from the **Visual Style:** drop-down list. All preset and saved visual styles are available in the drop-down list.

Figure 7-11.
The selected viewport configuration has been applied to the active viewport within the original configuration.

New configuration is applied to the viewport

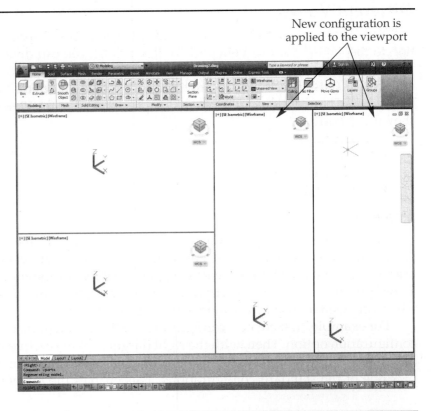

AutoCAD and Its Applications—Advanced

Exercise 7-2

Complete the exercise on the companion website.
www.g-wlearning.co m/CAD

Drawing in Multiple Viewports

When used with 2D drawings, viewports allow you to display a view of the entire drawing, plus views showing portions of the drawing. This is similar to using the **VIEW** command, except you can have several views on screen at once. You can also adjust the zoom magnification in each viewport to suit different areas of the drawing.

Viewports are also a powerful aid when constructing 3D models. You can specify different viewpoints in each viewport and see the model take shape as you draw. A model can be quickly constructed because you can switch from one viewport to another while drawing and editing. For example, you can draw a line from a point in one viewport to a point in another viewport simply by changing viewports while inside of the **LINE** command. The result is shown in each viewport.

In Chapter 6, you constructed a solid object. It was a base that had an angled surface from which a cylinder projected. See **Figure 7-12.** Now, you will construct the same object, but using two viewports. First, create a vertical configuration of two viewports. In the **Viewports** dialog box, set the right-hand viewport to display the southeast isometric view. Also, select the Conceptual visual style. Set the left-hand viewport to display the front view. Remember, this will also set the UCS to the front preset orthographic UCS in that viewport. Also, select the Wireframe visual style for the left-hand viewport. Close the dialog box and make the left-hand viewport active. You may also want to set the parallel projection current in both viewports.

Figure 7-12.
You will construct the object from Chapter 6 using multiple viewports.

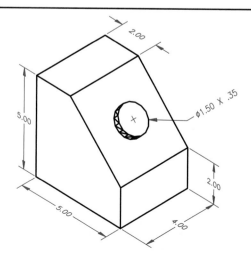

Next, draw a polyline using the coordinates shown in **Figure 7-13**. Remember, to enter absolute coordinates with dynamic input turned on, type a pound sign (#) before the coordinate. This temporarily overrides the dynamic input of relative coordinate values. Be sure to use the **Close** option for the last segment. As you construct the side view, you can clearly see its true size and shape in the left-hand viewport. At the same time, you can see the construction in 3D in the right-hand viewport. Notice that each viewport has a different UCS, as indicated by the UCS icon.

The next step is to extrude the shape to create the base. The **EXTRUDE** command is used to do so, as was the case in Chapter 6. With the left-hand viewport current, enter the **EXTRUDE** command, pick the polyline, and enter an extrusion height of –4 units. The base of the object is now complete, **Figure 7-14**.

Now, the cylinder needs to be created on the angled face. First, split the left-hand viewport into two horizontal viewports (top and bottom) using the **New Viewports** tab of the **Viewports** dialog box. Set both of the new viewports to display the current view. Pick the **OK** button to close the dialog box. Then, make the upper-left viewport current and set it up to always display a plan view of the current UCS by setting the **UCSFOLLOW** system variable to 1.

Using the UCS icon grips, as discussed in Chapter 6, create a new UCS on the angled face in the right-hand viewport. You can move the UCS icon dynamically using the origin grip, or you can use the **3 Point** option of the **UCS** command. If you are using the **3 Point** option, use the pick points shown in **Figure 7-14**. To quickly access the **3 Point** option, right-click on the UCS icon and select **3 Point** from the UCS icon shortcut menu.

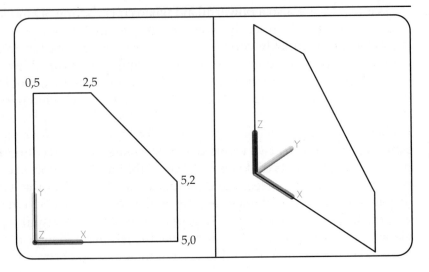

Figure 7-13.
The screen is divided into two viewports. A side view of the object appears in the left-hand viewport and a 3D view appears in the right-hand viewport. Notice the UCS icons.

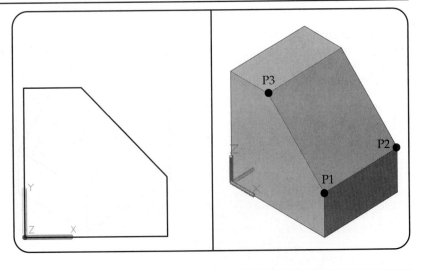

Figure 7-14.
The base of the object is now complete. A new UCS will be created on the angled face. If you are using the **3 Point** option of the UCS command, use the pick points shown here.

Notice in your drawing that after making the right-hand viewport active, the view in the upper-left viewport automatically changes to a plan view of the new current UCS.

With the right-hand viewport current, turn on 3D object snap. Turn off the dynamic UCS function, if it is on. Then, enter the **CYLINDER** command. When locating the center of the cylinder base, move the pointer to the angled face until the **Center of face** object snap is displayed. Then, pick the point. Next, enter a diameter of 1.5 units and a height of .35 units.

The object is now complete, **Figure 7-15**. Notice how the lower-left and right-hand viewports have different UCSs. Each viewport can have its own UCS. The view in the upper-left viewport is the plan view of the current UCS. If the lower-left viewport is made active, the plan view will be of the UCS in that viewport. This is because of the **UCSFOLLOW** system variable setting of 1 in the upper-left viewport. The UCS orientation in one viewport is not affected by a change to the UCS in another viewport unless the **UCSFOLLOW** system variable is set to 1 in a viewport. If a viewport arrangement is saved with several different UCS configurations, every named UCS remains intact and is displayed when the viewport configuration is restored.

The **REGEN** command affects only the current viewport. To regenerate all viewports at the same time, use the **REGENALL** command. This command can be entered by typing REGENALL.

The Quick Text mode is controlled by the **REGEN** command. Therefore, if you are working with text displayed with the Quick Text mode in viewports, be sure to use the **REGENALL** command in order for the text to be regenerated in all viewports.

 The UCS configuration in each viewport is controlled by the **UCSVP** system variable. When **UCSVP** is set to 1 in a viewport, the UCS is independent from all other UCSs, which is the default. If **UCSVP** is set to 0 in a viewport, its UCS will change to reflect any changes to the UCS in the current viewport.

 Exercise 7-3

Complete the exercise on the companion website.
www.g-wlearning.com/CAD

Figure 7-15.
The cylinder is drawn to complete the object. Notice the plan view in the upper-left viewport.

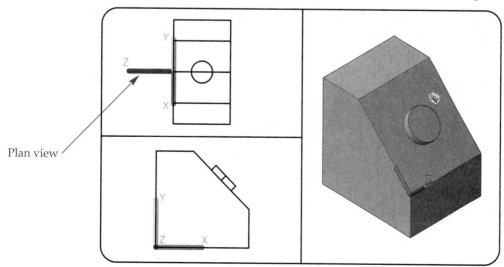

Plan view

Chapter Review

Answer the following questions. Write your answers on a separate sheet of paper or complete the electronic chapter review on the companion website.
www.g-wlearning.com/CAD

1. What is the purpose of *viewports*?
2. How do you name a configuration of viewports?
3. What is the purpose of saving a configuration of viewports?
4. Explain the difference between *tiled* and *floating* viewports.
5. Name the system variable controlling the maximum number of viewports that can be displayed at one time.
6. How can a named viewport configuration be redisplayed on screen?
7. How can a list of named viewport configurations be displayed?
8. What relationship must two viewports have before they can be joined?
9. What is the significance of the dominant viewport when two viewports are joined?
10. When creating a new viewport configuration, how can you set a visual style in a viewport?

Drawing Problems

1. Construct seven template drawings, each with a preset viewport configuration. Use the following configurations and names. Save each template under the same name as the viewport configuration.

Number of Viewports	Configuration	Name
2	Horizontal	TWO-H
2	Vertical	TWO-V
3	Right	THREE-R
3	Left	THREE-L
3	Above	THREE-A
3	Below	THREE-B
3	Vertical	THREE-V

2. Construct one of the problems from Chapter 4 using viewports. Use one of your template drawings from Problem 7-1. Save the drawing as P07_02.

3. This is an orthographic drawing of a light fixture bracket. Create it as a solid model. If appropriate, use geometric constraints on the base 2D shapes. Use solid primitives and Boolean commands as needed. Use the dimensions given. Similar holes have the same offset dimensions. Use multiple viewports to construct the drawing. Begin with a four-viewport layout, then switch viewports as needed to work on specific areas of the model. Create new UCSs as needed. Display an appropriate pictorial view of the drawing in the upper-right viewport. Plot the 3D view of the drawing to scale on a B- or C-size sheet of paper. Save the drawing as P07_03.

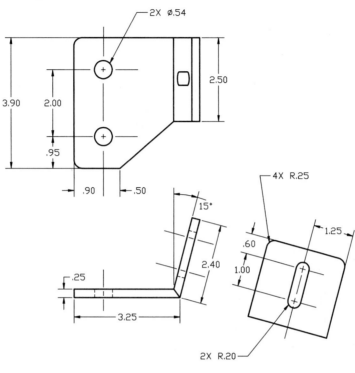

Light Fixture Bracket

4. This is an orthographic drawing of an angle bracket. Create it as a solid model. If appropriate, use geometric constraints on the base 2D shapes. Use solid primitives and Boolean commands as needed. Use the dimensions given. Similar holes have the same offset dimensions. Use multiple viewports to construct the drawing. Begin with a four-viewport layout, then switch viewports as needed to work on specific areas of the model. Create new UCSs as needed. Display an appropriate pictorial view of the drawing in the upper-right viewport. Plot the 3D view of the drawing to scale on a B- or C-size sheet of paper. Save the drawing as P07_04.

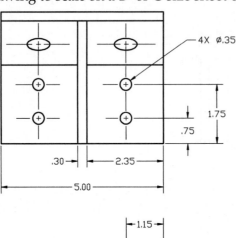

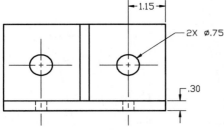
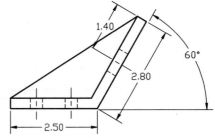

Angle Bracket

Drawing Problems - Chapter 7

Model Extrusions and Revolutions

Learning Objectives

After completing this chapter, you will be able to:

- ✓ Create solids and surfaces by extruding 2D profiles.
- ✓ Extrude surfaces.
- ✓ Create symmetrical 3D solids and surfaces by revolving 2D profiles.
- ✓ Revolve surfaces.
- ✓ Extrude and revolve objects using mathematical expressions and constraints.
- ✓ Use solid extrusions and revolutions as construction tools.

A complex shape can be created by applying a thickness to a two-dimensional profile. This is called *extruding* the shape. You have been introduced to the operation in previous chapters. Two or more profiles can be extruded to intersect. The resulting union can form a new shape by performing a Boolean operation. Symmetrical objects can be created by revolving a 2D profile about an axis to create a new shape.

Creating Extruded Models

An *extrusion* is a two-dimensional shape that has been "extruded" into a 3D solid or surface. The **EXTRUDE** command allows you to create extrusions from lines, arcs, elliptical arcs, 2D polylines, 2D splines, circles, ellipses, 2D solids, 3D solid faces and edges, regions, planar and nonplanar surfaces, meshes, and donuts. Objects in a block cannot be extruded. By default, closed objects, such as circles, polygons, closed poly-lines, and donuts, are converted to a *solid extrusion* when they are extruded. A solid extrusion represents a solid object and has mass properties. Open-ended objects, such as lines, arcs, polylines, elliptical arcs, and splines, are converted to a *surface extrusion* when they are extruded. Surface extrusions have no mass properties.

Extrusions can be created along a straight line or along a path curve. A taper angle can also be applied as you extrude an object. **Figure 8-1** illustrates a polygon extruded into a solid.

Ribbon

Home
> Modeling
Solid
> Solid
Surface
> Create

Extrude

Type

EXTRUDE
EXT

EXTRUDE

Figure 8-1.
The **EXTRUDE**
command creates a
3D solid or surface
from a 2D profile.
A—The initial,
closed 2D profile.
B—The extruded
solid object shown
with hidden lines
removed.

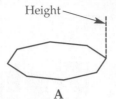
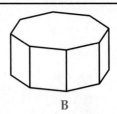

A

B

When the **EXTRUDE** command is selected, you are prompted to select the objects to extrude. Select the objects and press [Enter]. You are then prompted for the extrusion height. The height is always applied along the Z axis of the selected object, not the current UCS. A positive value extrudes above the XY plane of the object. A negative height value extrudes below the XY plane. If a pictorial view is displayed, you can drag the mouse to set the extrusion above or below the XY plane and then enter the height value.

Before entering a height, you can specify a taper angle. The taper angle can be any value *between* +90° and –90°. A positive angle tapers to the inside of the object from the base. A negative angle tapers to the outside of the object from the base. See Figure 8-2. However, the taper angle cannot result in edges that "fold into" the extruded object.

If the command is selected from the **Solid** tab in the ribbon, the mode is automatically set to solid. However, if an open profile is selected, a surface is still created. If the command is selected from the **Surface** tab in the ribbon, the mode is automatically set to surface. In either case, you can use the **Mode** option to change the output type.

PROFESSIONAL TIP

Objects such as polylines, lines, and arcs that have a thickness can be converted to surfaces using the **CONVTOSURFACE** command. Circles and closed polylines with a thickness can be converted to solids using the **CONVTOSOLID** command. The thickness property for an object is controlled by the Thickness setting in the **Properties** palette.

Extrusions along a Path

A 2D shape can be extruded along a path to create a 3D solid or surface. The path can be a line, circle, arc, ellipse, polygon, polyline, mesh edge, 3D solid edge, surface

Figure 8-2.
A—A positive
angle tapers to the
inside of the object
from the base.
B—A negative angle
tapers to the outside
of the object.

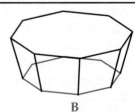

A

B

edge, or spline. Multiple line segments and other objects can be first joined to form a polyline path. The corners of angled segments on the extruded object are mitered, while curved segments are smooth. See **Figure 8-3**.

When open objects, such as lines, arcs, polylines, elliptical arcs, and splines, are used as the profile, they are converted to a swept surface when extruded along a path. A *sweep* is a solid or surface that is created when an open or closed curve is pulled, or swept, along a 2D or 3D path. An extrusion is really a form of a sweep. Sweeps are discussed in detail in Chapter 9.

The **Mode** option of the **EXTRUDE** command allows you to select the type of object that is created: surface or solid. Keep in mind, objects that are not closed, such as lines and polylines, *always* result in extruded surfaces. A solid can only be created by the **EXTRUDE** command if the original object is closed.

To extrude along a path, enter the **EXTRUDE** command and select the objects to extrude. When prompted for the height of the extrusion, select the **Path** option. If needed, first enter a taper angle. Then, pick the object to be used as the extrusion path.

Objects can also be extruded along a line at an angle to the base object, **Figure 8-4**. Notice that the plane at the end of the extruded object is parallel to the original object. Also notice that the length of the extrusion is the same as that of the path. The path does not need to be perpendicular to the object.

If the path begins perpendicular to the profile, the cross section of the resulting extrusion is perpendicular to the path, regardless if the path is a straight line, curve, or spline. See **Figure 8-5**. If the path is a spline or curve that does not begin perpendicular to the profile, the profile may not remain perpendicular to the path as it is extruded.

If one of the endpoints of the path is not on the plane of the object to be extruded, the path is temporarily moved to the center of the profile. The extrusion is then created as if the path were connected to the original object, as shown in **Figure 8-4**.

Figure 8-3.
A—Angled segments are mitered when extruded.
B—Curves are smoothed when extruded.

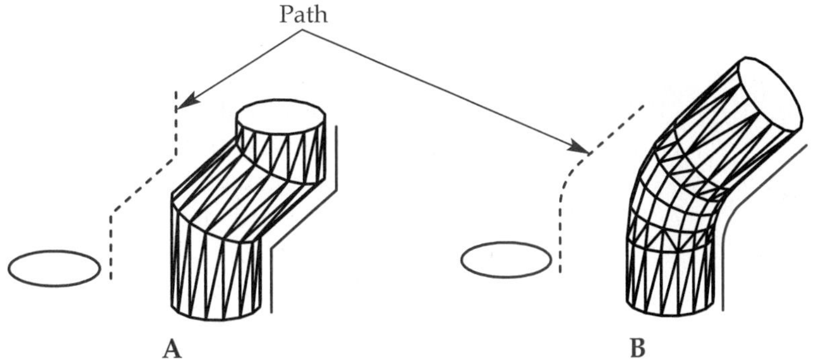

Figure 8-4.
A—An object extruded along a path.
B—The end of an object extruded along an angled path is parallel to the original object.

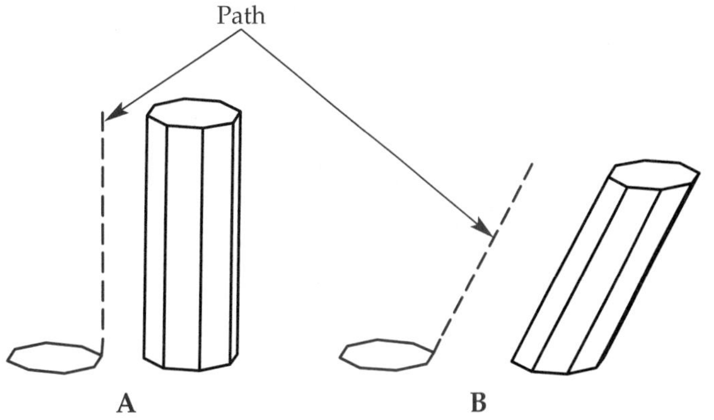

Figure 8-5.
A—Splines can be used as extrusion paths. Notice that the profile on the right is not perpendicular to the start of the path. B—The resulting extrusions.

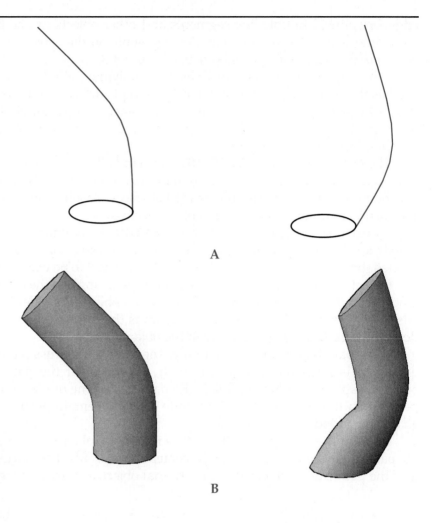

A

B

The **DELOBJ** system variable allows you to delete or retain the original extruded objects and path definitions. The settings are:

0   All original geometry and path definitions are retained.
1   Objects used for extrusion (profile curves) are deleted.
2   All geometry used to define the extrusion, including path definitions, is deleted.
3   All profile-defining and path-defining geometry is deleted. This includes such geometry used with the **SWEEP** and **LOFT** commands, should these actions create a solid. This is the default.
–1 You are prompted to delete objects used for the extrusion (profile curves).
–2 You are prompted to delete all geometry used to define the extrusion, including path and curve definitions.
–3 You are prompted to delete all defining geometry if the extruded entity is a surface. If the extruded entity is a solid, all defining geometry is deleted.

The **DELOBJ** system variable also affects the **REVOLVE**, **SWEEP**, and **LOFT** commands.

Extruding Regions

In Chapter 2, you learned how to create 2D regions. As an example, you created the top view of the base shown in **Figure 8-6A** as a region. Regions can be extruded to create 3D solids. The base you created in Chapter 2 can be extruded to create the final solid shown in **Figure 8-6B**. Any features of the region, such as holes, are extruded the same thickness as the rest of the object. If the profile was created as polylines, the holes must be separately extruded and then subtracted from the solid. Using this method, you can construct a fairly complex 2D region that includes curved profiles, holes, slots, etc. Then, a complex 3D solid can be quickly created. Additional details can be added using editing commands or Boolean operations.

Exercise 8-1

Complete the exercise on the companion website.
www.g-wlearning.com/CAD

Extruding a Surface

A planar surface can be extruded into a solid object in the same manner as a region. Nonplanar (curved) surfaces can be extruded, but only into extruded surfaces. Whereas both surfaces and regions have no thickness, the surface is an object composed of a mesh and the region is actually a solid that possesses mass properties. A planar surface can be quickly converted to a solid using the **EXTRUDE** command. Simply select the surface when prompted to select objects. The surface can be extruded in a specific direction, along a path, or at a taper angle.

Any closed object, such as a circle, rectangle, polygon, or polyline, can be converted into a surface with the **Object** option of the **PLANESURF** command. Pick the **Planar Surface** button on the **Create** panel of the **Surface** tab in the ribbon. Then, enter the **Object** option and select the closed object. The resulting surface can then be extruded into a 3D solid.

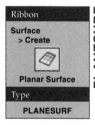

Ribbon

Surface
> Create

Planar Surface

Type

PLANESURF

PLANESURF

PROFESSIONAL TIP

You can also extrude a face on an existing solid into a new solid. When prompted to select objects, press the [Ctrl] key and pick the face to extrude. A face is a *subobject* of a solid. Subobject editing is covered in detail in Chapter 12.

Figure 8-6.
A—The 2D region that will be extruded.
B—The solid object created by extruding the region, shown in a 3D wireframe display.

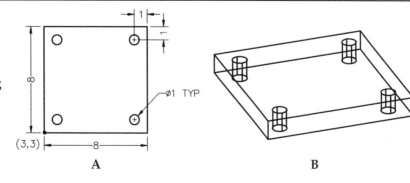

Creating Revolved Models

REVOLVE

Ribbon

Home
> Modeling
Solid
> Solid
Surface
> Create

Revolve

Type

REVOLVE
REV

The **REVOLVE** command allows you to create solids and surfaces by revolving a shape about an axis. Shapes that can be revolved include lines, arcs, circles, ellipses, polygons, polylines, closed splines, regions, planar surfaces, and donuts. The selected object can be revolved at any angle up to 360°. Open curves and line segments create revolved surfaces. Closed curves can be used to create either solids or surfaces. Surface revolutions have no mass properties.

When the command is selected, you are prompted to pick the objects to revolve. Then, you must define the axis of revolution. The default option is to pick the two endpoints of an axis of revolution. This is shown in **Figure 8-7**. You can also revolve about an object or the X, Y, or Z axis of the current UCS. Once the axis is defined, you are prompted to enter the angle through which the profile will be revolved. When the angle is specified by either moving the pointer or entering a value, the revolution is created.

If the command is selected from the **Solid** tab in the ribbon, the mode is automatically set to solid. However, if an open profile is selected, a surface is still created. If the command is selected from the **Surface** tab in the ribbon, the mode is automatically set to surface. In either case, you can use the **Mode** option to change the output type.

PROFESSIONAL TIP

When creating solid models, keep in mind that the final part will most likely need to be manufactured. Be aware of manufacturing processes and methods as you design parts. It is easy to create a part in AutoCAD with internal features that may be impossible to manufacture, especially when revolving a profile.

Revolving about an Object

You can select an object, such as a line, as the axis of revolution. **Figure 8-8** shows a solid created using the **Object** option of the **REVOLVE** command. Both a full-circle

Figure 8-7.
Points P1 and P2 are selected as the axis of revolution for the profile.

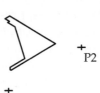

Figure 8-8.
An axis of revolution can be selected using the **Object** option of the **REVOLVE** command. Here, the line is selected as the axis.

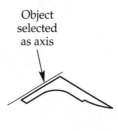

Object selected as axis

Full Circle　　　　　**270°**

(360°) revolution and a 270° revolution are shown. Enter the **Object** option when prompted for the axis of revolution. Then, pick the axis object and enter the angle through which the profile will be rotated. You can use the **Start Angle** option before entering an angle of revolution. This allows you to specify the point at which the revolution starts and then the angle of revolution.

Revolving about the X, Y, or Z Axis

The X axis of the current UCS can be used as the axis of revolution by selecting the **X** option of the **REVOLVE** command. The origin of the current UCS is used as one end of the X axis line. Notice in **Figure 8-9** that two different shapes can be created from the same 2D profile by changing the UCS origin. No hole appears in the object in **Figure 8-9B** because the profile was revolved about an edge that coincides with the X axis. The Y or Z axis can also be used as the axis of revolution. See **Figure 8-10**.

Exercise 8-2

Complete the exercise on the companion website.
www.g-wlearning.com/CAD

Figure 8-9.
A—A solid is created using the X axis as the axis of revolution. B—A different object is created with the same profile by changing the UCS origin.

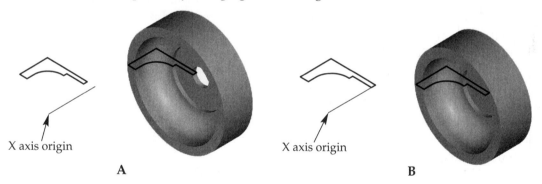

Figure 8-10.
A—A solid is created using the Y axis as the axis of revolution. B—A different object is created by changing the UCS origin.

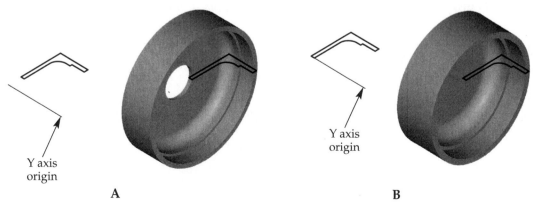

Revolving Regions

Earlier in this chapter, you learned that regions can be extruded. In this manner, holes, slots, keyways, etc., can be created. Regions can also be revolved. A complex 2D shape can be created using Boolean operations on regions. Then, the region can be revolved. One advantage of this method is it may be easier to create a region than trying to create a complex 2D profile as a single, closed polyline.

Revolving Surfaces

Just as planar and nonplanar surfaces can be extruded, they can also be revolved. When the **REVOLVE** command is selected, simply pick the surface when prompted to select objects. The surface can be revolved about an axis defined by two pick points, an object, or the X, Y, or Z axis of the current UCS.

PROFESSIONAL TIP

You can also revolve a face on an existing solid into a new solid. When prompted to select objects, turn on selection cycling or press the [Ctrl] key and pick the face to revolve. Subobject editing is covered in detail in Chapter 12.

Using Extrude and Revolve to Create Surfaces

The **Mode** option of the **EXTRUDE** and **REVOLVE** commands allows you to set the output of the command to a surface or solid. If the command is selected from the **Solid** tab in the ribbon, the mode is automatically set to solid. If the command is selected from the **Surface** tab in the ribbon, the mode is automatically set to surface. The **Mode** option can be used at the "select objects" prompt in the command to change the output type.

The **SURFACEMODELINGMODE** system variable controls the type of surface created. The default setting of 0 creates a *procedural surface*. This is a standard surface composed of multiple flat polygons, but it has no control vertices. By default, this type of surface is associated with the object used to create it. The **SURFACEASSOCIATIVITY** system variable controls this setting.

A NURBS surface can be created by setting the **SURFACEMODELINGMODE** system variable to a value of 1. A *NURBS surface* is composed of splines and contains control vertices that enable the control of the curve shape with great precision.

Surface modeling variables can be set on the **Create** panel in the **Surface** tab of the ribbon. See **Figure 8-11**. When the **Surface Associativity** button is on, the **SURFACEASSOCIATIVITY** system variable is set to 1. When the **NURBS Creation** button is on, the **SURFACEMODELINGMODE** system variable is set to 1.

Figure 8-11.
Surface modeling variables can be set on the **Create** panel in the **Surface** tab of the ribbon.

Pick to associate the surface with the original object

Pick to create a NURBS surface

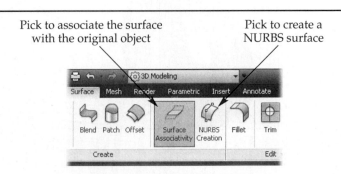

Extruding Surfaces from Objects with Dimensional Constraints

When the **Mode** option is set to create a surface, you can use the **Expression** option to create a parametric relationship between a dimensional constraint and the extruded surface. For example, the circle in **Figure 8-12** has been given a dimensional constraint named dia1. To create an extruded surface cylinder with a height that is always one-half of the diameter, enter the Expression option. Then, at the **Enter expression:**prompt, enter:

    dia1/2

In this manner, using appropriate formulaic expressions, you can create an extruded surface that has a mathematical relationship to a 2D object that has a dimensional constraint.

CAUTION

You can enter an expression when creating a solid. The end result is calculated, but the dimensional constraint is removed. Therefore, the parametric aspect of using an expression is lost.

Revolving Surfaces from Objects with Dimensional Constraints

The **Expression** option of the **REVOLVE** command can be used with dimensional constraints, in the same manner as the **EXTRUDE** command, in order to construct a revolved surface. Examples of mathematical expressions used with constraints are shown in **Figures 8-13** and **8-14**. These examples illustrate how linear and angular dimensional constraints can be used to calculate an angular value for a revolved surface.

The 2D profile in **Figure 8-13A** is to be revolved at an angle that is one-half that of the angular dimension (ang2). This is achieved by selecting the **Expression** option and then entering:

    ang2/2

The result is shown in **Figure 8-13B**.

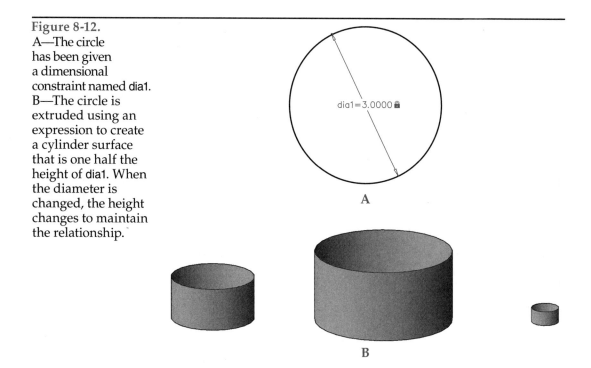

Figure 8-12.
A—The circle has been given a dimensional constraint named dia1.
B—The circle is extruded using an expression to create a cylinder surface that is one half the height of dia1. When the diameter is changed, the height changes to maintain the relationship.

dia1=3.0000

A

B

Figure 8-13.
A—The 2D profile is to be revolved at an angle that is one-half of the angular dimension ang2. B—The revolved surface is created using an expression that references the angular dimension ang2.

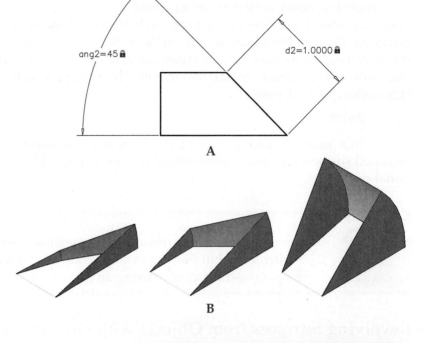

ang2=45

d2=1.0000

A

B

Figure 8-14.
The surface is created using a negative value based on the dimensional constraint d2 to revolve the object below the XY plane.

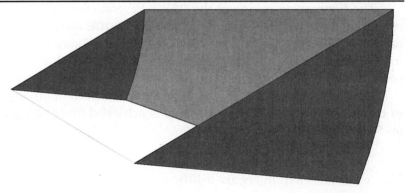

In **Figure 8-14**, the same 2D geometry is used. However, the aligned dimension d2 is used to create a negative value to revolve the object below the XY plane. The following formula is entered:

> d2–31

The result of subtracting 31 from the value of d2 is processed as the angular value for the revolution. The result is shown in **Figure 8-14**.

Using Extrude and Revolve as Construction Tools

It is unlikely that an extrusion or revolution will result in a finished object. Rather, these operations will be used with other solid model construction methods, such as Boolean operations, to create the final object. The next sections discuss how to use **EXTRUDE** and **REVOLVE** with other construction methods to create a finished solid object.

Creating Features with Extrude

You can create a wide variety of features with the **EXTRUDE** command. Study the shapes shown in Figure 8-15. These detailed solid objects were created by drawing a profile and then using the **EXTRUDE** command. The objects in Figures 8-15C and 8-15D must be constructed as regions before they are extruded. For example, the five holes (circles) in Figure 8-15D must be removed from the base region using the **SUBTRACT** command.

Look at Figure 8-16. This is part of a clamping device used to hold parts on a mill table. There is a T-slot milled through the block to receive a T-bolt and one side is stair-stepped, under which parts are clamped. If you look closely at the end of the object, most of the detail can be drawn as a 2D region and then extruded. However, there are also two holes in the top of the block to allow for bolting the clamp to the mill table. These features must be added to the extruded solid.

First, change the UCS to the front preset orthographic UCS. Display a plan view of the UCS. Then, draw the profile shown in Figure 8-17 using the **PLINE** command. You can draw it in stages, if you like, and then use the **PEDIT** command to join all segments into a single polyline.

Next, use the **EXTRUDE** command to create the 3D solid. Extrude the profile a distance of −6 units with a 0° taper. This will extrude the object away from you. Display the object from the southeast isometric preset viewpoint or use the view cube to display a pictorial view. The object should look similar to Figure 8-16 without the holes in the top. Set the Conceptual visual style current, if you like.

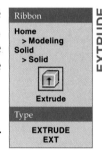

Ribbon

Home
> Modeling
Solid
> Solid

Extrude

Type

EXTRUDE
EXT

EXTRUDE

Figure 8-15.
Detailed solids can be created by extruding the profile of an object. The profiles are shown here in color.

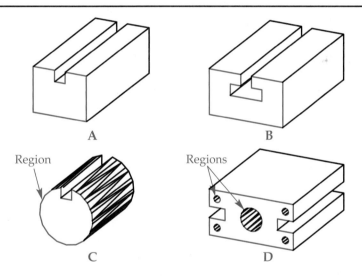

Region

Regions

A   B

C   D

Figure 8-16.
Most of this object can be created by extruding a profile. However, the holes must be added after the extruded solid is created.

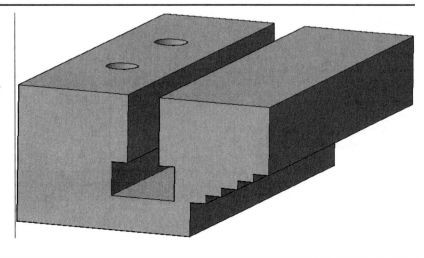

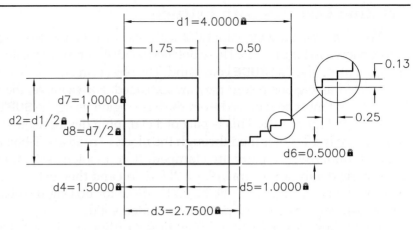

Figure 8-17.
This is the profile that will be extruded for the clamping block. Notice the dimensional constraints that have been applied.

The two holes are ∅.5 units and evenly spaced on the surface through which they pass. Change to the WCS and draw a construction line from midpoint to midpoint, as shown in **Figure 8-18**. Then, set **PDMODE** to an appropriate value, such as 3, and use the **DIVIDE** command to divide the construction line into three parts. The two points created by the **DIVIDE** command are equally spaced on the surface and can be used to locate the two holes.

There are two ways to create a hole. You can draw a circle and extrude it to create a cylinder or you can draw a solid cylinder. Either way, you need to subtract the cylinder to create the hole. Drawing a solid cylinder is probably easiest. When prompted for a center, use the **Node** object snap to select the point. Then, enter the diameter. Finally, enter a negative height so that the cylinder extends into the solid or drag the cylinder down in the 3D view so it extends all of the way through the block. The actual height is not critical, as long as it extends through the block.

You can either copy the first cylinder to the second point or draw another cylinder. When both cylinders are located, use the **SUBTRACT** command to remove them from the solid. The object is now complete and should look like **Figure 8-16**.

Creating Features with Revolve

The **REVOLVE** command is very useful for creating symmetrical, round objects. However, many times the object you are creating is not completely symmetrical. For example, look at the camshaft in **Figure 8-19**. For the most part, this is a symmetrical, round object. However, the cam lobes are not symmetrical in relation to the shaft and bearings. The **REVOLVE** command can be used to create the shaft and bearings. Then, the cam lobes can be created and added.

Start a new drawing and make sure the WCS is the current UCS. Using the **PLINE** command, draw the profile shown in **Figure 8-20A**. This profile will be revolved through 360°, so you only need to draw half of the true plan view of the cam profile. The profile represents the shaft and three bearings.

REVOLVE

| Ribbon |
| Home |
| > Modeling |
| Solid |
| > Solid |

Revolve

| Type |
| REVOLVE |
| REV |

Figure 8-18.
Draw a construction line (shown here in color) and divide it into three parts.

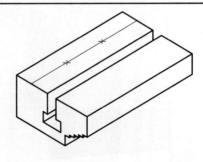

Figure 8-19.
For the most part, this object is symmetrical about its center axis. However, the cam lobes are not symmetrical about the axis.

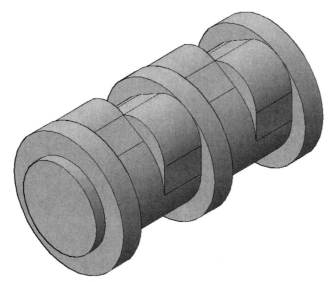

Next, display the drawing from the southwest isometric preset viewpoint. Then, use the **REVOLVE** command to create the base camshaft as a 3D solid. Pick the endpoints shown in **Figure 8-20A** as the axis of revolution. Revolve the profile through 360°. Perform a zoom extents and set the Conceptual visual style current to clearly see the object.

Now, you need to create one cam lobe. Change the UCS to the left orthographic preset. Then, draw a construction point in the center of the left end of the camshaft. Use the **Center** object snap and an appropriate **PDMODE** setting. Next, draw the profile shown in **Figure 8-20B**. Use the construction point as the center of the large radius. You may want to create a new layer and turn off the display of the base camshaft.

Once the cam lobe profile is created, use the **REGION** command to create a region. Then, use the **EXTRUDE** command to extrude the region a height of –.5 units (into the camshaft). The extrusion should have a 0° taper. If you turned off the display of the base camshaft, turn it back on now.

One cam lobe is created, but it is not in the proper position. With the left UCS current, move the cam lobe –.375 units on the Z axis. If a different UCS is current, the axis of movement will be different. This places the front surface of the cam lobe on the back surface of the first bearing. Now, make a copy of the lobe that is located –.5 units on the Z axis. Finally, copy the first two cam lobes –1.25 units on the Z axis.

Figure 8-20.
A—This profile will be revolved to create the shaft and bearings. Notice the parallel constraints that have been applied. B—This is the profile of one cam lobe, which will be extruded.

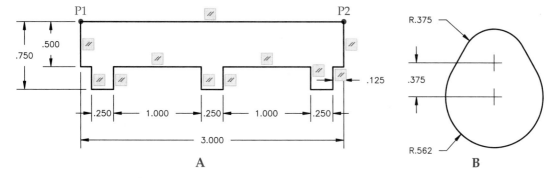

You now need to rotate the four cam lobes to their correct orientations. Make sure the left UCS is still current. Then, rotate the first and third cam lobes 30°. If a different UCS is current, you can use the **3DROTATE** command. The center of rotation should be the center of the shaft. There are many points on the shaft to which the **Center** object snap can snap; they are all acceptable. You can also use the construction point as the center of rotation. Rotate the second and fourth cam lobes –30° about the same center.

Finally, use the **UNION** command to join all objects. The final object should appear as shown in **Figure 8-19**. Use the view cube to see all sides of the object. You can also create a rotating display using the **3DORBIT** command.

Multiple Intersecting Extrusions

Many solid objects have complex curves and profiles. These can often be constructed from the intersection of two or more extrusions. The resulting solid is a combination of only the intersecting volumes of the extrusions. The following example shows the construction of a coat hook.

1. Construct the first profile, **Figure 8-21A**.
2. Construct the second profile located on a common point with the first, **Figure 8-21B**.
3. Construct the third profile located on the common point, **Figure 8-21C**.
4. Extrude each profile the required dimension into the same area. Be careful to specify positive or negative heights for each extrusion, **Figures 8-21D** and **8-21E**.
5. Use the **INTERSECT** command to create a composite solid from the volume shared by the three extrusions, **Figure 8-21F**.

Figure 8-21.
Constructing a coat hook. A—Draw the first profile. B—Draw the second profile. C—Draw the third profile. All three profiles should have a common origin. D—Extrude each profile so that the extruded objects intersect. E—The extruded objects after the Conceptual visual style is set current. F—Use the **INTERSECT** command to create the composite solid. The final solid is shown here with the Conceptual visual style set current.

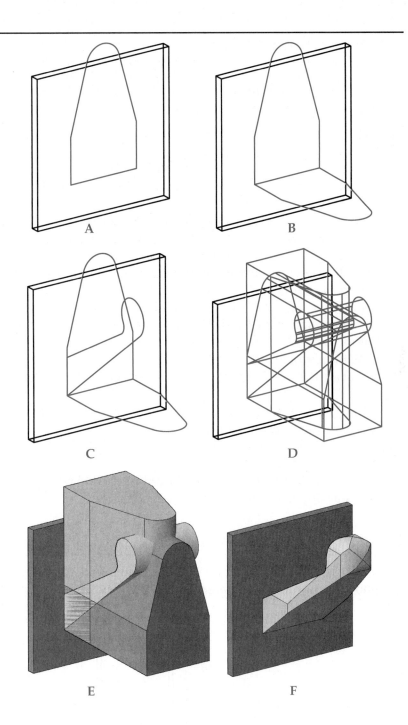

Chapter Review

Answer the following questions. Write your answers on a separate sheet of paper or complete the electronic chapter review on the companion website.
www.g-wlearning.com/CAD

1. What is an *extrusion*?
2. How do you create a surface extrusion?
3. Briefly describe how to create a solid extrusion.
4. Which command can be used to convert circles and closed polylines with a thickness to solids?
5. How can an extrusion be constructed to extend below the XY plane of the current UCS?
6. What is the range in which a taper angle can vary?
7. How can a curved extrusion be constructed?
8. Which system variable allows you to delete or retain the original extruded objects and path definitions?
9. How is the height of an extrusion applied in relation to the original object?
10. Which option of the **PLANESURF** command can be used to convert a closed object into a surface?
11. What is a *surface revolution*?
12. What are the five different options for selecting the axis of revolution for a revolved solid?
13. How can a given profile be revolved twice (or more) about the same axis in order to create different shaped solids?
14. Which option of the **REVOLVE** command controls the type of object created when the command is used on a circle?
15. What mathematical expression would be used to revolve a profile 90° less than the angular dimensional constraint named ang4?

Drawing Problems

1. Construct a 12′ long section of wide flange structural steel with the cross section shown below. Use the dimensions given. Save the drawing as P08_01.

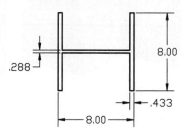

Problems 2–7. These problems require you to use a variety of solid modeling methods to construct the objects. Use **EXTRUDE**, **REVOLVE**, *solid primitives, new UCSs, Boolean commands, and editing tools such as sweeps, extrusions, and revolutions to assist in construction. If appropriate, apply geometric constraints to the base 2D drawing. Do not create section views. Save each drawing as* P08_*(problem number).*

2.

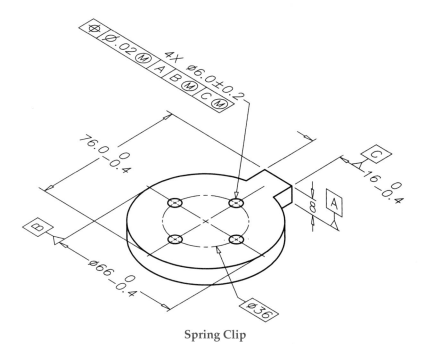

Spring Clip

3.

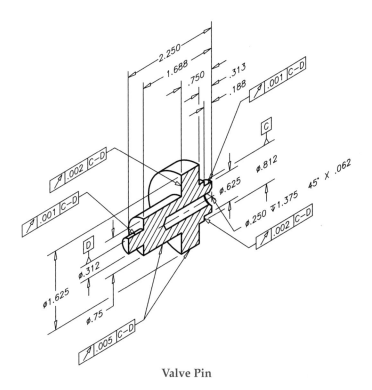

Valve Pin

4.

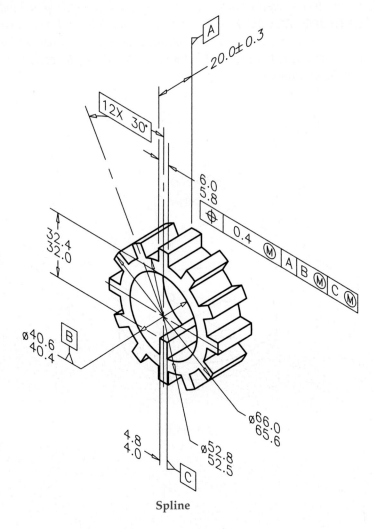

A

20.0± 0.3

12X 30°

6.0
5.8

32.4
32.0

⌀40.6
40.4 **B**

⌀66.0
65.6

4.8
4.0

⌀52.8
52.5 **C**

| ⊕ | 0.4 Ⓜ | A | B Ⓜ | C Ⓜ |

Spline

5.

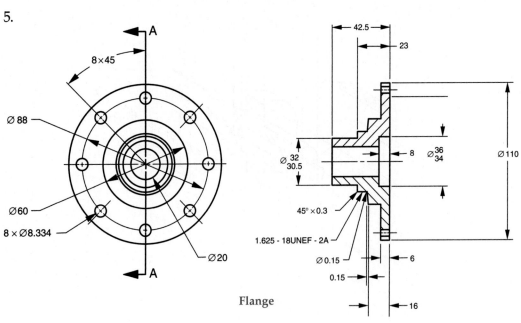

A

8×45

⌀ 88

⌀ 60

8 × ⌀8.334

⌀ 20

A

42.5

23

⌀ 32
30.5

8

⌀ 36
34

⌀ 110

45° × 0.3

1.625 - 18UNEF - 2A

⌀ 0.15

0.15

6

16

Flange

6.

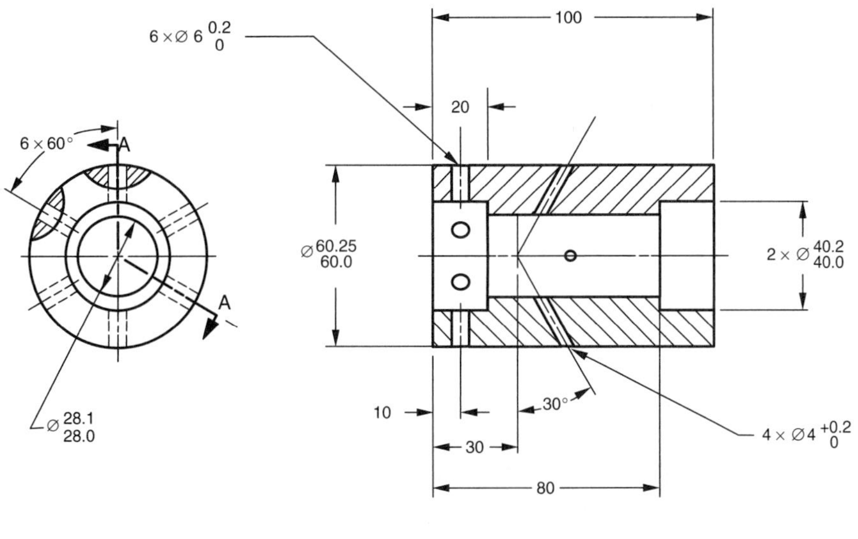

6 × ∅ 6 0.2 0
100
20
6 × 60°
A
A
∅60.25 60.0
2 × ∅ 40.2 40.0
∅ 28.1 28.0
10
30°
4 × ∅4 +0.2 0
30
80
SECTION A-A

Nozzle

7.

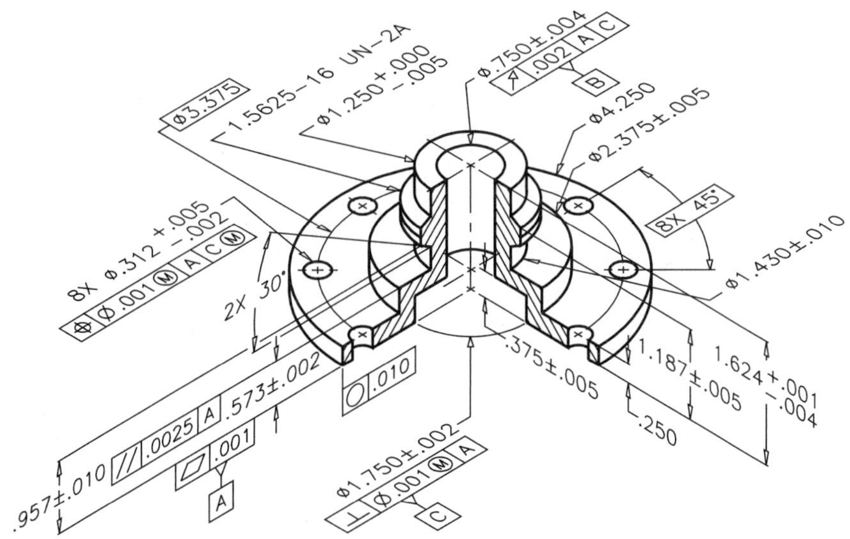

Φ3.375
1.5625–16 UN–2A
Φ1.250+.000 −.005
Φ.750±.004
.002 A C
B
Φ4.250
Φ2.375±.005
8X 45°
Φ1.430±.010
8X Φ.312 +.005 −.002
Φ.001 Ⓜ A C Ⓜ
2X 30°
.010
.375±.005
1.187±.005
1.624 +.001 −.004
.250
.957±.010
.0025 A
.573±.002
.001
A
Φ1.750±.002
Φ.001 Ⓜ A
C

Hub

8. Create the stairway shown below using the following parameters. Save the file as P08_08.
   A. Use the detail for the riser and tread dimensions.
   B. There are 13 risers.
   C. The stairs are 42″ wide.
   D. The landing at the top of the stairs is 48″ long from the face of the last riser.
   E. The vertical wall is 15′-1″ high on the inside and 15′-7″ long.
   F. The floor is 8′ wide on the inside and 15′-7″ long.
   G. Draw the floor and the wall as 1″ thick.
   H. The center of the banister is 3″ away from the wall and 34″ above the steps.
   I. The ends of the banister are directly above the face of the first and last riser.
   J. Use the detail for the profile of the banister. Use **EXTRUDE** as needed.

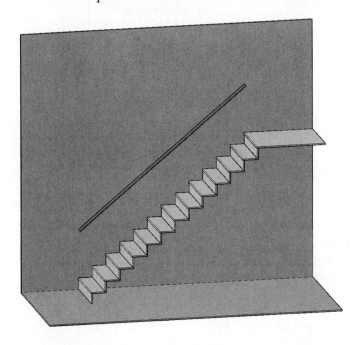

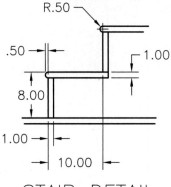

STAIR DETAIL

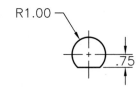

BANISTER DETAIL

9. Construct picture frame moldings using the profiles shown below.
   A. Draw each of the closed profiles shown. Use your own dimensions for the details of the moldings.
   B. The length and width of A and B should be no larger than 1.5″ × 1″.
   C. The length and width of C and D should be no larger than 3″ × 1.5″.
   D. Construct an 8″ × 12″ picture frame using moldings A and B.
   E. Construct a 12″ × 24″ picture frame using moldings C and D.
   F. Save the drawing as P08_09.

10. In this problem, you will refine the seat of the kitchen chair that you started in Chapter 2. You will use an extrusion following a path to create a curved, receding edge under the seat.
   A. Open P02_11 from Chapter 2. If you have not yet completed this model, do so now.
   B. Create a path for the extrusion by drawing a polyline that exactly matches either the upper or lower edge of the seat. Refer to the drawing shown.
   C. Change the view and UCS as needed to display a plan view of the edge of the seat. Draw the profile shown.
   D. Extrude the profile along the path and then subtract the extrusion from the seat.
   E. Save the drawing as P08_10.

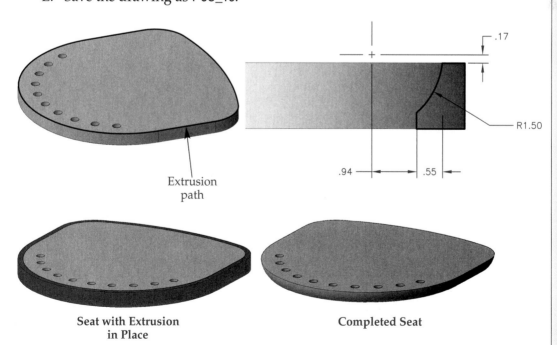

Extrusion path

Seat with Extrusion in Place

Completed Seat

11. In this problem, you will be taking a manually drawn layout from an archive. You are to create a 3D model of the garage to update the archive.

   A. Review the manually drawn layout. Make note of the construction details shown.

   B. Research any additional details needed to construct the model. For example, the thickness of the doors is not listed as these are purchased items. However, you need to know these dimensions to draw the 3D model.

   C. Using what you have learned, create the garage as a solid model. Be sure to create all components, including the studs in the walls, the footings, and the anchor bolts.

   D. Create layers as needed. For example, you may wish to place the wall sheathing on a layer so it can be hidden to show the studs.

   E. Save the drawing as P08_11.

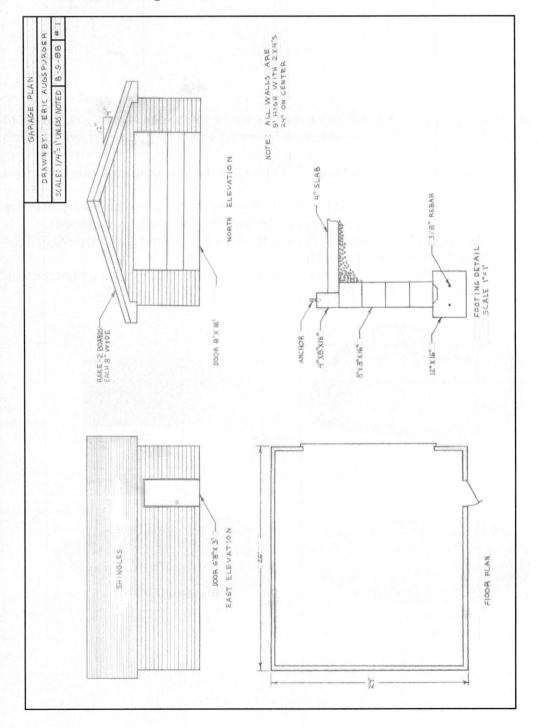

Sweeps and Lofts

Learning Objectives

After completing this chapter, you will be able to:

✓ Sweep 2D shapes along a 2D or 3D path to create a solid or surface object.

✓ Create 3D solid or surface objects by lofting a series of cross sections.

In the previous chapter, you learned about extruded solids and surfaces. Sweeps and lofts are similar to extrusions. In fact, an extrusion is really just a type of sweep. A *sweep* is an object created by extruding a single 2D profile along a path object. Sweeping an open shape along the path results in a surface object. If a closed shape is swept, a solid or surface object can be created. A *loft* is an object created by extruding between two or more 2D profiles. The shape of the loft object blends from one cross-sectional profile to the next. The profiles can control the loft or it can be controlled by one path or multiple guide curves. As with a sweep, open shapes result in surfaces and closed shapes give you solids. Open and closed shapes cannot be used together in the same loft.

Creating Swept Surfaces and Solids

The **SWEEP** command is used to create swept surfaces and solids. The command requires at least two objects:

- 2D shape to be swept.
- 2D or 3D shape to be used as the sweep path.

The profile can be aligned with the path, you can specify the base point, a scale factor can be applied, and the profile can be twisted as it is swept. The command procedure and options are the same for both swept solids and surfaces.

Sweeping an open shape creates a surface. See **Figure 9-1**. The objects that can be swept to create surfaces include lines, arcs, elliptical arcs, 2D polylines, 2D and 3D splines, traces, and 3D edge subobjects. Sweeping a closed shape creates a solid by default, but surfaces may be created as well. See **Figure 9-2**. Closed shapes that can be swept include circles, ellipses, closed 2D polylines, closed 2D splines, regions, 2D solids, traces, and 3D solid face subobjects. The sweep path can be a line, arc, circle, ellipse, elliptical arc, 2D polyline, 2D spline, 3D polyline, 3D spline, helix, or 3D edge subobject.

Ribbon

Home
  > Modeling
Solid
  > Solid
Surface
  > Create

Sweep

Type

SWEEP

SWEEP

Figure 9-1.
A—This open shape
will be swept along
the path (shown
in color). B—The
resulting surface.

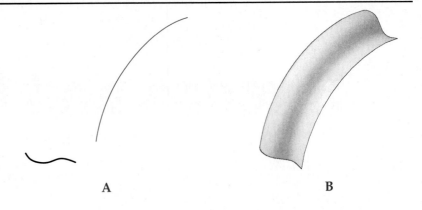

A   B

Figure 9-2.
A—This closed
shape will be swept
along the path
(shown in color).
B—The resulting
solid.

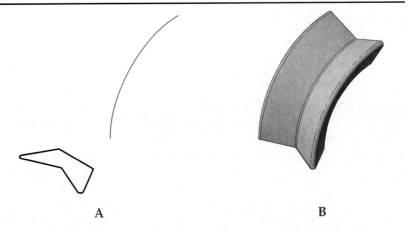

A   B

When the command is initiated, you are prompted to select the objects to sweep. Select the profile(s) and press [Enter]. Planar faces of solids may be selected by holding the [Ctrl] key as you select. Multiple profiles can be selected. They are swept along the same path, but separate objects are created.

Next, you are prompted to select the path. The path and profile can lie on the same plane. Select the object to be used as the sweep path and press [Enter]. To select the edge of a surface or solid as the path, press the [Ctrl] key and then select the edge. The profile is then moved to be perpendicular to the path and extruded along the path. The sweep starts at the endpoint of the path nearest to where you selected it.

The **Mode** option that is available when the **SWEEP** command is initiated controls the closed profiles creation mode. This allows you to change the way in which closed shapes are handled. Normally, the **SWEEP** command produces a solid object when a closed shape is swept, but surfaces may be created by changing this mode. For example, a circle swept with a line as the path will normally create a solid cylinder. Setting the **Mode** option to **Surface** will result in a tube created as a surface model.

 Exercise 9-1

Complete the exercise on the companion website.
www.g-wlearning.com/CAD

Changing the Alignment of the Profile

By default, the profile is aligned perpendicular to the sweep path. However, you can create a sweep where the profile is not perpendicular to the path. See **Figure 9-3**. After the **SWEEP** command is initiated, select the profile and press [Enter]. Before

Figure 9-3.
A—The profile and path for the sweep. B—By default, the profile is aligned perpendicular to the path when swept. C—Using the **Alignment** option, the profile can be swept so it is not perpendicular to the path.

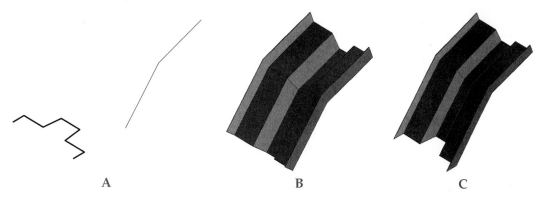

A   B   C

selecting the path, enter the **Alignment** option. The default setting of **Yes** means that the profile will be moved so it is perpendicular to the path. If you select **No**, the profile is kept in the same position relative to the path as it is swept. The position of the 2D shape determines the alignment.

Changing the Base Point

The base point is the location on the shape that will be moved along the path to create the sweep. By default, if the 2D shape intersects the path, the profile is swept along the path at the point of intersection. If the 2D shape does not intersect the path, the default base point depends on the type of object being swept. When lines and arcs are swept, the default base point is their midpoint. Open polylines have a default base point at the midpoint of their total length.

The base point can be any point on the 2D shape or anywhere in the drawing. See **Figure 9-4**. To change the base point, use the **Base point** option of the **SWEEP** command. When the command is initiated, select the profile and press [Enter]. Before selecting the path, enter the **Base point** option. Next, pick the new base point. It does not have to be on an existing object. Once the new base point is selected, pick the path to create the sweep.

Figure 9-4.
A—The profile and path for the sweep. B—The sweep is created with the default base point. C—The end of the path is selected as the base point. Notice the difference in this sweep and the one shown in B.

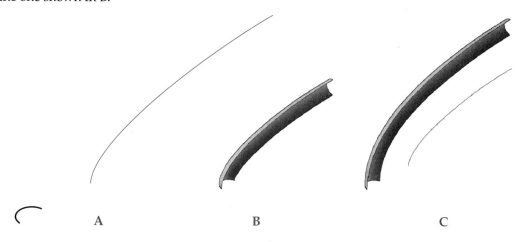

A   B   C

Scaling the Sweep Profile

By default, the size of the profile remains uniform from the beginning of the path to the end. However, using the **Scale** option of the **SWEEP** command, you can change the scale of the profile at the end of the path. This, in effect, tapers the sweep. **Figure 9-5** shows a .25 scale applied to a sweep object. A 2D polyline with multiple segments must be edited using the **Fit** or **Spline** option in order to be used as a path. Sharp corners will not work with the **Scale** option. A 3D polyline path must be a spline.

Once the **SWEEP** command is initiated, select the profile and press [Enter]. Before selecting the path, enter the **Scale** option. You are prompted for the scale. Enter the scale value and press [Enter]. The scale value must be greater than zero. You can also enter the **Reference** option. With this option, pick two points for the first reference line and then two points for the second reference line. The difference in scale between the two distances is the scale value. Once the scale is set, pick the path to create the sweep. The **Expression** option allows you to enter a mathematical expression to constrain the scaling of the sweep. This option only works when creating a surface sweep.

Twisting the Sweep

The profile can be rotated as it is swept along the length of the path by using the **Twist** option of the **SWEEP** command. The angle that you enter indicates the rotation of the shape along the path of the sweep. The higher the number, the more twists in the sweep. **Figure 9-6** shows how a simple, closed profile and a straight line can be used to create a milling tool. The profile was swept with a 270° twist.

Once the **SWEEP** command is initiated, select the profile and press [Enter]. Then, before selecting the path, enter the **Twist** option. You are prompted for the twist angle or to enter the **Bank** option.

Banking is the natural rotation of the profile on a 3D sweep path, similar to a banked curve on a racetrack. See **Figure 9-7**. The path must be 3D (nonplanar) to set banking. The banking option is disabled for a 2D path, although you can go through the process of turning it on when creating the sweep. Once you use the **Bank** option to turn banking on, it is on by default the next time the **SWEEP** command is used. To turn it off, enter a twist angle of zero (or the twist angle you wish to use). The **Expression** option allows you to enter a mathematical expression to constrain the number of rotations in the sweep. This option only works when creating a surface sweep.

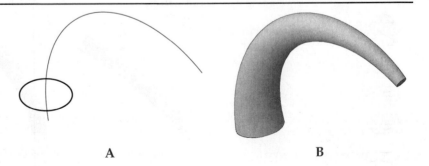

Figure 9-5.
A—The profile and path for the sweep. B—The resulting sweep. Notice how the .25 scale results in a tapered sweep.

A

B

Figure 9-6.
A—The profile and path for creating the end mill.
B—The resulting end mill model. Notice how the profile is twisted (rotated) as it is swept.

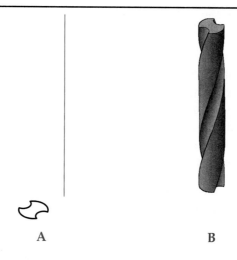

A   B

Figure 9-7.
A—The profile and path for the sweep are shown in color. B—Banking is off for this sweep. When viewed from the side, you can see that the profile does not bank through the curve. Look at the upper-right corner. C—Banking is on for this sweep. Notice how the profile banks, or leans, through the curve. Compare this to B.

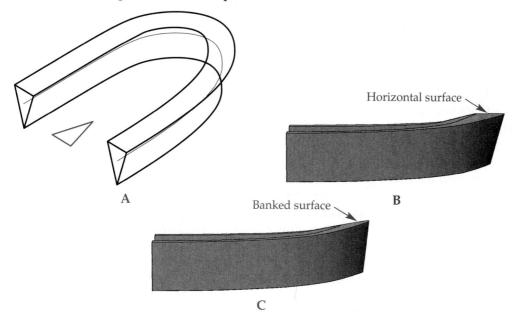

Horizontal surface

A   B

Banked surface

C

PROFESSIONAL TIP

The sweep options can be changed after the sweep is created using the **Properties** palette. In the **Geometry** section, you will find Profile rotation (alignment), Bank along path (banking), Twist along path (twist angle), and Scale along path (scale) settings.

Exercise 9-2

Complete the exercise on the companion website.
www.g-wlearning.com/CAD

Creating Lofted Objects

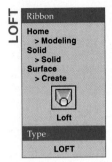

The **LOFT** command is used to create lofted surfaces and solids based on a series of cross-sectional profiles. **Figure 9-8** shows an example of a loft formed from a rectangle, circle, and polygon. The loft may be guided by only the cross sections, as shown in the figure, by a path, or by guide curves. Lofting open shapes results in a surface object, while lofting closed shapes creates a solid. Open and closed shapes cannot be combined in the same loft.

Objects that can be used as cross sections include lines, circles, arcs, points, ellipses, elliptical arcs, 2D polylines, 2D splines, regions, edge subobjects, surfaces, face subobjects, 2D solids, helices, and traces. Points may be used for the first and last cross sections only. The loft path may be a line, circle, arc, ellipse, elliptical arc, spline, helix, 2D or 3D polyline, or edge subobject. Guide curves may be composed of lines, arcs, elliptical arcs, 2D or 3D splines, 2D or 3D polylines, and edge subobjects. However, 2D polylines are limited to only one segment.

Once the command is initiated, you are prompted to select the cross-sectional profiles. Pick each profile in the order in which it should appear in the loft and press [Enter]. Be sure to individually select the cross sections in the order of the loft creation. You may not get the desired loft if you randomly select them or use a window selection. As the profiles are selected, the loft is previewed in a semitransparent state, allowing the user to make adjustments.

The **Mode** option that is available when you start the command controls the closed profiles creation mode and allows you to change the way that closed shapes are handled. It behaves exactly as it does in the **SWEEP** command.

To avoid having to draw point objects for cross sections, the **Point** option allows you to pick any point as either the start point or the end point of the loft. If you pick a point first, it is the start point and you can pick as many shapes as you want as the other cross sections. If you pick the other cross sections first, then the point must be the endpoint of the loft. The other cross sections must be closed shapes. Open shapes will not work with the **Point** option.

If you need to use the edges of existing 3D objects as cross sections, the **Join multiple edges** option works well. The edges must be touching at their end points and form a cross section. See **Figure 9-9**. This option is used to define a single cross section. When you press [Enter], you are again prompted to select cross sections. The option can be used again to select additional cross sections.

After selecting cross sections, you are prompted to select how the loft is to be controlled. As mentioned earlier, you can control the loft by the cross sections, a path, or guide curves. These options are discussed in the next sections.

Figure 9-8.
A—The three profiles will be lofted to create a solid. B—The resulting loft with the default settings.

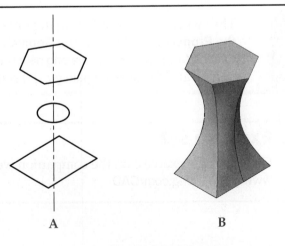

A

B

Figure 9-9.
The **Join multiple edges** option allows you to make a quick transition from the edges of one 3D object to another. A—Use the option twice to select two cross sections. B—The resulting loft is a transition between the two objects.

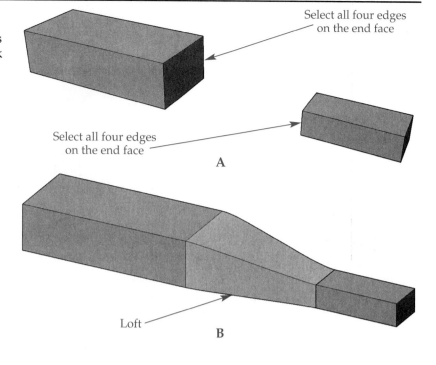

Select all four edges on the end face

Select all four edges on the end face

A

Loft

B

Controlling the Loft with Cross Sections

The **Cross sections only** option of the **LOFT** command is useful when the 2D cross sections are drawn in their proper locations in space. The command determines the transition from one cross section to the next. The cross sections are not moved by the command.

When you select the **Settings** option, the **Loft Settings** dialog box appears, **Figure 9-10**. The settings in this dialog box control the transition or contour between cross sections. As settings are changed, the preview is updated. When all settings have been made, pick the **OK** button to close the dialog box and create the loft.

When the **Ruled** option is selected in the dialog box, the loft has straight transitions between the cross sections. Sharp edges are created at each cross section. **Figure 9-11** shows the same cross sections in **Figure 9-8A** lofted with the **Ruled** option on. Compare this to **Figure 9-8B**.

Figure 9-10.
The **Loft Settings** dialog box is used to control the transition between profiles.

Select a contour setting

Check to connect the first and last cross sections

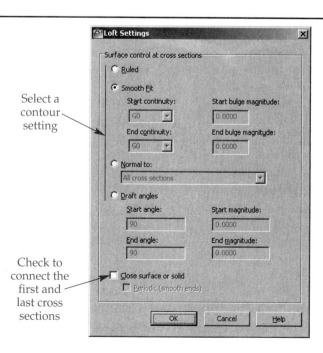

Figure 9-11.
The profiles in Figure 9-8A are lofted with the **Ruled** option selected in the **Loft Settings** dialog box. Compare this to Figure 9-8B.

The **Smooth Fit** option creates a smooth transition between the cross sections. Sharp edges are only created at the first and last cross sections. This is the default setting and the one used to create the loft shown in **Figure 9-8B**. The **Start continuity:** and **End continuity:** settings control the tangency and curvature of the first and last cross sections. The **Start bulge magnitude:** and **End bulge magnitude:** settings control the size of the curve at the first and last cross sections. These options only apply if the first and/or last cross sections are regions. See **Figure 9-12**. The start and end continuity can be set to G0, G1, or G2. The G0 (positional continuity) setting creates a sharp edge. The loft transitions precisely at the location of the profile. The G1 (tangential continuity) setting is the default. It creates the largest bulge radius. The G2 (curvature or continuous continuity) setting creates a smaller bulge radius. For all three settings, bulge magnitude is initially set to 0.5. Altering the value will make the bulge either larger or smaller but has no effect when continuity is set to G0.

When the **Normal to:** option is selected in the **Loft Settings** dialog box, you can choose how the normal of the transition is treated at the cross sections. A *normal* is a vector extending perpendicular to the cross section. When the transition is normal to a cross section, it is perpendicular to the cross section. You can set the transition normal to the first cross section, last cross section, both first and last cross sections, or all cross sections. See **Figure 9-13**. Select the normal setting in the drop-down list. You will have to experiment with these settings to get the desired loft shape.

Figure 9-12.
Region profiles are lofted with the **Smooth Fit** option and different continuity settings selected in the **Loft Settings** dialog box. Cross sections were selected from bottom to top. Compare these results with Figure 9-8B. A—G0 continuity at the start and end of the loft. Bulge magnitude has no effect at this setting. B—G1 continuity at the start and end of the loft. This is the default setting. C—G2 continuity at the start and end of the loft. Compare the bulge radius to B.

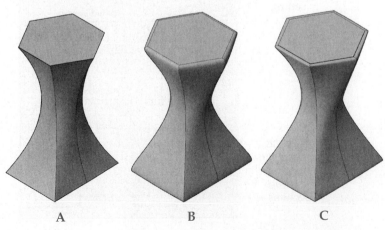

A   B   C

Figure 9-13.
The profiles in Figure 9-8A are lofted with the **Normal to:** option on in the **Loft Settings** dialog box. Cross sections were selected from bottom to top. Compare these results with Figure 9-8B and Figure 9-12. A—Start cross section. B—End cross section. C—Start and End cross sections. D—All cross sections.

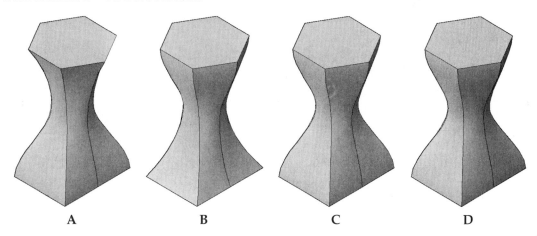

In manufacturing, plastic or metal parts are sometimes formed in a two-part mold. A slight angle is designed into the parts on the inside and outside surfaces to make removing the part from the mold easier. This taper is called a *draft angle*. The **Draft angles** option allows you to add a taper to the beginning and end of the loft.

When setting the draft angle, you can set the angle and magnitude. See **Figure 9-14**. The default draft angle is 90°, which means the transition is perpendicular to the cross section. The magnitude represents the relative distance from the cross section, in the same direction as the draft angle, before the transition starts to curve toward the next cross section. Magnitude settings depend on the size of the cross sections, the draft angle values, and the distance between the cross sections. You may have to experiment with different magnitude and angle settings to get the desired loft shape.

The **Close surface or solid** option is used to connect the last cross section to the first cross section. See **Figure 9-15**. This option "closes" the loft, similar to the **Close** option of the **LINE** or **PLINE** command. The shapes from **Figure 9-8** are shown in **Figure 9-16** with the **Close surface or solid** option on. This option is only available when the **Ruled** or **Smooth Fit** option is selected.

If the **Smooth Fit** option is selected and the **Close surface or solid** check box is checked, the **Periodic (smooth ends)** check box is available. If a closed-loop loft is created

Figure 9-14.
When setting the draft angle, you can set the angle and the magnitude. A—Draft angle of 90° and a magnitude of zero. B—Draft angle of 30° and a magnitude of 180. C—Draft angle of 60° and a magnitude of 180.

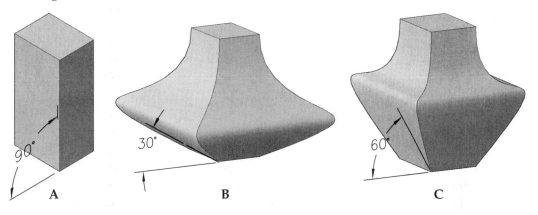

Figure 9-15.
A—These profiles will be used to create a sealing ring. They should be selected in a counterclockwise direction starting with the first cross section. B—The resulting loft with the default settings. Notice the gap between the first and last cross sections. C—By checking the **Close surface or solid** check box in the **Loft Settings** dialog box, the loft continues from the last cross section to the first cross section.

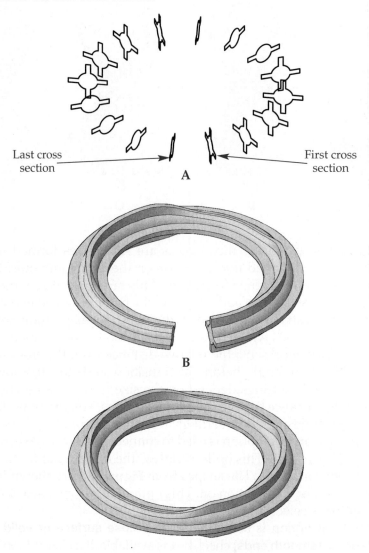

Last cross section

First cross section

A

B

Figure 9-16.
The same cross sections from Figure 9-8 are used in this loft. However, the **Close surface or solid** option has been applied. Notice how the loft is inside out.

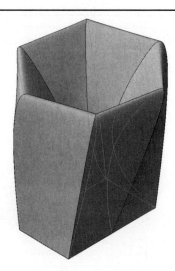

similar to the loft in **Figure 9-15**, the seam may kink if the loft is reshaped in some way. Checking the **Periodic (smooth ends)** check box will help alleviate this problem.

PROFESSIONAL TIP

It may be easier to experiment with the loft settings after the loft is created. Select the loft and change the settings in the **Properties** palette. In addition, special grips and handles appear at the locations of the profiles. Picking on a grip will open a shortcut menu and dragging a handle will alter the loft dynamically.

Exercise 9-3

Complete the exercise on the companion website.
www.g-wlearning.com/CAD

Controlling the Loft with Guide Curves

Guide curves are lines that control the shape of the transition between cross sections. They do not have to be *curves*. They can be lines, arcs, elliptical arcs, splines (2D or 3D), or polylines (2D or 3D). There are four rules to follow when using guide curves:

- The guide curve should start on the first cross section.
- The guide curve should end on the last cross section.
- The guide curve should intersect all other cross sections.
- The surface control in the **Loft Settings** dialog box must be set to **Smooth Fit** (**LOFTNORMALS** = 1).

When the **Guides** option of the **LOFT** command is entered, you are prompted to select the guide curves. Select all of the guide curves and press [Enter]. The loft is created. The order in which guide curves are selected is not important.

For example, **Figure 9-17A** shows two circles that will be lofted. If the **Cross sections only** option is used, a cylinder is created, **Figure 9-17B**. However, if the **Guides** option is used and the two guide curves shown in **Figure 9-17A** are selected, one side of the cylinder is deformed similar to a handle or grip. See **Figure 9-17C**.

Lofting is used to create open-contour shapes such as fenders, automobile interior parts, fabrics, and other ergonomic consumer products. **Figure 9-18** shows the use of open 2D splines in the construction of a fabric covering. Notice how each cross section

---

Figure 9-17.
A—These two circles will be lofted. The lines shown in color will be used as guide curves. B—When the circles are lofted using the **Cross sections only** option, a cylinder is created. C—When the **Guides** option is used and the guide curves shown in A are selected, the resulting loft is shaped like a handle or grip.

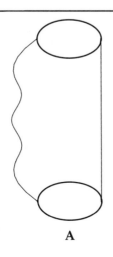

A B C

Figure 9-18.
A—The open profiles shown in black and the guide curve shown in color will be used to create a fabric covering for the three solid objects. B—The resulting fabric covering. This is a surface because the profiles were open.

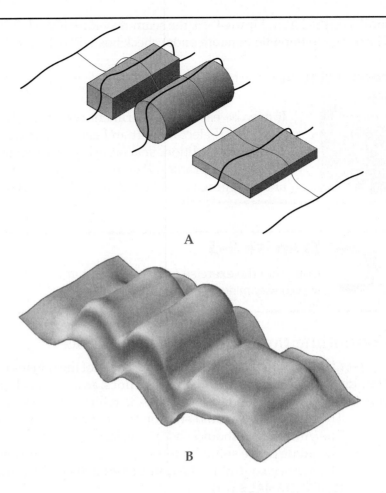

A

B

is intersected by the guide curve. There is a cross section at the beginning of the guide curve and one at the end. These conditions fulfill the rules outlined earlier.

CAUTION

Guide curves only work well when the surface control is set to **Smooth Fit** (**LOFTNORMALS** = 1). If you get an error message when using guide curves or the curves are not reflected in the end result, make sure **LOFTNORMALS** is set to 1 and try it again.

Controlling the Loft with a Path

The **Path** option of the **LOFT** command places the cross sections along a single path. The path must intersect the planes on which each of the cross sections lie. However, the path does *not* have to physically touch the edge of each cross section, as is required of guide curves. When the **Path** option is entered, you are prompted to select the path. Once the path is picked, the loft is created. The cross sections remain in their original positions.

Figure 9-19 shows how 2D shapes can be positioned at various points on a path to create a loft. The rectangular shape does not cross the path. However, as long as the path intersects the plane of the rectangle, which it does, the shape will be included in the loft definition. The last shape at the top of the helix is a point object, causing the loft to taper.

Figure 9-19.
A—The profiles shown in black will be lofted along the path shown in color. Notice how the rectangular profile is not intersected by the path, but the path does intersect the plane on which the rectangle lies.
B—The resulting loft.

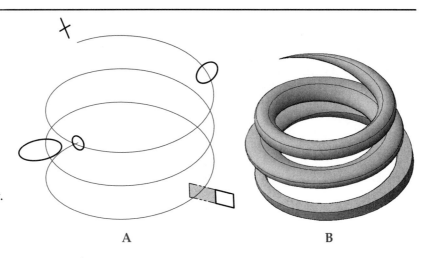

A   B

 Exercise 9-4

Complete the exercise on the companion website.
www.g-wlearning.com/CAD

 # Chapter Review

Answer the following questions. Write your answers on a separate sheet of paper or complete the electronic chapter review on the companion website.
www.g-wlearning.com/CAD

1. What is a *loft*?
2. What option of the **SWEEP** command determines whether the sweep will be a solid or a surface?
3. When using the **SWEEP** command, on which endpoint of the path does the sweep start?
4. What is the purpose of the **Base point** option of the **SWEEP** command?
5. After the sweep or loft is created, how may the creation options be changed?
6. Which objects may be used as a sweep path?
7. How is the alignment of a sweep set to be perpendicular to the start of the path?
8. Which **SWEEP** command option is used to taper the sweep?
9. What is the difference between the **Ruled** and **Smooth Fit** options in the **LOFT** command?
10. What does the **Bank** option of the **LOFT** command do?
11. How can you close a loft?
12. List five objects that may be used as guide curves in a loft.
13. What are the four rules that must be followed when using guide curves?
14. When using the **Path** option of the **LOFT** command, what must the path intersect?
15. How can a loft be created so it tapers to a point at its end?

Drawing Problems

1. Create the lamp shade shown. Create two separate loft objects for the top and the bottom. Then, union the two pieces. Finally, scale a copy and hollow out the lamp shade. Save the drawing as P09_01.

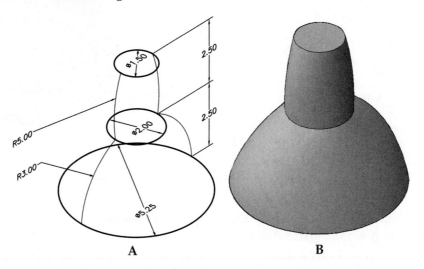

A   B

2. Create the two shampoo bottles shown. One design uses cross sections only and the other uses a guide curve. Each bottle is made up of two loft objects. Join the pieces so each bottle is one solid. Save the drawing as P09_02.

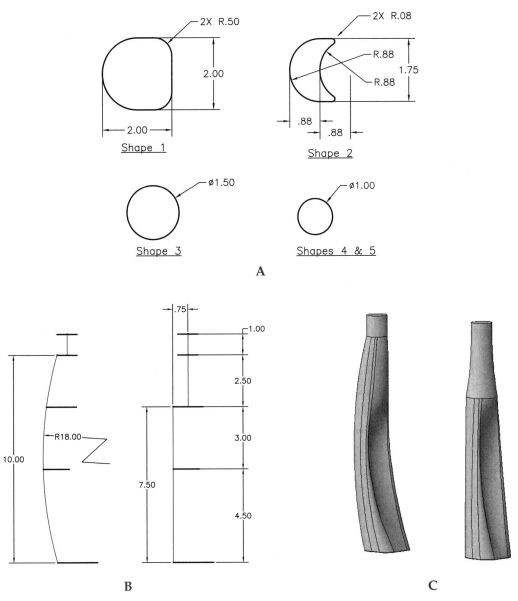

Shape 1

Shape 2

Shape 3

Shapes 4 & 5

A

B

C

3. Create as a loft the automobile fender shown below. Use either the **Guide** or the **Path** option and the line shown in color. Save the drawing as P09_03.

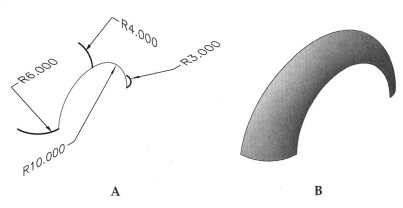

A

B

4. Draw as a loft the C-clamp shown. Use the shapes (A, B, C, and D) as the cross sections and the polyline (in color) as the guide curve. Add Ø1 unit cylinders to the ends. Make one cylinder .125H and the other 1.125H. The cylinders should be centered on profile D and located at the ends of the loft as shown. Make a Ø.625 hole through the larger cylinder. Save the drawing as P09_04.

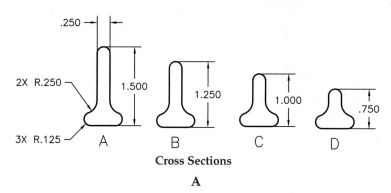

.250

2X R.250

1.500

1.250

1.000

.750

3X R.125   A   B   C   D

Cross Sections

A

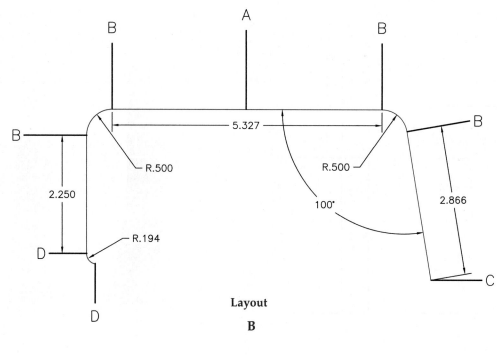

A

B

B

B

5.327

R.500   R.500

B

2.250

100°

2.866

R.194

D

D

C

Layout

B

C

5. In this problem, you will draw a racetrack for toy cars by sweeping a 2D shape along a polyline path.
   A. Draw the polyline path shown with the coordinates given. Turn it into a spline.
   B. Draw the 2D profile shown using the dimensions given. Turn it into a region or a polyline.
   C. Use the **SWEEP** command to create the racetrack, as shown in the shaded view.
   D. You may have to use the **Properties** palette to adjust the sweep after it is drawn.
   E. Save the drawing as P09_05.

@0,50,10

@23<135

@28<91

@18<20

@0,23,0

Start here

Polyline Path

A

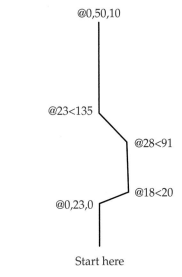

2D Profile

B

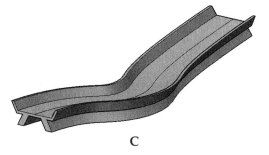

C

6. In this problem, you will cut a UNC thread in a cylinder by sweeping a 2D shape around a helix and subtracting it.
   A. Draw a ⌀.25 cylinder that is 1.00 in height.
   B. Draw the thread cutter profile shown below. The long edge of the cutter should be aligned with the vertical edge of the cylinder.
   C. Draw a helix centered on the cylinder with base and top radii of .125, a turn height of .050, and a total height of 1.000.
   D. Sweep the 2D shape along the helix. Then, subtract the resulting solid from the cylinder. Refer to the shaded view.
   E. If time allows, create another cutter profile to cut a .0313 × 45° chamfer on the end of the thread. Use a circle as a sweep path or revolve the profile about the center of the cylinder.
   F. Save the drawing as P09_06.

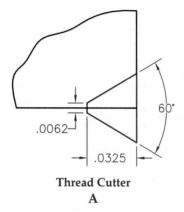

Thread Cutter
A

B

7. Create the furniture leg shown. Either the **LOFT** command or the **SWEEP** command may be used. However, one command may work better than the other. The profile dimensions refer to the bottom end (small end) of the leg. The top is twice the size of the bottom. Note that the top and the bottom of the leg are parallel. Save the drawing as P09_07.

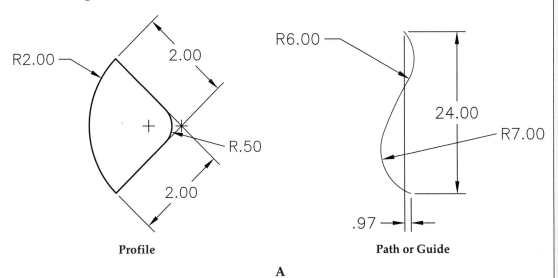

Profile   Path or Guide

A

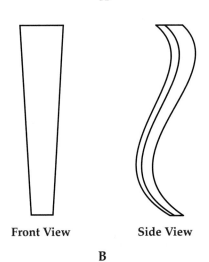

Front View   Side View

B

C

8. In this problem, you will add a seatback to the kitchen chair you started modeling in Chapter 2. In Chapter 8, you refined the seat.
   A. Open P08_10 from Chapter 8.
   B. Draw an arc for the top of the bow. Using a 14.25″ length of the arc, divide it into seven equal segments.
   C. Position eight ∅.50 circles at the division points.
   D. Draw circles at the top of each hole in the seat.
   E. Create a loft between each lower circle and each upper circle.
   F. Using the information in the drawings, create the outer bow for the seatback.
   G. Save the drawing as P09_08.

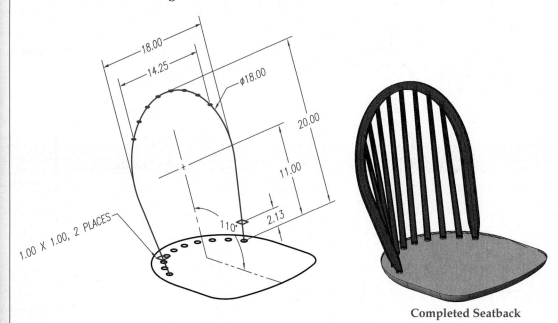

Completed Seatback

AutoCAD and Its Applications—Advanced

Creating and Working with Solid Model Features

Learning Objectives

After completing this chapter, you will be able to:

✓ Change properties on solids.
✓ Align objects.
✓ Rotate objects in three dimensions.
✓ Mirror objects in three dimensions.
✓ Create 3D arrays.
✓ Fillet solid objects.
✓ Chamfer solid objects.
✓ Slice a solid using various methods.
✓ Construct features on solid models.
✓ Remove features from solid models.

Changing Properties Using the Properties Palette

Properties of 3D objects can be modified using the **Properties** palette, which is thoroughly discussed in *AutoCAD and Its Applications—Basics*. This palette is displayed using the **PROPERTIES** command. You can also double-click on a solid object or select the solid, right-click, and pick **Properties** from the shortcut menu.

The **Properties** palette lists the properties of the currently selected object. For example, Figure 10-1 lists the properties of a selected solid sphere. Some of its properties are parameters, which is why solid modeling in AutoCAD can be considered parametric. You can change the sphere's radius; diameter; X, Y, Z coordinates; linetype; linetype scale; color; layer; lineweight; and visual settings. The categories and properties available in the **Properties** palette depend on the selected object.

To modify an object property, select the property. Then, enter a new value in the right-hand column. The drawing is updated to reflect the changes. You can leave the **Properties** palette open as you continue with your work.

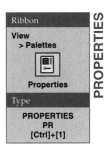

Ribbon

View
> Palettes

Properties

Type

PROPERTIES
PR
[Ctrl]+[1]

PROPERTIES

Figure 10-1.
The **Properties** palette can be used to change many of the properties of a solid.

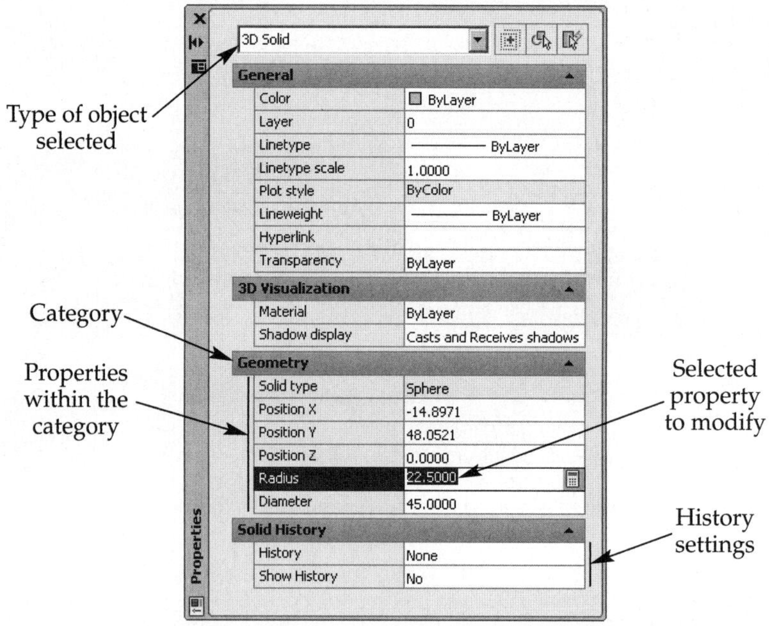

Type of object selected

Category

Properties within the category

Selected property to modify

History settings

Solid Model History

AutoCAD can automatically record a history of a composite solid model's construction. A *composite solid* is created by a Boolean operation or by using the **SOLIDEDIT** command. The retention of a composite solid's history is controlled by the History property setting in the **Solid History** category of the **Properties** palette. By default, the History property is set to None. Refer to **Figure 10-1**. This means that the history is not saved and that edges, vertices, and faces of solids can be directly edited. If the History property in the **Properties** palette is set to Record, the solid history is "recorded," or retained. This allows you to work with the geometry used in modeling operations to create the composite solid. Depending on your modeling preferences, it is generally a good idea to have the history recorded. Then, at any time, you can graphically display all of the geometry that was used to create the model.

The retention of solid model history for newly created composite solids is controlled by the **SOLIDHIST** system variable. By default, the **SOLIDHIST** system variable is set to 0. This means that all new solids have their History property set to None and solid history is not recorded. If the **SOLIDHIST** system variable is set to a value of 1, all new solids have their History property set to Record and solid history is preserved. With either setting of the system variable, 0 (None) or 1 (Record), the **Properties** palette can be used to change the setting for individual solids. The current setting of the **SOLIDHIST** system variable is indicated by the **Solid History** button located in the **Primitive** panel of the **Solid** tab on the ribbon. See **Figure 10-2**. With a default setting of 0 (None), the **Solid History** button has a white background. Picking this button when it has a white background changes the background color to blue and sets the **SOLIDHIST** system variable to 1 (Record).

To view the graphic history of a composite solid, set the **Show History** property in the **Properties** palette to Yes. All of the geometry used to construct the model is displayed. If the **SHOWHIST** system variable is set to 0, the Show History property is set to No for all solids and cannot be changed. If this system variable is set to 2, the Show History property is set to Yes for all solids and cannot be changed. A **SHOWHIST** setting of 1 allows the Show History property to be individually set for each solid. This is the default setting.

Figure 10-2.
The **Solid History** button, located in the **Primitive** panel of the **Solid** tab on the ribbon, is used to control the **SOLIDHIST** system variable setting. This button has a white background when the **SOLIDHIST** system variable is set to 0 (the default value). If picked, the button has a blue background.

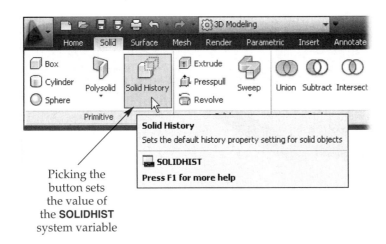

Picking the button sets the value of the **SOLIDHIST** system variable

An example of showing the history on a composite solid is provided in **Figure 10-3**. In **Figure 10-3A**, the model appears in its current edited format. The History property setting was initially set to Record before performing two Boolean subtraction operations. The Show History property is currently set to No and the Conceptual visual style is set current. In **Figure 10-3B**, the Show History property is set to Yes. Isolines have also been turned on. You can see the geometry that was used in the Boolean subtraction operations. Using subobject editing techniques, the individual geometry can be selected and edited. In **Figure 10-3C**, the sphere (the subtracted object) is selected while pressing the [Ctrl] key. Selecting the sphere in this manner selects the solid primitive *subobject*. Subobject editing is discussed in detail in Chapter 12.

PROFESSIONAL TIP

If the Show History property is set to Yes to display the components of the composite solid, as seen in **Figure 10-3**, the components will appear when the drawing is plotted. Be sure to set the Show History property to No before you print or plot.

Selection Cycling

When selecting objects that are on top of each other or occupy the same space, *selection cycling* is the preferred method to select one of the objects. When editing, you may need to erase, move, or copy one of the objects that overlap in order to select the correct object. The [Shift] key and spacebar can be pressed at the same time to cycle through objects at a pick point. When you need to cycle through objects:

1. At the "select objects" prompt, hold down the [Shift] key and spacebar, then click to select the object you want.
2. Keep clicking until the object you want to select is highlighted.
3. Press the [Enter] key.

Figure 10-3.
A—The object appears in its current state with the Show History property turned off. The History property was initially set to Record before subtracting two solid primitives.
B—The Show History property is set to Yes and the display of isolines has been turned on.
C—When the sphere is selected while pressing the [Ctrl] key, grips are displayed to indicate the sphere primitive subobject is selected. The move grip tool can be used to edit the original sphere primitive.

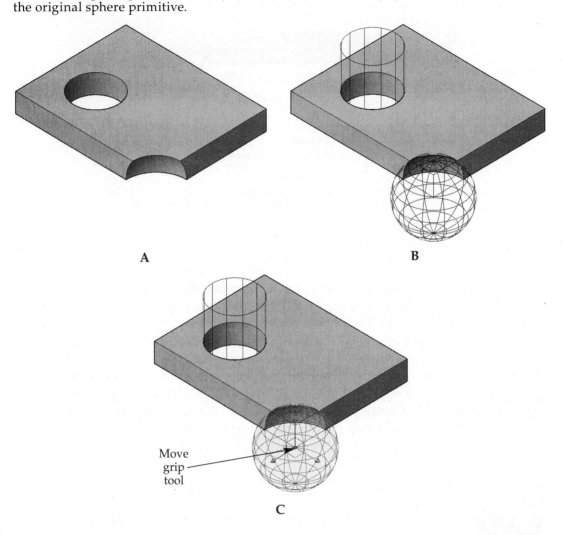

A   B

Move
grip
tool

C

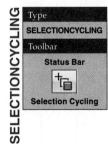

You can use the **SELECTIONCYCLING** system variable to turn on selection cycling instead of using the [Shift] key and spacebar. The **Selection Cycling** button on the status bar is used for toggling selection cycling. There are three settings for the system variable:

- Off (0).
- On, but the list dialog box does not display (1).
- On and the list dialog box displays the selected objects that can be cycled through (2).

It is recommended that you turn on selection cycling. With selection cycling turned on, you can cycle and select the faces that may overlap one another.

In **Figure 10-4**, the two 3D objects occupy the same space. The tapered 3D object needs to be moved up using the **3DMOVE** command. Turn on selection cycling. When you select the tapered object, the **Selection** dialog box lists the objects overlapping at the pick point. Select which object you want to work with by picking it in the dialog box. As your cursor is over an object in the list, the object is highlighted in the drawing area.

Figure 10-4.
Using selection
cycling to choose
which object to select.

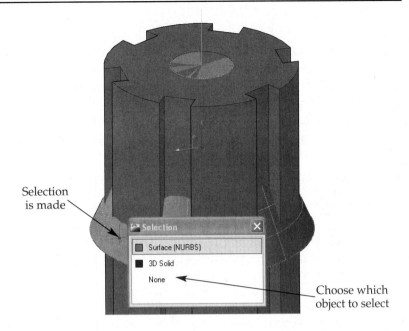

Selection
is made

Choose which
object to select

Aligning Objects in 3D

AutoCAD provides two different methods with which to move and rotate objects in a single command. This is called *aligning* objects. The simplest method is to align 3D objects by picking source points on the first object and then picking destination points on the object to which the first one is to be aligned. This is accomplished with the **3DALIGN** command, which allows you to both relocate and rotate the object. The second method is possible with the **ALIGN** command. The **ALIGN** command aligns 2D or 3D objects by selecting three sets of alignment pairs. You first select the source object, then the first source point, and finally the first destination point on the destination object. This technique is repeated two more times to align one object to another. This method of aligning allows you to not only move and rotate an object, but scale the object being aligned.

Ribbon

Home
> Modify

3D Align

Type

3DALIGN
3AL

Move and Rotate Objects in 3D Space

The basic function of moving and rotating an object relative to a second object or set of points is done with the **3DALIGN** command. It allows you to reorient an object in 3D space. Using this command, you can correct errors of 3D construction and quickly manipulate 3D objects. The **3DALIGN** command requires existing points (source) and the new location of those existing points (destination).

For example, refer to **Figure 10-5**. The wedge in **Figure 10-5A** is aligned in its new position in **Figure 10-5B** as follows. Set the **Intersection** or **Endpoint** running object snap to make point selection easier. Refer to the figure for the pick points.

Select objects: *(pick the wedge)*
1 found
Select objects: ↵
   Specify source plane and orientation…
Specify base point or [Copy]: *(pick P1)*
Specify second point or [Continue] <C>: *(pick P2)*
Specify third point or [Continue] <C>: *(pick P3)*
   Specify destination plane and orientation…
Specify first destination point: *(pick P4)*
Specify second destination point or [eXit] <X>: *(pick P5)*
Specify third destination point or [eXit] <X>: *(pick P6)*

Figure 10-5.
The **3DALIGN**
command can be
used to properly
orient 3D objects.
A—Before aligning.
Note the pick points.
B—After aligning.

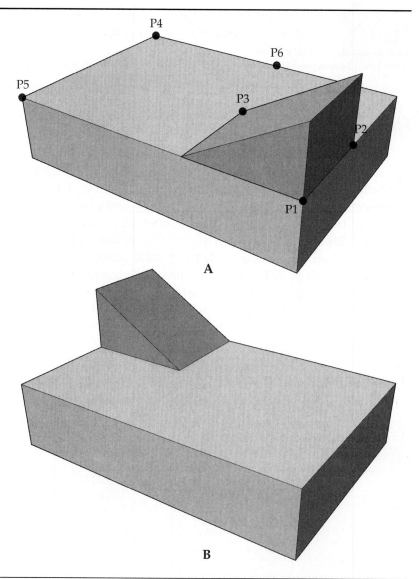

You can also use the **3DALIGN** command to align cylindrical 3D objects. The procedure is similar to aligning planar objects, which require three pick points. However, you only need two pick points per object.

For example, the socket head cap screw in **Figure 10-6** is aligned to a new position using the **3DALIGN** command. Set the center running object snap to make point selection easier. Refer to the figure for the pick points.

> Select objects: *(pick the socket head cap screw)*
> 1 found
> Select objects: ↵
>   Specify source plane and orientation…
> Specify base point or [Copy]: *(pick P1)*
> Specify second point or [Continue] <C>: *(pick P2)*
> Specify third point or [Continue] <C>: ↵
>   Specify destination plane and orientation…
> Specify first destination point: *(pick P3)*
> Specify second destination point or [eXit] <X>: *(pick P4)*
> Specify third destination point or [eXit] <X>: ↵

The idea with this command is you are aligning a plane defined by three points with another plane defined by three points. The planes do not need to correspond to actual planar faces. In the example shown in **Figure 10-5**, the alignment planes coincide with planar faces. However, in the example shown in **Figure 10-6**, the alignment planes define planes on which axes lie.

PROFESSIONAL TIP

You can also use the **3DALIGN** command to copy an object, rather than move it, and realign it at the same time. Just select the **Copy** option at the Specify base point or [Copy]: prompt. Then, continue selecting the points as previously discussed.

Figure 10-6.
Using the **3DALIGN** command to align cylindrical objects. A—Before aligning. Note the pick points. B—After aligning.

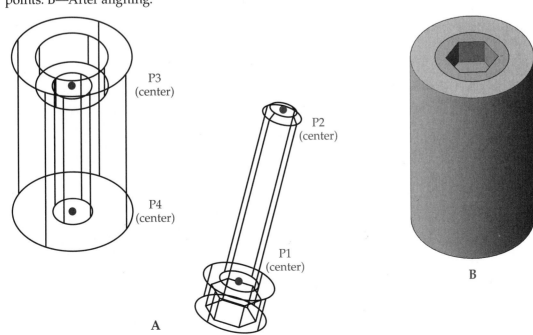

Exercise 10-1

Complete the exercise on the companion website.
www.g-wlearning.com/CAD

Move, Rotate, and Scale Objects in 3D Space

ALIGN

Ribbon
Home > Modify
Align

Type
ALIGN AL

The **ALIGN** command has the same functions of the **3DALIGN** command, but adds the ability to scale an object. See **Figure 10-7**. The 90° bend must be rotated and scaled to fit onto the end of the HVAC assembly. Two source points and two destination points are required, **Figure 10-7A**. Then, you can choose to scale the object.

Select objects: *(pick the 90° bend)*
1 found
Select objects: ↵
Specify first source point: *(pick P1)*
Specify first destination point: *(pick P2; a line is drawn between the two points)*
Specify second source point: *(pick P3)*
Specify second destination point: *(pick P4; a line is drawn between the two points)*
Specify third source point or <continue>: ↵
Scale objects based on alignment points? [Yes/No] <N>: Y↵

The 90° bend is aligned and uniformly scaled to meet the existing ductwork object. See **Figure 10-7B**. You can also align using three source and three destination points. However, when doing so, you cannot scale the object.

PROFESSIONAL TIP

Before using 3D editing commands, set running object snaps to enhance your accuracy and speed.

Figure 10-7.
Using the **ALIGN** command. A—Two source points and two destination points are required. Notice how the bend is not at the proper scale. B—You can choose to scale the object during the operation. Notice how the aligned bend is also properly scaled.

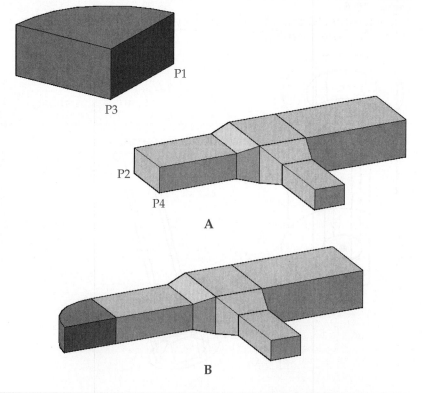

A

B

AutoCAD and Its Applications—Advanced

Complete the exercise on the companion website.
www.g-wlearning.com/CAD

3D Moving

The **3DMOVE** command allows you to quickly move an object along any axis or plane of the current UCS. When the command is initiated, you are prompted to select the objects to move. If the 2D Wireframe visual style is current, the visual style is temporarily changed to the Wireframe visual style because the gizmo is not displayed in 2D mode. After selecting the objects, press [Enter]. The *move gizmo*, also called a *grip tool*, is displayed in the center of the selection set. By default, the move gizmo is also displayed when a solid is selected with no command active.

The move gizmo is a tripod that appears similar to the shaded UCS icon. See **Figure 10-8A**. You can relocate the tool by right-clicking on the gizmo and selecting **Relocate Gizmo** from the shortcut menu. Then, move the gizmo to a new location and pick. You can also realign the gizmo using the shortcut menu.

If you move the pointer over the X, Y, or Z axis of the gizmo, the axis changes to yellow. To restrict movement along that axis, pick the axis. If you move the pointer over one of the right angles at the origin of the gizmo, the corresponding two axes turn yellow. Pick to restrict the movement to that plane. You can complete the movement by either picking a new point or by direct distance entry.

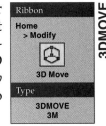

If the **GTAUTO** system variable is set to 1, the move gizmo is displayed when a solid is selected with no command active. If the **GTLOCATION** system variable is set to 0, the gizmo is placed on the UCS icon (not necessarily the UCS *origin*). Both variables are set to 1 by default.

3D Rotating

As you have seen in earlier chapters, the **ROTATE** command can be used to rotate 3D objects. However, the command can only rotate objects in the XY plane of the current UCS. This is why you had to change UCSs to properly rotate objects. The **3DROTATE** command, on the other hand, can rotate objects on any axis regardless of the current UCS. This is an extremely powerful editing and design tool.

When the command is initiated, you are prompted to select the object(s) to rotate. After selecting the objects, press [Enter]. The *rotate gizmo* is displayed in the center of the selection set. See **Figure 10-8B**. The gizmo provides you with a dynamic, graphic representation of the three axes of rotation. After selecting the objects, you must specify a location for the gizmo, which is the base point for rotation.

Now, you can use the gizmo to rotate the objects about the tool's local X, Y, or Z axis. As you hover the cursor over one of the three circles in the gizmo, a vector is displayed that represents the axis of rotation. To rotate about the tool's X axis, pick the red circle on the gizmo. To rotate about the Y axis, pick the green circle. To rotate about

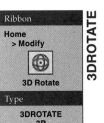

Figure 10-8.
A—The move gizmo is a tripod that appears similar to the shaded UCS icon. B—This is the rotate gizmo. The three axes of rotation are represented by the circles. The origin of the rotation is where you place the center grip.

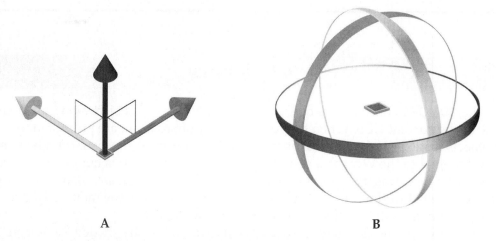

A   B

the Z axis, pick the blue circle. Once you select a circle, it turns yellow and you are prompted for the start point of the rotation angle. You can enter a direct angle at this prompt or pick the first of two points defining the angle of rotation. When the rotation angle is defined, the object is rotated about the selected axis.

The following example rotates the bend in the HVAC assembly shown in **Figure 10-9A**. Set the center of face 3D object snap. Then, select the command and continue:

> Current positive angle in UCS: ANGDIR=*(current)* ANGBASE=*(current)*
> Select objects: *(pick the bend)*
> 1 found
> Select objects: ↵
> Specify base point: *(acquire the center of the face and pick)*
> Pick a rotation axis: *(pick the green circle)*
> Specify angle start point or type an angle: **180.**↵

Note that the rotate gizmo remains visible through the base point and the angle of rotation selections. The rotated object is shown in **Figure 10-9B**.

If you need to rotate an object on an axis that is not parallel to the current X, Y, or Z axes, use a dynamic UCS with the **3DROTATE** command. Chapter 6 discussed the benefits of using a dynamic UCS when creating objects that need to be parallel to a surface other than the XY plane. With the object selected for rotation and the dynamic UCS option active (pick the **Allow/Disallow Dynamic UCS** button on the status bar), right-click and select **Relocate Gizmo**. Then, move the rotate gizmo over a face of the object. The gizmo aligns itself with the surface so that the Z axis is perpendicular to the face. Carefully place the gizmo over the point of rotation using object snaps. Make sure the tool is correctly positioned before picking to locate it. Then, enter an angle or use polar tracking to rotate the object about the appropriate axis on the gizmo.

PROFESSIONAL TIP

By default, the move gizmo is displayed when an object is selected with no command active. To toggle between the move gizmo and the rotate gizmo, select the grip at the tool's origin and press the spacebar. Then, pick a location for the tool's origin. You can toggle back to the move gizmo using the same procedure.

Figure 10-9.
A—Use object tracking or object snaps to place the grip tool in the middle of the rectangular face. Then, select the axis of rotation. B—The completed rotation.

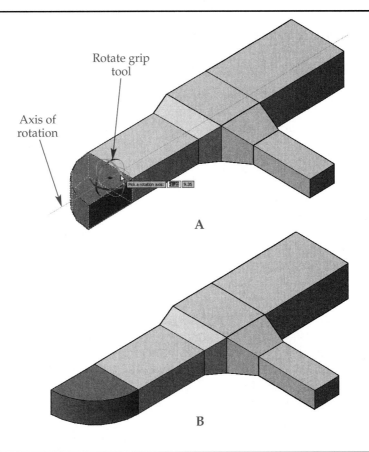

Rotate grip tool

Axis of rotation

Pick a rotation axis:

A

B

Exercise 10-3

Complete the exercise on the companion website.
www.g-wlearning.com/CAD

3D Mirroring

The **MIRROR** command can be used to rotate 3D objects. However, like the **ROTATE** command, the **MIRROR** command can only work in the XY plane of the current UCS. Often, to properly mirror objects with this command, you have to change UCSs. The **MIRROR3D** command, on the other hand, allows you to mirror objects about any plane regardless of the current UCS.

The default option of the command is to define a mirror plane by picking three points on that plane, **Figure 10-10A**. Object snaps should be used to accurately define the mirror plane. To mirror the wedge in **Figure 10-10A**, set the midpoint object snap, select the command, and use the following sequence. The resulting drawing is shown in **Figure 10-10B**.

Ribbon

Home
> Modify

3D Mirror

Type

MIRROR3D

MIRROR3D

> Select objects: *(pick the wedge)*
> 1 found
> Select objects: ↵
> Specify first point of mirror plane (3 points) or
> [Object/Last/Zaxis/View/XY/YZ/ZX/3points] <3points>: *(pick P1, which is the midpoint of the box's top edge)*
> Specify second point on mirror plane: *(pick P2)*
> Specify third point on mirror plane: *(pick P3)*
> Delete source objects? [Yes/No] <N>: ↵

Figure 10-10.
The **MIRROR3D** command allows you to mirror objects about any plane regardless of the current UCS. A—The mirror plane defined by the three pick points is shown here in color. Point P1 is the midpoint of the top edge of the base. B—A copy of the original is mirrored.

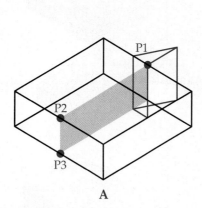

A

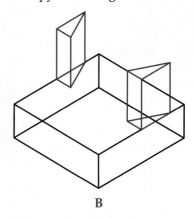

B

There are several different ways to define a mirror plane with the **MIRROR3D** command. These are:

- **Object.** The plane of the selected circle, arc, or 2D polyline segment is used as the mirror plane.
- **Last.** Uses the last mirror plane defined.
- **Zaxis.** Defines the plane with a pick point on the mirror plane and a point on the Z axis of the mirror plane.
- **View.** The viewing direction of the current viewpoint is aligned with a selected point to define the plane.
- **XY, YZ, ZX.** The mirror plane is placed parallel to one of the three basic planes of the current UCS and passes through a selected point.
- **3points.** Allows you to pick three points to define the mirror plane, as shown in **Figure 10-10.**

Exercise 10-4

Complete the exercise on the companion website.
www.g-wlearning.com/CAD

Creating 3D Arrays

An *array* is an arrangement of objects in a 2D or 3D pattern. An array can be created as a rectangular, polar, or path array. You probably used arrays to complete some of the problems in previous chapters. A 2D array is created on the XY plane of the current UCS. A 3D array is an arrangement of objects in 3D space. The **ARRAYRECT, ARRAYPOLAR,** and **ARRAYPATH** commands can be used to create both 2D and 3D arrays. These commands provide the same functions as the **Rectangular, Polar,** and **Path** options of the **ARRAY** command and are available in the drop-down menu located in the **Modify** panel of the **Home** tab on the ribbon. See **Figure 10-11.**

An array can be created as an associative or non-associative array. Creating an associative array creates an *array object*, which can be modified as a single entity.

AutoCAD and Its Applications—Advanced

Figure 10-11.
The **ARRAYRECT**,
ARRAYPOLAR,
and **ARRAYPATH**
commands can be
accessed from the
drop-down menu
located in the **Modify**
panel of the **Home**
tab on the ribbon.

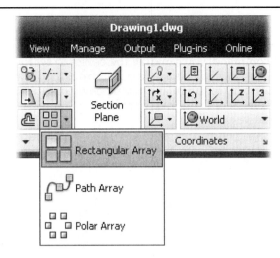

For example, you can edit the source object to change all of the items in the array at once. You can also perform other modifications, such as deleting one or more items in the array, while maintaining the associativity of the arrayed items.

When creating a 3D array, the information you specify depends on the type of array being created. Many of the options are similar to those used when creating a 2D array. The following sections discuss 3D rectangular, polar, and path arrays.

The legacy **3DARRAY** command can also be used to create 3D arrays, but it is limited to creating rectangular and polar arrays and cannot create associative arrays. Using the **ARRAYRECT**, **ARRAYPOLAR**, or **ARRAYPATH** command is the preferred method to create a 3D array.

3D Rectangular Arrays

In a *3D rectangular array*, as with a 2D rectangular array, you must enter the number of rows and columns. However, you must also specify the number of *levels*, which represents the third (Z) dimension. The command sequence is similar to that used when creating a 2D array.

An example of where a 3D rectangular array may be created is the layout of structural columns on multiple floors of a commercial building. In **Figure 10-12A**, you can see two concrete floor slabs of a building and a single structural column. It is now a simple matter of arraying the column in rows, columns, and levels.

To draw a 3D rectangular array, select the **ARRAYRECT** command. Select the object to array and press [Enter]. Then, specify the number of rows and columns. As when creating a 2D array, you can drag the cursor and pick to set the number of rows, number of columns, and item spacing dynamically. As you drag the cursor, the array will populate based on your cursor movement, **Figure 10-12B**. Without exiting the command, you can then adjust the row, column, and spacing values if needed using the appropriate command options.

In **Figure 10-12C**, there are three rows, five columns, and two levels. Drag the cursor and pick points to locate an initial arrangement for the array. Then, continue as follows. Note that for the following command sequence, the drawing units have been set to architectural.

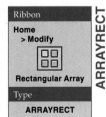

Figure 10-12.
A—Two floors and one structural column are drawn. Creating a 3D rectangular array will place all of the required columns on both floors at the same time. B—After entering the **ARRAYRECT** command, you can drag the cursor to set the number of rows and columns dynamically. The floor objects are removed from the view for illustration purposes only. C—An associative 3D rectangular array made up of three rows and five columns on two levels. D—An arrangement of three rows and seven columns after using the **Properties** palette to edit the associative array.

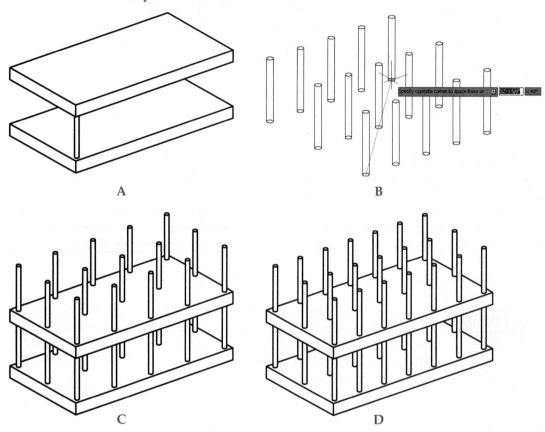

A
B
C
D

Press Enter to accept or [ASsociative/Base point/Rows/Columns/Levels/eXit]
    <eXit>: **R**↵
Enter the number of rows or [Expression] <*current*>: **3**↵
Specify the distance between rows or [Total/Expression] <*current*>: **10'**↵
Specify the incrementing elevation between rows or [Expression] <0">: ↵
Press Enter to accept or [ASsociative/Base point/Rows/Columns/Levels/eXit]
    <eXit>: **C**↵
Enter the number of columns or [Expression] <*current*>: **5**↵
Specify the distance between columns or [Total/Expression] <*current*>: **10'**↵
Press Enter to accept or [ASsociative/Base point/Rows/Columns/Levels/eXit]
    <eXit>: **L**↵
Enter the number of levels or [Expression] <1>: **2**↵
Specify the distance between levels or [Total/Expression] <*current*>: **12'8"**↵
Press Enter to accept or [ASsociative/Base point/Rows/Columns/Levels/eXit]
    <eXit>: ↵

The result is shown in Figure 10-12C. By default, an associative array is created. You can create a non-associative array by selecting the **Associative** option.

As shown in the previous sequence, when setting the distance between rows, you can also define an elevation increment between rows when the Specify the incrementing elevation between rows or [Expression] <0">: prompt appears. The elevation increment is different from the distance between levels. The elevation increment sets the spacing between rows along the Z axis so that each successive row is drawn on a higher or lower plane. This option can also be used when creating a 3D polar array and is addressed in the next section.

The **Base point** option is used to define the base point for the array. By default, the base point defined by AutoCAD is the centroid of the object(s) selected. You may want to select a more logical base point, such as an endpoint of an edge or the center point of a circular face. In **Figure 10-12B**, the center of the cylindrical base of the column has been selected as the base point of the array.

If any of the properties of an associative array require changes, use the **Properties** palette to edit the array. The arrayed items will update based on the changes. In **Figure 10-12D**, the array is shown after changing the number of columns to 7 and the column spacing to 6'-8".

 Grips are available for editing an associative array after the array is created. The grip options allow you to change the row and column spacing and other array parameters. Editing with grips is discussed in Chapter 12.

 ### Exercise 10-5

Complete the exercise on the companion website.
www.g-wlearning.com/CAD

3D Polar Arrays

A *3D polar array* is similar to a 2D polar array. However, the axis of rotation in a 2D polar array is parallel to the Z axis of the current UCS. In a 3D polar array, you can define a centerline axis of rotation that is not parallel to the Z axis of the current UCS. In other words, you can array an object in a UCS different from the current one. In addition, as with a 3D rectangular array, you can array the object in multiple "levels" along the Z axis. The **ARRAYPOLAR** command can be used to create a 3D arrangement in rows, levels, or both rows and levels.

To create a 3D polar array, select the **ARRAYPOLAR** command. Select the object to array and press [Enter]. Then, pick the center point of the array or use the **Axis of rotation** option to select a centerline axis. The **Axis of rotation** option allows you to select a center-line axis that is different from the Z axis of the current UCS. Using this option requires you to pick two points to define the axis. In **Figure 10-13A**, the leg attached to the hub must be arrayed about the center axis. The center axis is drawn as a construction line. Use the **Axis of rotation** option to pick the two endpoints of the axis. Once you define the axis, items in the array generate dynamically. You can then use the command options to adjust the array.

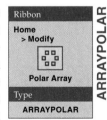

```
Specify center point of array or [Base point/Axis of rotation]: A↵
Specify first point on axis of rotation: (pick one endpoint of the axis)
Specify second point on axis of rotation: (pick the other endpoint of the axis)
Enter number of items or [Angle between/Expression] <4>: 6↵
Specify the angle to fill (+=ccw, -=cw) or [EXpression] <360>: ↵
Press Enter to accept or [ASsociative/Base point/Items/Angle between/Fill angle/
    ROWs/Levels/ROTate items/eXit] <eXit>: AS↵
Create associative array [Yes/No] <Yes>: N↵
Press Enter to accept or [ASsociative/Base point/Items/Angle between/Fill angle/
    ROWs/Levels/ROTate items/eXit] <eXit>: ↵
```

Figure 10-13.
A— Six new legs need to be arrayed about the center axis of the existing part. B—The model after using the **ARRAYPOLAR** command.

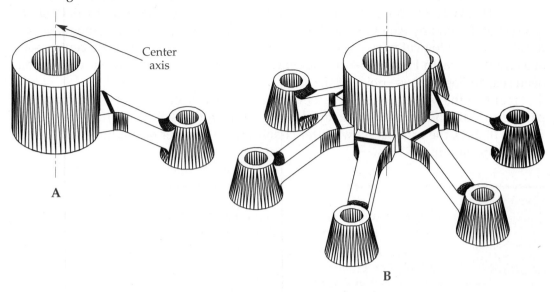

The result is shown in **Figure 10-13B**. In this case, a non-associative array is created. The legs are arrayed as individual objects and the resulting arrayed objects can be selected individually. The **Associative** option setting is maintained by AutoCAD the next time an array command is accessed. You can verify the current setting of this option after entering the command and selecting objects. The Associative = Yes or Associative = No prompt appears. If the Associative = No prompt appears and you want to create an associative array, select the **Associative** option.

The **Rows** option is used to create multiple rows of arrayed objects. After selecting this option, enter the number of rows and the distance between rows. Then, set an elevation increment value to control the spacing along the Z axis between each successive row. An array of seats in a theater can be created in this manner. **Figure 10-14** shows an example of arraying a single seat to create multiple rows of seats. To create this array, select the **ARRAYPOLAR** command, select the first seat, and use the following command sequence. If needed, use the **Associative** option to create an associative array. For this example, the default **Center point** option is used to set the center point of the array. Using this option is sufficient because the Z axis of the current UCS is parallel to the required axis of rotation.

Specify center point of array or [Base point/Axis of rotation]: *(Use the* **Center** *object snap to pick the center point of the top arc of the first platform)*
Enter number of items or [Angle between/Expression] <4>: **8**↵
Specify the angle to fill (+=ccw, −=cw) or [EXpression] <360>: **80**↵
Press Enter to accept or [ASsociative/Base point/Items/Angle between/Fill angle/ ROWs/Levels/ROTate items/eXit] <eXit>: **ROWS**↵
Enter the number of rows or [Expression] <1>: **6**↵
Specify the distance between rows or [Total/Expression] <*current*>: **96**↵
Specify the incrementing elevation between rows or [Expression] <0.0000>: **7**↵
Press Enter to accept or [ASsociative/Base point/Items/Angle between/Fill angle/ ROWs/Levels/ROTate items/eXit] <eXit>: ↵

The result is shown in **Figure 10-14B**. Notice that each successive row of seats is situated on a higher plane. In more complex models, the **Levels** option can be used to create multiple levels of rows.

AutoCAD and Its Applications—Advanced

Figure 10-14.
Using the **Rows** option of the **ARRAYPOLAR** command to create a polar array of seats arranged in multiple rows. A—The first seat is modeled on the bottom platform. B—The result after using the **ARRAYPOLAR** command.

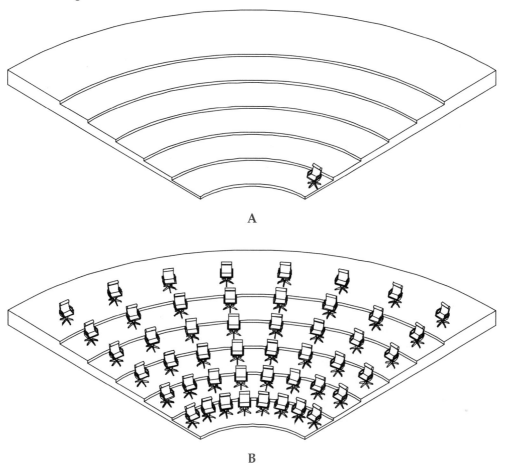

A

B

The **Expression** option allows you to enter a mathematical expression to calculate one of the array parameters. For example, you can enter an expression to calculate the number of items, rows, or levels. The **Expression** option can also be used to define the spacing between rows and the incrementing elevation.

Exercise 10-6

Complete the exercise on the companion website.
www.g-wlearning.com/CAD

3D Path Arrays

A *3D path array* is similar to a 2D path array. Objects can be arrayed along a path or a segment of a path. The path, also called a *path curve*, can be a line, circle, arc, ellipse, spline, polyline, helix, or 3D polyline. See **Figure 10-15**. As with a 3D polar array, you can create a 3D arrangement in rows, levels, or both rows and levels. A 3D path array can be created as an associative or non-associative array.

Figure 10-15.
Examples of 3D path arrays used to array posts.

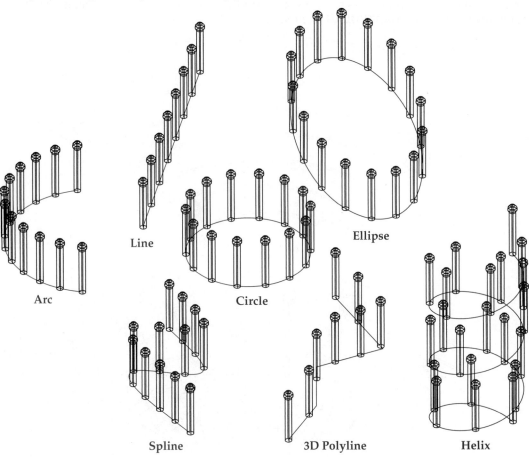

Line

Ellipse

Arc

Circle

Spline

3D Polyline

Helix

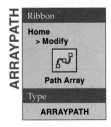

To create a 3D path array, select the **ARRAYPATH** command. Select the object to array and press [Enter]. You are prompted to select the path curve. The object to be arrayed does not have to intersect the path curve. Once you select the path curve, items in the array generate dynamically as you drag the cursor. As an option, you can select the **Orientation** option to set the array's base point, orientation, or both. The default base point of the array is the endpoint of the path curve closest to where you select it. This point serves as the start point of the array. Depending on the result you want, you can select a different base point (start point), such as a point on the object. The default orientation of the array is the current orientation of the object. The **2 Points** option can be used to pick two points to define a different orientation. The **Normal** option can be used to align the object "normal" to the path. Using this option aligns the Z axis of the object perpendicular to the path.

After specifying the base point and orientation of the array or using the defaults, continue as follows. You can use the command options to adjust the number of items to be arrayed and the distance between items. As with a 2D path array, you can distribute the object along the path evenly by using the **Divide** option. You can also specify the total distance between the first and last objects, or the distance between each object. In **Figure 10-16A**, the **Divide** option is used:

> Enter number of items along path or [Orientation/Expression] <Orientation>: **10**↵
> Specify the distance between items along path or [Divide/Total/Expression] <Divide evenly along path>: **D**↵
> Press Enter to accept or [ASsociative/Base point/Items/Rows/Levels/Align items/Z direction/eXit] <eXit>: ↵

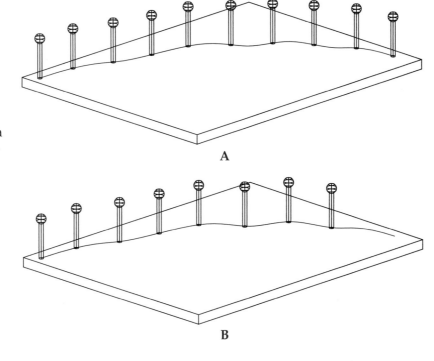

Figure 10-16.
Using the
ARRAYPATH
command.
A—Specifying
10 items to array
and using the
Divide option.
B—Specifying 8
items to array and a
distance of 25 units
between objects.

A

B

In **Figure 10-16B**, the number of items and the distance between items are specified:

Enter number of items along path or [Orientation/Expression] <Orientation>: **8**↵
Specify the distance between items along path or [Divide/Total/Expression] <Divide evenly along path>: **25**↵
Press Enter to accept or [ASsociative/Base point/Items/Rows/Levels/Align items/Z direction/eXit] <eXit>: ↵

The **Align items** option is used to align the arrayed objects tangent to the direction of the array path. The **Z direction** option is used to change the Z axis direction of the arrayed objects. By default, the Z axis of each object is aligned in the same orientation used by the original object. If the **Z direction** option is set to **No**, the Z axis of the object changes direction to follow the path as the path changes direction.

The **Rows** and **Levels** options allow you to create an arrangement in rows and levels. The options are similar to those used with the **ARRAYPOLAR** command. In **Figure 10-17**, an oval table is created using the **Levels** option. The table feet and middle supports are created as a 3D path array. One of the feet is arrayed along an elliptical path to create the first level of the array, and the second level is created during the same command sequence. In this example, a non-associative array is created. This allows the middle supports to be edited as individual objects after creating the array. The middle supports have a smaller diameter than the feet and a longer length. To complete the table, the bottom shelf is copied along the Z axis to create the top.

In **Figure 10-17A**, the bottom shelf has been created as an extruded solid from an ellipse. The shelf is .375″ thick and is centered on the UCS origin. The underside of the shelf rests on the XY plane. A second ellipse is drawn on the XY plane and rests on the underside of the shelf. This ellipse serves as the path for the 3D path array. A construction cylinder has also been created with its base resting on the XY plane. The construction cylinder is 2″ in diameter and 2″ in height, the dimensions of the table feet. The construction cylinder is the item to be arrayed. During the operation, the base point of the array is specified as the top center point of the construction cylinder. This aligns the top of the cylinder with the start of the path. Since the path is an ellipse, AutoCAD defines the start of the path as one of the quadrant points on the major axis. When the base point is selected, the cylinder is then aligned with the quadrant point. This is a sufficient start point for the array.

Figure 10-17.
Using the **ARRAYPATH** command to create the cylindrical feet and supports for an oval table. A—The bottom shelf is created as an extruded ellipse. A construction ellipse is created to serve as the path. A construction cylinder is the object to be arrayed. B—A 3D path array consisting of two levels with four items on each level is created. The array is created as a non-associative array. C—The supports are edited to the correct diameter and height. D—The bottom shelf is copied along the Z axis to create the top. Shown is a shaded version of the completed model.

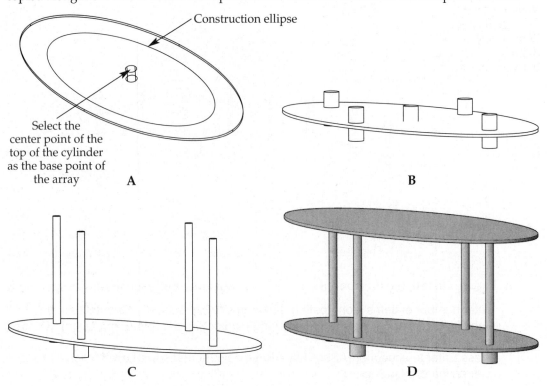

After selecting the **ARRAYPATH** command, the construction cylinder is selected as the item to array and the construction ellipse is selected as the path. The command sequence continues as follows.

> Enter number of items along path or [Orientation/Expression] <Orientation>: **O**↵
> Specify base point or [Key point] <end of path curve>: *(Use the **Center** object snap to pick the center of the top of the cylinder)*
> Specify direction to align with path or [2Points/NORmal] <current>: ↵
> Enter number of items along path or [Expression] <*current*>: **4**↵
> Specify the distance between items along path or [Divide/Total/Expression] <Divide evenly along path>: **D**↵
> Press Enter to accept or [ASsociative/Base point/Items/Rows/Levels/Align items/Z direction/eXit] <eXit>: **AS**↵
> Create associative array [Yes/No] <*current*>: **N**↵
> Press Enter to accept or [ASsociative/Base point/Items/Rows/Levels/Align items/Z direction/eXit] <eXit>: **L**↵
> Enter the number of levels or [Expression] <1>: **2**↵
> Specify the distance between levels or [Total/Expression] <*current*>: **2-3/8**↵
> Press Enter to accept or [ASsociative/Base point/Items/Rows/Levels/Align items/Z direction/eXit] <eXit>: ↵

The resulting array is shown in **Figure 10-17B**. The four cylinders on each level of the array are evenly divided along the elliptical path. Because a non-associative array was created, the cylinders can be selected and edited as independent objects. Notice that the four cylinders in the second level of the array rest on the top of the shelf at the correct position along the Z axis. Also, notice the appearance of the construction cylinder in the center. The construction cylinder rests on the XY plane, not on the top of the shelf. In **Figure 10-17C**, the arrayed cylinders have been changed to 1″ in diameter and 14″ in height using the **Properties** palette. In addition, the construction cylinder has been placed on a frozen layer. A shaded version of the completed table is shown in **Figure 10-17D** after copying the shelf along the Z axis to create the top.

PROFESSIONAL TIP

If you edit the path shape used for an associative 3D path array, the objects follow the new edited path.

Exercise 10-7

Complete the exercise on the companion website.
www.g-wlearning.com/CAD

Filleting Solid Objects

A *fillet* is a rounded interior edge on an object, such as a box. A *round* is a rounded exterior edge. The **FILLET** command is used to create both fillets and rounds in 2D and 3D. Additionally, the **FILLETEDGE** command can be used to fillet 3D objects.

Before a fillet or round is created at an intersection, the solid objects that intersect need to be joined using the **UNION** command. Then, use the **FILLET** command. See **Figure 10-18**. Since the object being filleted is actually a single solid and not two objects, only one edge is selected. In the following sequence, the fillet radius is first set at .25, then the fillet is created.

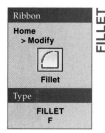

Ribbon

Home
> Modify

Fillet

Type

FILLET
F

Current settings: Mode = *current*, Radius = *current*
Select first object or [Undo/Polyline/Radius/Trim/Multiple]: **R**↵
Specify fillet radius <*current*>: **.25**↵
Select first object or [Undo/Polyline/Radius/Trim/Multiple]: (*pick the edge to be filleted or rounded*)
Enter fillet radius or [Expression] <0.2500>: ↵
Select an edge or [Chain/Loop/Radius]: ↵ (*this fillets the selected edge, but you can also select other edges at this point*)
1 edge(s) selected for fillet.

Examples of fillets and rounds are shown in **Figure 10-19**.

Figure 10-18.
A—Pick the edge where two unioned solids intersect to create a fillet. B—The fillet after rendering.

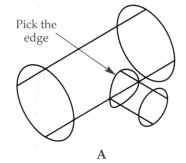

Pick the edge

A

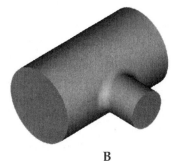

B

Figure 10-19.
Examples of fillets
and rounds. The
wireframe displays
show the objects
before the **FILLET**
command is used.

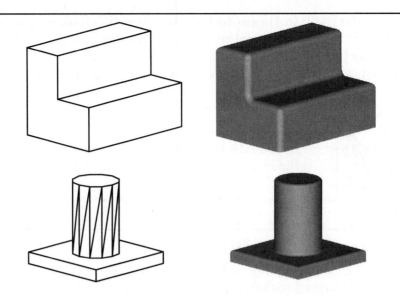

Ribbon
Solid
> Solid Editing

Fillet Edge

Type
FILLETEDGE

The **FILLETEDGE** command works in a similar manner. Once the command is entered, select the edges to fillet. You can continue to select edges or enter the **Chain**, **Loop**, or **Radius** option. The **Chain** option is used to select a chain of continuous edges that have rounded corners, **Figure 10-20A**. The **Loop** option is similar to the **Chain** option and is used to select a loop of edges, **Figure 10-20B**. When using the **Loop** option, the **Next** option can be used to select the adjacent loop of edges. After using the **Chain** or **Loop** option, you can select individual edges by entering the **Edge** option. Once all edges are selected, press [Enter]. You are prompted to either accept the fillet or enter a radius. If the current radius is acceptable, press [Enter]. If not, enter the **Radius** option and set the new value. The advantages of using this command are 1) a preview is shown and 2) a linear stretch grip is associated with the fillet. The linear stretch grip allows for subobject editing, which is discussed in Chapter 12.

Figure 10-20.
Using the **FILLETEDGE** command. A—The **Chain** option is used to select a chain of continuous edges with rounded corners to fillet. B—The **Loop** option is used to select a loop of edges to fillet.

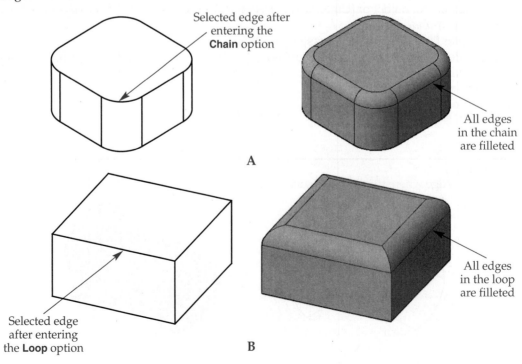

Selected edge after entering the **Chain** option

All edges in the chain are filleted

A

Selected edge after entering the **Loop** option

All edges in the loop are filleted

B

You can construct and edit solid models while the object is displayed in a shaded view. If your computer has sufficient speed and power, it is often much easier to visualize the model in a 3D view with a shaded visual style set current. This allows you to realistically view the model. If an edit or construction does not look right, just undo and try again.

Chamfering Solid Objects

A *chamfer* is a small square edge on the edges of an object. The **CHAMFER** command can be used to create a chamfer on a 2D or 3D object. Just as when chamfering a 2D line, there are two chamfer distances. Therefore, you must specify which surfaces correspond to the first and second distances. The feature to which the chamfer is applied must be constructed before chamfering. For example, if you are chamfering a hole, the object (cylinder) must first be subtracted to create the hole. If you are chamfering an intersection, the two objects must first be unioned.

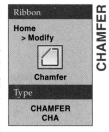

After you enter the command, you must pick the edge you want to chamfer. The edge is actually the intersection of two surfaces of the solid. One of the two surfaces is highlighted when you select the edge. The highlighted surface is associated with the first chamfer distance. This surface is called the *base surface*. If the highlighted surface is not the one you want as the base surface, enter N at the [Next/OK] prompt and press [Enter]. This highlights the next surface. An edge is created by two surfaces. Therefore, when you enter N for the next surface, AutoCAD cycles through only two surfaces. When the proper base surface is highlighted, press [Enter].

The **CHAMFEREDGE** command works in a similar manner. Once the command is entered, select an edge to chamfer. Then set the distance of the chamfer. When you select an edge with a distance setting, a preview of the chamfer will appear. You can continue to select edges or enter the **Loop** option. The **Loop** option is similar to the **Loop** option used with the **FILLETEDGE** command and allows you to select a loop of edges. Once all edges are selected, press [Enter]. You are prompted to accept the chamfer. If the current chamfer distances are acceptable, press [Enter]. If not, enter the **Distance** option and set new distance values. The advantages of using this command are 1) a preview is shown and 2) a linear stretch grip is associated with the chamfer. The linear stretch grip allows for subobject editing, which is discussed in Chapter 12.

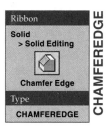

Using the **CHAMFER** command to chamfer a hole is shown in **Figure 10-21A**. In **Figure 10-21B**, the cylinder is unioned to the base in order to create the chamfer at the intersection. The ends of the cylinder in **Figure 10-21B** are chamfered by first picking a vertical isoline on the cylindrical face to define the base surface. Then, the top edge is selected, followed by the intersection edge. The following command sequence is illustrated in **Figure 10-21A**.

```
(TRIM mode) Current chamfer Dist1 = 1.0000, Dist2 = 1.0000
Select first line or [Undo/Polyline/Distance/Angle/Trim/mEthod/Multiple]: (select a
    top edge)
Base surface selection...
Enter surface selection option [Next/OK (current)] <OK>: (select Next if the top
    surface is not selected or press [Enter])
Specify base surface chamfer distance or [Expression] <1.0000>: .125↵
Specify other surface chamfer distance or [Expression] <1.0000>: .125↵
Select an edge or [Loop]: (select the hole diameter edge)
Select an edge or [Loop]: ↵
```

Figure 10-21.
A—A hole is chamfered by picking the top surface, then the edge of the hole. B—The top edge of the cylinder is chamfered by first picking the side, then the top edge. Both edges can be chamfered at the same time, as shown here.

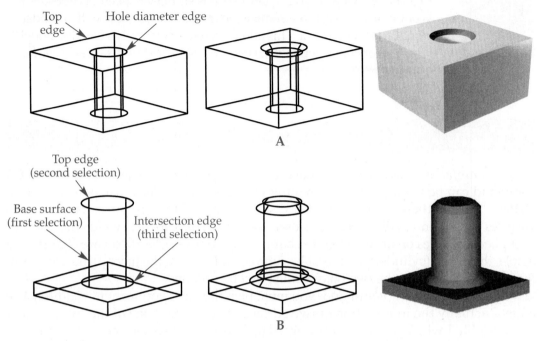

Top edge
Hole diameter edge

A

Top edge
(second selection)

Base surface
(first selection)

Intersection edge
(third selection)

B

Editing a fillet or chamfer is discussed in Chapter 12. The **SOLIDEDIT** command discussed in Chapter 13 can also be used to edit a fillet or chamfer.

Exercise 10-8

Complete the exercise on the companion website.
www.g-wlearning.com/CAD

Converting to Solids

CONVTOSOLID

| Ribbon |
| Home |
| > Solid Editing |
| Convert to Solid |
| Type |
| CONVTOSOLID |

Additional flexibility in creating solids is provided by the **CONVTOSOLID** command. This command allows you to directly convert certain closed objects into solids. You can convert:

- Circles with thickness.
- Wide, uniform-width polylines with thickness. This includes polygons and rectangles.
- Closed, zero-width polylines with thickness. This includes polygons, rectangles, and closed revision clouds.
- Mesh primitives and other watertight mesh objects. Keep in mind that the smoothness level applied to the mesh primitive appears on the object when it is converted to a solid.
- Watertight surface models.

AutoCAD and Its Applications—Advanced

First, select the command. Then, select the objects to convert and press [Enter]. The objects are instantly converted with no additional input required. **Figure 10-22** shows the three different objects before and after conversion to a solid.

If an object that appears to be a closed polyline with a thickness does not convert to a solid and the command line displays the message Cannot convert an open curve, the polyline was not closed using the **Close** option of the **PLINE** command. Use the **PEDIT** or **PROPERTIES** command to close the polyline and use the **CONVTOSOLID** command again.

PROFESSIONAL TIP

To quickly create a straight section of pipe, draw a donut with the correct ID and OD of the pipe. Then, use the **Properties** palette to give the donut a thickness equal to the length of the section you are creating. Finally, use the **CONVTOSOLID** command to turn the donut into a solid.

Slicing a Solid

A 3D solid can be sliced at any location by using existing objects such as circles, arcs, ellipses, 2D polylines, 2D splines, or surfaces. Why slice a solid? To create complex, angular, contoured, or organic shapes that traditionally cannot be created by just using solids. Good modeling techniques incorporate slicing of solids. For example, think of a computer mouse. This would be difficult to create with solids. But, you can create a basic solid shape, then slice the solid using contoured surfaces as slicing tools. After slicing the solid, you can choose to retain either or both sides of the model. The slices can then be used for model construction or display and presentation purposes.

The **SLICE** command is used to slice solids. When the command is initiated, you are asked to select the solids to be sliced. Select the objects and press [Enter]. Next,

Ribbon
Solid
> Solid Editing
Slice
Type
SLICE

SLICE

Figure 10-22.
A—From left to right, two polylines and an edited mesh sphere primitive that will be converted into solids. B—The resulting solids.

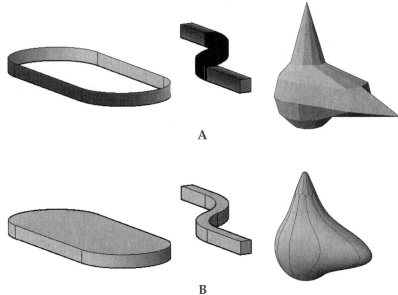

A

B

you must define the slicing path. The default method of defining a path requires you to specify two points on a slicing plane. The plane passes through the two points and is perpendicular to the XY plane of the current UCS. Refer to **Figure 10-23** as you follow this sequence:

1. Select the command and pick the object to be sliced.
2. Pick the start point of the slicing plane. See **Figure 10-23A**.
3. Pick the second point on the slicing plane.
4. You are prompted to specify a point on the desired side to keep. Select anywhere on the back half of the object. The point does not have to be *on* the object. It must simply be on the side of the cutting plane that you want to keep.
5. The object is sliced and the front half is deleted. See **Figure 10-23B**.

When prompted to select the side to keep, you can press [Enter] to keep both sides. If both sides are retained, two separate 3D solids are created. Each solid can then be used for additional modeling purposes.

There are several additional options for specifying a slicing path. These options are listed here and described in the following sections.

- **Planar object**
- **Surface**
- **Zaxis**
- **View**
- **XY**
- **YZ**
- **ZX**
- **3points**

Once the **SLICE** command has been used, the history of the solid to that point is removed. If a history of the work is important, then save a copy of the file or place a copy of the object on a frozen layer prior to performing the slice.

Figure 10-23.
Slicing a solid by picking two points. A—Select two points on the cutting plane. The plane passes through these points and is perpendicular to the XY plane of the current UCS. B—The sliced solid.

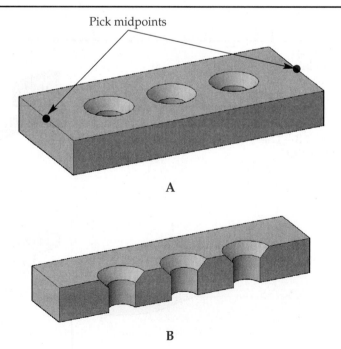

Pick midpoints

A

B

Planar Object

A second method to create a slice through a 3D solid is to use an existing planar object. Planar objects include circles, arcs, ellipses, 2D polylines, and 2D splines. See **Figure 10-24A**. The plane on which the planar object lies must intersect the object to be sliced. The current UCS has no effect on this option.

Be sure that the planar object has been moved to the location of the slice. Then, select the **SLICE** command, pick the object to slice, and press [Enter]. Next, enter the **Planar Object** option and select the slicing path object (the circle, in this case). Finally, specify which side is to be retained. See **Figure 10-24B**. Again, if both sides are kept, they are separate objects and can be individually manipulated.

Surface

A surface object can be used as the slicing path. The surface can be planar or nonplanar (curved). Surface modeling is discussed in detail in Chapter 11. Using a surface as a slicing path is a technique that you can use to quickly create a mating die. For example, refer to **Figure 10-25**. First, draw the required surface. The surface should exactly match the stamped part that will be manufactured, **Figure 10-25A**. Then, draw a box that encompasses the surface. Next, select the **SLICE** command, pick the box, and press [Enter]. Then, enter the **Surface** option and select the surface. You may need to turn on selection cycling or work in a wireframe display. Finally, when prompted to select the side to keep, press [Enter] to keep both sides. The two halves of the die can now be moved and rotated as needed, **Figure 10-25B**.

Figure 10-24.
Slicing a solid with a planar object.
A—The circle is drawn at the proper orientation and in the correct location.
B—The completed slice.

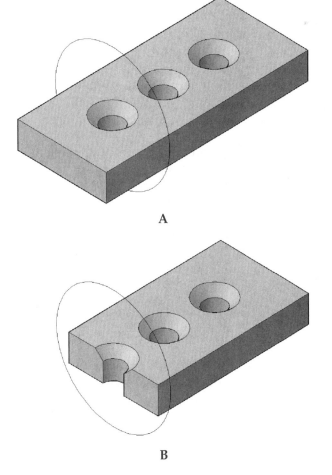

A

B

Figure 10-25.
Slicing a solid with a surface. A—Draw the surface and locate it within the solid to be sliced. The solid is represented here by the wireframe. B—The completed slice with both sides retained. The top can now be moved and rotated as shown here.

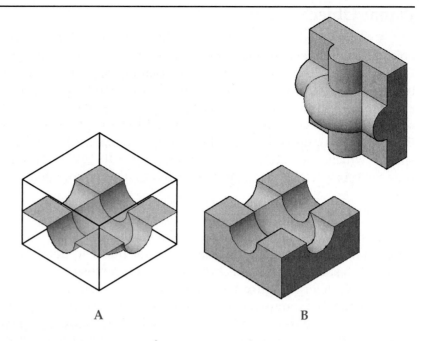

A   B

Exercise 10-9

Complete the exercise on the companion website.
www.g-wlearning.com/CAD

Z Axis

You can specify one point on the cutting plane and one point on the Z axis of the plane. See **Figure 10-26.** This allows you to have a cutting plane that is not parallel to the current UCS XY plane. First, select the **SLICE** command, pick the object to slice, and press [Enter]. Next, enter the **Zaxis** option. Then, pick a point on the XY plane of the cutting plane followed by a point on the Z axis of the cutting plane. Finally, pick the side of the object to keep.

View

A cutting plane can be established that is aligned with the viewing plane of the current viewport. The cutting plane passes through a point you select, which sets the depth along the Z axis of the current viewing plane. First, select the **SLICE** command, pick the object to slice, and press [Enter]. Next, enter the **View** option. Then, pick a point in the viewport to define the location of the cutting plane on the Z axis of the viewing plane. Use object snaps to select a point on an object. The cutting plane passes through this point and is parallel to the viewing plane. Finally, pick the side of the object to keep.

Figure 10-26.
Slicing a solid using the **Zaxis** option.
A—Pick one point on the cutting plane and a second point on the Z axis of the cutting plane.
B—The resulting slice.

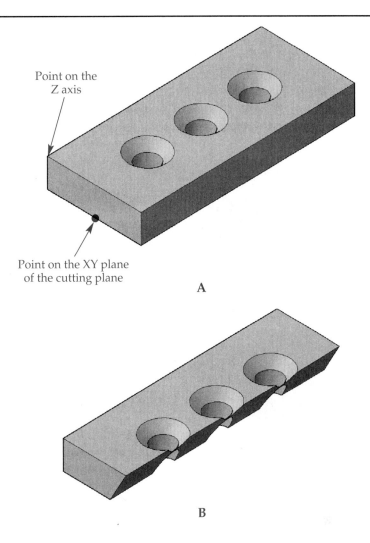

Point on the Z axis

Point on the XY plane of the cutting plane

A

B

XY, YZ, and ZX

You can slice an object using a cutting plane that is parallel to any of the three primary planes of the current UCS. See **Figure 10-27**. The cutting plane passes through the point you select and is aligned with the primary plane of the current UCS that you specify. First, select the **SLICE** command, pick the object to slice, and press [Enter]. Next, enter the **XY**, **YZ**, or **ZX** option, depending on the primary plane to which the cutting plane will be parallel. Then, pick a point on the cutting plane. Finally, pick the side of the object to keep.

Three Points

Three points can be used to define the cutting plane. This allows the cutting plane to be aligned at any angle, similar to using the **Zaxis** option. See **Figure 10-28**. First, select the **SLICE** command, pick the object to be sliced, and press [Enter]. Then, enter the **3points** option. Pick three points on the cutting plane and then select the side of the object to keep.

Exercise 10-10

Complete the exercise on the companion website.
www.g-wlearning.com/CAD

Figure 10-27.
Slicing a solid using the **XY**, **YZ**, and **ZX** options. A—The object before slicing. The UCS origin is in the center of the first hole and at the midpoint of the height. B—The resulting slice using the **XY** option. C—The resulting slice using the **YZ** option. D—The resulting slice using the **ZX** option.

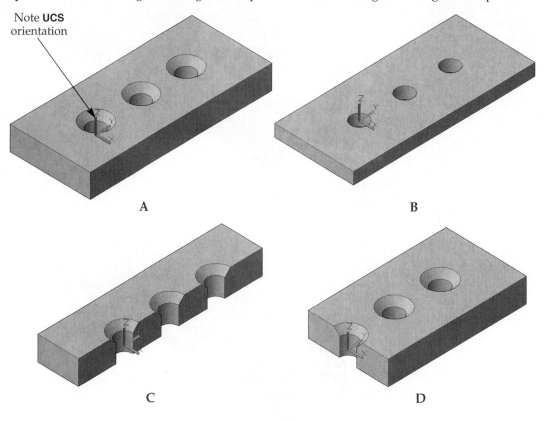

Note **UCS** orientation

A

B

C

D

Figure 10-28.
Slicing a solid using the **3points** option. A—Specify three points to define the cutting plane. B—The resulting slice.

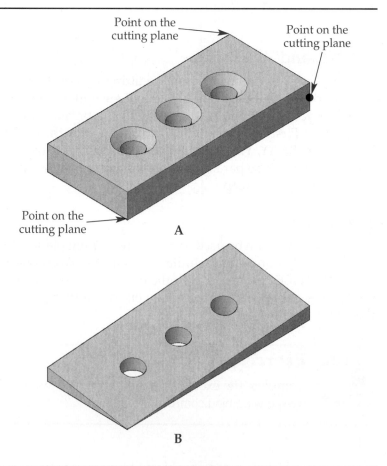

Point on the cutting plane

Point on the cutting plane

Point on the cutting plane

A

B

Removing Features

Sometimes, it may be necessary to remove a feature that has been constructed. For example, suppose you placed a R.5 fillet on an object based on an engineering sketch. Then, the design is changed to a R.25 fillet. Subobject editing is the best technique to accomplish this. Subobject editing is covered in Chapter 12.

Constructing Features on Solid Models

A variety of machining, structural, and architectural features can be created using some basic solid modeling techniques. The features discussed in the next sections are just a few of the possibilities.

Counterbore and Spotface

A *counterbore* is a recess machined into a part, centered on a hole, that allows the head of a fastener to rest below the surface. Create a counterbore as follows.
1. Draw a cylinder representing the diameter of the hole, **Figure 10-29A**.
2. Draw a second cylinder that is the diameter of the counterbore and center it at the top of the first cylinder. Move the second cylinder so it extends below the surface of the object to the depth of the counterbore, **Figure 10-29B**.
3. Subtract the two cylinders from the base object, **Figure 10-29C**.

A *spotface* is similar to a counterbore, but is not as deep. See **Figure 10-30**. It provides a flat surface for full contact of a washer or underside of a bolt head. Construct it in the same way as a counterbore.

Figure 10-29. Constructing a counterbore. A—Draw a cylinder to represent a hole. B—Draw a second cylinder to represent the counterbore. C—Subtract the two cylinders from the base object.

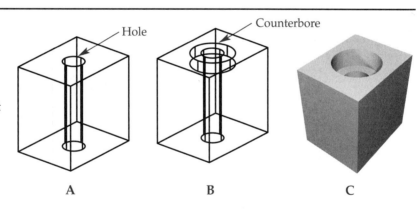

A   B   C

Figure 10-30. Constructing a spotface. A—The bottom of the second, larger-diameter cylinder should be located at the exact depth of the spotface. However, the height may extend above the surface of the base. Then, subtract the two cylinders from the base. B—The finished solid.

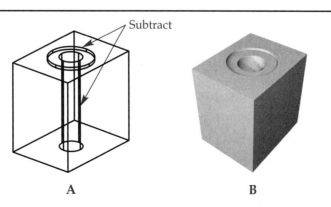

A   B

Countersink

A *countersink* is like a counterbore with angled sides. The sides allow a flat-head machine screw or wood screw to sit flush with the surface of an object. A countersink can be drawn in one of two ways. You can draw an inverted cone centered on a hole and subtract it from the base or you can chamfer the top edge of a hole. Chamfering is the quickest method.

1. Draw a cylinder representing the diameter of the hole, **Figure 10-31A**.
2. Subtract the cylinder from the base object.
3. Select the **CHAMFER** or **CHAMFEREDGE** command.
4. Select the top edge of the base object.
5. Enter the chamfer distance(s).
6. Pick the top edge of the hole, **Figure 10-31B**.

Boss

A *boss* serves the same function as a spotface. However, it is an area raised above the surface of an object. Draw a boss as follows.

1. Draw a cylinder representing the diameter of the hole. Extend it above the base object higher than the boss is to be, **Figure 10-32A**.
2. Draw a second cylinder the diameter of the boss. Place the base of this cylinder above the top surface of the base object a distance equal to the height of the boss. Give the cylinder a negative height value so that it extends inside of the base object, **Figure 10-32B**.
3. Union the base object and the second cylinder (boss). Subtract the hole from the unioned object, **Figure 10-32C**.
4. Fillet the intersection of the boss with the base object, **Figure 10-32D**.

Figure 10-31.
Constructing a countersink. A—Subtract the cylinder from the base to create the hole. B—Chamfer the top of the hole to create a countersink.

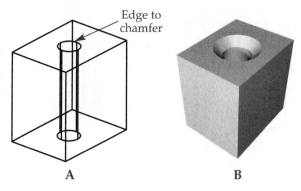

Figure 10-32.
Constructing a boss. A—Draw a cylinder for the hole so it extends above the surface of the object. B—Draw a cylinder the height of the boss on the top surface of the object. C—Union the large cylinder to the base. Then, subtract the small cylinder (hole) from the unioned objects. D—Fillet the edge to form the boss.

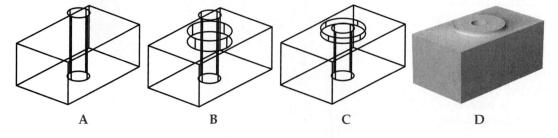

O-Ring Groove

An *O-ring* is a circular seal that resembles a torus. It sits inside of a groove constructed so that part of the O-ring is above the surface. An *O-ring groove* can be constructed by placing the center of a circle on the outside surface of a cylinder. Then, revolve the circle around the cylinder. Finally, subtract the revolved solid from the cylinder.

1. Construct the cylinder to the required dimensions, **Figure 10-33A**.
2. Rotate the UCS on the X axis (or appropriate axis).
3. Draw a circle with a center point on the surface of the cylinder, **Figure 10-33B**.
4. Revolve the circle 360° about the center of the cylinder, **Figure 10-33C**.
5. Subtract the revolved object from the cylinder, **Figure 10-33D**.

PROFESSIONAL TIP

In many cases, you will draw the O-ring as a torus. A copy of the torus can be used to create the O-ring groove instead of revolving a circle as described in the previous section.

Architectural Molding

Architectural molding features can be quickly constructed using extrusions. First, construct the profile of the molding as a closed shape, **Figure 10-34A**. Then, extrude the profile the desired length, **Figure 10-34B**.

Figure 10-33.
Constructing an O-ring groove. A—Construct a cylinder; this one has a round placed on one end. B—Draw a circle centered on the surface of the cylinder. C—Revolve the circle 360° about the center of the cylinder. D—Subtract the revolved object from the cylinder. E—The completed O-ring groove.

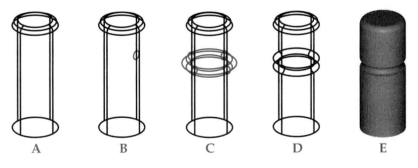

Figure 10-34.
A—The molding profile. B—The profile extruded to the desired length.

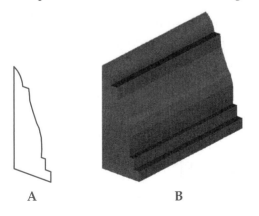

Figure 10-35.
Constructing corner molding. A—Copy and rotate the molding profile. B—Extrude the profiles to the desired lengths. C—Union the two extrusions to create the mitered corner. Note: The view has been rotated. D—The completed corner.

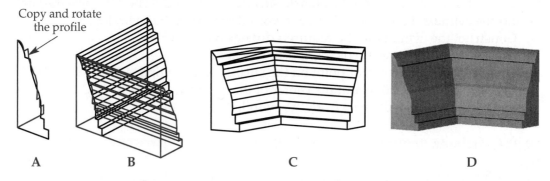

Copy and rotate the profile

A   B   C   D

Corner intersections of molding can be quickly created by extruding the same shape in two different directions and then joining the two objects. First, draw the molding profile. Then, copy and rotate the profile to orient the local Z axis in the desired direction, **Figure 10-35A.** Next, extrude the two profiles the desired lengths, **Figure 10-35B.** Finally, union the two extrusions to create the mitered corner molding, **Figure 10-35C.**

Exercise 10-11

Complete the exercise on the companion website.
www.g-wlearning.com/CAD

Chapter Review

Answer the following questions. Write your answers on a separate sheet of paper or complete the electronic chapter review on the companion website.
www.g-wlearning.com/CAD

1. Which properties of a solid can be changed in the **Properties** palette?
2. What does the History property control?
3. What is the preferred command for aligning objects to create an assembly of parts?
4. How does the **3DALIGN** command differ from the **ALIGN** command?
5. How does the **3DROTATE** command differ from the **ROTATE** command?
6. How does the **MIRROR3D** command differ from the **MIRROR** command?
7. Which command allows you to create a rectangular array by defining rows, columns, and levels?
8. How does a 3D polar array differ from a 2D polar array?
9. Which option of the **ARRAYPATH** command can be used to evenly distribute the arrayed object along the path?
10. Which command is used to fillet a 3D object?
11. Which command is used to chamfer a 3D object?
12. What is the preferred command to convert a mesh into a solid?
13. Name four types of surfaces that can be used to slice objects.
14. Briefly describe the function of the **SLICE** command.
15. Which **SLICE** command option would be used to create a contoured solid object like a computer mouse?
16. Which two commands can be used to create a countersink?

Drawing Problems

1. Construct an 8″ diameter tee pipe fitting using the dimensions shown below. Hint: Extrude and union two solid cylinders before subtracting the cylinders for the inside diameters.
   A. Use **EXTRUDE** to create two sections of pipe at 90° to each other, then use **UNION** to union the two pieces together.
   B. Fillet and chamfer the object to finish it. The chamfer distance is .25″ × .25″.
   C. The outside diameter of all three openings is 8.63″ and the pipe wall thickness is .322″.
   D. Save the drawing as P10_01.

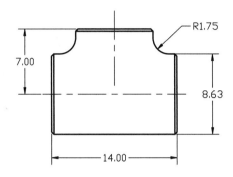

2. Construct an 8″ diameter, 90° elbow pipe fitting using the dimensions shown below.
   A. Use **EXTRUDE** or **SWEEP** to create the elbow.
   B. Chamfer the object. The chamfer distance is .25″ × .25″. Note: You cannot use the **CHAMFER** command.
   C. The outside diameter is 8.63″ and the pipe wall thickness is .322″.
   D. Save the drawing as P10_02.

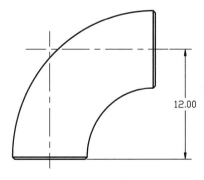

Problems 3–6. These problems require you to use a variety of solid modeling functions to construct the objects. Use all of the solid modeling and editing commands you have learned so far to assist in construction. Create new UCSs as needed and use a dynamic UCS when practical. Use **SOLIDHIST** *and* **SHOWHIST** *to record and view the steps used to create the solid models. Create copies of the completed models and split them as required to show the internal features visible in the section views. Save each drawing as* P10_*(problem number).*

3.

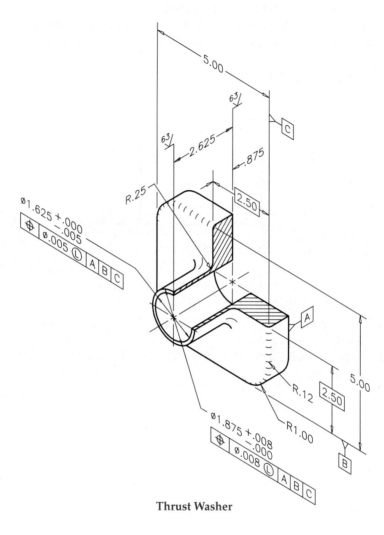

Thrust Washer

4.

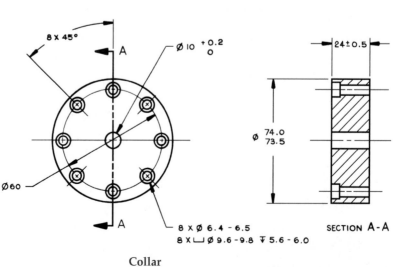

Collar

5.

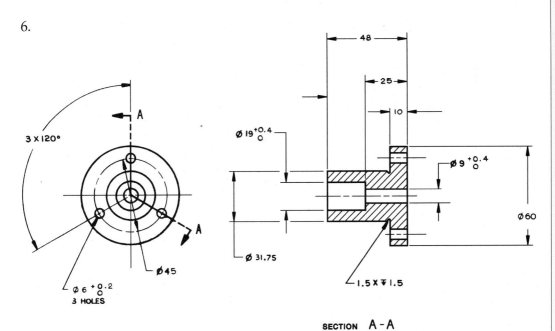

Diffuser

6.

SECTION A-A

Bushing

7. In this problem, you will add legs to the kitchen chair you started in Chapter 2. In Chapter 8, you refined the seat, and in Chapter 9, you added the seatback. In this chapter, you complete the model. In Chapter 18, you will add materials to the model and render it.

   A. Open **P09_08** from Chapter 9.

   B. Using the **LINE** command, create the framework for lofting the legs and crossbars as shown. The crossbars are at the midpoints of the legs. Position the lines for the double crossbar about 4.5" apart.

   C. Using a combination of the **3DROTATE** and **3DMOVE** commands in conjunction with new UCSs, draw and position circles as shapes for lofting the legs and crossbars. The legs transition from ∅1.00 at the ends to ∅1.25 at the midpoints. The crossbars transition from ∅.75 at the ends to ∅1.00 at a position 2" from each end.

   D. Create the legs and crossbars. The completed model is shown.

   E. Save the drawing as **P10_07**.

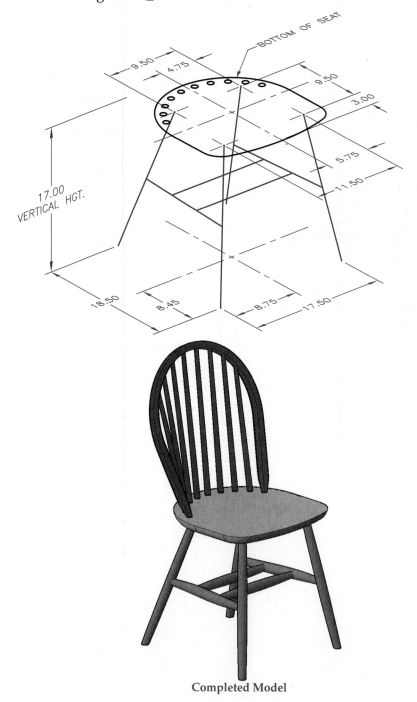

Completed Model

Advanced Surface Modeling

Learning Objectives

After completing this chapter, you will be able to:

✓ Understand and work with different types of surface models.
✓ Create procedural surfaces.
✓ Create NURBS surfaces.
✓ Create network surfaces.
✓ Create surface models from existing surfaces.
✓ Blend and patch surfaces.
✓ Offset, fillet, extend, and trim surfaces.
✓ Convert existing models to NURBS surfaces.
✓ Edit NURBS surface control vertices.
✓ Convert 2D objects to surfaces.
✓ Thicken a surface into a solid.
✓ Sculpt watertight surfaces into solids.

Overview

This chapter describes advanced surface modeling techniques and workflows used in AutoCAD. Surface modeling provides the ability to create a more freeform shape with tools that solid modeling cannot provide. You have been introduced to basic surface modeling techniques in previous chapters. As you have learned, one way to create surface models is to extrude, revolve, sweep, or loft profiles. This chapter builds on those techniques. In this chapter, you will develop an understanding of surface modeling techniques that can stand alone in the design process or work in combination with other modeling techniques.

As discussed in previous chapters, a *solid model* is created with a closed and bounded profile and has mass and volume properties. A *mesh* consists of vertices, edges, and faces that define the 3D mesh shape. A mesh does not have mass or volume. A *surface model* can be thought of as a thin-walled object with no "Z" depth. A surface model does not have mass or volume.

There are a number of workflows in AutoCAD available to the 3D designer. The following approaches can be considered depending on the nature of the work or the requirements of a specific application:

- Creating 3D models as solids, meshes, procedural surfaces, or NURBS surfaces (procedural surfaces and NURBS surfaces are discussed in the next section).
- Using Boolean operations on solids to create composite solids.
- Slicing composite solids using surfaces.
- Converting solids to mesh models.
- Converting solids to surface models.
- Converting surface models to NURBS surfaces.

These are just a few of the possible workflows. Editing techniques are also available and often play a significant role in surface modeling.

Understanding Surface Model Types

There are two basic types of surface models in AutoCAD: procedural surfaces and NURBS surfaces. A *procedural surface* is a standard surface object without control vertices. By default, a procedural surface, when created, is an *associative surface*. This means that the surface maintains associativity to the defining geometry or to other surrounding surfaces. Editing the defining geometry of an associative surface, or an adjacent surface in a "chain" of associative surfaces, modifies the surface.

A *NURBS surface* is based on splines or curves. The acronym *NURBS* stands for non-uniform rational B-spline. NURBS surfaces are based on a mathematical model and are used to create organic, freeform shapes. NURBS surfaces have control vertices that can be manipulated to edit the shape of the surface with great precision. Unlike a procedural surface, a NURBS surface cannot be created as an associative surface.

A third type of surface in AutoCAD is a *generic surface*. A generic surface has no associative history and no control vertices.

The type of surface model created is controlled by the **SURFACEMODELINGMODE** system variable. The default setting, 0, creates procedural surfaces. If the **SURFACEMODELINGMODE** system variable is set to 1, NURBS surfaces are created.

When creating a procedural surface, the **SURFACEASSOCIATIVITY** system variable setting determines whether an associative surface is created. The default setting, 1, creates associative surfaces. This system variable has no effect when creating NURBS surfaces.

Surface models can be created from either closed and bounded geometry or open profile geometry. When using modeling commands such as **EXTRUDE**, **REVOLVE**, **SWEEP**, and **LOFT**, the **Mode** option determines whether a surface model or solid model is created. Open profile curves always create surfaces, regardless of the **Mode** option setting.

The advantage to a procedural surface is the ease with which the surface can be created based on common shapes. In addition, working with procedural surfaces allows the designer to take advantage of associative modeling. Based on the design intent, the designer can use profile curves such as lines, circles, arcs, ellipses, helices, points, polylines, 3D polylines, and splines as the basis for the model. A procedural surface model can then be created using commands such as **EXTRUDE**, **REVOLVE**, **SWEEP**, **LOFT**, or **PLANESURF** (as discussed in previous chapters). If created as an associative surface, the model is linked to the defining geometry and can be modified by editing the geometry.

The commands used to create surface models are located in the **Surface** tab of the ribbon. See **Figure 11-1**. As discussed in Chapter 8, selecting a command from

Figure 11-1.
The **Surface** tab of the ribbon.

Picking this button
sets the value of
the **SURFACEASSOCIATIVITY**
system variable

Picking this button
sets the value of
the **SURFACEMODELINGMODE**
system variable

the **Create** panel in the **Surface** tab automatically sets the model creation mode to surface. Referring to **Figure 11-1**, notice the **Surface Associativity** and **NURBS Creation** buttons in the **Create** panel. The status of the **Surface Associativity** button indicates the **SURFACEASSOCIATIVITY** system variable setting. By default, the button has a blue background to indicate the system variable is turned on. The status of the **NURBS Creation** button indicates the **SURFACEMODELINGMODE** system variable setting. By default, this button does not have a blue background. This indicates that procedural surfaces are created by default.

PROFESSIONAL TIP

Procedural surfaces are also referred to as *explicit surfaces*. The terms *procedural surface* and *explicit surface* are interchangeable. This text uses procedural surfaces to illustrate various methods of surface creation. NURBS surfaces, as discussed later in this chapter, are used in creating models that represent sophisticated freeform shapes. Procedural surfaces can be converted to NURBS surfaces in order to model more sophisticated surface model shapes.

Planar surface creation is discussed in Chapter 2. Extruded and revolved surfaces are discussed in Chapter 9. Swept and lofted surfaces are discussed in Chapter 10.

 Exercise 11-1

Complete the exercise on the companion website.
www.g-wlearning.com/CAD

Working with Associative Surfaces

Procedural surfaces, when created, are associative by default. An associative surface changes shape or adjusts to the modifications made to the defining profile geometry (or other adjoining surfaces). This provides flexibility in the design. However, it is important to remember that when modifying the shape of an associative surface, you modify the profile geometry, *not* the surface. Modifying the profile geometry maintains the associative relationship. If you pick on the surface and then attempt to modify it, AutoCAD issues a warning that the surface will lose its associativity with the defining curve, surface, or parametric equation. If you choose to continue with the operation, the associativity is lost. You can cancel the operation to preserve the associativity.

Set the **SURFACEASSOCIATIVITY** system variable to 1 to create associative surfaces. If the system variable is set to 0, surfaces that are created have no associativity to defining profile curves or other surfaces.

An example of creating a procedural surface model is shown in **Figure 11-2**. In this example, an associative surface is created from a series of cross-sectional profiles using the **LOFT** command. As discussed in Chapter 9, a loft is created by selecting the cross-sectional profiles in order. In **Figure 11-2A**, four open profile curves are selected to create the loft. By default, this creates a surface model. In **Figure 11-2B**, the object is shown after lofting. In **Figure 11-2C**, the object is shown after moving and scaling the third open profile. Because the surface model is associative, the model updates and adjoining surfaces adjust to conform to the new curve shape.

In **Figure 11-3**, a loft surface is created from three closed cross-sectional profiles (circles). The surface is created by using the **Mode** option or by selecting the command from the **Surface** tab in the ribbon. In **Figure 11-3B**, the object is shown after moving and scaling the third closed profile at the top. Because the surface is associative, the model updates and conforms to the new curve shape.

Figure 11-2.
Creating a loft surface as an associative surface model. A—The original cross-sectional profiles. B—The model after using the **LOFT** command. C—The model after editing one of the cross-sectional profiles.

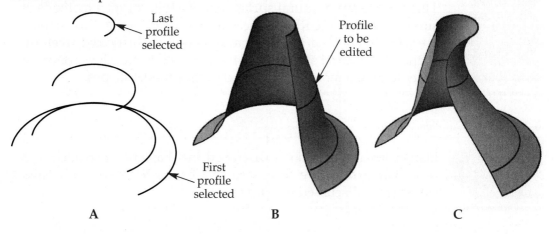

A   B   C

Figure 11-3.
Modifying a loft surface created from closed profiles (circles). A—The original loft surface. B—The model after editing the closed profile at the top.

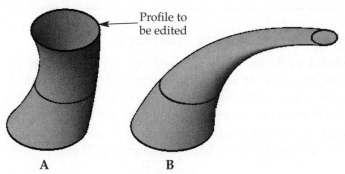

A   B

PROFESSIONAL TIP

The **DELOBJ** system variable is ignored if the **SURFACEASSOCIATIVITY** system variable is set to 1. Also, when creating a NURBS surface, the surface is a NURBS surface and is not associative.

Removing Surface Associativity

The surface associativity can be removed from a surface once the surface has been created. This can be done by selecting the surface and opening the **Properties** palette. The Maintain Associativity property in the **Surface Associativity** category controls the associativity of the surface. See **Figure 11-4**. By default, the property is set to Yes. Selecting **Remove** from the drop-down list removes the associativity and changes the property to None. This converts the surface to a generic surface.

The **Show Associativity** property in the **Surface Associativity** category controls whether adjoining associative surfaces are highlighted when a surface is selected in order to indicate dependency. When this property is set to Yes and a surface is selected, AutoCAD highlights other surfaces to which the surface is dependent. This can be useful for identifying associative relationships in a chain of surfaces.

PROFESSIONAL TIP

When moving, scaling, or rotating an associative surface, be sure to select the underlying curve geometry defining the surface. Failure to select the underlying geometry will result in the loss of the associativity.

Determining Modeling Workflows for Procedural Surfaces and NURBS Surfaces

When the design of a 3D model requires a freeform shape that would be difficult to create using solids, start by creating a procedural surface. You can convert the surface as required. A practical application is creating a surface model of a car fender. Start with a lofted surface based on four guide curves. Finish creating the fender by creating several procedural surfaces or patches, as discussed later in this chapter. Then, convert the fender surfaces to NURBS surfaces as needed and add further editing techniques for a more freeform sculpted shape.

Different factors determine when to use procedural surface modeling and NURBS surface modeling. For example, create procedural surfaces when it is important to maintain associativity and you plan to edit the original geometry. On the other hand, NURBS surfaces have control vertices that typically permit greater flexibility when editing. NURBS surfaces are often very useful for modeling organic shapes. The extent to which the design will require further editing can serve as a guideline for determining the best modeling approach.

Figure 11-4.
Associativity settings for a surface are accessed in the **Properties** palette.

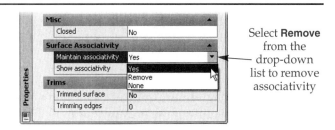

Select **Remove** from the drop-down list to remove associativity

The following sections discuss surface modeling commands and techniques available in AutoCAD. Procedural surfaces are shown in examples as the default creation method. NURBS surface creation and editing techniques are covered later in this chapter.

Creating Network Surfaces

A *network surface* is a surface model created by a group or "network" of profile curves or edges. A network surface is similar to a loft surface. As with a loft, the defining profiles can be open or closed curves, such as splines. The defining profiles can also be the edges of existing objects, including region edges, surface edge subobjects, and solid edge subobjects. The curves or edges selected can intersect at coincident points, but do not have to intersect.

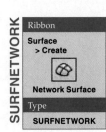

SURFNETWORK

Ribbon
Surface
> Create

Network Surface

Type
SURFNETWORK

The **SURFNETWORK** command is used to create network surfaces. After selecting this command, select the curves or edges defining the first direction of the surface. Make sure to select the curves in the order of surface creation. Then, press [Enter]. Next, select the curves or edges defining the second direction of the surface. Press [Enter] when you are done selecting the profiles. This creates the surface and ends the command. See **Figure 11-5**. A network surface is created as an associative surface by default.

The curves selected for the two directions define the U and V directions of the surface. The U and V directions can be thought of as the local directions of the surface and can be defined in either order. The U and V directions define the "flow" of the surface.

Figure 11-6 shows examples of creating network surfaces from similar sets of profile curves. In **Figure 11-6A**, a series of connected profile curves defines the network surface. In **Figure 11-6B**, two of the curves do not intersect with other profiles. Notice the differences between the resulting surface models.

Creating a network surface from region edges, surface subobject edges, and solid subobject edges is shown in **Figure 11-7**. Creating a network surface in this manner may result in some unexpected surface shapes. To select the profile edges, press and hold the [Ctrl] key. You can also use the edge subobject filter by selecting **Edge** from the **Selection** panel in the **Home** tab of the ribbon. In **Figure 11-7**, the four objects are located at different "Z" heights. The network surface is created from the four nonintersecting edges.

Figure 11-5.
Creating a network surface. Splines are used as the profiles in this example. A—The profiles used to define the surface are selected in the numbered order shown. Profiles 1–3 define the first direction and are shown in color. Profiles 4–8 define the second direction. B—The resulting surface model.

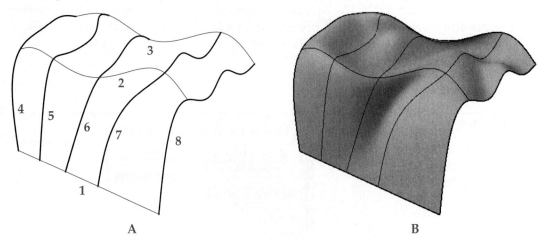

A   B

Figure 11-6.
Network surfaces. A—Profiles 1–3 define the first direction of the surface and are shown in color. Profiles 4 and 5 define the second direction of the surface. The resulting surface model is shown in wireframe and shaded form. B—Profiles 1–4 define the first direction of the surface and are shown in color. Profiles 5 and 6 define the second direction of the surface. The resulting surface model is shown in wireframe and shaded form.

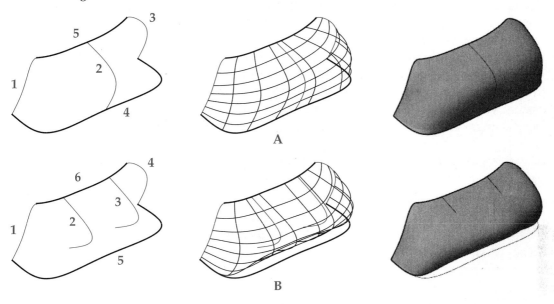

Figure 11-7.
Creating a network surface from the edges of existing objects. A—Two region edges (1 and 2) are selected to define the first direction of the surface. A surface subobject edge (3) and solid subobject edge (4) are selected to define the second direction of the surface. B—The resulting surface model.

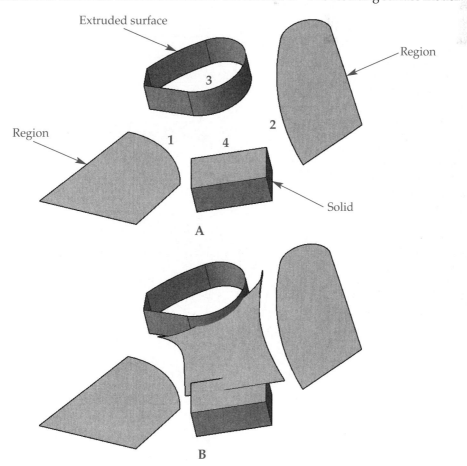

Referring to **Figure 11-6**, when working in a wireframe display, isolines appear to represent the curved surfaces of the surface model. The **SURFU** and **SURFV** system variables control the number of isolines displayed in the U and V directions of the surface. The number of isolines does not include the lines defining the object's boundary. The default setting for both system variables is 6. These settings can be changed for a surface by selecting the surface and opening the **Properties** palette. The U isolines and V isolines properties in the **Geometry** category control the number of isolines displayed in the U and V directions. See **Figure 11-8**. Setting a higher value can give you a better understanding of the curvature of the model. When working in a shaded display instead of a wireframe display, you can view the isoline representation by hovering the cursor over the model. The default settings for the **SURFU** and **SURFV** system variables can be changed in the **3D objects** section of the **Options** dialog box. The settings can range from 0 to 200.

PROFESSIONAL TIP

Surface associativity plays an important role in the creation of network surfaces. When editing a network surface with associativity, select a curve that forms the basis for the surface and modify it as needed. The result will be a new network surface shape.

Exercise 11-2

Complete the exercise on the companion website.
www.g-wlearning.com/CAD

Figure 11-8.
Isolines defining the curvature of the surface model appear when working in a wireframe display.
A—The **Properties** palette is used to set the number of isolines in the U and V directions of the surface.
B—A network surface with the default number of isolines in the U and V directions.
C—The surface after increasing the values of the U isolines and V isolines properties from 6 to 10.

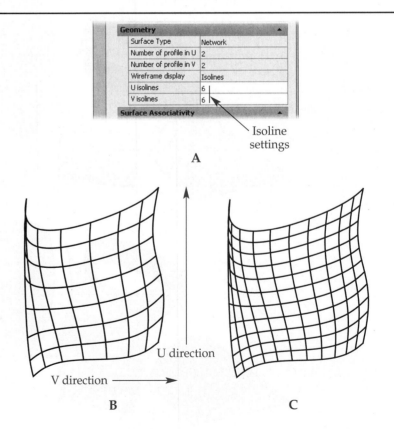

Creating Surfaces from Existing Surfaces

In addition to creating surfaces from profile curves, you can create surfaces from existing surfaces using the **SURFBLEND**, **SURFPATCH**, **SURFOFFSET**, **SURFFILLET**, and **SURFEXTEND** commands. These commands are discussed in the following sections. Surface models created with these commands are created as associative surfaces by default. This maintains associativity between the surfaces used to create the surface and the resulting surface.

Blend Surfaces

When working with surface models, there are situations when you need to "blend" together surfaces that do not meet or touch. The **SURFBLEND** command is used to create a *blend surface* between two surface edges or two solid edges. When blending surfaces, you select the surface edges to blend, *not* the surfaces.

Select the **SURFBLEND** command and then select the first edge to blend. Press and hold the [Ctrl] key to select the first edge or use the edge subobject filter. You can select multiple edges or use the **Chain** option to select a chain of continuous edges. Press [Enter] after defining the first edge. Next, select the second edge. Select a single edge or multiple edges, or use the **Chain** option to select a chain of continuous edges. When you press [Enter], a preview of the surface appears. You can press [Enter] to accept the default settings, as shown in **Figure 11-9**, or you can use the **Continuity** and **Bulge magnitude** options to specify the continuity and bulge magnitude settings at the edges. Different settings can be applied at each edge. The settings are similar to those when creating a loft, as discussed in Chapter 9.

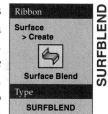

Continuity defines how the surfaces blend together at the starting and ending edges. The following options are available:

- **G0 (positional continuity).** This option creates a sharp transition between surfaces. The position of the surfaces is maintained continuously at the surface edges. This option is used for creating flat surfaces.
- **G1 (tangential continuity).** This option forms surfaces so that the end tangents match at the edges. The two surfaces blend together tangentially. This is the default option, as shown in **Figure 11-9B**.
- **G2 (curvature).** This option creates a curvature blend between the surfaces. The surfaces share the same curvature.

Figure 11-9.
A blend surface is created between two existing surfaces by selecting a starting and ending edge. A—Two existing loft surfaces. B—The model after creating a blend surface with the default settings of the **SURFBLEND** command.

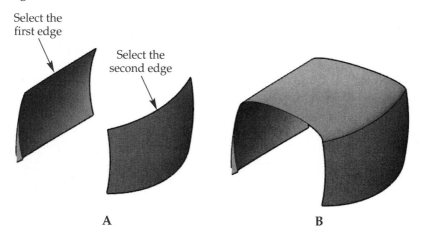

Select the first edge

Select the second edge

A

B

Examples of blend surfaces with different continuity settings are shown in **Figure 11-10**.

Bulge magnitude defines the size or "bulge" of the radial transition where the surfaces meet. See **Figure 11-11**. The default setting is 0.5. Valid values range from 0 to 1. A greater value is valid, but results in a larger roundness to the blend. Using different surface modeling techniques instead of entering a value greater than 1 is recommended. If the surface is set to G0 (positional continuity), changing the default bulge magnitude value has no effect.

Using different continuity and bulge magnitude settings modifies the surface and provides a way to create different blend surface shapes. You can change the continuity by using the grips that appear when creating the blend surface. Picking on a grip displays a menu with the continuity options. The same grips appear when selecting a blend surface after it has been created. You can also use the **Properties** palette to edit the settings of a blend surface.

Blend surfaces can also be created between the edges of regions or solid objects. An example of creating a blend surface between two solid subobject edges to form a cap is shown in **Figure 11-12**.

PROFESSIONAL TIP

Continuity settings are retained when exporting a 3D model to other 3D CAD modeling applications.

Figure 11-10.
Blend surfaces created with different continuity settings. A—G0 continuity. B—G1 continuity. C—G2 continuity.

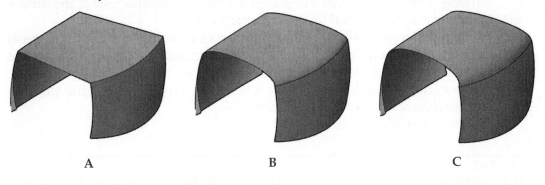

A B C

Figure 11-11.
Bulge magnitude determines the size of the radial transition at the edges where surfaces meet. A—The original model consists of two surfaces created from extruded arcs. B—The model after capping the ends with blend surfaces. For each surface, the continuity is set to G1 and the bulge magnitude is set to 0. C—For each surface, the continuity is set to G1. The bulge magnitude at the left edge is set to 0.5. The bulge magnitude at the right edge is set to 1.

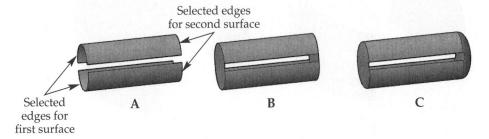

Selected edges for second surface

Selected edges for first surface

A B C

Figure 11-12.
Creating a blend surface between two solid subobject edges. A—The original model consists of two solid boxes. B—A blend surface is created between two subobject edges. The continuity is set to G2 and the bulge magnitude is set to 1.

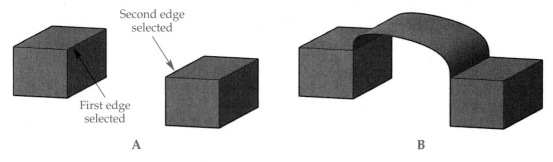

Second edge selected

First edge selected

A

B

Exercise 11-3

Complete the exercise on the companion website.
www.g-wlearning.com/CAD

Patch Surfaces

A *patch surface* is used to create a "patch" over an opening in an existing surface. A patch surface is used when it is necessary to close an opening or gap in the model. You can think of a patch surface as one of the many squares making up a quilt.

The **SURFPATCH** command is used to create a surface patch based on one or more edges forming a closed loop. You can select one or more surface edges or a series of curves. As when using the **SURFBLEND** command, you can specify the continuity and bulge magnitude to define the curvature of the surface.

Select the **SURFPATCH** command and then select one or more surface edges defining a closed loop. You can use the **Chain** option to select a chain of continuous surface edges. You can also use the **Curves** option to select multiple curves forming a closed loop. After selecting the edges or curves, press [Enter]. A preview appears and you can press [Enter] to create the surface using the default settings. The **Continuity** and **Bulge** magnitude options can be used to change the default settings as previously discussed. The default continuity setting is G0. The default bulge magnitude setting is 0.5.

The **Guides** option allows you to use a guide curve to constrain the shape of the surface patch. You can select one or more curves to define the guide curve. You can also select points to define the guide curve. When selecting points, use object snaps as needed.

Examples of creating patch surfaces are shown in **Figure 11-13**. In **Figure 11-13A**, the top of the tent requires a patch. To create the patch, the single edge representing the opening in the model is selected with the default **Surface edges** option. In **Figure 11-13B**, the patch surface is created with the continuity set to G1. This is the appropriate setting for the patch surface. In this case, you would want the patch to be tangent to the existing surface. The default bulge magnitude (0.5) is used. In **Figure 11-13C**, the patch surface is created with the continuity set to G1, but the bulge magnitude is set to 1. Notice the different result.

When using the **Guides** option, draw a curve to serve as the guide curve prior to selecting the **SURFPATCH** command. In **Figure 11-14A**, the top of the tent requires a patch. A new middle post has been added to the model, and a spline is drawn to serve as a guide curve. After selecting the **SURFPATCH** command, select the surface edge. Then, select the **Guides** option, select the spline, and press [Enter]. Press [Enter] to create the patch surface, or adjust the continuity and bulge magnitude settings as needed. See **Figure 11-14B**.

Figure 11-13.
Creating a patch surface to close an opening in a surface model. A—The original model. The single surface edge indicated forms a closed loop and is selected to generate the patch surface. B—A patch surface created with the continuity set to G1 and the bulge magnitude set to 0.5. C—A patch surface created with the continuity set to G1 and the bulge magnitude set to 1.

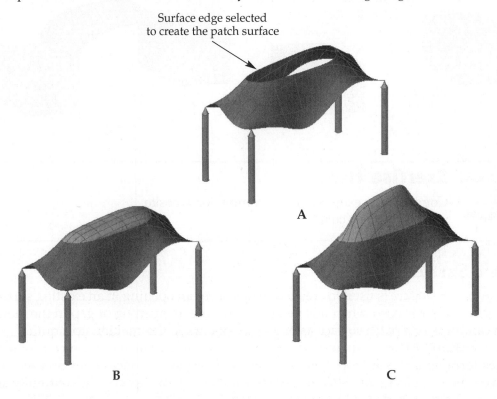

Surface edge selected
to create the patch surface

A

B

C

Figure 11-14.
Using a guide curve to constrain the shape of a patch surface. A—The original model. The middle post has been added. The spline is drawn for use with the **Guides** option. B—The patch surface created after selecting the spline as the guide curve. Notice the resulting shape.

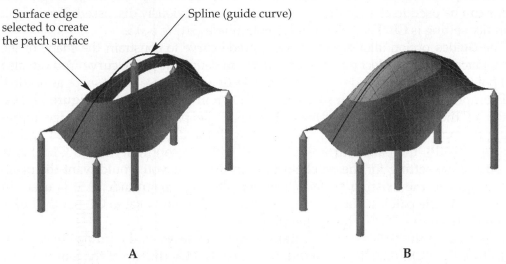

Surface edge
selected to create
the patch surface

Spline (guide curve)

A

B

In Figure 11-15, the game controller is to be redesigned with a new top shape. The model has been converted from a mesh model to a surface. In the original model (the mesh model), the top faces were deleted. The surface opening has eight continuous surface edges. The **Chain** option is used to assist in selecting edges to create the surface patch. After selecting the **SURFPATCH** command, select the **Chain** option and select one of the edges. The remaining edges are automatically selected. Next, press [Enter]. You can adjust the continuity and bulge magnitude settings or press [Enter] to create the patch surface. See **Figure 11-15B**.

Using the **Curves** option of the **SURFPATCH** command is shown in Figure 11-16. In Figure 11-16A, multiple curves have been created to design a sophisticated freeform shape. First, the **LOFT** command is used to create six loft surfaces defining the sides. Refer to the shaded surfaces shown in Figure 11-16B. Then, the **Curves** option of the **SURFPATCH** command is used to create three surface patches forming the top of the model. Refer to Figure 11-16C. To use the **Curves** option, enter it after selecting the **SURFPATCH** command. Then, select the curves defining the patch surface.

Surfaces created using the **SURFPATCH** command may differ from those created with the **LOFT** command. In addition, you may come across design situations where curves used with the **LOFT** command will not create a surface, but will when used with the **SURFPATCH** command.

PROFESSIONAL TIP

The **PREVIEWCREATIONTRANSPARENCY** system variable controls the transparency of surface previews when using the **SURFBLEND**, **SURFPATCH**, and **SURFFILLET** commands. The default setting is 60. Setting a higher value increases the transparency of the surface preview.

Exercise 11-4

Complete the exercise on the companion website.
www.g-wlearning.com/CAD

Figure 11-15.
Using the **Chain** option to select multiple surface edges to define the patch surface. A—The original model is a surface converted from a mesh model. The top opening includes eight surface edges. B—The patch surface created after using the **Chain** option to select the edges. The continuity is set to G2 and the bulge magnitude is set to 0.7.

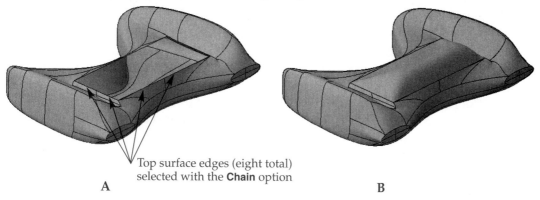

Top surface edges (eight total) selected with the **Chain** option

A   B

Figure 11-16.
Using the **LOFT** and **SURFPATCH** commands to create a freeform design from a series of curves.
A—The original model. The **LOFT** command is used to create six separate loft surfaces to form the sides. Each loft surface consists of two cross-sectional curves (shown in color).
B—The **Curves** option of the **SURFPATCH** command is used to create three separate surface patches forming the top of the model. Each surface is created from four curves (shown in color).
C—The model after creating surface patches. For each surface patch, the continuity is set to G1 and the bulge magnitude is set to 0.

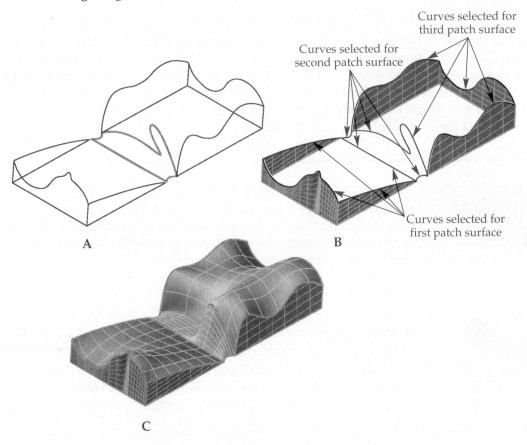

Offsetting Surfaces

The **SURFOFFSET** command allows you to offset a surface to create a new, parallel surface at a specified distance. You can offset a surface in one direction or in both directions from an existing surface. You can also offset a region to create a new surface.

Select the **SURFOFFSET** command and select the surface or region to offset. Then, press [Enter]. A preview of the offset surface appears with offset arrows indicating the direction of the offset. See **Figure 11-17A**. Next, specify the offset distance. If the design calls for the offset surface to be located on the opposite side, select the **Flip direction** option. You can offset the surface to both sides by selecting the **Both sides** option. After specifying the offset distance, press [Enter] to create the offset surface. In **Figure 11-17B**, the hair dryer housing is offset to the inside of the existing surface at a distance of .125.

The **Solid** option allows you to create a new solid based on the specified offset distance. In **Figure 11-17C**, the **Solid** option is used to create a solid. This is similar to using the **THICKEN** command, as discussed later in this chapter.

The **Connect** option can be used when you have more than one surface to offset and you need to maintain connection between the surfaces. When using this option, the original surfaces must be connected. See **Figure 11-18**.

The **Expression** option allows you to enter an expression to constrain the offset distance. Surface associativity must be enabled in order to use this option.

Figure 11-17.
Using the **SURFOFFSET** command. A—After selecting the surface to offset, offset arrows are displayed to indicate the direction of the offset. B—The surface is offset to the inside of the hair dryer housing at a distance of .125. C—The **Solid** option is used to create a new solid using the specified offset distance from the base surface.

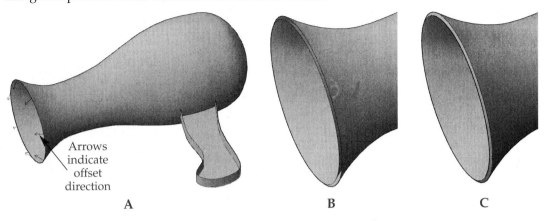

Arrows indicate offset direction

A   B   C

Figure 11-18.
The **Connect** option of the **SURFOFFSET** command is used to maintain the connection between surfaces when offsetting multiple surfaces. A—The original model. B—The vertical surfaces are offset using the **Connect** option.

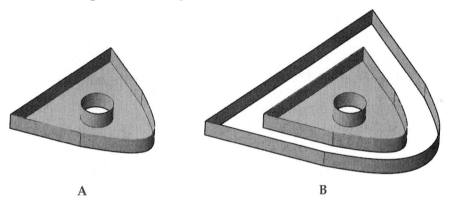

A   B

Exercise 11-5

Complete the exercise on the companion website.
www.g-wlearning.com/CAD

Creating Fillet Surfaces

The **SURFFILLET** command is used to create a fillet between two existing surfaces. The fillet created is a rounded surface that is tangent to the existing surfaces. You can create fillet surfaces from existing surfaces or regions. Using the **SURFFILLET** command is similar to using other fillet commands in AutoCAD. Commands used to fillet solids are introduced in Chapter 10.

To create a fillet surface, select the **SURFFILLET** command. First, set a radius using the **Radius** option. Then, select two surfaces. See **Figure 11-19A**. By default, the existing surfaces are trimmed to form the new surface. The surface trimming mode can be set by selecting the **Trim surface** option. AutoCAD stores the radius you specify as the setting for the **FILLETRAD3D** system variable. If you do not specify a radius, the current **FILLETRAD3D** system variable setting is used.

Ribbon
Surface
> Edit

Surface Fillet

Type
SURFFILLET

SURFFILLET

Figure 11-19.
Fillet surfaces created with the **SURFFILLET** command. For each example, the **Trim surface** option is set to **Yes**. A—A fillet surface created to fill the area between two surfaces. B—A fillet surface created between two surfaces meeting at an edge. C—A fillet surface created between two intersecting surfaces.

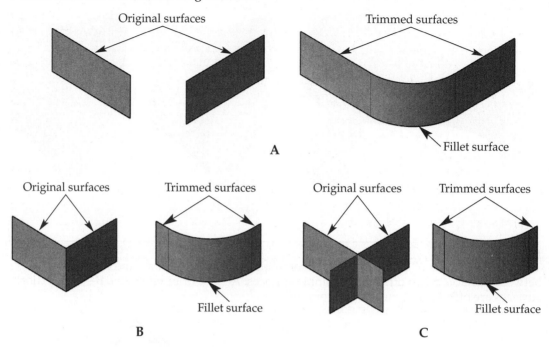

Creating a fillet surface between surfaces that do not meet is shown in **Figure 11-19A**. You can also create a fillet surface between two surfaces that share an edge or intersect, as shown in **Figure 11-19B** and **Figure 11-19C**.

After selecting the two surfaces, a preview appears and you can drag the fillet grip to adjust the radius dynamically. If dynamic input is turned on, you can type a value. If the fillet radius is too large, AutoCAD displays a message stating that the fillet surface cannot be created. When using the **Radius** option to set the radius, you can select the **Expression** option to enter a mathematical expression for the radius value.

After creating the fillet surface, you can use the **Properties** palette to edit the fillet surface radius. A radius of zero is not permissible.

PROFESSIONAL TIP

You can use the **UNION** command to union surfaces. However, it is not recommended. You will lose the surface associativity between the surfaces and the defining profile curves. Use surface editing commands instead.

Exercise 11-6

Complete the exercise on the companion website.
www.g-wlearning.com/CAD

Extending Surfaces

Ribbon

Surface
> Edit

Surface Extend

Type

SURFEXTEND

SURFEXTEND

You can add length to an existing surface using the **SURFEXTEND** command. When extending a surface, you can specify whether the new surface is created as a continuation of the existing surface or as a new surface. The surface extends to a new length using the specified distance.

Select the **SURFEXTEND** command and select one or more surface edges to extend. After selecting an edge, press [Enter]. A preview of the extended surface appears and you can drag the cursor dynamically to set the distance. You can also enter a distance by typing a value. The **Expression** option allows you to enter a mathematical expression for the extension distance. If you specify a distance and press [Enter], the surface is extended using the default settings. See **Figure 11-20**.

Before specifying the extension distance, you can use the **Modes** option to specify the extension mode. The two options are **Extend** and **Stretch**. The default **Extend** option is used to extend the surface in the same direction as the existing surface and attempt to maintain the surface shape based on the surface contour. The **Stretch** option is also used to extend the surface in the same direction as the existing surface. However, the resulting extension may not have the same surface contour.

After specifying the extension mode, the **Creation type** option can be used to set the type of surface created. The two options are **Merge** and **Append**. The default **Merge** option is used to extend the surface as one surface. The **Append** option is used to create a new surface extending from the original surface. This option results in two surfaces instead of one merged surface. After creating the new surface, you can use the **Properties** palette to edit the extension distance.

Exercise 11-7

Complete the exercise on the companion website.
www.g-wlearning.com/CAD

Trimming Surfaces

The **SURFTRIM** command can be used to trim surfaces or regions using other existing surfaces. You can trim any part of a surface where the surface intersects with another surface, region, or curve. In addition, you can project an existing object onto

---

Figure 11-20.
Extending an existing surface using the **SURFEXTEND** command. A—The original model. B—The extended surface.

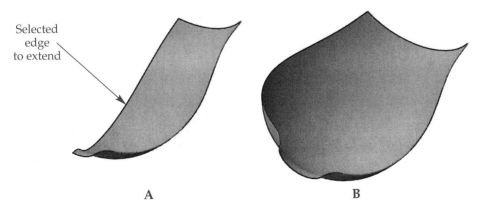

Selected
edge
to extend

A

B

Ribbon

Surface
> Edit

Surface Trim

Type

SURFTRIM

a surface to serve as a trimming boundary. The object to be trimmed and the cutting object do not have to intersect. When an associative surface is trimmed, it remains associative and retains the ability to be modified by editing the cutting object.

Select the **SURFTRIM** command and then select one or more surfaces or regions to trim. After selecting the objects to trim, press [Enter]. Next, you are prompted to select the cutting objects. Select one or more curves, surfaces, or regions. Then, press [Enter]. You are then prompted to select the surface areas to be trimmed. Select one or more areas. As you select each area, it is trimmed by the cutting object(s). If you trim an area that you wish to restore, use the **Undo** option. When you are finished trimming, press [Enter] to end the command. See **Figure 11-21**.

The **Extend** and **Projection direction** options are available after selecting the **SURFTRIM** command. The **Extend** option determines whether a surface used as a cutting edge is extended to meet the surface to be trimmed. By default, this option is set to **Yes**. The **Projection direction** option specifies the projection method used for projected geometry, as discussed in the next section.

Exercise 11-8

Complete the exercise on the companion website.
www.g-wlearning.com/CAD

Using Projected Geometry to Trim Surfaces

With the **SURFTRIM** command, you can trim surfaces using cutting objects other than existing surfaces. Objects used in this manner are referred to by AutoCAD as *curves*. Selecting a curve, such as an arc or circle, allows you to project the geometry onto the surface and use it as the cutting edge. See **Figure 11-22**. In the example shown, the cell phone case is selected as the surface to trim. The arcs located above the cell phone case are selected as the cutting edges. When selected, the arcs are projected onto the surface by AutoCAD. See **Figure 11-22B**. The areas to be trimmed are then selected to complete the sides. See **Figure 11-22C**.

To use a curve instead of an existing surface for trimming, select the curve after selecting the object to trim. You can select lines, arcs, circles, ellipses, polylines, splines, and helices. The **Projection direction** option of the **SURFTRIM** command can be used

Figure 11-21.
Trimming surfaces with the **SURFTRIM** command. A—The original model consists of three loft surfaces. Two of the loft surfaces are to be trimmed. B—The model after trimming.

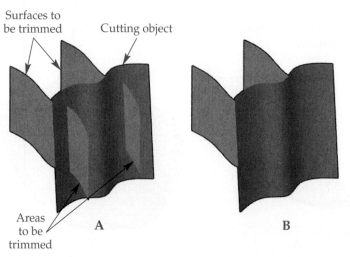

Surfaces to be trimmed

Cutting object

Areas to be trimmed

A

B

Figure 11-22.
Using projected geometry with the **SURFTRIM** command. A—The cell phone case model is a surface created from a mesh. The arcs drawn above the model are used for trimming the sides. B—Selecting the two curves and pressing [Enter] projects the curves to the model surface. The areas within the projected geometry on each side are selected as the areas to trim. C—The model after trimming.

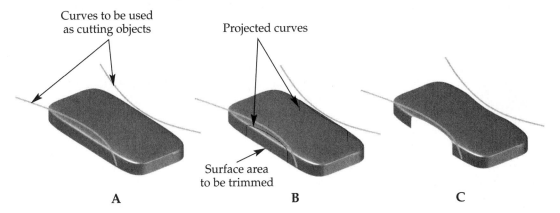

Curves to be used as cutting objects

Projected curves

Surface area to be trimmed

A   B   C

to set the projection method used by AutoCAD when projecting curves onto a surface. The following settings are available:

- **Automatic.** The cutting object is projected onto the surface to be trimmed. The projection is based on the current viewing direction. In a plan view, the projection of the cutting object is in the viewing direction. In a 3D view, the projection of a planar curve is normal to the curve, and the projection of a 3D curve is parallel to the direction of the Z axis of the current UCS. The **Automatic** option is set by default.
- **View.** The cutting object is projected in a direction based on the current view.
- **UCS.** The cutting object is projected in the positive or negative direction of the Z axis of the current UCS.
- **None.** The cutting object is not projected and must lie on the surface in order to perform the trim.

PROFESSIONAL TIP

The **SURFTRIM** command defaults to the **Automatic** option. Automatically projected geometry is used in most trim situations.

When using the **PROJECTGEOMETRY** command, the **SURFACEAUTOTRIM** system variable controls automatic trimming. The default setting is 0. A setting of 1 enables automatic trimming of surfaces. The current setting is indicated by the **Auto Trim** button in the **Project Geometry** panel of the **Surface** tab on the ribbon.

Exercise 11-9

Complete the exercise on the companion website.
www.g-wlearning.com/CAD

Ribbon
Surface
> Edit

Surface Untrim
Type
SURFUNTRIM

Untrimming Surfaces

If you need to restore a trimmed surface back to its original shape, use the **SURFUNTRIM** command. After selecting this command, select the edge of the surface area to untrim. If the surface has multiple trimmed edges, you can use the **Surface** option. The **SURFUNTRIM** command untrims surfaces trimmed by the **SURFTRIM** command. It does not untrim surfaces trimmed using the **PROJECTGEOMETRY** command.

PROFESSIONAL TIP

Open the **Properties** palette if you are unsure if a surface has been trimmed. The Trimmed surface property setting indicates the surface trim status.

Exercise caution if you are trimming an associative surface using another surface as a cutting object. When using another surface as the cutting object, turn off surface associativity (set the **SURFACEASSOCIATIVITY** system variable to 0) before trimming to avoid potential problems in future edits.

NURBS Surfaces

When creating a NURBS surface, you use many of the same commands that you would use to create a procedural surface. You can use splines and various curve shapes to create NURBS surfaces. In addition, you can convert procedural surfaces into NURBS surfaces.

As discussed earlier in this chapter, a NURBS surface is created when the **SURFACEMODELINGMODE** system variable is set to 1. In addition, NURBS surfaces are non-associative. The setting of the **SURFACEASSOCIATIVITY** system variable has no effect when creating a NURBS surface.

The advantage of working with NURBS surfaces is that you use control vertices to control or influence the shape of the surface. The ability to edit control vertices provides significant flexibility in creating and sculpting freeform, organic shapes. For example, in computer animation work, NURBS surface modeling techniques are commonly used for modeling characters to produce the organic shape desired. In AutoCAD, you can create highly sophisticated, freeform shapes using NURBS surface models. This chapter introduces the NURBS surface modeling tools available in AutoCAD.

NURBS Surface Modeling Workflows

There are two common workflows used in NURBS surface modeling. You can begin by creating procedural surfaces and then convert them to NURBS surfaces, or you can create the initial surfaces as NURBS surfaces.

When you start the modeling process by working from procedural surfaces, the following workflow is common:

- Create procedural surfaces using commands such as **EXTRUDE, REVOLVE, SWEEP, LOFT, PLANESURF,** and **SURFNETWORK.** The **SURFACEMODELINGMODE** system variable should be set to 0.

- Create other surfaces, such as blend surfaces, patches, fillets, and offset surfaces. Use the commands presented in this chapter.
- Convert the surfaces into NURBS surfaces.
- Edit the NURBS surfaces as needed to create the desired sculpted shape.

When you start the modeling process by creating NURBS surfaces, the following workflow is common:

- Set the **SURFACEMODELINGMODE** system variable to 1 (on). With this setting, NURBS surfaces are created.
- Create the surfaces needed to create the desired model shape. When using this approach, you use splines or curves to define the surface profile. Splines are created using the **SPLINE** command. Splines used for NURBS surface models are typically created with the **Method** option of the **SPLINE** command set to **CV**. This creates splines with control vertices (CVs), also known as *CV splines*. See **Figure 11-23**. Control vertices play a major role in editing NURBS surfaces.
- Edit the NURBS surfaces as needed to create the desired shape.

There are several important points to keep in mind when you are working with NURBS surfaces. Once a procedural surface is converted to a NURBS surface, the NURBS surface *cannot* be converted back to a procedural surface. In addition, once a NURBS surface is created, it *cannot* be converted to a procedural surface. The design workflow you use is important. Make sure to plan ahead so that your modeling process is suitable for the design.

Figure 11-23.
Splines used to create NURBS surfaces are typically created as CV splines. A—A CV spline used for creating an extruded NURBS surface. B—When you select a CV spline, the control vertices are displayed for editing. C—A NURBS surface model after using the **EXTRUDE** command to extrude the spline.

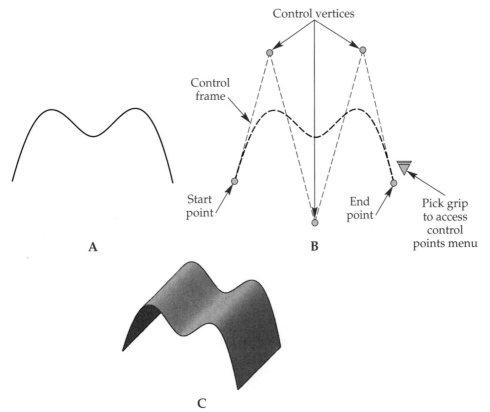

Creating and Editing NURBS Surfaces

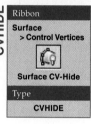

If the surface model you are creating will have only slight modifications, a simple approach is to create a procedural surface and then convert it to a NURBS surface. The **CONVTONURBS** command is used to convert a procedural surface to a NURBS surface. After entering this command, select the surface to convert. When you press [Enter], the surface is converted.

If the intention is to create more complex, freeform shapes, you can create NURBS surfaces without first converting procedural surfaces. In this case, set the **SURFACEMODELINGMODE** system variable to 1. Create profile geometry using splines and use the appropriate surface modeling commands.

The model in **Figure 11-24** shows an example of editing a NURBS surface converted from a procedural surface. The original model shown in **Figure 11-24A** is a loft surface created from open profiles. In **Figure 11-24B**, the model is shown after using the **CONVTONURBS** command. The model is shown selected with control vertices displayed (the wireframe view is shown for reference only). When you select a NURBS surface, control vertices do not appear by default. The display of control vertices is controlled by the **CVSHOW** command. To display control vertices, select the **CVSHOW** command and then select the NURBS surface. Then, press [Enter]. You can also select the command after initially selecting the surface. To remove the display of control vertices, select the **CVHIDE** command. Using this command removes the display of control vertices from all objects in the drawing.

Editing control vertices is similar to editing grips. Pick on the control vertex grip and pull or drag. You can press and hold the [Shift] key to select multiple control vertices. You can also use the gizmo that appears. As you pull or drag, the shape of the surface is modified. In **Figure 11-24C**, the model is shown after editing a control vertex on the back end of the surface. A shaded view of the model after editing and using the **CVHIDE** command is shown in **Figure 11-24D**.

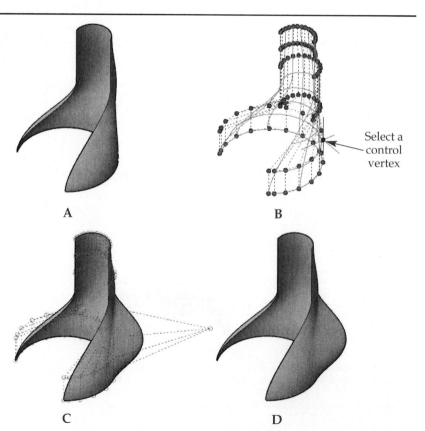

Figure 11-24.
Editing a NURBS surface converted from a procedural surface. A—The original model is a loft surface created as a procedural surface. B—The model after using the **CONVTONURBS** and **CVSHOW** commands. A wireframe view is shown for reference only. C—The model after editing one of the control vertices on the back end. D—The model after using the **CVHIDE** command.

Select a control vertex

A

B

C

D

For greater control when editing the control vertices of a NURBS surface, you can use the **3DEDITBAR** command. Select this command and select the NURBS surface to edit. You are then prompted to select a point on the surface. When you select a point, the *3D edit bar gizmo* appears, **Figure 11-25A**. This gizmo is similar to the move gizmo that appears when working with 3D objects, as discussed in Chapter 10. However, it contains additional grips for setting the tool options and modifying the tangencies of the surface. See **Figure 11-25B**. The grips include a square grip, triangle grip, and tangent arrow grip. The square grip represents the initial base point of the edit. Picking on the grip and dragging reshapes the surface from the base point. The triangle grip is used to specify the method for reshaping the surface. Picking on the grip displays a shortcut menu with the **Move Point** and **Tangent Direction** options. The **Move Point** option is used to reshape the surface by moving the base point. The **Tangent Direction** option is used to adjust the magnitude or bulge of the tangency at the base point. The tangent arrow grip is used to dynamically modify the tangency.

Use the 3D edit bar gizmo in the same manner as other gizmo tools. Pull or drag on a grip and then pick to set a distance or use direct distance entry. The movement can be restricted to the X, Y, or Z axis or the XY, XZ, or YZ plane by picking the appropriate axis or plane on the tool. The surface updates dynamically as you drag a grip.

Right-clicking on the tool displays a shortcut menu with additional options for relocating the base point, setting the tangency direction, and aligning the gizmo. The **Move Point Location** and **Move Tangent Direction** options serve the same functions as the **Move Point** and **Tangent Direction** options previously discussed.

Ribbon
Surface > Control Vertices
Surface CV-Edit Bar
Type
3DEDITBAR

3DEDITBAR

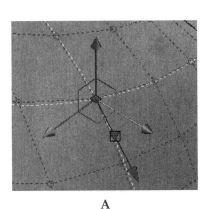

Exercise 11-10

Complete the exercise on the companion website.
www.g-wlearning.com/CAD

Additional Surface Modeling Methods

Additional methods are available for working with surface models. These methods include converting existing objects to surfaces and using the **THICKEN** and **SURFSCULPT** commands to create solid models from surfaces. These methods are discussed in the following sections.

Figure 11-25.
Using the **3DEDITBAR** command to edit a NURBS surface. A—The 3D edit bar gizmo appears after selecting a surface and a point on the surface. B—The grips available on the 3D edit bar gizmo.

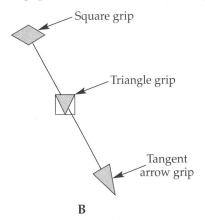

A   B

Converting Objects to Surfaces

AutoCAD provides a great deal of flexibility in converting and transforming objects. For example, a simple line can be quickly turned into a surface. Refer to **Figure 11-26**.

1. Use the Thickness property in the **Properties** palette to give the line a thickness. Notice that the object is still a line object, as indicated in the drop-down list at the top of the **Properties** palette.
2. Select the **CONVTOSURFACE** command.
3. Pick the line. Its property type is now listed in the **Properties** palette as a surface extrusion.

Other objects that can be converted to surfaces using the **CONVTOSURFACE** command are 2D solids, 3D solids, arcs with thickness, open polylines with a thickness and no width, regions, and planar 3D faces.

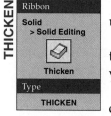

> The **THICKNESS** system variable can be used to assign a default Thickness property to new 2D objects that you create, such as lines, polylines, polygons, and circles. The value of the **THICKNESS** system variable does not affect the thickness of a planar surface or 3D surfaces.

Thickening a Surface into a Solid

A surface has no thickness. But, a surface can be quickly converted to a 3D solid using the **THICKEN** command.

To add thickness to a surface, enter the **THICKEN** command. Then, pick the surface(s) to thicken and press [Enter]. Next, you are prompted for the thickness. Enter a thickness value or pick two points on screen to specify the thickness. See **Figure 11-27**.

The **THICKEN** command can be used in conjunction with the **CONVTOSURFACE** command to quickly convert a 2D line into a solid. For example, create the line and then use the **Properties** palette to give the line a thickness. Convert the line into a surface using the **CONVTOSURFACE** command, and then use the **THICKEN** command to create a solid from the surface.

By default, the original surface object is deleted when the 3D solid is created with **THICKEN**. This is controlled by the **DELOBJ** system variable. To preserve the original surface, change the **DELOBJ** value to 0.

Figure 11-26.
Converting a line into a surface. First, draw the line. Next, give the line a thickness using the **Properties** palette. Then, convert the line to a surface using the **CONVTOSURFACE** command.

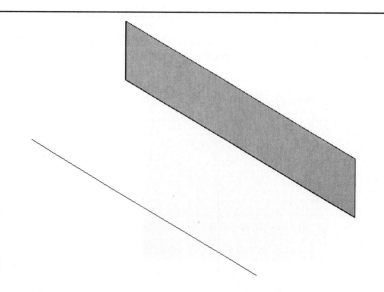

Figure 11-27.
A—This surface
will be thickened
into a solid. B—The
thickened surface is
a 3D solid.

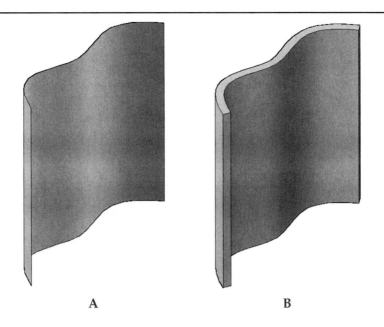

A   B

Exercise 11-11

Complete the exercise on the companion website.
www.g-wlearning.com/CAD

Sculpting: Surface to Solid

When designing using surfaces, you can sculpt a surface into a solid using the **SURFSCULPT** command. This is similar to using the **CONVTOSOLID** command as discussed in Chapter 10. The main use for sculpting is to create a solid from a watertight area by trimming and combining multiple surfaces. The command can also be used on solid and mesh objects.

Notice the surfaces in **Figure 11-28** create a watertight volume where trimming of the surface can occur. A single-surface enclosed area or multiple surfaces that create an enclosed area are considered to be watertight objects. Watertight objects can be converted to a solid.

Which command should you use, **CONVTOSOLID** or **SURFSCULPT**? The difference is very subtle.

- **CONVTOSOLID** works the best when you have a watertight mesh and want to convert it to a solid. Also, it works well on polylines and circles with thickness.
- **SURFSCULPT** works the best when you have watertight surfaces or solids that completely enclose a space (no gaps). This is shown in **Figure 11-28**.
- As a best practice, use **CONVTOSOLID** for converting watertight meshes to solids. Use **SURFSCULPT** for converting surfaces or solids.

Ribbon
Surface > Edit
Sculpt
Type
SURFSCULPT

SURFSCULPT

The watertight surface area must have a G0 continuity (positional continuity) for the **SURFSCULPT** command to work properly.

Figure 11-28.
A—These six surfaces form a watertight volume. B—The solid is sculpted using the **SURFSCULPT** command.

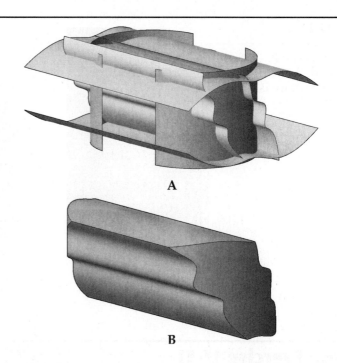

A

B

Chapter Review

Answer the following questions. Write your answers on a separate sheet of paper or complete the electronic chapter review on the companion website.
www.g-wlearning.com/CAD

1. Name the two basic types of surface models in AutoCAD.
2. Which type of surface is created when the **SURFACEMODELINGMODE** system variable is set to 0?
3. What is an *associative surface*?
4. Which system variable determines whether a surface model is associative when created?
5. When editing the shape of an associative surface, what should be selected to maintain the surface associativity?
6. What is a *network surface*?
7. What two system variables set the number of isolines displayed in the U and V directions of a surface model?
8. What is the purpose of the **SURFBLEND** command?
9. What are the three options used to define surface continuity? What is the result of using each option?
10. Define *bulge magnitude*.
11. What is the purpose of the **SURFPATCH** command?
12. What is the purpose of the **SURFOFFSET** command?
13. How do you create a new solid when using the **SURFOFFSET** command?
14. What are the two creation type options available when using the **SURFEXTEND** command? What is the purpose of each option?
15. What are the three object types that can be used as cutting objects when trimming a surface?
16. What are the two basic ways to create a NURBS surface?
17. What command is used to display control vertices on a NURBS surface?
18. What command can be used to convert a 2D line or polyline into a surface model?
19. What is the purpose of the **THICKEN** command and which type of object does it create?
20. What is the preferred command to convert a watertight series of surfaces into a solid?

Drawing Problems

1. Create a loft surface from the three cross sections shown. The spacing between cross sections is 5 units. The circle cross section is a Ø6 circle. Use your own coordinates for the circle center point and the ellipse center points. To draw the two ellipse cross sections, refer to the dimensions given. The ellipses are centered on the same center point along the Z axis (refer to the top view). The major axis of the top ellipse is parallel to the minor axis of the lower ellipse. Use your own orientation for the ellipse axes relative to the circle (the exact orientation is not important). When creating the loft, create a procedural surface with associativity. After creating the loft, edit it by changing the dimensions of the ellipse cross sections. Refer to the dimensions given. Save the drawing as P11_01.

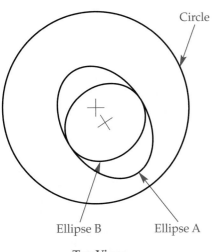

Top View

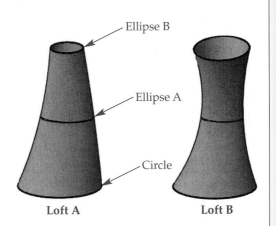

Cross Section	Dimensions (Loft A)	Edited Dimensions (Loft B)
Ellipse A	Major diameter = 3.80	Major diameter = 3.00
	Minor diameter = 2.70	Minor diameter = 2.00
Ellipse B	Major diameter = 2.70	Major diameter = 6.00
	Minor diameter = 2.30	Minor diameter = 4.00

2. Create the cell phone case shown. Create the case as an extruded surface. Create a patch surface for the top surface. Then, use the arcs to trim the sides.
   A. Draw the base profile in the top view using the dimensions given.
   B. Draw the arcs in the top view. The arcs should extend past the perimeter of the cell phone case. Draw the first arc on the right side of the profile and then mirror it to the other side. Move the arcs along the Z axis so they are located .75 units above the bottom of the cell phone case.
   C. Extrude the base profile to a height of .5 units.
   D. Create the top of the cell phone case by creating a patch surface with C2 continuity and a bulge magnitude of .125.
   E. Trim the sides by projecting the arcs and selecting areas to trim.
   F. Save the drawing as P11_02.

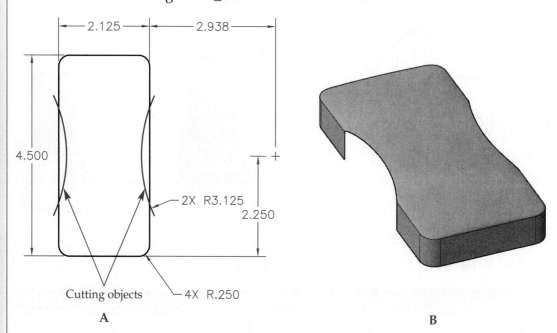

A

B

3. Create the hair dryer handle shown. Create the base profile using a spline and a straight line segment. Use the profile to create a planar surface. Then, extrude the planar surface and fillet the bottom end. Finally, convert the model to a NURBS surface and edit the top surface.

A. Draw a spline and a line using the profile shown. Use the **CV** option of the **SPLINE** command to create a CV spline. Dimensions are not important. Create the general shape by picking points to define the control vertices as indicated. Draw a line to form the straight segment at the end of the handle. Then, use the **JOIN** command to join the line to the spline. The resulting object should be a single, closed spline.

B. Create a planar surface from the profile.

C. Extrude the planar surface to a height of .85 units.

D. Use the **SURFFILLET** command to fillet the bottom end of the handle. Use a radius of .375 units. Set the trimming mode to **Yes**.

E. Using the **CONVTONURBS** command, convert the model to a NURBS surface. Edit the control vertices of the top surface to create a different shape.

F. Save the drawing as P11_03.

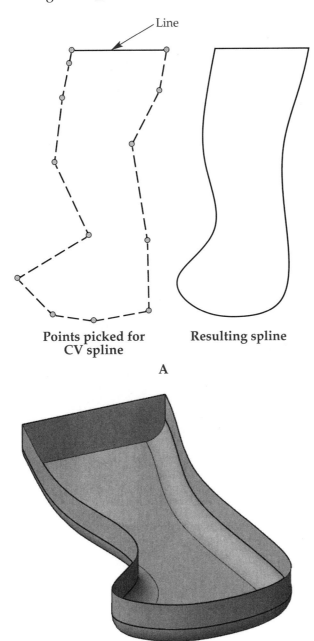

Line

Points picked for
CV spline

Resulting spline

A

B

4. Create the computer speaker using surface modeling commands. Use associative surfaces and edit the height from 4.5 units to 6 units, as shown. Save the drawing as P11_04.

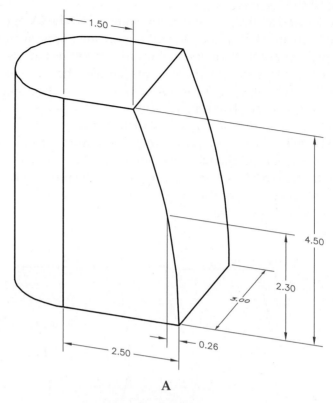

A

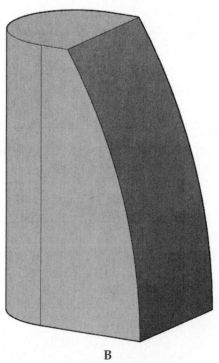

B

5. Create the kitchen chair shown. In earlier chapters, you created this model as a solid model. In this problem, use surface modeling commands to create the model. Use the drawings shown to create the seat, seatback, legs, and crossbars.

   A. Create the top of the seat using the profile shown. Use the edge profile to create the curved, receding edge extending to the bottom of the seat. The bottom of the seat is 1″ below the top.

   B. Create the profile geometry for the seatback. Loft the circular cross sections to create the supports. Loft the square and circular cross sections to create the bow.

   C. Create the framework for lofting the legs and crossbars as shown on the next page. The crossbars are at the midpoints of the legs. Position the lines for the double crossbar about 4.5″ apart. The legs transition from ∅1.00 at the ends to ∅1.25 at the midpoints. The crossbars transition from ∅.75 at the ends to ∅1.00 at a position 2″ from each end.

   D. Save the drawing as P11_05.

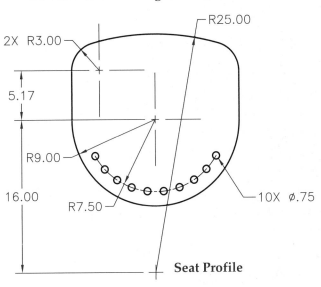

Seat Profile

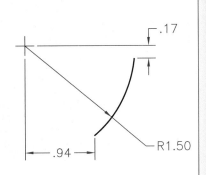

Edge Profile

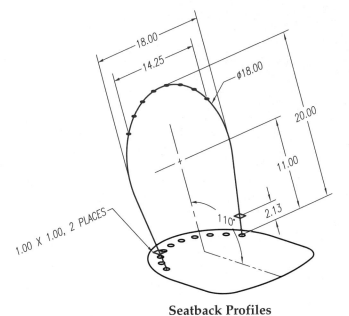

Seatback Profiles

Completed Seatback

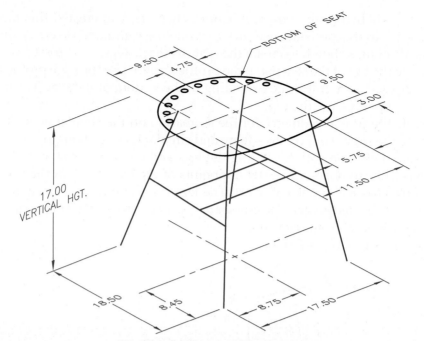

Profiles for Legs and Crossbars

Completed Model

Subobject Editing

Learning Objectives

After completing this chapter, you will be able to:

- ✓ Select subobjects (faces, edges, and vertices).
- ✓ Edit solids using grips.
- ✓ Edit fillet and chamfer subobjects.
- ✓ Edit composite solid model subobjects.
- ✓ Edit face subobjects.
- ✓ Edit edge subobjects.
- ✓ Edit vertex subobjects.
- ✓ Extrude a closed boundary using the **PRESSPULL** command.
- ✓ Offset planar surface edges using the **OFFSETEDGE** command.

Grip Editing

There are three basic types of 3D solids in AutoCAD. The commands **BOX, WEDGE, PYRAMID, CYLINDER, CONE, SPHERE,** and **TORUS** create 3D solid *primitives. Swept objects* are 2D open or closed profiles given thickness by the **EXTRUDE, REVOLVE, SWEEP,** and **LOFT** commands to create a 3D solid or surface. Finally, 3D solid *composites* are created by a Boolean operation (**UNION, SUBTRACT,** or **INTERSECT**) or by using the **SOLIDEDIT** command. The **SOLIDEDIT** command is discussed in Chapter 13. Smooth-edged solid primitives are achieved by converting mesh objects to solids.

There are two basic types of grips—base and parameter—that may be associated with a solid object. These grips provide an intuitive means of modifying solids. Base grips are square and parameter grips are typically arrows. The editing techniques that can be performed with these grips are discussed in the next sections.

3D Solid Primitives

The 3D solid primitives all have basically the same types of grips (base and parameter). However, not all grips are available on all primitives. All primitives have a base grip at the centroid of the base. This grip functions like a standard grip in 2D work. It can be used to stretch, move, rotate, scale, or mirror the solid.

Boxes, wedges, and pyramids have square base grips at the corners that allow the size of the base to be changed. See **Figure 12-1.** The object dynamically changes in the viewport as you select and move the grip or after you type the new coordinate location for the grip and press [Enter]. If ortho is off, the length and width can be changed at the same time by dragging the grip, except in the case of a pyramid. The triangular parameter grips on the base allow the length or width to be changed. The height can be changed using the top parameter grip. Each object has one parameter grip for changing the height of the apex and one for changing the height of the plane on which the base sits. A pyramid also has a parameter grip at the apex for changing the radius of the top.

Cylinders, cones, and spheres have four parameter grips for changing the radius of the base, or the cross section in the case of a sphere. See **Figure 12-2.** Cylinders and cones also have parameter grips for changing the height of the apex and the height of the plane on which the base sits. Additionally, a cone has a parameter grip at the apex for changing the radius of the top.

A torus has a parameter grip located at the center of the tube. See **Figure 12-3.** This grip is used to change the radius of the torus. Parameter grips at each quadrant of the tube are used to change the radius of the tube.

A polysolid does not have parameter grips. Instead, a base grip appears at each corner and edge midpoint of the starting face of the solid. See **Figure 12-4.** Use these

Figure 12-1.
Boxes, wedges, and pyramids have square base grips at the corners and parameter grips on the sides of the base and center of the top face, edge, or vertex.

Figure 12-2.
Cylinders, cones, and spheres have four parameter grips for changing their radius. Cylinders and cones have parameter grips for changing their height. Cones also have a parameter grip for changing the radius of the top.

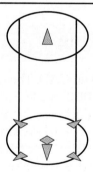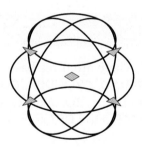

Figure 12-3.
A torus has a parameter grip located at the center of the tube for changing the radius of the torus. There are also parameter grips for changing the radius of the tube.

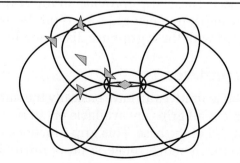

Figure 12-4.
A polysolid has a base grip at each corner and edge midpoint of the starting face of the solid and one at the endpoint and midpoint of each segment.

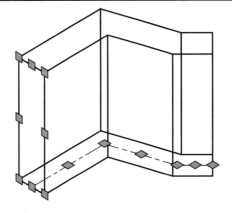

grips to change the cross-sectional shape or height of the polysolid. The corners do not need to remain square. Base grips also appear at the endpoint and midpoint of each segment centerline. Use these grips to change the location of each segment's midpoint or endpoints.

Grips in AutoCAD are multifunctional. For example, if dynamic input is turned on, you can hover over the parameter grips of an object to display various design parameters. In **Figure 12-5A**, the box primitive is selected to display grips. Hovering over the parameter grip at the side of the base displays the length dimension. Hovering over parameter grips on other types of primitives, such as a cylinder or sphere, displays the corresponding dimension of the shape. See **Figure 12-5B** and **Figure 12-5C**. You can quickly subobject edit a primitive by selecting a parameter grip and typing a delta value to add or remove geometry. For example, suppose a cylinder has a height of 10 units. The new height requirement is 15. With dynamic input turned on, select the cylinder, pick the parameter grip, and drag it so the dynamic input changes to delta entry. In the dynamic input box, type 5 for the new height and press [Enter].

Swept Solids

Extrusions, revolutions, sweeps, and lofts are considered swept solids and typically have base grips located at the vertices of their profiles. These can be used to change the size of the profile and, thus, the solid. Other grips that appear include:

- A parameter grip appears on the upper face of an extrusion for changing the height.
- A base grip appears on the axis of a revolved solid for changing the location of the axis in relation to the profile.
- Base grips appear on the vertices of a sweep path for modifying the swept object.

Composite Solids

Composite solids are created by using one of the Boolean commands (**UNION**, **SUBTRACT**, or **INTERSECT**) on a solid. Solids that have been modified using any of the options of the **SOLIDEDIT** command also become composite solids, as do meshes converted into solids. The solid may still look like a primitive, sweep, loft, etc., but it is a composite. The grips available with the previous objects are no longer available, unless performing subobject editing on a composite created with a Boolean command (as discussed later in this chapter). Composite solids have a base grip located at the centroid of the base surface. This grip can be used to stretch, move, rotate, scale, or mirror the solid when the 2D Wireframe visual style is current. See **Figure 12-6A**. This grip also appears if no gizmo is selected to display when a 3D visual style is current.

Figure 12-5.
Using parameter grips to display design data for a box, cylinder, and sphere. A—Hovering over the parameter grip on the side of the base displays the box length. B—Hovering over the parameter grip at the apex of the cylinder displays the cylinder height. C—Hovering over the parameter grip at the cross section of the sphere displays the sphere radius.

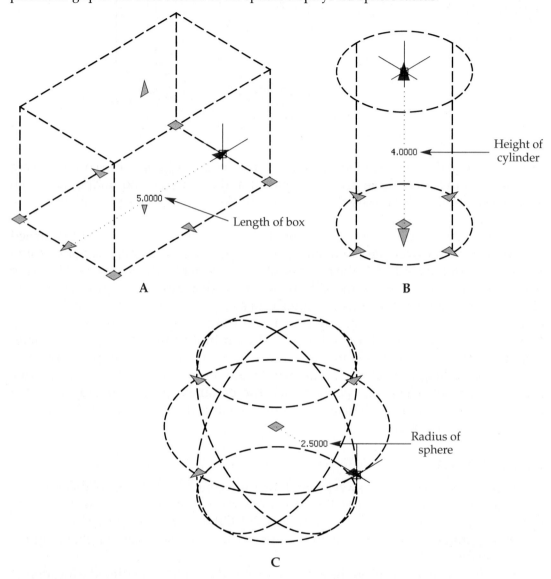

The gizmo drop-down list selection in the **Selection** panel on the **Home** tab of the ribbon controls the display of the default gizmo in a 3D visual style. If the move gizmo is set to appear in a 3D visual style (the default option), it appears when you select a composite solid. See **Figure 12-6B**. You can use the move, rotate, or scale gizmo to make modifications.

The grip options for composite solids converted from meshes are similar to those for other composite solids. Composite solids converted from meshes have several grips, but these all act as base grips.

Exercise 12-1

Complete the exercise on the companion website.
www.g-wlearning.com/CAD

Figure 12-6.
Grip editing options available for composite solids. A—A base grip appears on the composite solid when the object is selected if the 2D Wireframe visual style is current. Right-clicking displays a shortcut menu with the grip editing options. B—The move gizmo appears when the object is selected if a 3D visual style is current and the move gizmo is selected in the **Selection** panel in the **Home** tab of the ribbon. Right-clicking on the gizmo displays the shortcut menu.

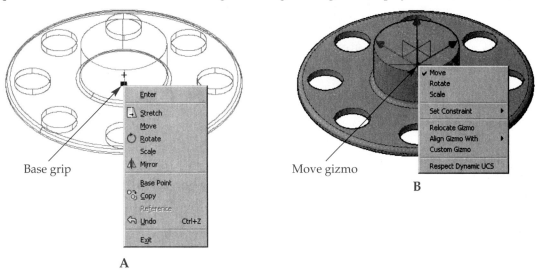

Using Grips with Surfaces

Surfaces can be edited using grips in the same manner previously discussed with solids. A planar surface created with **PLANESURF** can be moved, rotated, scaled, and mirrored, but not stretched. Base grips are located at each corner.

As you learned in Chapter 8, a variety of AutoCAD objects can be extruded to create a surface. Three of these objects—arc, line, and polyline—are shown extruded into surfaces in **Figure 12-7**. Notice the location and type of grips on the surface extrusions. Base grips are located on the original profile that was extruded to make the surface. These grips enable you to alter the shape of the surface. A parameter grip located on the top of the surface is used to change the height of the extrusion.

Surfaces that have been extruded, or swept, along a path can be edited with grips. Also, the grips located on the path allow you to change the shape of the surface extrusion. See **Figure 12-8**.

Figure 12-7.
Surfaces extruded from an arc, line, and polyline. Notice the grips.

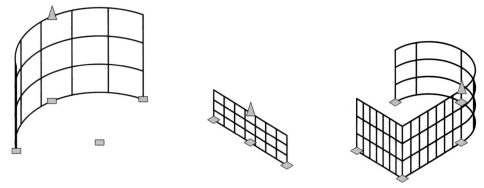

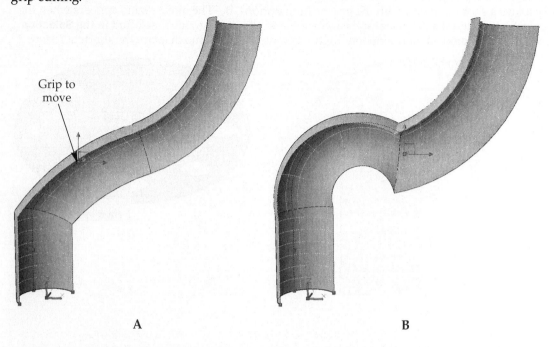

Grip to move

A   B

Overview of Subobject Editing

AutoCAD solid primitives, such as cylinders, wedges, and boxes, are composed of three types of subobjects: faces, edges, and vertices. As discussed in Chapter 3, mesh models are also made up of subobjects. In addition, the objects that are used with Boolean commands to create a composite solid are considered subobjects, if the solid history is recorded. The primitive subobjects can be edited. See **Figure 12-9.** Once selected, the primitive subobjects can be modified or deleted as needed from the composite solid. **Figure 12-10** illustrates the difference between a composite solid model, the solid primitives used to construct it, and an individual subobject of one of the primitives.

Subobjects can be easily edited using grips, which provide an intuitive and flexible method of solid model design. For example, suppose you need to rotate a face

Figure 12-9.
A—If the solid history is recorded, selecting a subobject solid primitive within a composite solid displays its grips. B—The grips on the primitive can be used to edit the primitive.

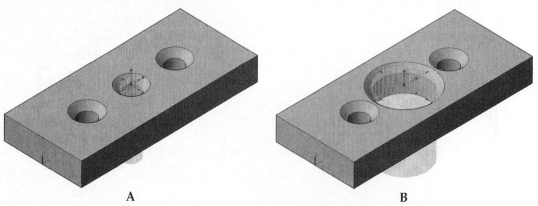

A   B

Figure 12-10.
A—The composite solid model is selected. Notice the single base grip. B—The wedge primitive subobject has been selected. Notice the grips associated with the primitive. C—An edge subobject within the primitive subobject is selected for editing.

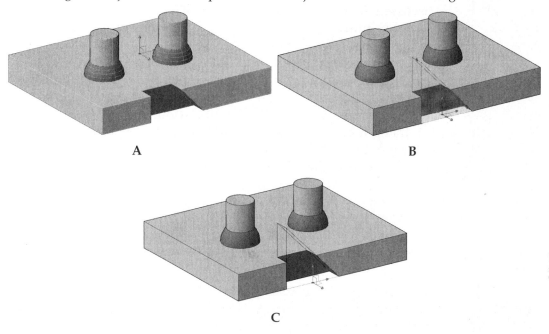

subobject in the current XY plane. You can select the subobject, pick its base grip, and then cycle through the editing functions to the rotate function. You can also use the **ROTATE** command on the selected subobject.

To select a subobject, press the [Ctrl] key and pick the subobject. You can select multiple subobjects and subobjects on multiple objects. To select a subobject that is hidden in the current view, first display the model as a wireframe. After creating a selection set, select a grip and edit the subobject as needed. Multiple objects can be selected in this manner. To deselect objects, press the [Shift]+[Ctrl] key combination and pick the objects to be removed from the selection set.

If objects or subobjects are overlapping, you have two options for cycling through the subobjects. The first and preferred method is to set the **SELECTIONCYCLING** system variable to 2. This is the same as picking the **Selection Cycling** button on the status bar to turn it on. Selection cycling is discussed in Chapter 10. The second method is available when selection cycling is turned off. Press the [Ctrl] key and the spacebar to turn on cycling and pick the subobject. Then, release the spacebar, continue holding the [Ctrl] key, and pick until the subobject you need is highlighted. Press [Enter] or the spacebar to select the highlighted subobject.

The [Ctrl] key method can be used to select subobjects for use with editing commands such as **MOVE**, **COPY**, **ROTATE**, **SCALE**, **ARRAY**, and **ERASE**. Some commands, like **ARRAY**, **STRETCH**, and **MIRROR**, are applied to the entire solid. Other operations may not be applied at all, depending on which type of subobject is selected. You can also use the **Properties** palette to change the color of edge and face subobjects or the material assigned to a face. The color of a subobject primitive can also be changed, but not the color of its subobjects.

Type
SELECTIONCYCLING
Toolbar
Status Bar
Selection Cycling

SELECTIONCYCLING

Use caution when using the **SOLIDEDIT** command with solid primitives and composite solids. The **SOLIDEDIT** command is discussed in Chapter 13. Using the **SOLIDEDIT** command with a solid primitive removes the history from the primitive. Do not use subobject editing methods if you do not plan on editing the solid primitive in the future. Once you use subobject editing on a primitive, the original geometric properties of the primitive *cannot* be edited with the **Properties** palette. The associated properties, such as the length, width, or height, are lost.

Using the **SOLIDEDIT** command with a composite solid may remove some of the history from the solid. Therefore, the original objects—some of the subobjects—are no longer available for subobject editing. However, you may still be able to perform some subobject edits that will accomplish your design intent.

PROFESSIONAL TIP

The subobject filters help to quickly select the specific type of subobject. They are located in the **Selection** panel on the **Home**, **Solid**, and **Mesh** tabs of the ribbon. The filters are discussed in Chapter 3.

Subobject Editing Fillets and Chamfers

If you improperly create a fillet or chamfer, edit it by holding down the [Ctrl] key and selecting the fillet or chamfer. Then, use editing methods to change the fillet or chamfer. A fillet has a parameter grip that can be used to change the radius of the fillet.

On some occasions during the editing process of fillets or chamfers, the fillet radius or the chamfer distances will not appear in the **Properties** palette. If this is the case and you only see a single face grip on the surface of the fillet or chamfer, erase the fillet or chamfer and reapply it. Fillet and chamfer subobject editing is a form of composite solid editing.

PROFESSIONAL TIP

The **Delete** option of the **SOLIDEDIT** command deletes selected faces. This is a quick way to remove features such as chamfers, fillets, holes, and slots. This technique is discussed in Chapter 13.

Exercise 12-2

Complete the exercise on the companion website.
www.g-wlearning.com/CAD

Face Subobject Editing

Faces of 3D solids can be modified using commands such as **MOVE**, **ROTATE**, and **SCALE** or by using grips and gizmos (grip tools). To select a face on a 3D solid, press the [Ctrl] key and pick within the boundary of the face. Do not pick the edge of the face. The face subobject filter can also be used to limit the selection to a face.

Face grips are circular and located in the center of the face, as shown in **Figure 12-11**. In the case of a sphere, the grip is located in the center of the sphere since there is only one face. The same is true of the curved face on a cylinder or cone.

If you select a solid primitive, all of the grips associated with that primitive are displayed. See **Figure 12-12A**. If you edit a solid primitive face, the history of the primitive is deleted and the object becomes a composite solid. Then, when the object is selected, a single base grip is displayed. See **Figure 12-12B**.

The same holds true for composite solids, if the solid history has been recorded. Solid model history is discussed in Chapter 10. If you select a primitive subobject in a composite solid with a recorded solid history, all of the grips associated with the primitive are displayed. If you edit a solid primitive face of the composite solid, the history of the primitive is deleted and the object becomes a new composite solid. Then, when the subobject is selected, a single base grip is displayed.

Figure 12-11.
Face grips are located in the center of face subobjects.

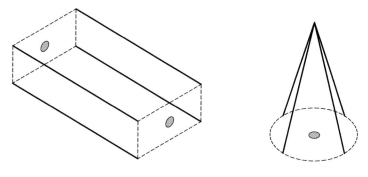

Figure 12-12.
A—This primitive is selected for editing. Notice the grips associated with the primitive.
B—If the primitive is edited, the history of that primitive is deleted and a single base grip is displayed.

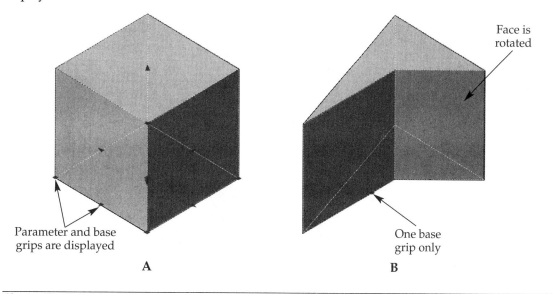

Face is rotated

Parameter and base grips are displayed

One base grip only

A

B

While pressing the [Ctrl] key and selecting a face, it may be difficult to select the face you want or to deselect faces you do not need. Use the view cube to transparently change the viewpoint. You can also press and hold the [Shift] key and press and hold the mouse wheel button at the same time to activate the transparent **3DORBIT** command.

The recorded history of a composite solid can be displayed by selecting the solid, opening the **Properties** palette, and changing the Show History property in the **Solid History** category to Yes.

Moving Faces

When a face of a 3D solid is moved, all adjacent faces are dragged and stretched with it. The shape of the original 3D primitive or solid determines the manner in which the face can be moved and how adjacent faces react. A face can be moved using the **MOVE** command, **3DMOVE** command, move gizmo, or by dragging the face's base grip. When moving a face, use the gizmo, polar tracking, or direct distance entry. Otherwise, the results may appear correct in the view in which the edit is made, but, when the view is changed, the actual result may not be what you wanted. See **Figure 12-13**.

The move gizmo, as discussed earlier in this chapter, is displayed by default when the face is selected. To use this gizmo, move the pointer over the X, Y, or Z axis of the gizmo; the axis changes to yellow. To restrict movement along that axis, pick the axis. If you move the pointer over one of the right angles at the origin of the gizmo, the corresponding two axes turn yellow. Pick to restrict the movement to that plane. You can complete the movement by either picking a new point or by using direct distance entry.

The face base grip is used to access options when dynamically moving a face. First, select the face. Then, hover over the face base grip to display the base grip shortcut menu. See **Figure 12-14A**. The options in this menu determine the effects of the edit on the selected face and adjacent faces. When using the **Extend Adjacent Faces** option, the moved face maintains

Figure 12-13.
A—The original solid primitives. B—The box and wedge are dynamically edited without using exact coordinates or distances. C—When the viewpoint is changed, you can see that dynamic editing has produced unexpected results.

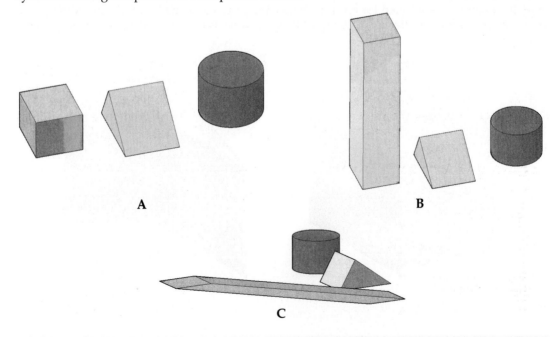

A

B

C

Figure 12-14.
A—The original solid primitive. B—Using the **Extend Adjacent Faces** option keeps the adjacent faces in their original planes, but alters the modified face. C—When using the **Move Face** option, the face maintains its shape and orientation.

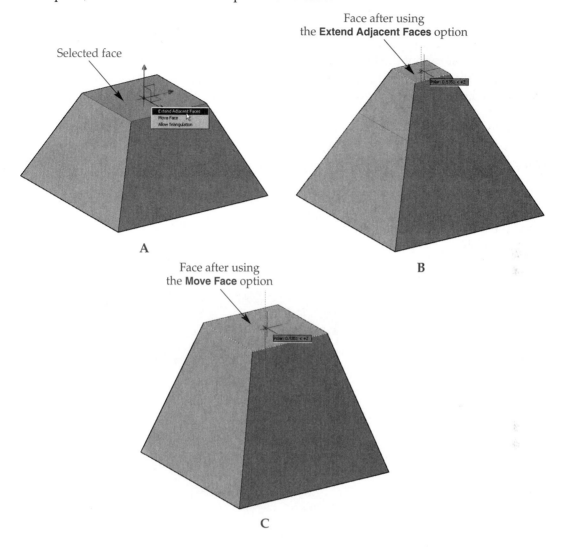

its shape and orientation. However, its size is modified because the planes of adjacent faces are maintained. See **Figure 12-14B**. If you select **Move Face**, the moved face maintains its size, shape, and orientation. The shape and plane of adjacent faces are changed. See **Figure 12-14C**. If you select **Allow Triangulation**, the moved face maintains its size, shape, and orientation. However, adjacent faces are subdivided into triangular faces, if needed.

During the edit, you can press the [Ctrl] key to cycle through the move options available in the base grip shortcut menu. For example, after selecting the face, pick the face grip or gizmo. Then, press and release the [Ctrl] key to cycle through the options.

PROFESSIONAL TIP

It is always important to keep the design intent of your solid model in mind. If you are creating a conceptual design, you may be able to use subobject grip editing and gizmos without entering precise coordinates. But, if you are working on a design for manufacturing or production, it is usually critical to use tools such as direct distance entry, gizmos with exact values, and polar tracking for greater accuracy.

Rotating Faces

Before rotating any primitive or subobject, you must know in which plane the rotation is to occur. The **ROTATE** command permits a rotation in the current XY plane. But, you can get around this limitation by using the **3DROTATE** command. This command allows you to select a rotation plane by means of the rotate gizmo discussed in Chapter 10.

The rotate gizmo provides a dynamic, graphic representation of the three axes of rotation. To rotate about the tool's X axis, pick the red circle on the gizmo. To rotate about the Y axis, pick the green circle. To rotate about the Z axis, pick the blue circle. Once you select a circle, it turns yellow and you are prompted for the start point of the rotation angle. You can enter a direct angle at this prompt or pick a point to define the angle of rotation. When the rotation angle is defined, the face is rotated about the selected axis.

For example, in **Figure 12-15A**, the top face is selected and the rotate gizmo is placed on a corner of the face. After picking the rotation axis on the gizmo, specify the

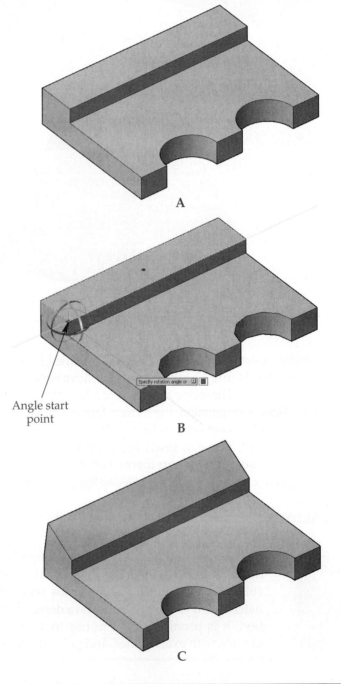

Figure 12-15.
Using the rotate gizmo to rotate a face. A—The top face is selected and the gizmo is placed on a base point of the face. B—The axis of rotation and a starting point for the angle are selected. C—The completed rotation.

Angle start point

A

B

C

angle start point and then the angle end point. Notice in **Figure 12-15B** that dynamic input can be used to enter an exact angle value. The result is shown in **Figure 12-15C**.

The [Ctrl] key is used to access options when dynamically rotating a face. **Figure 12-16A** shows a rotation without pressing [Ctrl]. The shape and size of the face being rotated is maintained, while the adjacent faces change. **Figure 12-16B** shows a rotation after pressing [Ctrl] once. The shape and size of the face being rotated changes, while the plane and shape of adjacent faces are maintained. Pressing the [Ctrl] key a second time maintains the shape and orientation of the selected face, but triangular faces may be created on adjacent faces. Pressing the [Ctrl] key a third time resets the function.

PROFESSIONAL TIP

When you are working in a 3D view, you may want to use the **3DMOVE** and **3DROTATE** commands exclusively during editing sessions. If this is the case, set the **GTDEFAULT** system variable to 1 (the default is 0). This automatically executes the **3DMOVE** and **3DROTATE** commands in a 3D view when you select the **MOVE** or **ROTATE** commands. If the 2D Wireframe visual style is current, the Wireframe visual style is set current for the duration of the command.

Figure 12-16.
A—Rotating a face without pressing the [Ctrl] key. The large, top face on the object has been selected for rotation. B—Pressing the [Ctrl] key once keeps the adjacent faces in their original planes. The shape and size of the face being rotated changes.

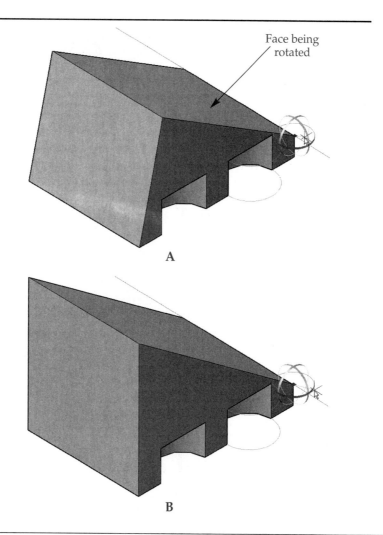

Face being rotated

A

B

Scaling Faces

Scaling a face is a simple procedure. First, select the face to be scaled. Then, select a base point and dynamically pick to change the scale or use a scale factor. See **Figure 12-17.** Pressing the [Ctrl] key has no effect on the scaling process, except to turn it off or on, if the base point is on the same plane as the face. However, if the base point is not on the same plane as the selected face, then pressing the [Ctrl] key has the same effect as for a rotate face-editing operation.

Coloring Faces

To change the color of a face, use the [Ctrl] key selection method to select the face. Next, open the **Properties** palette. See **Figure 12-18.** In the **General** category, pick the

Figure 12-17.
Scaling a face. A—The original solid. B—The dark face is scaled down. C—The dark face is scaled up.

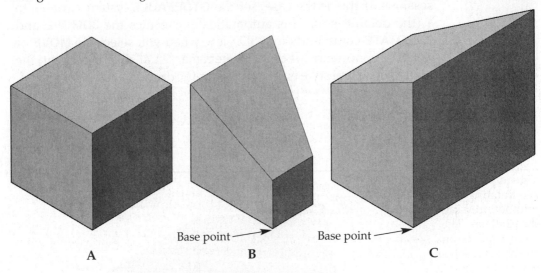

Base point
Base point

A   B   C

Figure 12-18.
Changing the color of a face or the material assigned to it.

A face
is selected

Pick to select
a color

Select a
material

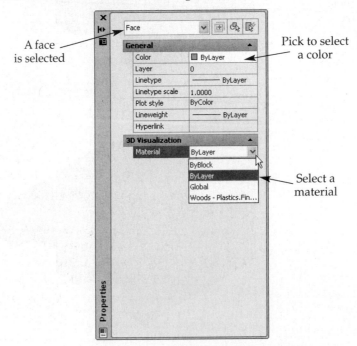

drop-down list for the Color property. Select the desired color or pick Select Color... and choose a color from the **Select Color** dialog box. To change the material applied to the face, pick the drop-down list for the Material property in the **3D Visualization** category. Select a material from the list. A material must be loaded into the drawing to be available in this drop-down list. Materials are discussed in detail in Chapter 17.

Extruding a Solid Face

Planar faces on 3D solids can be extruded into new solids. Refer to the HVAC duct assembly shown in **Figure 12-19A**. A new, reduced trunk needs to be created on the left end of the assembly. This requires two pieces: a reducer and the trunk.

First, select the **EXTRUDE** command. At the "select objects" prompt, press the [Ctrl] key and pick the face subobject to be extruded. Next, since this is a reduced trunk, specify a taper angle. Enter the **Taper angle** option and specify the angle. In this case, a 15° angle is used. Finally, specify the extrusion height. The height of the reducer is 12". See **Figure 12-19B**.

Now, the new trunk needs to be created. Select the **EXTRUDE** command. Press the [Ctrl] key and pick the face to extrude. Since this piece is not tapered, enter the extrusion height, which in this case is 44". See **Figure 12-19C**. The two new pieces are separate solid objects. If the assembly is to be one solid, use the **UNION** command and join the two new solids to the assembly.

Figure 12-19.
A—A new, reduced trunk needs to be created on the left end of the HVAC assembly. The face shown in color will be extruded. B—The **Taper angle** option of the **EXTRUDE** command is used to create the reducer. The face shown in color will be extruded to create the extension. C—The **EXTRUDE** command is used to create an extension from the reducer.

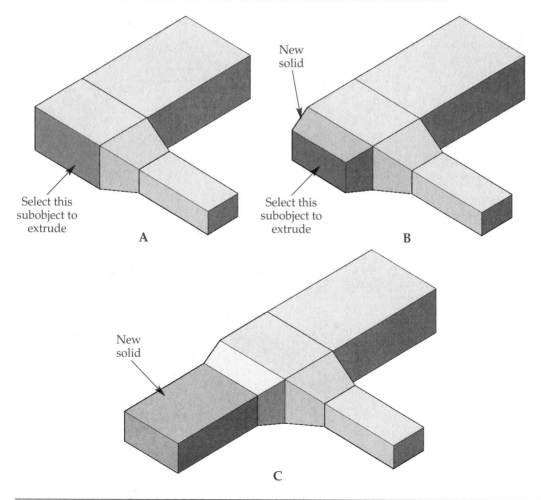

Revolving a Solid Face

Planar faces on 3D solids can be revolved in the same manner as other AutoCAD objects to create new solids. Refer to **Figure 12-20A**. The face on the left end of the HVAC duct created in the last section needs to be revolved to create a 90° bend. First, select the **REVOLVE** command. At the "select objects" prompt, press the [Ctrl] key and pick the face subobject to be revolved.

Next, the axis of revolution needs to be specified. You can pick the two endpoints of the vertical edge, but you can also pick the edge subobject. Enter the **Object** option of the command, press the [Ctrl] key, and select the edge subobject.

Finally, the 90° angle of revolution needs to be specified. **Figure 12-20B** shows the face revolved into a new solid. The bend is a new, separate solid. If necessary, use the **UNION** command to join the bend to the assembly.

Exercise 12-3

Complete the exercise on the companion website.
www.g-wlearning.com/CAD

Figure 12-20.
A—The face on the left end of the HVAC duct (shown in color) needs to be revolved to create a 90° bend. B—Use the **REVOLVE** command and pick the face subobject to be revolved.

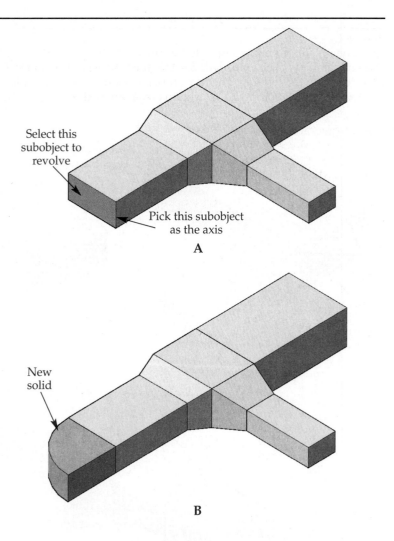

Select this subobject to revolve

Pick this subobject as the axis

A

New solid

B

Edge Subobject Editing

Individual edges of a solid can be edited using grips and gizmos in the same manner as faces. To select an edge subobject, press the [Ctrl] key and pick the edge. The edge subobject filter can also be used to limit the selection to an edge.

Grips on linear edges are rectangular and appear in the middle of the edge, **Figure 12-21**. In addition to solid edges, the edges of regions can be altered using **MOVE**, **ROTATE**, and **SCALE**, but grips are not displayed on regions as they are on solid subobjects.

Remember, editing subobjects of a primitive removes the primitive's history. This should always be a consideration if it is important to preserve the solid primitives that were used to construct a 3D solid model. Instead of editing the primitive subobjects at their subobject level, it may be better to add or remove material with a Boolean operation, thus preserving the solid's history.

There are several ways to perform a 3D edit. Keep the following options in mind when working with subobject editing.
- Selecting the **MOVE**, **ROTATE**, or **SCALE** command and picking a subobject will not display a gizmo, regardless of the current gizmo button displayed in the **Selection** panel.
- Selecting the **3DMOVE**, **3DROTATE**, or **3DSCALE** command and picking a subobject will display the appropriate gizmo, regardless of the current gizmo button displayed in the **Selection** panel.
- Picking a subobject without having selected a command first will display the gizmo corresponding to the current gizmo button in the **Selection** panel.

Moving Edges

To move an edge, select it using the [Ctrl] key, as previously discussed. See **Figure 12-22A**. By default, the gizmo corresponding to the gizmo button in the **Selection** panel appears. If the move gizmo is not current, select the **Move Gizmo** button in the **Selection** panel. Select the appropriate axis handle and dynamically move the edge or use direct distance entry, **Figure 12-22B**. The move gizmo remains active until the [Esc] key is pressed to deselect the edge.

Figure 12-21.
Edge grips are rectangular and displayed in the middle of the edge.

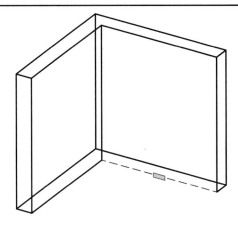

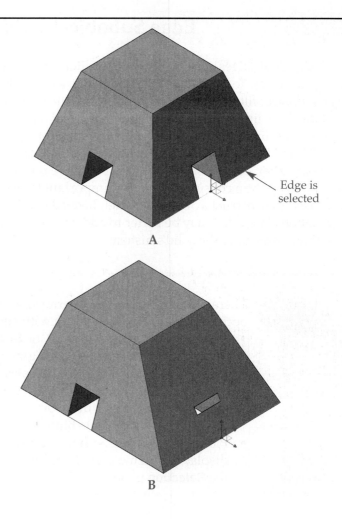

Edge is
selected

A

B

If you pick an edge grip to turn it hot, the gizmo is bypassed. This places you in the standard grip editing mode. You can stretch, move, rotate, scale, and mirror the edge. In this case, the stretch function works in the same manner as the move gizmo, but less reliably. You must be careful to use either ortho, polar tracking, or direct distance entry, but the possibility for error still exists.

The options available when dynamically moving an edge are similar to those used when moving a face. See Figure 12-23A. First, select the edge. Then hover over the base edge grip to display the base grip shortcut menu. Selecting the **Extend Adjacent Faces** option and moving the edge maintains the orientation of the moved edge, but its length is modified. See Figure 12-23B. This is because the planes and orientation of adjacent faces are maintained. Selecting the **Move Edge** option and moving the edge maintains the length and orientation of the moved edge. However, the shape and planes of adjacent faces are changed. See Figure 12-23C. Selecting the **Allow Triangulation** option and moving the edge maintains the length and orientation of the moved edge. But, if the move alters the planes of adjacent faces, those faces may become *nonplanar*. In other words, the face may now be located on two or more planes. If this happens, adjacent faces are divided into triangles, Figure 12-23D. This is visible when the object in Figure 12-23D is displayed in two orthographic views. See Figure 12-24.

During the edit, you can press the [Ctrl] key to cycle through the move options available in the base grip shortcut menu. For example, after selecting the edge, pick the face grip or gizmo. Then, press and release the [Ctrl] key to cycle through the options.

Figure 12-23.
Moving an edge. A—Selecting the edge and then hovering over the base grip displays the base grip shortcut menu. B—Using the **Extend Adjacent Faces** option maintains the orientation of the moved edge, but its length is modified because the planes of adjacent faces are maintained. C—When using the **Move Face** option, the edge maintains its length and orientation, but the shape and planes of adjacent faces are changed. D—When using the **Allow Triangulation** option, the adjacent faces may be triangulated.

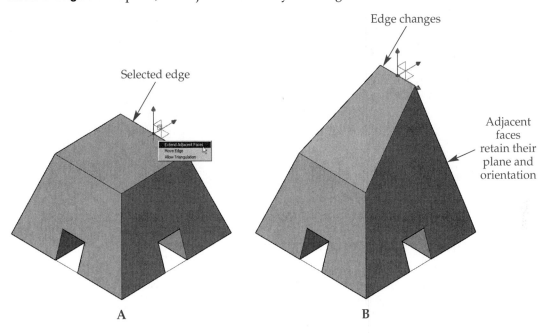

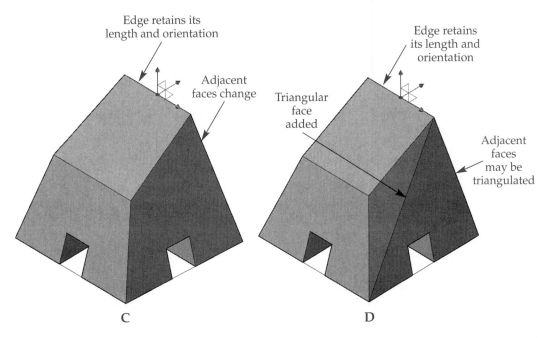

Figure 12-24.
Triangulated faces are clear in plan views. A—Front plan view. B—Side plan view.

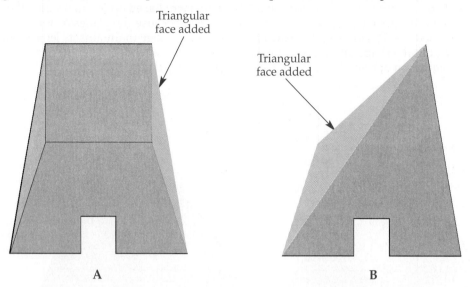

Triangular face added

Triangular face added

A

B

Rotating Edges

Before you select an edge to rotate, do a little planning. Since there are a wide variety of edge rotation options, it will save time if you first decide on the location of the base point about which the edge will rotate. Next, determine the direction and angle of rotation. Based on these criteria, choose the option that will accomplish the task the quickest.

To rotate an edge, enter the **ROTATE** or **3DROTATE** command. Then, pick the edge using the [Ctrl] key. Select a base point and then enter the rotation. You can also select the edge, pick the edge grip, and cycle to the **ROTATE** mode.

Edges are best rotated using the rotate gizmo. It provides a graphic visualization of the axis of rotation. If you select a dynamic UCS while using the **3DROTATE** command, you have a variety of rotation axes to use because the gizmo can be located on a temporary plane.

There are a few options to achieve different results when dynamically rotating an edge. The [Ctrl] key is used to access these options. First, select the edge. Then, pick the edge grip or gizmo and press and release the [Ctrl] key to cycle through the options. If the [Ctrl] key is not pressed, the rotated edge maintains its length, but the shape and planes of adjacent faces are changed. See **Figure 12-25A**. If the [Ctrl] key is pressed once, the length of the rotated edge is modified because the planes of adjacent faces are maintained. See **Figure 12-25B**. If the [Ctrl] key is pressed twice, the rotated edge maintains its length, but if the rotation causes faces to become nonplanar, the adjacent faces may be triangulated. See **Figure 12-25C**. Pressing the [Ctrl] key a third time resets the function.

Figure 12-25.
Rotating an edge. A—The [Ctrl] key is not pressed. B—The [Ctrl] key is pressed once. Notice the top edge of the dark face. C—The [Ctrl] key is pressed twice.

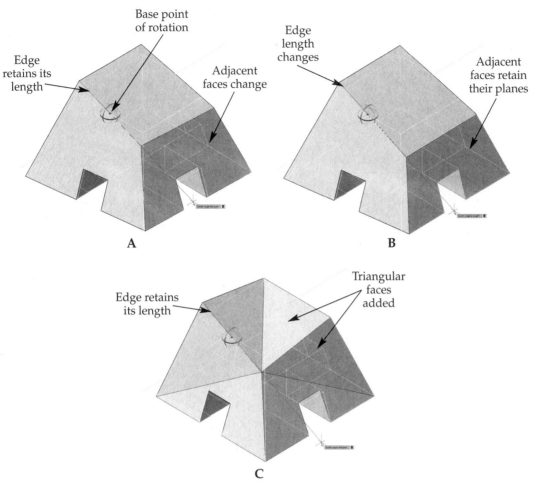

Scaling Edges

Only linear (straight-line) edges can be scaled. Circular edges, such as the ends of cylinders, can be modified using grips or the **SOLIDEDIT** command. These tools can be used to change the diameter or establish taper angles. See Chapter 13 for a complete discussion of the **SOLIDEDIT** command.

To scale a linear edge, enter the **SCALE** command. Select the edge using the [Ctrl] key. Pick a base point for the operation and enter a scale factor. You can also select the edge, pick the edge grip, and cycle to the **SCALE** mode. However, using the scale gizmo may be the best option.

The direction of the scaled edge is related to the base point you select. The base point remains stationary, while the vertices in either direction are scaled. If you enter the **SCALE** command, you are prompted for the base point. If you select the edge grip, the grip becomes the base point. The differences in opposite end and midpoint scaling of an edge are shown in **Figure 12-26**.

There are a few options to achieve different results when dynamically scaling an edge. The [Ctrl] key is used to access these options. First, select the edge. Then, pick the edge grip and cycle to the **SCALE** mode or use the scale gizmo and press and release the [Ctrl] key to cycle through the options.

If the [Ctrl] key is not pressed, the edge is scaled. The shape and planes of adjacent faces are changed to match the scaled edge. See **Figure 12-27A**.

Figure 12-26.
The differences in opposite end and midpoint scaling of an edge. A—The original object.
B—The edge is scaled down with a base point on the left corner. C—The edge is scaled down
to the same scale factor, but the base point is on the right corner. D—The edge is scaled down
to the same scale factor with the base point at the middle of the edge.

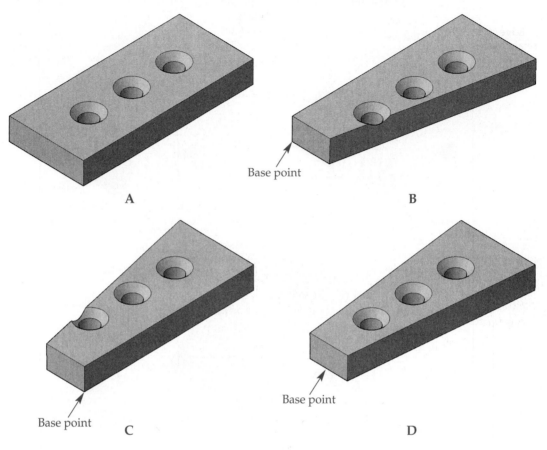

Base point

A

B

Base point

C

Base point

D

Figure 12-27.
Scaling the front edge with the base point at the middle of the edge. The original object is
shown in Figure 12-26A. A—If the [Ctrl] key is not pressed, the edge is scaled and the shape
and planes of adjacent faces are changed. B—If the [Ctrl] key is pressed twice, the edge
is scaled, as are edges attached to it. If the scaling causes faces to become nonplanar, the
adjacent faces may be triangulated.

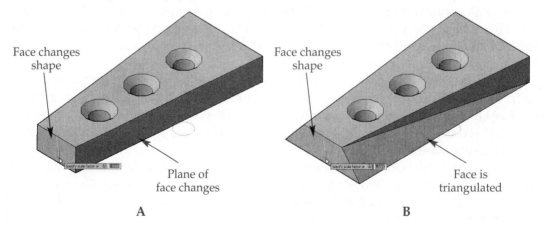

Face changes
shape

Face changes
shape

Plane of
face changes

Face is
triangulated

A

B

If the [Ctrl] key is pressed once, the edge is, in effect, not scaled. This is because the planes of adjacent faces are maintained.

If the [Ctrl] key is pressed twice, the edge is scaled, as are edges attached to the modified edge. However, if the scaling causes faces to become nonplanar, they may be triangulated. See **Figure 12-27B**.

Coloring Edges

To change the color of an edge, use the [Ctrl] key selection method to select the edge. Next, open the **Properties** palette. In the **General** category, pick the drop-down list for the Color property. Select the desired color or pick Select Color... and choose a color from the **Select Color** dialog box. Edges cannot have materials assigned to them.

Deleting Edges

Edges can be deleted in certain situations. In order for an edge to be deleted, it must completely divide two faces that lie on the same plane. If this condition is met, the **ERASE** command or the [Delete] key can be used to remove the edge. The two faces become a single face.

Exercise 12-4

Complete the exercise on the companion website.
www.g-wlearning.com/CAD

Vertex Subobject Editing

The modification of a single vertex involves moving the vertex and stretching all edges and planar faces attached to it. Vertex grips are circular and located on the vertex, as shown in **Figure 12-28**. Use the vertex subobject filter to assist in selecting vertices. A single vertex cannot be rotated or scaled, but you can select multiple vertices and perform rotating and scaling edits. When editing multiple vertices in this manner, you are, in effect, editing edges.

As with other subobject editing functions performed on a 3D solid primitive, the solid's history is removed when a vertex is modified. The solid can no longer be edited using the primitive grips; only a single base grip is displayed. Further editing of the solid must be done with the **SOLIDEDIT** command, discussed in Chapter 13, or through subobject editing.

Figure 12-28.
Vertex grips are
circular and placed
on the vertex.

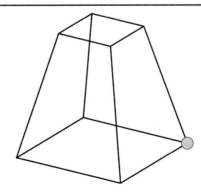

Moving Vertices

To move a vertex, select it using the [Ctrl] key method. If the move gizmo is not displayed, select it from the gizmo drop-down list in the **Selection** panel. You can use the move gizmo, the **MOVE** command, or standard grip editing modes to move the vertex. Hovering over the base vertex grip displays a shortcut menu with two move options. See **Figure 12-29A**. The **Move Vertex** option allows the vertex to be moved without triangulating adjacent faces, but the faces may change shape. See **Figure 12-29B**. In some cases, AutoCAD may deem it necessary to triangulate faces. Using the **Allow Triangulation** option results in the triangulation of adjacent faces when the vertex is moved. See **Figure 12-29C**. Pressing the [Ctrl] key during the edit cycles through the move options.

PROFESSIONAL TIP

If you are dragging a vertex and faces become triangulated, you can transparently change your viewpoint to see the effect of the triangulation. Press and hold the [Shift] key. At the same time, press and hold the mouse wheel button. Now, move the mouse to change the viewpoint. This is a transparent instance of the **3DORBIT** command.

Figure 12-29.
Moving a vertex. A—The original object. B—Using the **Move Vertex** option moves the vertex and changes some of the adjacent faces. B—When using the **Allow Triangulation** option, adjacent faces are triangulated.

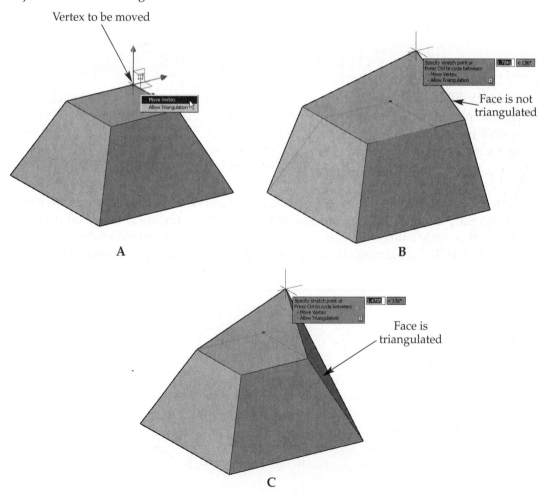

Rotating Vertices

As previously stated, a single vertex cannot be rotated or scaled, but two or more vertices can be. Since two vertices define a line, or edge, any edit is an edge modification. However, the process is slightly different than the edge modifications described earlier in this chapter.

To rotate an edge by selecting its endpoints, press the [Ctrl] key and select each vertex. See **Figure 12-30A**. You may need to use the [Ctrl]+spacebar option to turn on cycling. Notice that grips appear at each selected vertex, but the edges between the vertices are not highlighted.

The **ROTATE** command can now be used to rotate the vertices (if **PICKFIRST** is set to 1). However, a more efficient method for rotating vertices is to use the **3DROTATE** command. The combination of the rotate gizmo and the UCS icon enable you to graphically view the rotation plane. Once the command is initiated, move the base point of the rotate gizmo if needed. See **Figure 12-30B**. Then, select the axis of revolution. Finally, pick the angle start point and enter the rotation. See **Figure 12-30C**.

If the [Ctrl] key is not pressed while dynamically rotating the vertices, the area of the selected vertices does not change and adjacent faces are triangulated. This is because the edges of the adjacent faces are attached to the selected vertices, so their edge length changes as the selected edge is rotated. If the [Ctrl] key is pressed once, the adjacent faces are not triangulated unless necessary, but the faces may change shape.

Figure 12-30.
To rotate or scale vertices, multiple vertices must be selected. In effect, the edges are modified. A—Vertices are selected to be rotated. B—The rotate gizmo is placed at the base of rotation and the rotation axis is selected. C—The vertices are rotated.

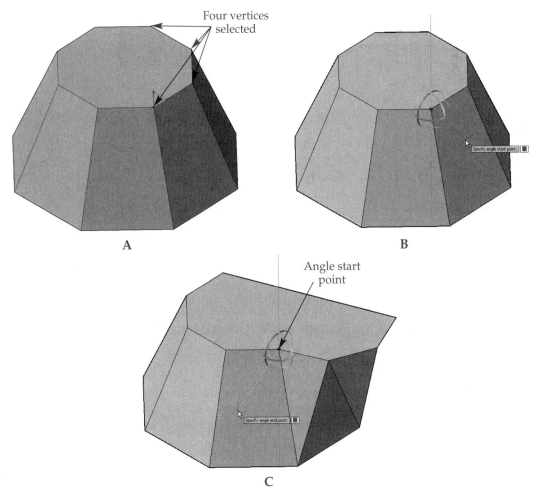

If the selected edge does not dynamically rotate at the "angle end point" prompt, then the desired rotation is not possible.

Scaling Vertices

As mentioned earlier, it is not possible to scale a single vertex. However, two or more vertices can be selected for scaling. This, in effect, scales edges. The selection methods are the same as discussed for rotating vertices and the use of the [Ctrl] key while dragging produces the same effects. As the pointer is dragged, the dynamic display of scaled edges may be difficult to visualize. Therefore, it is best to use a scale factor or the **Reference** option to achieve properly scaled edges.

Exercise 12-5

Complete the exercise on the companion website.
www.g-wlearning.com/CAD

Using Subobject Editing as a Construction Tool

This section provides an example of how subobject editing can be used to not only make changes to existing solids and composites, but as a powerful construction tool. Some of the procedures of subobject editing, such as editing faces, edges, and vertices, are used to construct an HVAC assembly. The entire model is constructed from a single, solid cube. This is the only primitive you will draw.

Editing Faces

1. Begin by setting the units to architectural and drawing a 24" cube. Display the model from the southeast isometric viewpoint.
2. Select the left-hand face and move it 60" to the left. Also, select the front face and move it out 12". This forms the first duct. See **Figure 12-31**.
3. Using the **EXTRUDE** command, select the left-hand face and extrude it 36" to create a new solid. This is a tee junction from which two branches will extend.
4. Select the front face of the new solid and extrude it 28" with a taper angle of 10° to create a new solid that is a reducer.
5. Select the left-hand face of the tee junction and extrude it 20" with a taper angle of 10° to create a new solid that is a second reducer. See **Figure 12-32**.
6. Select the left-hand face of the 20" reducer and move it 3 17/32" along the positive Z axis. This places the top surface of the reducer level with the trunk of the duct. Next, extrude the left-hand face of this reducer 60" into a new solid.
7. Use the **REVOLVE** command to turn the left-hand face of the 60" extension into a new solid that is a 90° bend. Your drawing should now look like **Figure 12-33**.

Figure 12-31.
The left-hand face of the cube is moved 60″. The front face is then moved 12″.

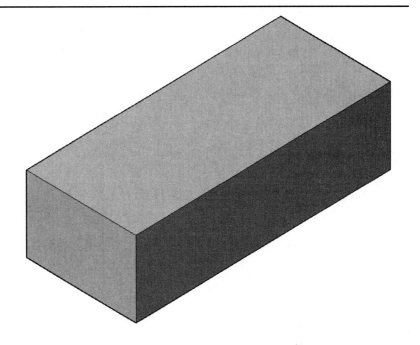

Figure 12-32.
Two reducers are created by extruding faces from the tee junction.

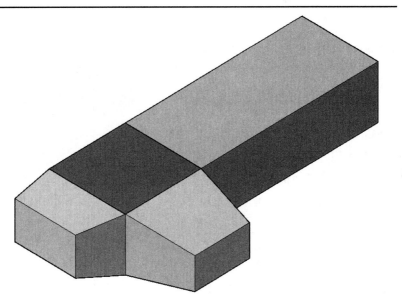

Figure 12-33.
The left end of the 60″ extrusion is revolved 90° to create an elbow.

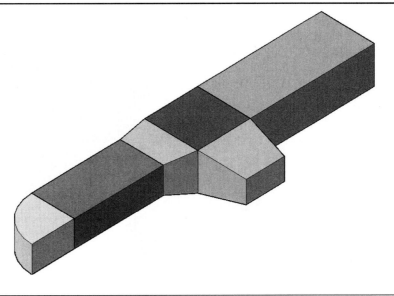

Editing Edges and Vertices

1. The bottom surface of the 28″ reducer must be level with the bottom of the tee junction and main trunk. Select the bottom edge of the reducer's front face and move it down 4 15/16″.
2. Select the two top vertices on the 28″ reducer's front face and move them down (negative Z) 3″.
3. Select the front, rectangular face of the 90° bend and extrude it 72″ into a new solid.
4. Select the front face of the 28″ reducer and extrude it 108″ into a new solid.
5. Select the front face of the new solid created in step 4 and extrude it 26″ to create a new solid that will be a tee junction.
6. Extrude the left-hand face of the tee junction 20″. See **Figure 12-34**.
7. Move the top edge of the left-hand face on the 20″ extrusion created in step 6 down 4″.
8. Move each vertical edge of the 20″ extrusion 6″ toward the center of the duct.
9. Mirror a copy of the 20″ extrusion to the opposite side of the tee junction. The completed drawing should look like **Figure 12-35**.

Figure 12-34.
The face of the revolved elbow is extruded 72″. The face of the right branch is extruded 108″. The right duct is then extruded 26″ and the left face of that extrusion is extruded by 20″.

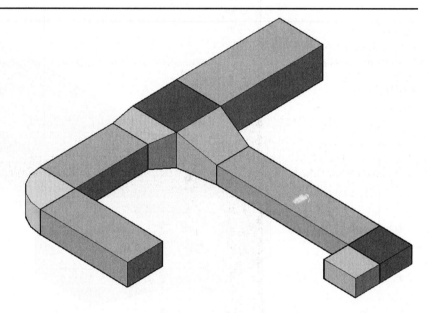

Figure 12-35.
The reducer is mirrored to create the final assembly.

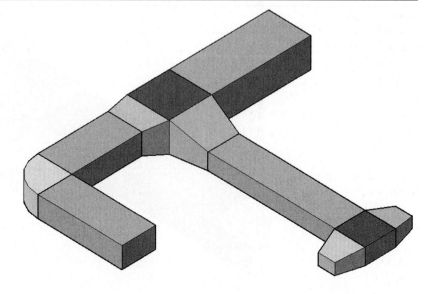

For many of the previously discussed subobject editing techniques, you may find it difficult to edit an object after a complex composite solid is created, especially if it has multiple fillets. You may find it necessary to select the subobject, delete it, and start the process over. Or, you may find it necessary to add geometry to the existing solid using the **UNION** command and then edit the new object as needed.

Other Solid Editing Tools

There are other tools that can be used in solid model editing. As you will learn in Chapter 13, the **SOLIDEDIT** command can be used to edit faces, edges, and vertices, much like subobject editing. In addition, you can extrude a closed boundary with the **PRESSPULL** command, offset edges with the **OFFSETEDGE** command, and explode a solid with the **EXPLODE** command. These methods are discussed in the next sections.

Presspull

The **PRESSPULL** command allows any closed boundary to be extruded. The boundary can be a flat surface, closed polyline, circle, or region. Also, you can use the **PRESSPULL** command to add or subtract to an existing solid by selecting and dragging a closed and bounded area of the solid. The extrusion is always applied perpendicular to the plane of the boundary, but can be in the positive or negative direction. When applied to the face of a solid, it is very similar to the **Extrude Face** option of the **SOLIDEDIT** command, though dynamic feedback is provided for the extrusion with **PRESSPULL**.

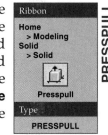

Once the command is initiated, you are prompted to pick inside of the bounded areas to extrude. Move the pointer inside of a boundary and pick. Drag the boundary to a new location and pick, or, if dynamic input is on, enter the distance to extrude the face. See **Figure 12-36**. By extruding the cylinder in a positive direction, the new cylinder is automatically unioned to the existing part, **Figure 12-36B**. Extruding in a negative direction subtracts the cylinder from the wedge, **Figure 12-36C**. The **PRESSPULL** command stays active so repetitive selections can occur in a single command sequence.

The entire boundary must be visible on the screen or the loop will not be found. Also, the **When a command is active** check box must be checked in the **Selection preview** area of the **Selection** tab in the **Options** dialog box.

Offsetting Edges

The **OFFSETEDGE** command allows you to offset the edges of a planar surface to create a new, closed object on the same plane as the existing surface. The new object is a closed polyline or spline that can then be used to create a new 3D object. The planar surface you offset can be on a 3D solid or surface model. After entering the **OFFSETEDGE** command, select the surface and then pick a point inside or outside of the existing edge. The point you pick determines where the offset is created. You can also use the **Distance** option to create the offset at a specific distance from the

Figure 12-36.
Using the **PRESSPULL** command. A—Pick inside of a boundary (shown in color) and drag the boundary to a new location. B—The completed operation with a positive distance, resulting in a union. C—The completed operation with a negative distance, resulting in a subtraction.

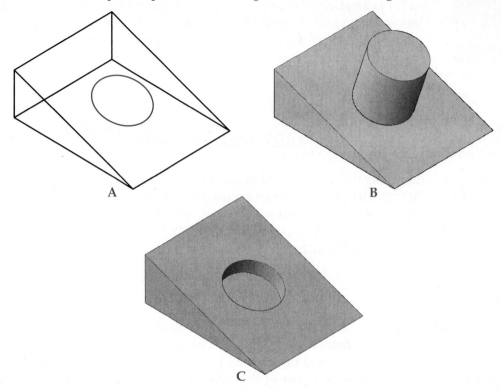

existing edge. The **OFFSETEDGE** command allows you to continue selecting faces to create offsets until you press [Esc] or [Enter]. The procedure is similar to that used with the **OFFSET** command.

An example of using the **OFFSETEDGE** command is shown in **Figure 12-37**. Before specifying the point through which the offset object is created, you can use the **Corner** option to create the offset object with round or square corners. Results after using the **Round** and **Corner** options are shown in **Figure 12-37B**. In both examples shown, the new offset object is a closed polyline.

Once the offset object is created, you can use the **EXTRUDE** or **PRESSPULL** command to create a new 3D feature. See **Figure 12-37C**. In each example shown, an extrusion is created with the **PRESSPULL** command.

Exercise 12-6

Complete the exercise on the companion website.
www.g-wlearning.com/CAD

Exploding a Solid

A solid can be exploded. This turns the solid into surfaces and/or regions. Flat surfaces on the solid are turned into regions. Curved surfaces on the solid are turned into surfaces. To explode a solid, select the **EXPLODE** command. Then, pick the solid(s) to explode and press [Enter].

Figure 12-37.
Using the **OFFSETEDGE** command. A—After entering the command, select a point on the planar surface with the edges to offset. B—The offset object can have rounded or square corners. In each example shown, a closed polyline is created. C—The new polyline object can be used for modeling purposes. Shown are results after using the **PRESSPULL** command. Extruding in a negative direction subtracts material. Extruding in a positive direction adds material.

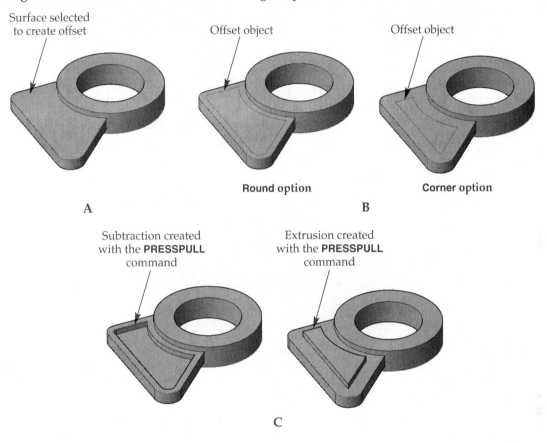

Surface selected to create offset

Offset object

Offset object

Round option

Corner option

A

B

Subtraction created with the **PRESSPULL** command

Extrusion created with the **PRESSPULL** command

C

Exercise 12-7

Complete the exercise on the companion website.
www.g-wlearning.com/CAD

Working with Associative Arrays

A 3D array constructed with the **ARRAYRECT**, **ARRAYPOLAR**, or **ARRAYPATH** command can be created as an associative or non-associative array. Arrays are introduced in Chapter 10. An associative array acts as a single object, much like a block. If you try to use a Boolean operation with an associative array, AutoCAD will not allow the operation to perform. As with a block, an array is a collection of objects that are defined as one object. Think of a 3D associative array as a group of solids in a plastic "wrapper." This type of object cannot be used in a Boolean operation. For example, if you attempt to subtract an associative polar array of cylinders from a plate, AutoCAD will display the following message:

No solids, surfaces, or regions selected.

To use arrayed objects for a Boolean operation, create the array as a non-associative array or explode the array using the **EXPLODE** command. See **Figure 12-38**. Exploding an array creates individual objects that can then be used for other modeling purposes.

Figure 12-38.
Using arrayed objects in a Boolean operation. A—A base plate with a single cylinder.
B—A polar array is created from the cylinder. If the array is created as an associative array,
it cannot be subtracted from the base plate in a Boolean operation. C—The result after
exploding the array and subtracting the individual cylinders from the base plate. Shown is a
shaded view of the model.

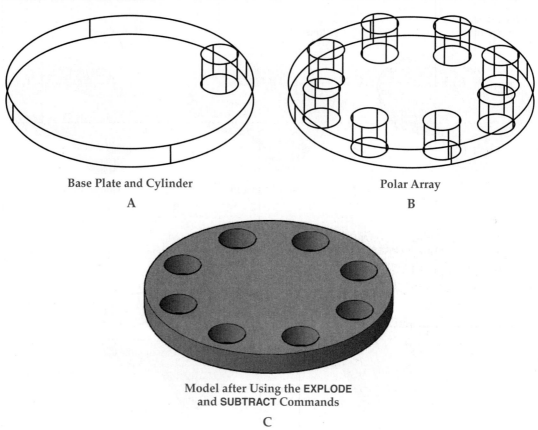

Base Plate and Cylinder

A

Polar Array

B

Model after Using the **EXPLODE**
and **SUBTRACT** Commands

C

Figure 12-38C shows the result after using the **SUBTRACT** command to subtract the
individual cylinders from the base plate.

Exercise 12-8

Complete the exercise on the companion website.
www.g-wlearning.com/CAD

Chapter Review

1. How do you select a subobject?
2. How do you deselect a subobject?
3. When grip editing, two types of grips appear on the object. Name the two types of grips.
4. If you have a cylinder primitive with a height of 10 units, but the height requirement has changed to 15 units, explain the procedure for adding 5 units to the cylinder height.
5. How can you change the radius of a fillet or the distances of a chamfer?
6. When moving a face on a solid primitive, how can you accurately control the axis of movement?
7. When moving a face on a solid primitive, which option maintains the planes of adjacent faces while modifying the size of the face?
8. Describe a major difference of function between the **ROTATE** and **3DROTATE** commands.
9. Which system variable enables you to use the **3DROTATE** command in a 3D view even if you select the **ROTATE** command?
10. How does the location and shape of an edge grip differ from a face grip?
11. What is the most efficient tool to use when rotating an edge and how is it displayed?
12. What is the only type of edge that can be scaled?
13. What is the only editing function that can be done when editing a single vertex?
14. How are two or more vertices selected for editing?
15. What is created when offsetting an edge of a solid with the **OFFSETEDGE** command?
16. Which option of the **OFFSETEDGE** command is used to create round corners on the resulting offset object?
17. What is the function of the **PRESSPULL** command?
18. On which types of surfaces or objects can the **PRESSPULL** command be used?
19. When a solid object is exploded, what happens to the flat surfaces of the solid?
20. When a solid object is exploded, what happens to the curved surfaces of the solid?

Drawing Problems

1. Draw the bookcase shown using the dimensions given. The final result should be a single solid object. Then, use grip and subobject editing procedures to edit the object as follows.
   A. Change the width of the bookcase to 3'.
   B. Change the height of the bookcase by eliminating the top section. The resulting height should be 3'-2".
   C. Save the drawing as P12_01.

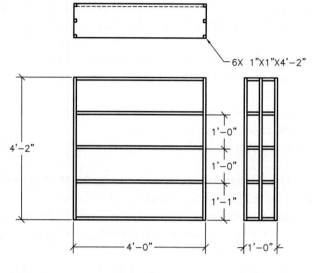

ALL WOOD THICKNESS IS 1"

2. Open problem P12_01. Save it as P12_02. Use primitive and subobject editing procedures to create the following edits.
   A. Change the depth of the top of the bookcase to 6-1/2".
   B. Change the depth of the bottom of the bookcase to 24".
   C. Reduce the height of the front uprights so they are flush with the top surface of the next lower shelf.
   D. Extend the front of the second lowest shelf to match the front of the bottom. Add two uprights at the front corners between the bottom and this shelf.
   E. Save the drawing.

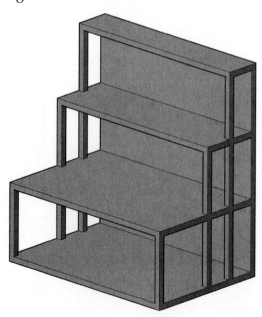

Drawing Problems - Chapter 12

3. Draw the mounting bracket shown below. Then, use primitive and subobject editing procedures to create the following edits.
   A. Change the 3.00″ dimension to 3.50″.
   B. Change the 2.50″ dimension in the front view to 2.75″.
   C. Change the location of the slot in the auxiliary view from .60″ to .70″ and change the length of the slot to 1.15″.
   D. Change the width of each foot in the top view from 2.00″ to 1.50″. The overall dimension (5.00″) should not change.
   E. Change the angle of the bend from 15° to 45°.
   F. Save the drawing as P12_03.

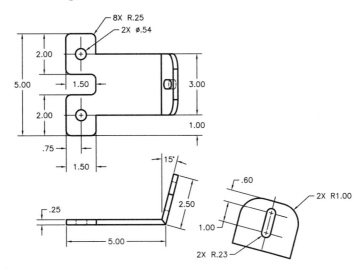

4. Draw as a single composite solid the desk organizer shown in the orthographic views. Then, use primitive and subobject editing procedures to create the following edits. The final object should look like the shaded view.
   A. Change the 3″ height to 3.25″.
   B. Change the 2″ height to 1.85″.
   C. Increase the thickness of the long compartment divider to .5″. The increase in thickness should be evenly applied along the centerline of the divider. Locate three evenly spaced, Ø5/16″ × 1.5″ holes in this divider.
   D. Angle the top face of the rear compartments by 30°. The height of the rear of the organizer should be approximately 4.5″ and all corners on the bottom of the organizer should remain square.
   E. Save the drawing as P12_04.

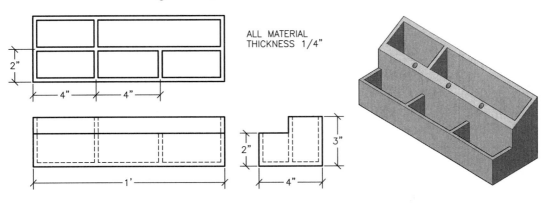

Drawing Problems - Chapter 12

5. Draw the pencil holder shown. Then, use primitive and subobject editing procedures to create the following edits.
   A. Change the depth of the base to 4.000". The base should be rectangular, not square, and the grooves should become shorter.
   B. Change the height of the top groove from .250" to .125".
   C. Change the diameter of two holes from ∅.450" to ∅.625".
   D. Change the diameter of the other two holes from ∅.450" to ∅1.000".
   E. Rotate the top face 15° away from the side with the grooves. The planes of the adjoining faces should not change. Refer to the shaded view.
   F. Save the drawing as P12_05.

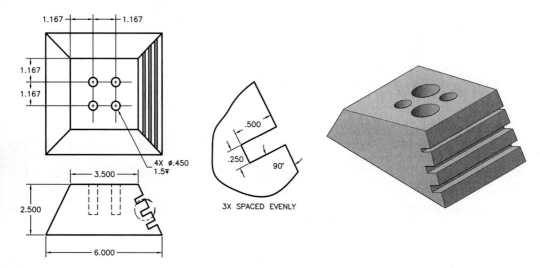

Learning Objectives

After completing this chapter, you will be able to:

✓ Change the shape and configuration of solid object faces.
✓ Copy and change the color of solid object edges and faces.
✓ Break apart a composite solid composed of physically separate entities.
✓ Extract a wireframe to project edges from a 3D solid using the **XEDGES** command.
✓ Use the **SOLIDEDIT** command to construct and edit a solid model.

AutoCAD provides expanded capabilities for editing solid models. As you saw in the previous chapter, grips can be used to edit a solid model. Also, the subobjects that make up a solid, such as faces, edges, and vertices, can be edited. A single command, **SOLIDEDIT**, allows you to edit faces, edges, or the entire body of the solid.

Mesh objects cannot be modified using the **SOLIDEDIT** command. The mesh object must be converted to a solid first. If you select a mesh for editing with the **SOLIDEDIT** command, you are given the option of converting it to a solid, as long as the display of the dialog box has not been turned off.

Overview of the SOLIDEDIT Command

The **SOLIDEDIT** command allows you to edit the faces, edges, and body of a solid. Many of the subobject editing functions discussed in Chapter 12 can also be performed with the **SOLIDEDIT** command. The options of the **SOLIDEDIT** command can be accessed in the **Solid Editing** panel on the **Home** or **Solid** tab of the ribbon or by typing SOLIDEDIT. See **Figure 13-1**. The quickest method of entering the command is by using the ribbon. For example, directly select a face editing option from the drop-down list in the **Solid Editing** panel, as shown in **Figure 13-1**.

Figure 13-1.
Accessing the **SOLIDEDIT** command options.

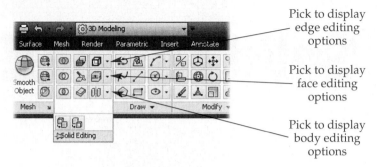

Pick to display edge editing options

Pick to display face editing options

Pick to display body editing options

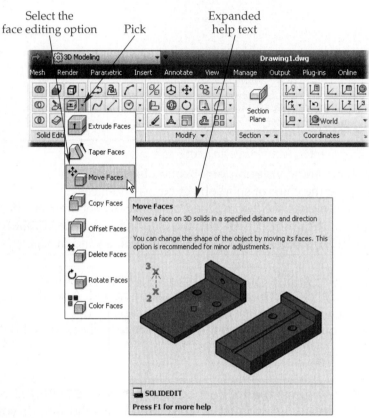

Select the face editing option

Pick

Expanded help text

When the **SOLIDEDIT** command is typed, you are first asked to select the component of the solid with which you wish to work. Specify **Face**, **Edge**, or **Body**. The editing options for the selected component are then displayed and are the same as those seen in **Figure 13-1**.

The editing function is directly entered when the option is selected from the ribbon. This is why using the ribbon is the most efficient method of entering the **SOLIDEDIT** command options.

The following sections provide an overview of the solid model editing functions of the **SOLIDEDIT** command. Each option is explained and the results of each are shown. A tutorial later in the chapter illustrates how these options can be used to construct a model.

AutoCAD displays a variety of error messages when invalid solid editing operations are attempted. Rather than trying to interpret the wording of these messages, just realize that what you tried to do will not work. Actions that may cause errors include trying to rotate a face into other faces or extruding and tapering an object at too great of an angle. When an error occurs, try the operation again with different parameters or determine a different approach to solving the problem in order to maintain the design intent.

Face Editing

The basic components of a solid are its faces and the greatest number of **SOLIDEDIT** options are for editing faces. All eight face editing options ask you to select faces. It is important to make sure you select the correct part of the model for editing. Remember the following three steps when using any of the face editing options.

1. First, select a face to edit. If you pick an edge, AutoCAD selects the two faces that share the edge. If this happens, use the **Remove** option to deselect the unwanted face. A more intuitive approach is to select the open space of the face as if you were touching the side of a part. AutoCAD highlights only that face.

2. Adjust the selection set at the Select faces or [Undo/Remove/ALL]: prompt. The following options are available.

   - **Undo.** Removes the previous selected face(s) from the selection set.
   - **Remove.** Allows you to select faces to remove from the selection set. This is only available when **Add** is current.
   - **All.** Adds all faces on the model to the selection set. This is only available after selecting at least one face. It can also be used to remove all faces if **Remove** is current.
   - **Add.** Allows you to add faces to the selection set. This is only available when **Remove** is current.

3. Press [Enter] to continue with face editing.

Extruding Faces

An extruded face is moved, or stretched, in a selected direction. The extrusion can be straight or have a taper. To extrude a face, select the command and pick the **Face**>**Extrude** option. Remember, the option is directly entered when picking the button in the ribbon. You are then prompted to select the face(s) to extrude. Nonplanar (curved) faces cannot be extruded. As you pick faces, the prompt verifies the number of faces selected. For example, when an edge is selected, the prompt reads 2 faces found. When done selecting faces, press [Enter] to continue.

Next, the height of the extrusion needs to be specified. A positive value adds material to the solid, while a negative value subtracts material from the solid. A taper can also be given.

Ribbon

Home
> Solid Editing
Solid
> Solid Editing

Extrude Faces

Type

SOLIDEDIT

> Specify height of extrusion or [Path]: *(enter the height)*
> Specify angle of taper for extrusion <0>: *(enter an angle or accept the default)*
> Solid validation started.
> Solid validation completed.
> Enter a face editing option
> [Extrude/Move/Rotate/Offset/Taper/Delete/Copy/coLor/mAterial/Undo/eXit] <eXit>: **X**↵
> Solids editing automatic checking: SOLIDCHECK=1
> Enter a solids editing option [Face/Edge/Body/Undo/eXit] <eXit>: **X**↵

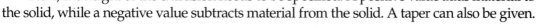

Figure 13-2.
Extruding faces on
an object. A—The
original object.
B—The top face is
extruded with a 0°
taper angle.
C—The top face of
the original object
is extruded with a
30° taper angle.
D—The top and
right-hand faces of
the original object
are extruded with
15° taper angles.

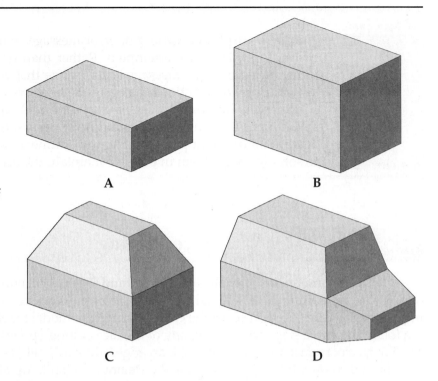

Figure 13-2 shows an original solid object and the result of extruding the top face with a 0° taper angle and a 30° taper angle. It also shows the original solid object with two adjacent faces extruded with 15° taper angles.

In addition to extruding a face perpendicular to itself, the extruded face can follow a path. Select the **Path** option at the Specify height of extrusion or [Path]: prompt. The path of extrusion can be a line, circle, arc, ellipse, elliptical arc, polyline, or spline. The extrusion height is the exact length of the path. See **Figure 13-3**.

Exercise 13-1

Complete the exercise on the companion website.
www.g-wlearning.com/CAD

Figure 13-3.
The path of
extrusion can be
a line, circle, arc,
ellipse, elliptical arc,
polyline, or spline.
Here, the paths are
shown in color.

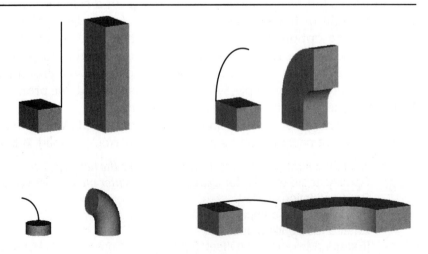

Moving Faces

The **Move** option moves a face in the specified direction and lengthens or shortens the solid object. In another application, a solid model feature (such as a hole) that has been subtracted from an object to create a composite solid can be moved with this option. Object snaps may interfere with the function of this option, so they may need to be toggled off during the operation.

To move a face, select the command and pick the **Face>Move** option. If the button is picked in the ribbon, the option is directly entered. You are then prompted to select the face(s) to move. When done selecting faces, press [Enter] to continue. Next, you are prompted to select a base point of the operation:

> Specify a base point or displacement: *(pick a base point)*
> Specify a second point of displacement: *(pick a second point or enter coordinates)*
> Solid validation started.
> Solid validation completed.
> Enter a face editing option
> [Extrude/Move/Rotate/Offset/Taper/Delete/Copy/coLor/mAterial/Undo/eXit] <eXit>: **X**↵
> Solids editing automatic checking: SOLIDCHECK=1
> Enter a solids editing option [Face/Edge/Body/Undo/eXit] <eXit>: **X**↵

When adjacent faces are perpendicular, the edited face is moved in a direction so the new position keeps the face parallel to the original. See **Figures 13-4A** and **13-4B**. Faces that are normal to the current UCS can be moved by picking a new location or entering a direct distance. If you are moving a face that is not normal to

Figure 13-4.
A—The hole will be moved using the **Move** option of the **SOLIDEDIT** command. B—The hole is moved. C—When the angled face is moved, a portion of it is altered to be coplanar with the vertical face. D—If the angled face is moved more, it becomes completely coplanar to the vertical face. This is a new, single face.

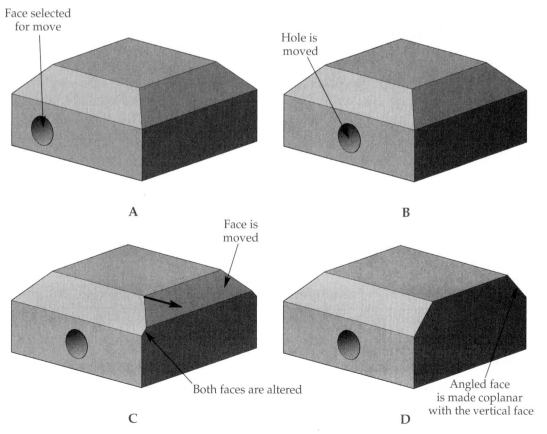

the current UCS, you can enter coordinates for the second point of displacement, but it may be easier to first use the **Face** option of the **UCS** command to align the UCS with the face to be moved.

When adjacent faces join at angles other than 90°, the moved face will be relocated as previously stated, but only if the movement is less than the dimensional offset of the two faces. For example, in **Figure 13-4B**, the top edge of the angled face is in .5″ from the vertical face. If the angled face is moved outward a distance of less than .5″, it is altered as shown in **Figure 13-4C**. A portion of the angled face becomes coplanar with the vertical face. If the angled face is moved outward a distance greater than .5″, it is altered so that it forms a single plane with the adjacent face. What has actually happened is that the angled face is moved beyond the adjacent face, while remaining parallel to its original position. Thus, in effect, it has disappeared because the adjacent, vertical face cannot be altered. See **Figure 13-4D**. In this example, the angled face was moved .75″. The new vertical face that is created can now be moved.

Exercise 13-2

Complete the exercise on the companion website.
www.g-wlearning.com/CAD

Offsetting Faces

SOLIDEDIT

Ribbon

Home
> Solid Editing
Solid
> Solid Editing

Offset Faces

Type

SOLIDEDIT

The **Offset** option may seem the same as the **Extrude** option because it moves faces by a specified distance or through a specified point. Unlike the **OFFSET** command in AutoCAD, this option moves all selected faces a specified distance. It is most useful when you wish to change the size of features such as slots, holes, grooves, and notches in solid parts. A positive offset distance increases the size or volume of the solid (adds material). A negative distance decreases the size or volume of the solid (removes material). Therefore, if you wish to make the width of a slot wider, provide a negative offset distance to decrease the size of the solid. Picking points to set the offset distance and direct distance entry are always taken as a positive value, so negative values must be entered using the keyboard.

To offset a face, select the command and pick the **Face>Offset** option. Remember, the option is directly entered when picking the button in the ribbon. You are then prompted to select the face(s) to offset. When done selecting faces, press [Enter] to continue. Next, enter the offset distance and press [Enter] and then exit the command. See **Figure 13-5** for examples of features edited with the **Offset** option.

PROFESSIONAL TIP

Nonplanar (curved) faces cannot be extruded, but can be offset. Using the **Offset** option, you can, in effect, "extrude" a nonplanar face.

Exercise 13-3

Complete the exercise on the companion website.
www.g-wlearning.com/CAD

Figure 13-5.
Offsetting faces. A—The original objects. The hole is selected to offset. The interior of the L is also selected to offset. B—A positive offset distance increases the size or volume of the solid. C—A negative offset distance decreases the size or volume of the solid.

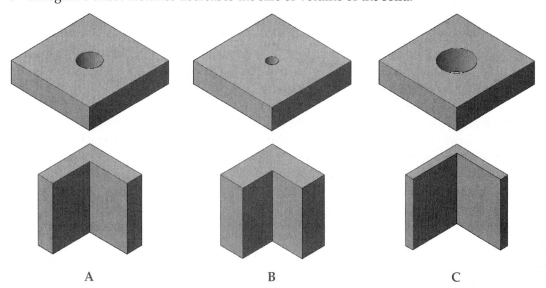

A   B   C

Deleting Faces

The **Delete** option deletes selected faces. This is a quick way to remove features such as chamfers, fillets, holes, and slots. To delete a solid face, select the command and pick the **Face>Delete** option. If the button is picked in the ribbon, the option is directly entered. You are then prompted to select the face(s) to delete. When done selecting faces, press [Enter] to continue and then exit the command. When a face is deleted, existing faces extend to fill the gap. No additional faces are created. For instance, the inclined surface of a wedge cannot be deleted as there are no existing faces that can be extended to fill the gap. When deleting the face that is a chamfered or filleted edge, the adjacent edges are extended to fill the gap. See **Figure 13-6**.

Ribbon
Home
> Solid Editing
Delete Faces
Type
SOLIDEDIT

Rotating Faces

The **Rotate** option rotates a face about a selected axis. To rotate a solid face, select the command and pick the **Face>Rotate** option. Remember, the option is directly entered when picking the button in the ribbon. You are then prompted to select the face(s) to rotate. When done selecting faces, press [Enter] to continue. There are several methods by which a face can be rotated.

Ribbon
Home
> Solid Editing
Rotate Faces
Type
SOLIDEDIT

Figure 13-6.
Deleting faces.
A—The original objects with three rounds. B—The faces of two rounds have been deleted.

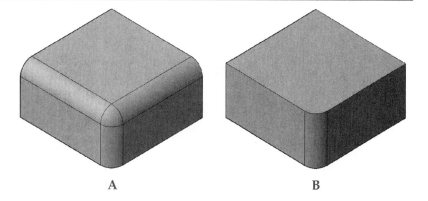

A   B

The **2points** option is the default. Pick two points to define the "hinge" about which the face will rotate. Then, provide the rotation angle and exit the command.

The **Axis by object** option allows you to use an existing object to define the axis of rotation. You can select the following objects.

- **Line.** The selected line becomes the axis of rotation.
- **Circle, arc, or ellipse.** The Z axis of the object becomes the axis of rotation. This Z axis is a line that passes through the center of the circle, arc, or ellipse and is perpendicular to the plane on which the 2D object lies.
- **Polyline or spline.** A line connecting the polyline or spline's start point and endpoint becomes the axis of rotation.

After selecting an object, enter the angle of rotation and exit the command.

When you select the **View** option, the axis of rotation is perpendicular to the current view, with the positive direction coming out of the screen. This axis is identical to the Z axis when the **UCS** command **View** option is used. Next, enter the angle of rotation and exit the command.

The **Xaxis**, **Yaxis**, and **Zaxis** options prompt you to select a point. The X, Y, or Z axis passing through that point is used as the axis of rotation. Then, enter the angle of rotation and exit the command.

Figure 13-7 provides several examples of rotated faces. Notice how the first and second pick points determine the direction of positive and negative rotation angles.

A positive rotation angle moves the face in a clockwise direction looking from the first pick point to the second. Conversely, a negative angle rotates the face counterclockwise. If the rotated face will intersect or otherwise interfere with other faces, an error message indicates that the operation failed or no solution was calculated. In this case, you may wish to try a negative angle if you previously entered a positive one. In addition, you can try using the opposite edge of the face as the axis of rotation by picking the appropriate points.

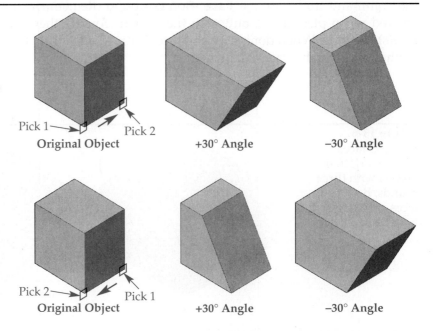

Figure 13-7.
When rotating faces, the first and second pick points determine the direction of positive and negative rotation angles.

Pick 1 — Pick 2
Original Object

+30° Angle

−30° Angle

Pick 2 — Pick 1
Original Object

+30° Angle

−30° Angle

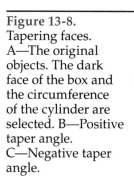

Tapering Faces

The **Taper** option tapers a face at the specified angle, from the first pick point to the second. To taper a solid face, select the command and pick the **Face>Taper** option. If the button is picked in the ribbon, the option is directly entered. You are then prompted to select the face(s) to taper. When done selecting faces, press [Enter] to continue:

Specify the base point: *(pick the base point)*
Specify another point along the axis of tapering: *(pick a point along the taper axis)*
Specify the taper angle: *(enter a taper value)*

Tapers work differently depending on whether the faces being tapered describe the outer boundaries of the solid, a cavity, or a removed portion of the solid. A positive taper angle always removes material. A negative taper angle always adds material. For example, if a positive taper angle is entered for a solid cylinder, the selected object is tapered in on itself from the base point along the axis of tapering, thus removing material. A negative angle tapers the object out away from itself to increase its size along the axis of tapering, thus adding material. See **Figure 13-8**.

On the other hand, if the face (or faces) of a feature such as a hole or slot are tapered, a positive taper angle increases the size of the feature along the axis of tapering. For example, if a round hole is tapered using a positive taper angle, its

Figure 13-8.
Tapering faces.
A—The original objects. The dark face of the box and the circumference of the cylinder are selected. B—Positive taper angle.
C—Negative taper angle.

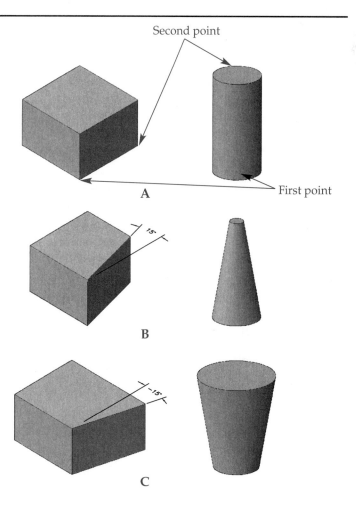

diameter increases from the base point along the axis of tapering, thus removing material from the solid. Conversely, if the same round hole is tapered using a negative taper angle, its diameter decreases from the base point along the axis of tapering, thus adding material to the solid. **Figure 13-9** shows some examples of tapering features.

Exercise 13-5

Complete the exercise on the companion website.
www.g-wlearning.com/CAD

Copying Faces

The **Copy** option copies a face to the location or coordinates given. The copied face is *not* part of the original solid model. It is actually a region, which can later be extruded, revolved, swept, etc., into a solid. This may be useful when you wish to construct a mating part in an assembly that has the same features on the mating faces or the same outline. This option is quick to use because you can pick a base point on the face, then enter a single direct distance value for the displacement. Be sure an appropriate UCS is set if you wish to use direct distance entry.

To copy a solid face, select the command and pick the **Face>Copy** option. Remember, the option is directly entered when picking the button in the ribbon. You are then prompted to select the face(s) to copy. When done selecting faces, press [Enter] to continue. You are prompted for a base point for the copy. Pick this point and then pick a second point of displacement or press [Enter] to use the first point as a displacement. See **Figure 13-10** for examples of copied faces.

PROFESSIONAL TIP

Copied faces can also be useful for creating additional views. For example, you can copy a face to create a separate plan view with dimensions and notes. A copied face can also be enlarged to show details and to provide additional notation for design or assembly.

Figure 13-9.
If a hole or slot is tapered using a positive taper angle, its diameter or width increases from the base point along the axis of tapering, thus removing material from the solid. A negative taper angle increases the volume of the solid.

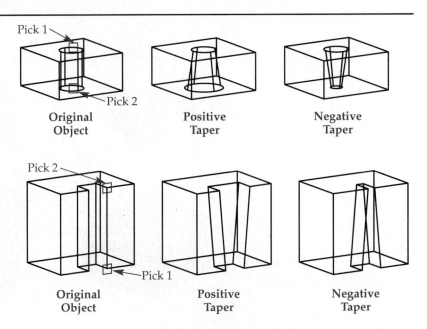

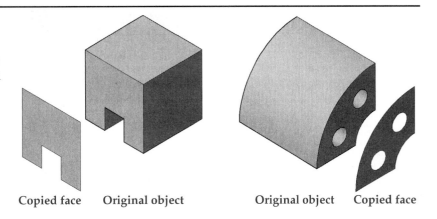

Figure 13-10.
A face can be quickly copied by picking a base point on the face and then entering a direct distance value for the displacement.

Copied face   Original object   Original object   Copied face

Coloring Faces

You can quickly change a selected face to a different color using the **Color** option. Select the command and pick the **Face>Color** option. If the button is picked in the ribbon, the option is directly entered. You are then prompted to select the face(s) to color. When done selecting faces, press [Enter] to continue. Next, choose the desired color from the **Select Color** dialog box that is displayed. Remember, the color of the object (or face) determines the shaded color.

PROFESSIONAL TIP

Use the **Color** option to enhance features on a solid. For example, if you have a model that has holes or slots, enhance just the color of these features.

The **Material** option of the **SOLIDEDIT** command allows you to assign a material to selected faces on a solid. Materials are discussed in Chapter 17.

Exercise 13-6

Complete the exercise on the companion website.
www.g-wlearning.com/CAD

Edge Editing

Edges can be edited in only two ways with the **SOLIDEDIT** command. They can be copied from the solid. Also, the color of an edge can be changed.

Copying an edge is similar to copying a face. To copy a solid edge, select the command and pick the **Edge>Copy** option. Remember, the option is directly entered when picking the button in the ribbon. See **Figure 13-11**. You are then prompted to select the edge(s) to copy. When done selecting edges, press [Enter] to continue. You are prompted for a base point for the copy. Pick this point and then pick a second point of displacement or press [Enter] to use the first point as a displacement. The edge is copied as a line, arc, circle, ellipse, or spline.

Figure 13-11.
Selecting
edge-editing
options.

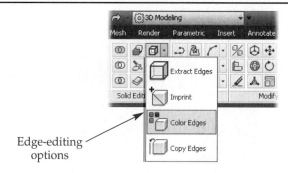

Edge-editing
options

SOLIDEDIT

Ribbon

Home
> Solid Editing

Color Edges

Type

SOLIDEDIT

To color a solid edge, select the command and pick the **Edge>Color** option. You are then prompted to select the edge(s) to color. When done selecting edges, press [Enter] to continue. Next, choose the desired color from the **Select Color** dialog box that is displayed and pick the **OK** button. The edges are now displayed with the new color. You may need to set a wireframe or hidden visual style current to see the change.

Extracting a Wireframe

XEDGES

Ribbon

Home
> Solid Editing
Solid
> Solid Editing

Extract Edges

Type

XEDGES

The **XEDGES** command creates copies of, or extracts, all of the edges on a selected solid. You can also select edges of a surface, mesh, region, or subobject. Once the command is initiated, you are prompted to select objects. Select one or more solids and press [Enter]. The edges are extracted and placed on top of the existing edges. See **Figure 13-12.** The new objects are created on the current layer.

Straight edges and the curved edges where cylindrical surfaces intersect with flat or other cylindrical surfaces are the only edges extracted. Spheres and tori have no edges that can be extracted. The round bases of cylinders and cones are the only edges of those objects that will be extracted.

Figure 13-12.
Extracting edges
with the **XEDGES**
command. A—The
original object.
B—The extracted
wireframe (edges).

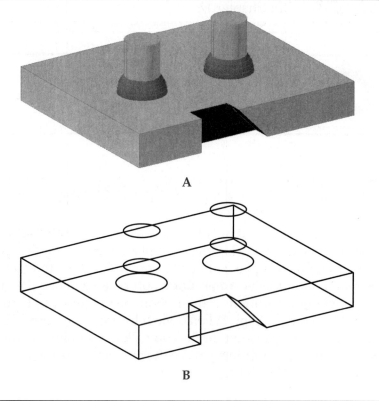

A

B

Exercise 13-7

Complete the exercise on the companion website.
www.g-wlearning.com/CAD

Body Editing

The body editing options of the **SOLIDEDIT** command perform editing operations on the entire body of the solid model. The body options are **Imprint**, **Separate**, **Shell**, **Clean**, and **Check**. The next sections cover these body editing options.

The **Imprint** option is a body editing function. However, since it modifies a body by adding edges, the option is located in the edges drop-down list in the **Solid Editing** panel on the **Home** tab. On the **Solid** tab, it is a separate button in the **Solid Editing** panel. In both places, the **Separate**, **Shell**, **Clean**, and **Check** options are found in the body drop-down list.

Imprint

Arcs, circles, lines, 2D and 3D polylines, ellipses, splines, regions, bodies, and 3D solids can be imprinted onto a solid, if the object intersects the solid. The imprint becomes a face on the surface based on the overlap between the two intersecting objects. Once the imprint has been made, the new face can be modified.

To imprint an object on a solid, select the **SOLIDEDIT** command and pick the **Body>Imprint** option. If IMPRINT is typed or the button is picked on the ribbon, the option is directly entered. Once the option is activated, you are prompted to select the solid. This is the object on which the other objects will be imprinted. Then, select the objects to be imprinted. You have the option of deleting the source objects.

The imprinted face can be modified using face editing options. **Figure 13-13** illustrates objects imprinted onto a solid model. Two of these are then extruded into the solid to create holes. The **PRESSPULL** command is used on the third object to create a cylindrical feature.

Figure 13-13.
Imprinted objects form new faces that can be extruded into the solid. A—A solid box with three objects on the plane of the top face. B—The objects are imprinted, then two of the new faces are extruded through the solid and used in subtraction operations. The **PRESSPULL** command is used on the third new face to create the cylindrical feature.

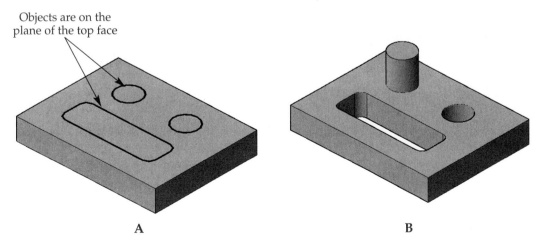

Objects are on the plane of the top face

A

B

Remember that objects are drawn on the XY plane of the current UCS unless you enter a specific Z value. Therefore, before you draw an object to be imprinted onto a solid model, be sure you have set an appropriate UCS for proper placement of the object by using the UCS icon grips or a dynamic UCS. Alternately, you can draw the object on the XY plane and then move the object onto the solid object.

Separate

SOLIDEDIT

Ribbon

Home
> Solid Editing
Solid
> Solid Editing

Separate

Type
SOLIDEDIT

The **Separate** option separates two objects that are both a part of a single solid composite, but appear as separate physical entities. This can happen when modifying solids using the Boolean commands. The **Separate** option may be seldom used, but it has a specific purpose. If you select a solid model and an object physically separate from the model is highlighted, the two objects are parts of the same composite solid. If you wish to work with them as individual solids, they must first be separated.

To separate a solid body, select the command and pick the **Body>Separate** option. If the button is picked in the ribbon, the option is directly entered. You are then prompted to select a 3D solid. After you pick the solid, it is automatically separated. No other actions are required and you can exit the command. However, if you select a solid in which the parts are physically joined, AutoCAD indicates this by prompting The selected solid does not have multiple lumps. A "lump" is a physically separate solid entity. In order to separate a solid, it must be composed of multiple lumps. See Figure 13-14.

Shell

SOLIDEDIT

Ribbon

Home
> Solid Editing
Solid
> Solid Editing

Shell

Type
SOLIDEDIT

A *shell* is a solid that has been "hollowed out." The **Shell** option creates a shell of the selected object using a specified offset distance, or thickness. To create a shell of a solid body, select the command and pick the **Body>Shell** option. Remember, the option is directly entered when picking the button in the ribbon. You are prompted to select the solid. Only one solid can be selected.

After selecting the solid, you have the opportunity to remove faces. If you do not remove any faces, the new solid object will appear identical to the old solid object when shaded or rendered. The thickness of the shell will not be visible. If you wish to create a hollow object with an opening, select the face to be removed (the opening).

Figure 13-14.
A—After the cylinder is subtracted from the box, the remaining solid is considered one solid.
B—Use the **Separate** option to turn this single solid into two solids.

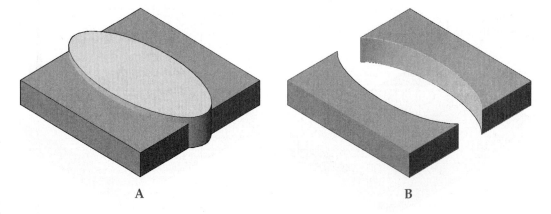

A   B

After selecting the object and specifying any faces to be removed, you are prompted to enter the shell offset distance. This is the thickness of the shell. A positive shell offset distance creates a shell on the inside of the solid body. A negative shell offset distance creates a shell on the outside of the solid body. See **Figure 13-15**. If you shell a solid that contains internal features, such as holes, grooves, and slots, a shell of the specified thickness is placed around those features. This is shown in **Figure 13-16**.

If the shell operation is not readily visible in the current view, you can rotate the view by pressing the [Shift] key and pressing the mouse wheel button at the same time to enter the transparent **3DORBIT** command. You can also rotate the view using the view cube, or you can see the results by picking the **X-Ray Effect** button in the **Visual Styles** panel on the **View** tab of the ribbon.

Figure 13-15.
A—The right-front, bottom, and left-back faces (marked here by gray lines) are removed from the shell operation. B—The resulting object after the shell operation.

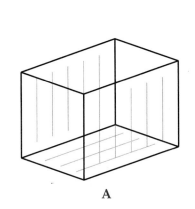
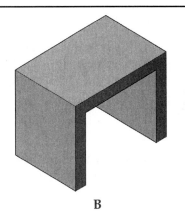

A   B

Figure 13-16.
If you shell a solid that contains internal features, such as holes, grooves, and slots, a shell of the specified thickness is also placed around those features. A—Solid object with holes subtracted. B—Wireframe display after shelling with a negative offset. C—The Conceptual visual style is set current.

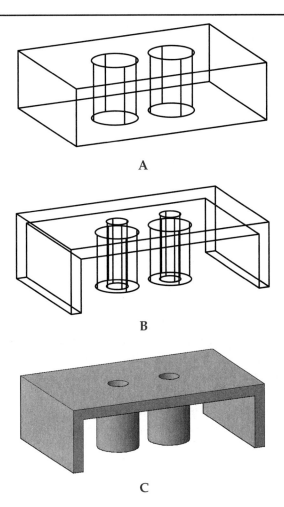

A

B

C

The **Shell** option of the **SOLIDEDIT** command is very useful in applications such as solid modeling of metal castings or injection-molded plastic parts.

Exercise 13-8

Complete the exercise on the companion website.
www.g-wlearning.com/CAD

Clean

The **Clean** option removes all unused objects and shared surfaces. Imprinted objects are not removed. Select the command and pick the **Body>Clean** option. If the button is picked in the ribbon, the option is directly entered. Then, pick the solid to be cleaned. No further input is required. You can then exit the command.

Check

The **Check** option simply determines if the selected object is a valid 3D solid. If a true 3D solid is selected, AutoCAD displays the prompt This object is a valid ShapeManager solid. You can then exit the command. If the object selected is not a 3D solid, the prompt reads A 3D solid must be selected. You are then prompted to select a 3D solid. To access the **Check** option, select the command and pick the **Body>Check** option. Remember, the option is directly entered when picking the button in the ribbon. Then, select the object to check.

Using SOLIDEDIT as a Construction Tool

This section provides an example of how the **SOLIDEDIT** command options can be used not only to edit, but also to construct a solid model. This approach makes it easy to design and construct a model without selecting a variety of commands. It also gives you the option of undoing a single editing operation or an entire editing session without ever exiting the command.

In the following example, **SOLIDEDIT** command options are used to imprint shapes onto the model body and then extrude those shapes into the body to create countersunk holes. Then, the model size is adjusted and an angle and taper are applied to one end. Finally, one end of the model is copied to construct a mating part.

Creating Shape Imprints on a Model

The basic shape of the solid model in this tutorial is drawn as a solid box. Then, shape imprints are added to it. Throughout this exercise, you may wish to change the UCS to assist in the construction of the part.
1. Draw a solid box using the dimensions shown in **Figure 13-17**.
2. Set the Wireframe visual style current.
3. On the top surface of the box, locate a single ∅.4 circle using the dimensions given. Then, copy or array the circle to the other three corners as shown in the figure.
4. Use the **Imprint** option to imprint the circles onto the solid box. Delete the source objects.

Figure 13-17.
The initial setup for the tutorial model.

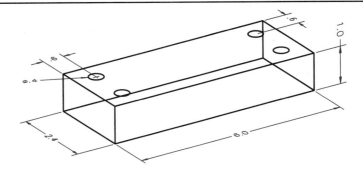

Extruding Imprints to Create Features

The imprinted 2D shapes can now be extruded to create new 3D solid features on the model. Use the **Face>Extrude** option to extrude all four imprinted circles.

1. When you select the edge of the first circle, all features on that face are highlighted, but only the circle you picked and the top face have actually been selected. If you pick inside the circle, only the circle is selected and highlighted. In either case, be sure to also pick the remaining three circles.
2. Remove the top face of the box from the selection set, if needed.
3. The depth of the extrusion is .16 units. Remember to enter –.16 for the extrusion height since the holes remove material. The angle of taper for extrusion should be 35°. Your model should look like **Figure 13-18A**.
4. Extrude the small diameter of the four tapered holes so the holes intersect the bottom of the solid body. Select the holes by picking the small diameter circles. Instead of calculating the distance from the bottom of the chamfer to the bottom surface, you can simply enter a value that is greater than this distance, such as the original thickness of the object. Again, since the goal is to remove material, use a negative value for the height of the extrusion. There is no taper angle. Your model should now look like **Figure 13-18B**.

Moving Faces to Change Model Size

The next step is to use the **Face>Move** option to decrease the length and thickness of the solid body.

1. Select either end face and the two holes nearest to it. Be sure to select the holes *and* the countersinks. Move the two holes and end face two units toward the other end, thus changing the object length to four units.
2. Select the bottom face and move it .5 units up toward the top face, thus changing the thickness to .5 units. See **Figure 13-19**.

Figure 13-18.
A—The imprinted circles are extruded with a taper angle of 35°. B—Holes are created by further extrusion with a taper angle of 0°.

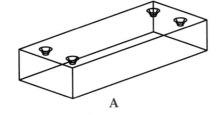

A

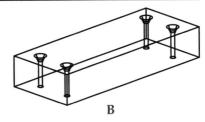

B

Figure 13-19.
The length of the object is shortened and the height is reduced.

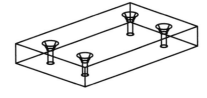

Offsetting a Feature to Change Its Size

Now, the **Face>Offset** option is used to increase the diameter of the four holes and to adjust a rectangular slot that will be added to the solid.

1. Using **Face>Offset**, select the four small hole diameters. Be sure to remove from the selection set any other faces that may be selected.
2. Enter an offset distance of –.05. This increases the hole diameter and decreases the solid volume. Exit the **SOLIDEDIT** command.
3. Select the **RECTANG** command. Set the fillet radius to .4 and draw a 2 × 1.6 rectangle centered on the top face of the solid. See **Figure 13-20A**.
4. Imprint the rectangle on the solid. Delete the source object.
5. Extrude the rectangle completely through the solid (.5 units). Remember to remove from the selection set any other faces that may be selected.
6. Offset the rectangular feature using an offset distance of .2 units. You will need to select all faces of the feature. This decreases the size of the rectangular opening and increases the solid volume. Your drawing should appear as shown in **Figure 13-20B**.

Tapering Faces

One side of the part is to be angled. The **Face>Taper** option is used to taper the left end of the solid.

1. Using **Face>Taper**, pick the face at the left end of the solid.
2. Pick point 1 in **Figure 13-21** as the base point and point 2 as the second point along the axis of tapering.
3. Enter a value of –10 for the taper angle. This moves the upper-left end away from the solid, creating a tapered end.

Figure 13-20.
A—The diameter of the holes is increased and a rectangle is imprinted on the top surface.
B—The rectangle is extruded to create a slot.

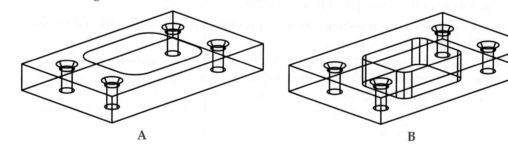

A   B

Figure 13-21.
The left end of the object is tapered. Notice the pick points. These points are also used when selecting an axis of rotation.

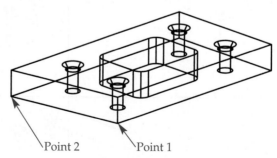

Point 2   Point 1

Rotating Faces

Next, use the **Face>Rotate** option to rotate the tapered end of the object. The top edge of the face will be rotated away from the holes, adding volume to the solid.

1. Using **Face>Rotate**, pick the face at the left end of the solid.
2. Pick point 1 in **Figure 13-21** as the first axis point and point 2 as the second point.
3. Enter a value of –30 for the rotation angle. This rotates the top edge of the tapered end away from the solid. See **Figure 13-22**.

Copying Faces

A mating part will now be created. This is done by first copying the face on the tapered end of the part.

1. Using the **Face>Copy** option, pick the angled face on the left end of the solid.
2. Pick one of the corners as a base point and copy the face one unit to the left. This face can now be used to create a new solid. See **Figure 13-23A**.
3. Draw a line four units in length on the negative X axis from the lower-right corner of the copied face. Use the **EXTRUDE** command on the copied face to create a new solid. Select the **Path** option and use the line as the extrusion path. See **Figure 13-23B**. If you do not use the **Path** option, the extrusion is projected perpendicular to the face.

The **Face>Extrude** option of the **SOLIDEDIT** command cannot be used to turn a copied face into a solid body.

Figure 13-22.
The tapered end of the object is modified by rotating the face.

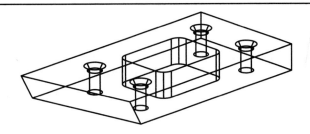

Figure 13-23.
Creating a mating part. A—The angled face is copied. B—The copied face is extruded into a solid.

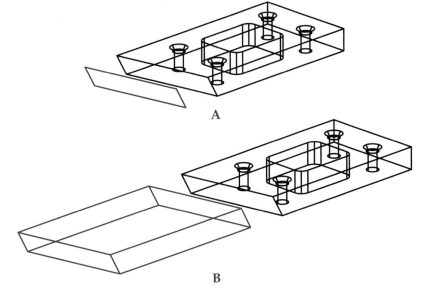

A

B

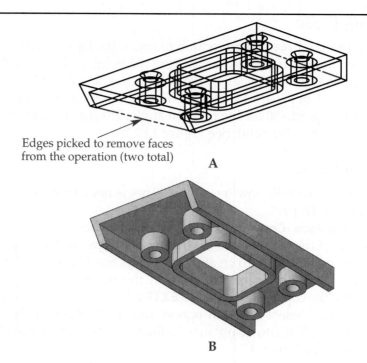

Figure 13-24.
A—The shelled object. B—The viewpoint is changed and the Conceptual visual style is set current.

Edges picked to remove faces from the operation (two total)

A

B

Creating a Shell

The bottom surface of the original solid will now be shelled out. Keep in mind that features such as the four holes and the rectangular slot will not be cut off by the shell. Instead, a shell will be placed around these features. This becomes clear when the operation is performed.

1. Select the **Shell** option and pick the original solid.
2. Pick the lower-left and lower-right edges of the solid. See **Figure 13-24A**. This removes the two side faces and the bottom face.
3. Enter a shell offset distance of .15 units. The shell is created and the model should appear similar to **Figure 13-24A**.
4. Use the view cube or **3DORBIT** command to view the solid from the bottom. Also, set the Conceptual visual style current. Your model should look like the one shown in **Figure 13-24B**.

Chapter Review

Answer the following questions. Write your answers on a separate sheet of paper or complete the electronic chapter review on the companion website.
www.g-wlearning.com/CAD

1. What three components of a solid model can be edited using the **SOLIDEDIT** command?
2. When using the **SOLIDEDIT** command, how many faces are highlighted if you pick an edge?
3. How do you deselect a face that is part of the selection set when using the **SOLIDEDIT** command?
4. How can you select a single face?
5. Name the objects that can be used as the path of extrusion when extruding a face.
6. What is one of the most useful aspects of the **Offset Faces** option?
7. How do positive and negative offset distance values affect the volume of the solid?

8. How is a single object, such as a cylinder, affected by entering a positive taper angle when using the **Taper Faces** option?
9. When a shape is imprinted onto a solid body, which component of the solid does the imprinted object become and how can it be used?
10. What is the purpose of the **XEDGES** command?
11. How does the **Shell** option affect a solid that contains internal features such as holes, grooves, and slots?
12. How can you determine if an object is a valid 3D solid?
13. Describe two ways to change the view of your model while you are inside of a command.
14. How can you extrude a face in a straight line, but not perpendicular to the face?

Drawing Problems

1. Complete the tutorial presented in this chapter. Then, perform the following additional edits to the original solid.
    A. Lengthen the right end of the solid by .5 units.
    B. Taper the right end of the solid with the same taper angle used on the left end, but taper it in the opposite direction.
    C. Fillet the two long, top edges of the solid using a fillet radius of .2 units.
    D. Rotate the face at the right end of the solid with the same rotation angle used on the left end, but rotate it in the opposite direction.
    E. Save the drawing as P13_01.

2. Construct the solid part shown using as many **SOLIDEDIT** options as possible. After completing the object, make the following modifications.
    A. Lengthen the 1.250" diameter feature by .250".
    B. Change the .750" diameter hole to .625" diameter.
    C. Change the thickness of the .250" thick flange to .375" (toward the bottom).
    D. Extrude the end of the 1.250" diameter feature .250" with a 15° taper inward.
    E. Save the drawing as P13_02.

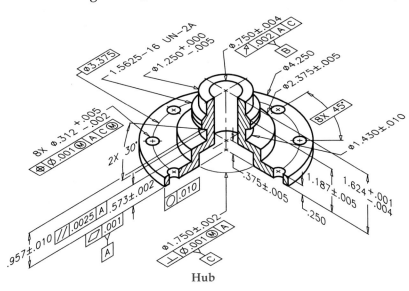

Hub

3. Construct the solid part shown. Then, perform the following edits on the solid using the **SOLIDEDIT** command.
    A. Change the diameter of the hole to 35.6/35.4.
    B. Add a 5° taper to each inner side of each tooth (the bottom of each tooth should be wider while the top remains the same).
    C. Change the width of the 4.8/4.0 key to 5.8/5.0.
    D. Save the drawing as P13_03.

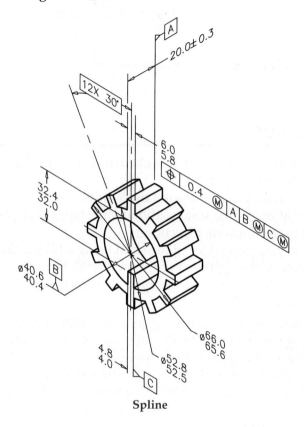

Spline

4. Construct the solid part shown using as many **SOLIDEDIT** options as possible. Then, perform the following edits on the solid.
    A. Change the depth of the counterbore to 10 mm.
    B. Change the color of all internal surfaces to red.
    C. Save the drawing as P13_04.

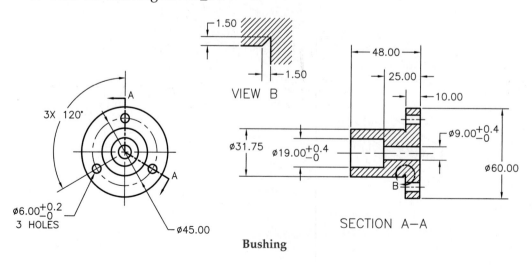

Bushing

AutoCAD and Its Applications—Advanced

5. Construct the solid part shown using as many **SOLIDEDIT** options as possible. Then, perform the following edits on the solid.
   A. Change the 2.625″ height to 2.325″.
   B. Change the 1.625″ internal diameter to 1.425″.
   C. Taper the outside faces of the .875″ high base at a 5° angle away from the part. Hint: The base cannot be directly tapered.
   D. Save the drawing as P13_05.

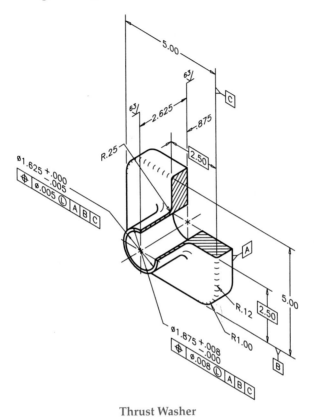

Thrust Washer

6. Construct the solid part shown using as many **SOLIDEDIT** options as possible. Then, change the dimensions on the model as follows. Save the drawing as P13_06.

Existing	New
100	106
80	82
Ø60	Ø94
Ø40	Ø42
30°	35°

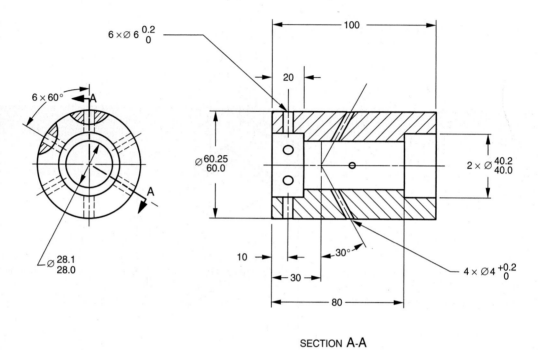

SECTION A-A

Nozzle

Text and Dimensions in 3D

Learning Objectives

After completing this chapter, you will be able to:

- ✓ Create text with a thickness.
- ✓ Draw text that is plan to the current view.
- ✓ Dimension a 3D drawing.

Creating Text with Thickness

A thickness can be applied to text after it is created. This is done using the **Properties** palette. The thickness setting is located in the **General** section. Once a thickness is applied, the hidden lines can be removed using the **HIDE** command or a shaded visual style. **Figure 14-1** shows six different fonts as they appear after being given a thickness and with the Conceptual visual style set current.

Ribbon
View
> Palettes
Properties Palette

Type
PROPERTIES
PR
[Ctrl]+[1]

PROPERTIES

Figure 14-1.
Six different fonts
with thickness after
the Conceptual visual
style is set current.

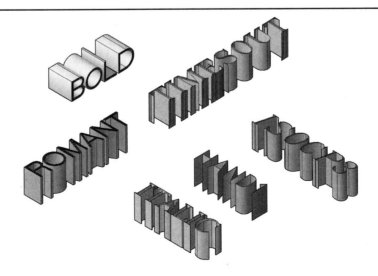

Only text created using the **TEXT** or **DTEXT** command (a text object) can be assigned thickness. Text created with the **MTEXT** command (an mtext object) cannot have thickness assigned to it. In addition, only AutoCAD SHX fonts can be given thickness. Therefore, when creating a text style to use for 3D purposes, select a text font in the **Text Style** dialog box with a .shx file extension. See **Figure 14-2**. Windows TrueType fonts *cannot* be used to create text with thickness.

Text and the UCS

Text is created parallel to the XY plane of the UCS in which it is drawn. Therefore, if you wish to show text appearing on a specific plane, establish a new UCS on that plane before placing the text. You can use the UCS icon grips to create a new UCS. **Figure 14-3** shows several examples of text on different UCS XY planes.

Changing the Orientation of a Text Object

If text is improperly placed or created using the wrong UCS, it can be edited using grips or editing commands. Editing commands and grips are relative to the current UCS. For example, if text is drawn with the WCS current, you can use the **ROTATE** command to change the orientation of the text in the XY plane of the WCS. However, to rotate the text so it tilts up from the XY plane of the WCS, you will need to change the UCS. Rotate the UCS as needed so the Z axis of the new UCS aligns with the axis about which you want to rotate. Then, the **ROTATE** command can be used to rotate the text. See **Figure 14-4**. The **3DROTATE** command can also be used instead of rotating the UCS.

Figure 14-2.
Only AutoCAD SHX fonts can be used for 3D text with thickness.

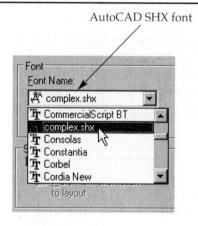

Figure 14-3.
Text located using three different UCSs.

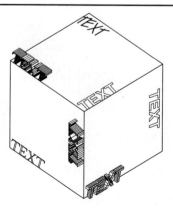

Figure 14-4.
Creating a new UCS to rotate text. A—Pick the UCS icon to display the grips and hover over the X axis grip. Select the **Rotate Around Y Axis** option from the shortcut menu. B—Rotate the UCS so the Z axis of the new UCS aligns with the axis about which you want to rotate. C—The **ROTATE** command is used to rotate the text 90°.

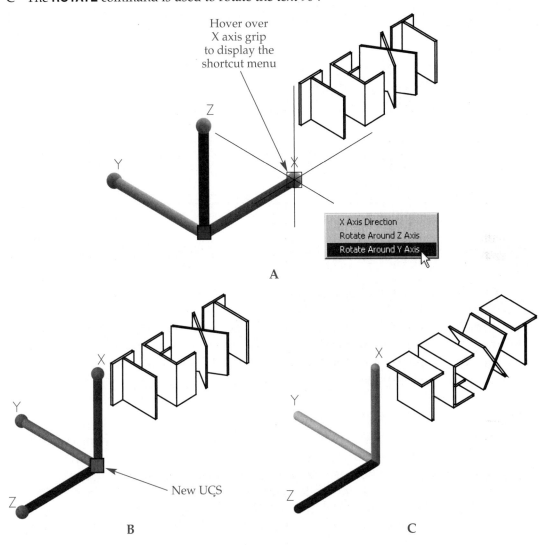

Using the UCS View Option to Create a Title

It is often necessary to create a pictorial view of an object, but with a note or title that is plan to your point of view. For example, you may need to insert the title of a 3D view. See **Figure 14-5.** This is done with the **View** option of the **UCS** command, which is discussed in Chapter 6. With this option, a new UCS is created perpendicular to your viewpoint. However, the view remains unchanged. Inserted text will be horizontal (or vertical) in the current view. Name and save the UCS if you will use it again.

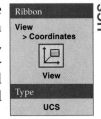

Exercise 14-1

Complete the exercise on the companion website.
www.g-wlearning.com/CAD

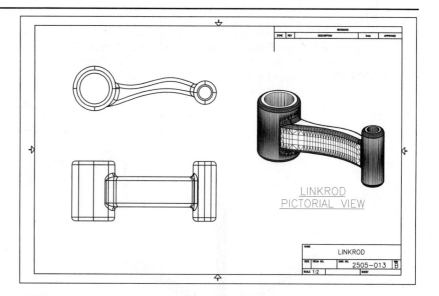

Figure 14-5.
This title (shown in color) has been correctly placed using the **View** option of the **UCS** command.

Dimensioning in 3D

Three-dimensional objects are seldom dimensioned for manufacturing, but may be used for assembly. Dimensioned 3D drawings are most often used for some sort of presentation, such as displays, illustrations, parts manuals, or training manuals. All dimensions, including those shown in 3D, must be clear and easy to read. The most important aspect of applying dimensions to a 3D object is planning. That means following a few basic guidelines.

Creating a 3D Dimensioning Template Drawing

If you often create dimensioned 3D drawings, make a template drawing containing a few 3D settings. Starting a drawing based on one of these templates will speed up the dimensioning process because the settings will already be made for you.

- Create named dimension styles with appropriate text heights. See *AutoCAD and Its Applications—Basics* for detailed information on dimensioning and dimension styles.
- Establish several named user coordinate systems that match the planes on which dimensions will be placed.
- If the preset isometric viewpoints will not serve your needs, establish and save several 3D viewpoints that can be used for different objects. These viewpoints will allow you to select the display that is best for reading dimensions.
- If 3D dimensioned views are to be used with a multiview 2D drawing, create a paper space drawing layout containing appropriate viewports and the required items listed above.

Multiview orthographic and pictorial drawing layouts can be created quickly and efficiently using the **VIEWBASE** command. This command can be accessed in the **Drawing Views** panel of the **Annotate** tab on the ribbon. Creating drawing views is discussed in Chapter 15.

Placing Dimensions in the Proper Plane

The location of dimensions and the plane on which they are placed are often a matter of choice. For example, **Figure 14-6** shows several options for placing a thickness dimension on an object. All of these are correct. However, several of the options can be eliminated when other dimensions are added. This illustrates the importance of planning.

The key to good dimensioning in 3D is to avoid overlapping dimension and extension lines in different planes. A freehand sketch can help you plan this. As you lay out the 3D sketch, try to group information items together. Dimensions, notes, and item tags should be grouped so that they are easy to read and understand. This technique is called *information grouping*.

Figure 14-7A shows the object from **Figure 14-6** fully dimensioned using the aligned technique. Notice that the location dimension for the hole is placed on the top surface. This avoids dimensioning to hidden points. **Figure 14-7B** shows the same object dimensioned using the unilateral technique.

To create dimensions that properly display, it may be necessary to modify the dimension text rotation. The dimension shown in **Figure 14-8A** is inverted because the positive X and Y axes are incorrectly oriented. Using the **Properties** palette, change the text rotation value to 180. The dimension text is then properly displayed, **Figure 14-8B**. Alternately, you can rotate the UCS before drawing the dimension, but this may be more time-consuming.

Figure 14-6.
A thickness dimension can be located in many different places. All locations shown here are acceptable.

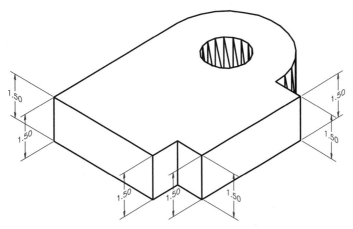

Figure 14-7.
A—An example of a 3D object dimensioned using the aligned technique. B—The object dimensioned with unilateral dimensions.

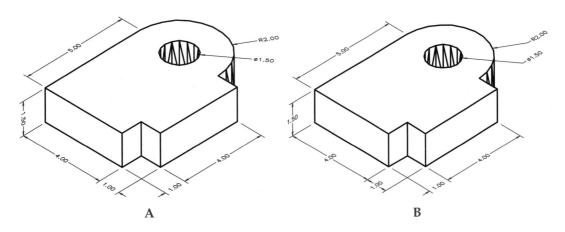

A   B

Figure 14-8.
A—This dimension text is inverted. B—The rotation value of the text is changed and the text reads correctly.

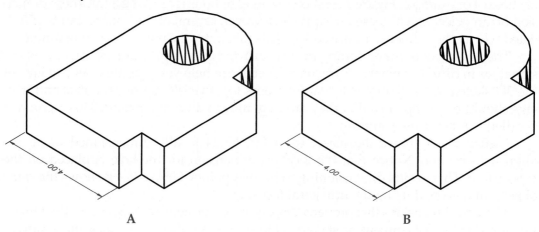

A   B

Prior to placing dimensions on a 3D drawing, you should determine the purpose of the drawing. For what will it be used? Just as dimensioning a drawing for manufacturing purposes is based on the function of the part, 3D dimensioning is based on the function of the drawing. This determines whether you use chain, datum, arrowless, architectural, or some other style of dimensioning. It also determines how completely the object is dimensioned.

Placing Leaders and Radial Dimensions in 3D

Although standards such as ASME Y14.5 should be followed when possible, the nature of 3D drawing and the requirements of the project may determine how dimensions and leaders are placed. Remember, the most important aspect of dimensioning a 3D drawing is its presentation. Is it easy to read and interpret?

Leaders and radial dimensions can be placed on or perpendicular to the plane of the feature. **Figure 14-9A** shows the placement of leaders and dimensions on the plane of the top surface. **Figure 14-9B** illustrates the placement of leaders and dimensions on two planes that are perpendicular to the top surface of the object. Remember that

Figure 14-9.
A—Leaders placed in the plane of the top surface. B—Leaders placed using two UCSs that are perpendicular to the top face.

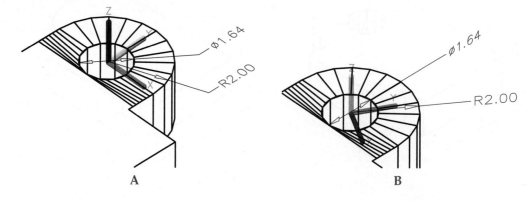

A   B

text, dimensions, and leaders are always created on the XY plane of the current UCS. Therefore, to create the layout in **Figure 14-9B**, you must use more than one UCS.

Exercise 14-2

Complete the exercise on the companion website.
www.g-wlearning.com/CAD

Chapter Review

Answer the following questions. Write your answers on a separate sheet of paper or complete the electronic chapter review on the companion website.
www.g-wlearning.com/CAD

1. How can you create 3D text with thickness?
2. If text is placed using the wrong UCS, how can it be edited to appear on the correct one?
3. How can text be placed horizontally based on your viewpoint if the object is displayed in 3D?
4. Name three items that should be a part of a 3D dimensioning template drawing.
5. What is *information grouping*?

Drawing Problems

For the following problems, create solid models as instructed. If appropriate, apply geometric constraints to the base 2D drawing. If you are using constraints and create profiles of a surface that can be converted into regions for solid modeling, apply geometric constraints and save the drawing with a different name than the final solid model.

1. This is a two-view orthographic drawing of a window valance mounting bracket. Create it as a solid model. Use solid primitives and Boolean commands as needed. Use the dimensions given. Similar holes have the same offset dimensions. Create new UCSs as needed. Display an appropriate pictorial view of the drawing. Then, add dimensions. Finally, add the material note so it is plan to the 3D view. Plot the drawing to scale on a C-size sheet of paper. Save the drawing as P14_01.

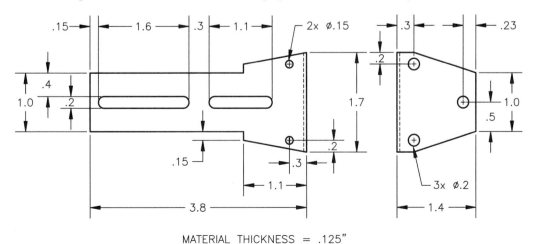

MATERIAL THICKNESS = .125"

2. Open P07_03. If you have not completed this problem, construct it using the directions for the problem in Chapter 7. Display the 3D view in a single viewport. Then, add dimensions to the 3D view, as shown in Chapter 7. Plot the drawing on a C-size sheet of paper. Save the drawing as P14_02.

3. Create the end table as a solid model using solid primitives and Boolean commands as needed. The end result should be a single object. As a test of your object-editing skills, try drawing the entire model by starting with only a single rectangle. You can copy, resize, extrude, and move objects as you create them from the single rectangle. Use the dimensions given and the following information to construct the model.

   A. Table height is 24″.
   B. Top of bottom shelf is 5″ off of the floor.
   C. Table legs must be located no less than 1/2″ from the tabletop edge.
   D. Shelf must be no closer than .75″ from the outside of table legs.
   E. Dimension the table as shown.
   F. Save the drawing as P14_03.

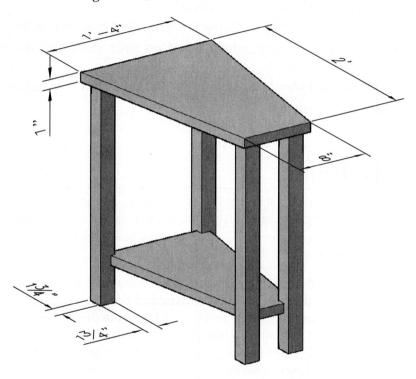

4. Shown are the profiles of a roof gutter (for the collection of rainwater) and a gutter downspout. Draw the profiles in 3D using the dimensions shown. Use the following additional information to construct a 3D model like the one shown in the shaded view.

    A. Offset the gutter profile to create a material thickness of .025″. Be sure to close the ends to create a closed polyline so a 3D solid is created when it is extruded.

    B. Extrude the gutter profile 12″ to create a one-foot section.

    C. Relocate the downspout profile on the underside of the gutter.

    D. Construct an extrusion path for the downspout. Refer to the shaded view, but use your own design.

    E. Extrude the downspout profile along the path.

    F. Dimension the end of the gutter profile in 3D.

    G. Save the drawing as P14_04.

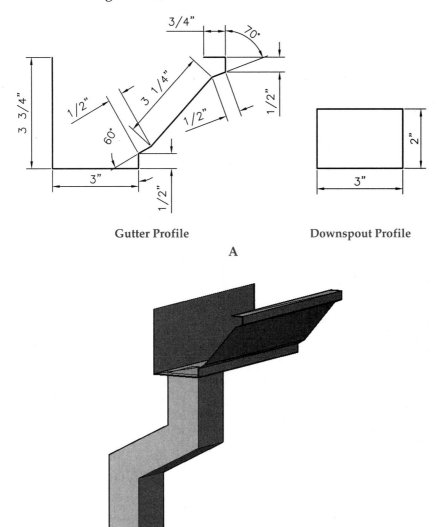

A

5. Open P07_04. If you have not completed this problem, construct it using the directions for the problem in Chapter 7. Display the 3D view in a single viewport. Then, add dimensions to the 3D view. Plot the drawing on a C-size sheet of paper. Save the drawing as P14_05.

Problems 6 and 7 are mechanical parts. Create a solid model of each part. Dimension each model. Place the title of each model so it is plan to the pictorial view. Plot the finished drawings on B-size paper. Save each drawing as P14_*(problem number).*

6.

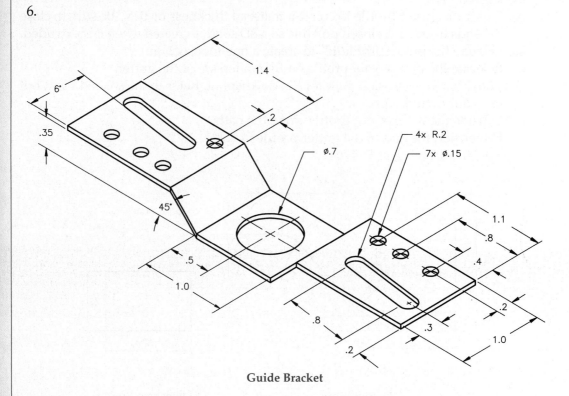

Guide Bracket

7.

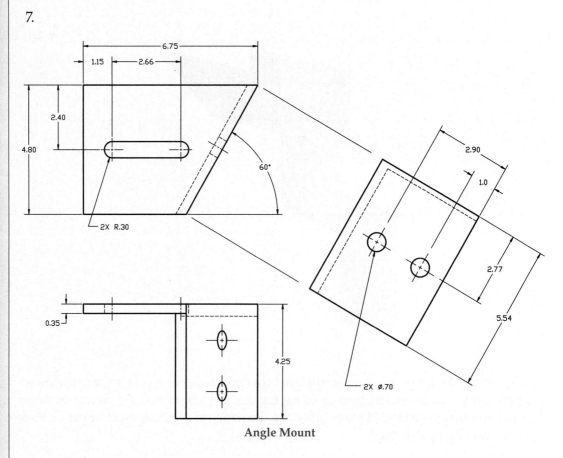

Angle Mount

Model Display, Documentation, and Analysis

Learning Objectives

After completing this chapter, you will be able to:

✓ Control the display of solid models.
✓ Create drawing views.
✓ Construct a 3D section plane through a 3D model.
✓ Adjust the size and location of section planes.
✓ Create a dynamic section of a 3D model.
✓ Construct 2D and 3D section blocks.
✓ Create a multiview layout of a solid model using **SOLVIEW** and **SOLDRAW**.
✓ Construct a profile of a solid using **SOLPROF**.
✓ Analyze solid and surface models.
✓ Exchange solid model file data with other programs.

Certain aspects of a solid model's appearance are controlled by the **ISOLINES**, **DISPSILH**, and **FACETRES** system variables. The **ISOLINES** system variable controls the number of lines used to define solids in wireframe displays. The **FACETRES** system variable controls the number of lines (smoothness) used to define solids in hidden and shaded displays. The **DISPSILH** system variable is used to display a silhouette edge. This chapter discusses options available for displaying 3D models. This chapter also discusses the tools available for creating drawing views, showing section views of objects, and conducting model analysis.

Controlling Solid Model Display

AutoCAD solid models can be displayed in wireframe form, with hidden lines removed, shaded, or rendered. A wireframe visual style named 2D Wireframe is the default display when a drawing is started based on the acad.dwt template. Wireframe displays are the quickest to display. When a drawing is started based on the acad3D.dwt or acadiso3D.dwt template, the default display is the Realistic visual style, which is a shaded display. The viewport controls, introduced in Chapter 1, provide quick access to visual styles. Visual styles are covered in detail in Chapter 16. The following sections introduce display options available in wireframe, hidden, and shaded displays.

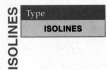

Isolines

The appearance of a solid model in a wireframe display is controlled by the **ISOLINES** system variable. *Isolines* represent the edges and curved surfaces of a solid model. This setting does *not* affect the final shaded or rendered object. However, if the Show property in the **Edge Setting** category in the **Visual Styles Manager** palette is set to Isolines, isolines are displayed when the visual style is set current. The default **ISOLINES** value is four. It can have a value from zero to 2047. All solid objects in the drawing are affected by changes to the **ISOLINES** value, as are all visual styles set to display isolines. **Figure 15-1** illustrates the difference between **ISOLINES** settings of four and 12.

The setting of the **ISOLINES** system variable can be changed in the **Visual Styles Manager** palette. You can also type **ISOLINES** and enter a new value, or you can change the **Contour lines per surface** setting in the **Display resolution** area of the **Display** tab in the **Options** dialog box. See **Figure 15-2**. The settings in the **Display** tab in the **Options** dialog box can also be used to control the **DISPSILH** and **FACETRES** system variables, discussed next.

Creating a Display Silhouette

In the 2D Wireframe visual style, a model can appear smooth with only a silhouette displayed, similar to the Hidden visual style. This is controlled by the **DISPSILH** (display silhouette) system variable. The **DISPSILH** system variable has two values, 0 (off) and 1 (on). **Figure 15-3** shows solids with **DISPSILH** set to 0 and 1 after setting the 2D Wireframe visual style current and then using **HIDE**.

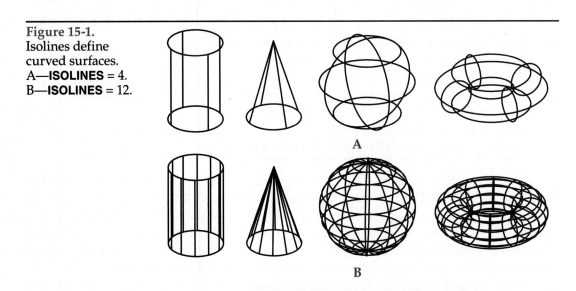

Figure 15-1.
Isolines define
curved surfaces.
A—**ISOLINES** = 4.
B—**ISOLINES** = 12.

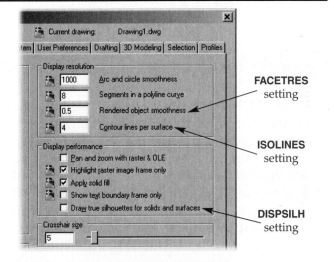

Figure 15-2.
The **ISOLINES**,
FACETRES, and
DISPSILH values can
be set in the **Options**
dialog box.

Figure 15-3.
A—The **HIDE** command used when **DISPSILH** is set to 0. Objects are displayed faceted. B—The **HIDE** command used when **DISPSILH** is set to 1. Facets are eliminated and only the silhouette is displayed.

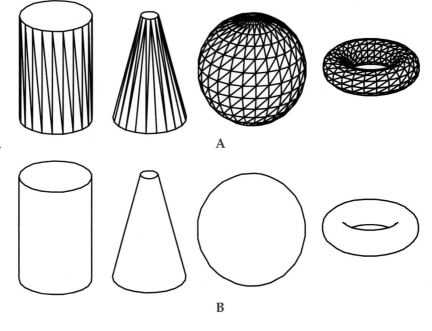

A

B

The setting can be changed by typing **DISPSILH** and entering a new value. You can also set the variable using the **Draw true silhouettes for solids and surfaces** check box in the **Display performance** area of the **Display** tab in the **Options** dialog box. Refer to **Figure 15-2.** A preferred technique is to set **ISOLINES** to 0 and **DISPSILH** to 1 to create a true display of the silhouette edge of a 3D model. For the 2D Wireframe visual style, the **DISPSILH** variable can also be set in the **Visual Styles Manager** palette. The variable is controlled by the Draw true silhouettes property in the **2D Wireframe options** category. Setting this property to Yes turns on silhouettes.

Controlling Surface Smoothness

The smoothness of curved surfaces in hidden, shaded, and rendered displays is controlled by the **FACETRES** system variable. This variable determines the number of polygon faces applied to the solid model. The value can range from .01 to 10.0 and the default value is .5. This system variable can be changed by typing **FACETRES** or by changing the **Rendered object smoothness** setting in the **Display resolution** area in the **Options** dialog box. Refer to **Figure 15-2.** For the 2D Wireframe visual style, the variable can be set in the **Visual Styles Manager** palette. The Solid smoothness property in the **Display resolution** category controls the variable. **Figure 15-4** shows the effect of two different **FACETRES** settings.

PROFESSIONAL TIP

When you use the **3DPRINT** command, **FACETRES** is set to 10 automatically.

Exercise 15-1

Complete the exercise on the companion website.
www.g-wlearning.com/CAD

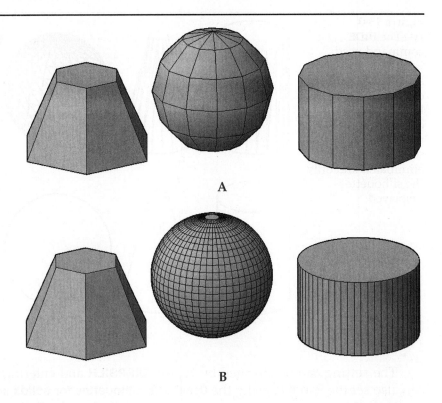

Figure 15-4.
A—The **FACETRES** setting is .5.
B—The **FACETRES** setting is 5.0.

A

B

Introduction to Model Documentation

There are multiple techniques available to create drawing views from 3D models in AutoCAD. Using these techniques, orthographic views and other standard drawing views can be created in order to produce multiview drawings. The techniques available are discussed in this chapter and include the following:

- Use the **VIEWBASE** and **VIEWPROJ** commands to create multiview drawings that are derived from and associated to a 3D solid or surface model. The drawing views are placed in a paper space layout. This is the most efficient method to create multiview drawings in AutoCAD.
- Use the **SECTIONPLANE** command to show the internal features of a 3D solid model. This command can create 2D and 3D section views on an object.
- Use the **SOLVIEW** command to create a multiview layout from a solid model. This command allows you to create a layout containing orthographic, section, and auxiliary views. The **SOLDRAW** command can then be used to complete profile and section views. **SOLDRAW** must be used after **SOLVIEW**.

Drawing Views

You can create 2D drawing views from 3D solid or surface models using the **VIEWBASE** command. This command allows you to quickly create a multiview drawing layout made up of orthographic views. Drawing views created with the **VIEWBASE** command are created in paper space (layout space). After creating a layout of views, you can dimension them using AutoCAD dimensioning commands.

Drawing views created with the **VIEWBASE** command are *associative*. This means that they are linked to the model from which they are created and can be updated to reflect changes to the model geometry. This capability allows you to keep drawing views up-to-date when design changes are required.

When using the **VIEWBASE** command, you create a *base view* of the model and then have the option to create additional views that are projected from the base view without exiting the command. You can create both orthographic and isometric views. A base view created with the **VIEWBASE** command is defined as a *parent view*. Views projected from the base view inherit the properties of the base view, such as the drawing scale and display properties, and are placed in orthographic alignment with the base view. If the base view is moved, any projected views are moved with it to maintain the parent-child relationship.

Drawing views can be created from sources other than AutoCAD 3D models. You can create views from Autodesk Inventor® part (IPT), assembly (IAM), or presentation (IPN) files. You can also create drawing views from models imported with the **IMPORT** command. The **IMPORT** command is discussed later in this chapter. Drawing views can also be created from an assembly or subassembly to create an assembly drawing. This chapter discusses creation of drawing views from part models.

Drawing View and Layout Setup

In order to create views with the **VIEWBASE** command, you must first activate paper space (layout space). Activate paper space by picking a layout tab. Then, before creating views, set up the page layout as required. If a default viewport appears in the layout, delete the viewport. Then, right-click on the active layout tab and select **Page Setup Manager** to access the layout settings. Select the desired paper size, drawing orientation, and other settings. If you have a template drawing already created, you can create a new layout based on the template by right-clicking on a layout tab and selecting **From template…**.

By default, drawing views created with the **VIEWBASE** command are generated using third-angle projection. This is the ANSI standard. The default projection used for drawing views can be set using the **VIEWSTD** command. This command and other drawing view commands are accessed from the **Drawing Views** panel on the **Annotate** tab of the ribbon. See **Figure 15-5**. Pick the dialog box launcher button at the lower-right corner of the **Drawing Views** panel to initiate the **VIEWSTD** command. This opens the **Drafting Standard** dialog box, **Figure 15-6**. The options in the **Projection type** area determine the type of projection used for new drawing views. By default, the **Third angle** button is selected. If the **First angle** button is selected, views are created using first-angle projection. Select this option to create views in accordance with ISO standards. The options in the **Thread style** area determine the type of thread representation used for views showing threaded features. The thread representation options are used when creating views from Autodesk Inventor models or imported models containing threaded features. A full circle thread representation conforms to the ANSI standard. This is the default option. A partial circle thread representation conforms to the ISO standard.

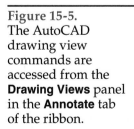

Figure 15-5.
The AutoCAD drawing view commands are accessed from the **Drawing Views** panel in the **Annotate** tab of the ribbon.

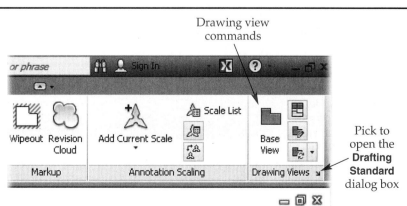

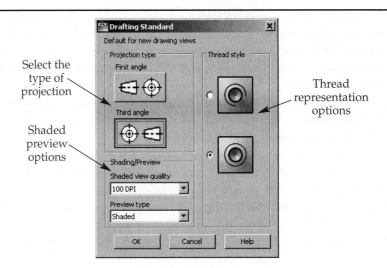

Figure 15-6.
The **Drafting Standard** dialog box.

Select the type of projection

Shaded preview options

Thread representation options

The options in the **Shading/Preview** area of the **Drafting Standard** dialog box determine the type of preview that appears when placing a drawing view. When **Shaded** is selected in the **Preview type** drop-down list, a shaded view appears. This is the default option. This option provides a preview image of the view orientation before the view is placed. When the **Bounding box** option is selected, a preview box representing the extents of the view appears. However, no preview image is displayed. The **Shaded view quality** option is set to 100 dpi by default. A higher setting will provide better resolution quality when a shaded preview is displayed. However, for all practical purposes, a setting of 150 dpi or higher does not produce a dramatic difference in resolution quality.

Creating Drawing Views

After setting up a layout and making the appropriate drawing view settings, you are ready to place the base view. As previously discussed, drawing views can be created from a 3D solid or surface model. Drawing views created in the layout are generated from the model you have created in model space. If no model exists in the drawing and you initiate the **VIEWBASE** command, the **Select File** dialog box appears. From this dialog box, you can select a model created in Autodesk Inventor.

When you select the **VIEWBASE** command, a preview representing a scaled orthographic view of the model appears. This is the *base view* of the model. You can then specify the location of the base view or select an option. To select an option, use dynamic input or make a selection from the **Drawing View Creation** contextual ribbon tab. See **Figure 15-7**. The **Orientation** option can be used to select a different view from the default base view. The **Orientation** options correspond to AutoCAD's six orthographic and four isometric preset views. These views are based on the WCS. By default, the front view is used as the base view by AutoCAD. Depending on the construction of your model, you may want to use a different view, such as the top view, to serve as the base view.

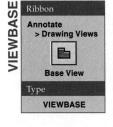

VIEWBASE

Ribbon
Annotate
> Drawing Views

Base View

Type

VIEWBASE

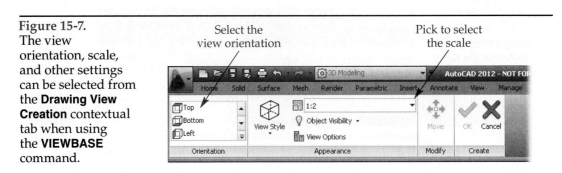

Figure 15-7.
The view orientation, scale, and other settings can be selected from the **Drawing View Creation** contextual tab when using the **VIEWBASE** command.

Select the view orientation

Pick to select the scale

In **Figure 15-8**, the top view of the model is selected as the base view. This view will establish the front view on the orthographic drawing. In most cases, the front orthographic view on the drawing describes the most critical contour of the model. You will typically use the front or top view of the 3D model for the front orthographic view on the drawing.

Creating a base view creates an AutoCAD *drawing view* object. A drawing view has a view border, a base grip, and properties that can be edited, such as the view scale. However, the content of the drawing view *cannot* be edited. When a base view is created, new layers are created by AutoCAD for the drawing view geometry. Object lines (visible lines) in the view are placed on a newly created layer named Visible. Hidden lines in the view are placed on a newly created layer named Hidden. The drawing view object is placed on a newly created layer named Base. Additional layers may be created by AutoCAD, depending on the type of model, display style used, and edges displayed in the view. However, the layers are only created to organize the drawing view geometry. The layer properties can be modified to change the appearance of the drawing view geometry, but the geometry cannot be otherwise modified.

The **Type** option of the **VIEWBASE** command is available before picking the initial location of the base view. This option is used to specify whether projected views are placed after placing the base view. By default, this option is set to **Base and Projected**. With this option, you can place additional, projected views while the **VIEWBASE**

Figure 15-8.
Placing a base view with the **VIEWBASE** command. A—The original model drawn in model space. B—The preview of the base view in paper space. The Top view option is selected for the base view because it is the most descriptive view of the model. C—The resulting base view.

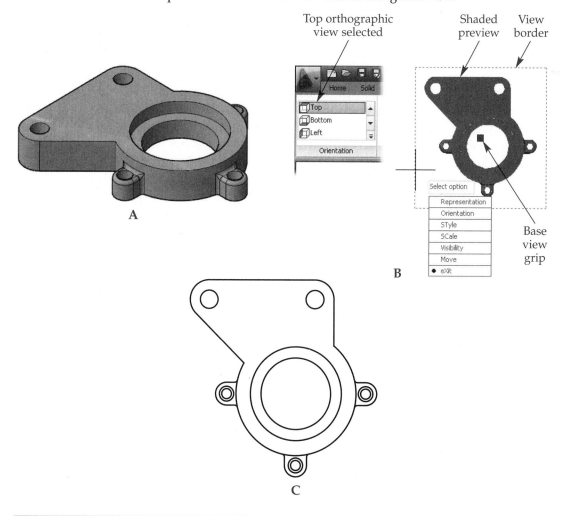

command is still active. Projected views that you place are *projected* from the base view and oriented in the proper orthographic alignment. As previously discussed, projected views have a parent-child relationship with the base view. The base view is the *parent* view and any projected views are *child* views. If the base view is moved, for example, the projected view is moved with it. Projected views are not created with the **VIEWBASE** command when the **Type** option is set to **Base only**. If you place a base view in this manner and later decide to create projected views from the base view, you can use the **VIEWPROJ** command. The **VIEWPROJ** command allows you to place projected views from any selected view.

After you pick the location for the base view, you can press [Enter] to place projected views (if the **Type** option is set to **Base and Projected**). Drag the cursor away from the base view and then pick to locate the projected view. The direction in which you move the cursor determines the orientation of the projected view. You can continue placing projected views, or you can press [Enter] to end the command. To end the command without placing projected views, press [Enter] twice.

When you place the base view, you can select an option to change the default style, scale, or visibility of the base view before placing projected views. Projected views inherit the properties of the base view. The **Style** option is used to set the display style of the base view. The four options available are **Wireframe**, **Wireframe with Hidden Edges**, **Shaded**, and **Shaded with Hidden Edges**. See Figure 15-9. These options can be selected using dynamic input or the **Drawing View Creation** contextual ribbon tab.

The **Scale** option is used to set the scale of the base view. You can select a scale from the drop-down list in the **Appearance** panel in the **Drawing View Creation** contextual ribbon tab. You can also specify the scale by typing a value.

The **Move** option allows you to move the base view after picking the initial location on screen. When you select the **Move** option, the view is reattached to the cursor so that you can move it to another location. After you pick a new location, the **VIEWBASE** options are again made available.

The **Visibility** option allows you to control the display of drawing geometry in the view. The **Interference edges** option controls whether both object and hidden lines are displayed for interference edges. By default, this option is set to **No**. The **Tangent edges** option controls whether tangential edges are displayed to show the intersection of surfaces. By default, this option is set to **No**. If this option is set to **Yes**, you can specify whether tangent edges are shortened to distinguish them from object lines that overlap. The **Bend extents** option is only available when working with a model that includes a view with sheet metal bends. The **Thread features** option controls thread displays on models with screw thread features. The **Presentation trails** option is used to control the display of trails in views created from presentation files.

Figure 15-9.
Display style options available for drawing views. A—Wireframe. B—Wireframe with Hidden Edges. C—Shaded. D—Shaded with Hidden Edges.

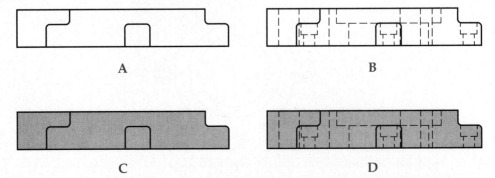

A

B

C

D

Additional drawing view options can be accessed by selecting **View Options** in the **Appearance** panel of the **Drawing View Creation** contextual ribbon tab. This option is not available on the command line. Selecting **View Options** opens the **View Options** dialog box. The options in the **View Justification** drop-down list determine the justification of the view. The *justification* refers to how the view is "anchored." When changes are made to the model, such as a change in size, the view updates based on the justification setting. If the justification is set to **Fixed**, geometry in the view unaffected by the edit does not change from the original location. If the justification is set to **Center**, the geometry updates about the center point of the view.

In **Figure 15-10**, a multiview drawing layout is created by placing three projected views after placing the base view. The projected views include two orthographic views and an isometric view. Notice that wireframe styles are used for the orthographic views and a shaded style is used for the isometric view. Drawing views can be edited after placing them to change properties as needed. In addition, as previously discussed, drawing views can be updated when changes are made to the model. Updating drawing views is discussed in the next section.

The **Representation** option of the **VIEWBASE** command is used to create a view based on a model representation and is only available when working with models created in Autodesk Inventor.

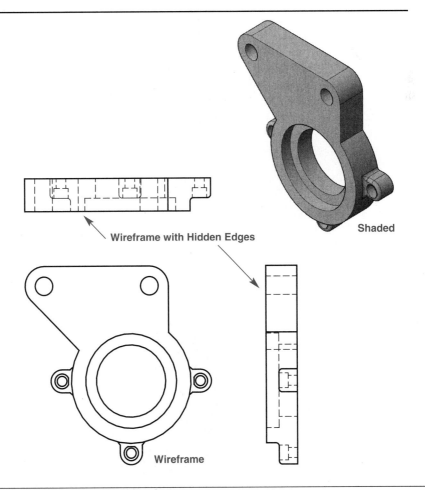

Figure 15-10.
A multiview drawing layout consisting of front, top, right-side, and isometric views. The top view of the model is used as the front view (base view) in the drawing. Notice the different style options that are used.

Wireframe with Hidden Edges

Shaded

Wireframe

You can create more than one base view in a layout. Additional base views can be used to create assembly or subassembly drawings.

Updating Drawing Views

Drawing views maintain an associative relationship with the model from which they are created. However, it is important to note that this associative relationship is controlled by the model, not the drawing view. When making design changes, you make changes to the model geometry, *not* the drawing view.

During the design process, it is often necessary to make modifications. If you have created drawing views from a model, and then make modifications to the model, the derived base view and any views projected from the base view will require updating. When this occurs, drawing views are considered *out-of-date*. AutoCAD does not automatically update the views when the model is modified. However, you can quickly update drawing views when needed.

When you make changes to a model in model space and then switch to a layout with drawing views, red markers appear to indicate that the views are out-of-date. See **Figure 15-11**. In addition, a balloon notification appears on the status bar. See **Figure 15-12**.

Figure 15-11.
Updating drawing views after editing the 3D model. A—The holes in the flanged coupler have been edited from ∅1.8 to ∅.9. B—After switching to the layout, markers appear to indicate that the drawing views are out-of-date. C—The views are updated to reflect the changes in the model.

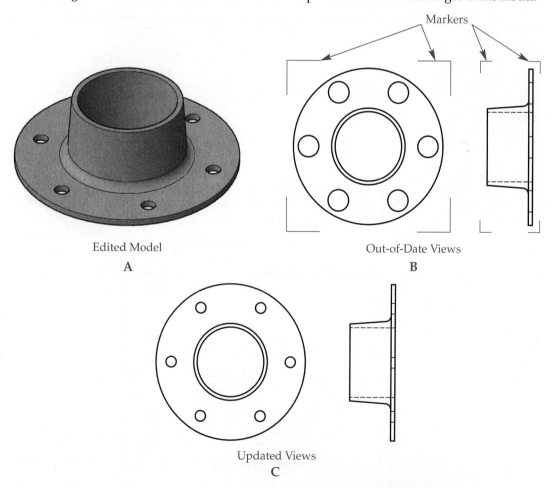

Markers

Edited Model
A

Out-of-Date Views
B

Updated Views
C

Figure 15-12.
AutoCAD displays
a notification on
the status bar when
drawing views
require updating.

> **A Model Has Changed** ⊠
> One or more drawing views are no longer up to date.
> Update all drawing views on this layout.

Pick to
update
views

You can pick the link in the balloon notification to update the views, or you can use one of the view update commands in the **Drawing Views** panel on the **Annotate** tab of the ribbon. In **Figure 15-11A**, the six arrayed holes have been modified by changing the diameter from ∅1.8 to ∅.9. In **Figure 15-11B**, the drawing views appear with markers to indicate they are out-of-date. To update the views, select **Update all Views** from the drop-down menu in the **Drawing Views** panel on the **Annotate** tab of the ribbon. After updating the views, the holes are updated to the modified diameter. See **Figure 15-11C**. In this example, all of the views are updated at once. This is the preferred method for updating views. However, you can also update views individually by selecting **Update View** from the drop-down menu in the **Drawing Views** panel on the **Annotate** tab of the ribbon. This requires you to select each view to update individually. Selecting **Update View** is the same as using the **VIEWUPDATE** command. When using this command, you are prompted to select each view to update.

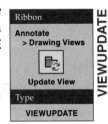

Ribbon
Annotate
> Drawing Views
Update All Views

Update All Views

Ribbon
Annotate
> Drawing Views
Update View
Type
VIEWUPDATE

VIEWUPDATE

Drawing views are different from viewports. You cannot pick inside of the view and edit or modify the drawing geometry by any standard AutoCAD editing technique. Drawing views can only be updated by changing the 3D model geometry.

PROFESSIONAL
TIP

In order to modify solid primitive subobjects in a composite solid model, the solid history must be recorded.

Editing Drawing Views

Drawing views can be edited after being created. Like other AutoCAD objects, drawing views can be moved or rotated. In addition, certain properties of drawing views, such as the display style and scale, can be modified.

The **VIEWEDIT** command can be used to edit the properties of a drawing view. You can quickly initiate this command by double-clicking on a view. The **Drawing View Editor** contextual ribbon tab appears, **Figure 15-13**. The editing options are similar to the options available when you create a base view.

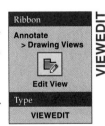

Ribbon
Annotate
> Drawing Views
Edit View
Type
VIEWEDIT

VIEWEDIT

Figure 15-13.
The **Drawing View**
Editor contextual
ribbon tab.

View property options

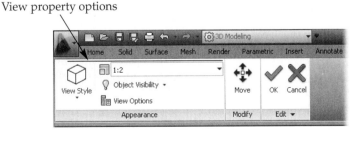

As previously discussed, projected views inherit the properties of the base view when created. If you select a projected view with the **VIEWEDIT** command, you can change the inherited properties, such as the display style, scale, and visibility.

Selecting **Move** in the **Modify** panel of the **Drawing View Editor** contextual ribbon tab allows you to move the drawing view. However, a more efficient way to move a drawing view is to select the view and then pick on the drawing view grip to move the view directly. When moving a parent view, any child views will move accordingly to maintain alignment. When moving a child view, you can move the view, but it cannot be moved out of alignment with the parent view. This applies to orthogonal views. If you move an isometric view, it is not aligned to other views and can be moved freely around the layout.

If you move the cursor over the drawing view grip to highlight it, a shortcut menu is displayed. The **Stretch** option is used to move the drawing view. The **Rotate** option is used to rotate the drawing view. You can rotate the view dynamically using the cursor or you can specify a rotation angle. If a drawing view is rotated, any parent-child relationships that exist between the view and other drawing views are broken.

The **Break Alignment** and **Repair Alignment** options are also available in the drawing view grip shortcut menu. The **Break Alignment** option allows you to remove the alignment constraint applied to a child view. If you select this option, the child view can be moved to a different location free of the base view. If a child view is moved, the alignment with the parent view can be restored by using the **Repair Alignment** option.

Exercise 15-2

Complete the exercise on the companion website.
www.g-wlearning.com/CAD

Dimensioning Drawing Views

After placing drawing views on a paper space layout, you can dimension each view as needed. When dimensioning drawing views, make sure that the **DIMASSOC** system variable is set to 2 so that associative dimensions are created. Associative dimensions are associated to the dimensions of the model and update when the physical model changes.

If you dimension drawing views with associative dimensions, make changes to the model, and then update the views, the dimensions may become disassociated. In this case, AutoCAD will ask if you want to reassociate the dimensions. See Figure 15-14. If you select **Try to reassociate annotations**, AutoCAD will attempt to reassociate the dimensions automatically. If AutoCAD is unable to reassociate all dimensions, you will be prompted to use the **DIMREASSOCIATE** command. Selecting **Run DIMREASSOCIATE** will allow you to pick points in the view to reassociate the dimensions. For more information on the **DIMREASSOCIATE** command and associative dimensioning, refer to *AutoCAD and Its Applications—Basics*.

Figure 15-14.
When updating views with associative dimensions, it may be necessary to reassociate dimensions. To continue with the update process, select **Try to reassociate annotations**.

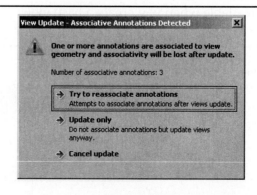

AutoCAD and Its Applications—Advanced

Using the EXPORTLAYOUT Command

You can use the **EXPORTLAYOUT** command to export a layout containing drawing views into a new drawing (DWG) file. However, this technique will break the associativity between the 3D model and the drawing views. The advantage to using this technique is that the drawing geometry can be edited in the same way as any other AutoCAD 2D geometry in model space. The disadvantage is that the 2D geometry has lost any associativity back to the 3D model. When using the **EXPORTLAYOUT** command, you save the exported layout as a DWG file. The drawing views become blocks in the new drawing file.

The **FLATSHOT** command can also be used to create a multiview orthographic drawing from a 3D model. However, the resulting drawing views do not have associative properties. The **FLATSHOT** command creates a flat projection of the 3D objects in the drawing from the current viewpoint. The view is created in model space. The view that is created is composed of 2D geometry and is projected onto the XY plane of the current UCS.

Creating Section Planes

The **SECTIONPLANE** command offers a powerful visualization and display tool. It allows you to construct a section plane, known as an AutoCAD *section object*, that can then be used as a plane to cut through a 3D model. Once the section object is drawn, it can be moved to any location, jogs can be added to it, and it can be rendered "live" so that internal features and sectioned material are dynamically visible as the cutting plane is moved. A variety of section settings allow you to customize the appearance of section features. Additionally, you can generate 2D sections/elevations or 3D sections that can be inserted into the drawing as a block. Once the command is initiated, you are prompted to select a face, the first point on the section object, or to enter an option. Sectioning options are located in the **Section** panel of the **Home** or **Solid** tab on the ribbon.

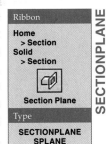

Ribbon

Home
> Section
Solid
> Section

Section Plane

Type

SECTIONPLANE
SPLANE

SECTIONPLANE

Pick a Face to Construct a Section Plane

The simplest way to create a section plane is to pick a flat face on the 3D object. Once the command is initiated, move the pointer until the face you wish to select is highlighted, then pick it. A transparent section object is placed on the face you selected and the model is cut at the plane. See **Figure 15-15**. The section plane can now be moved to create a section anywhere along the 3D model.

Figure 15-15.
Creating a section object on a face. A—The object before the face is selected. B—The section object is created.

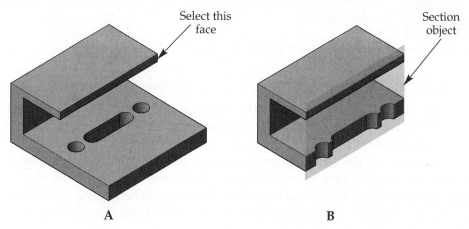

Select this face

Section object

A   B

Pick Two Points to Construct a Section Plane

A second method for defining a section plane is to pick two points through which the section object passes. The section object is perpendicular to the XY plane of the current UCS. When the command is initiated, pick the first point, which cannot be on a face. See P1 in **Figure 15-16.** It may be best to turn off dynamic UCSs or you could end up picking a face as the first point instead of a point. After picking the first point, move the pointer and notice that the section plane rotates about the first point. Next, pick the second point (P2) to define a line that cuts through the model. After the second point is picked, the section object is created. The section plane extends just beyond the edges of the model.

When the section object is created by picking two points, notice that the model is not automatically cut as it is when a face is selected. This is because *live sectioning* is not turned on when picking two points (when picking a face, this feature is turned on). To turn live sectioning on or off, select the section object, right-click, and select **Activate live sectioning** from the shortcut menu. Live sectioning is discussed later in this chapter.

Figure 15-16.
Creating a section object by selecting two points.

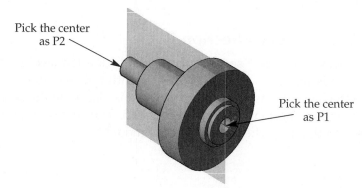

Pick the center as P2

Pick the center as P1

Pick Multiple Points to Construct a Section Plane

The previous method accepts only two points to construct a single section plane. Using the **Draw section** option, you can specify multiple points in order to create section plane *jogs*. In engineering drawing terminology, a section object drawn in this manner can represent an *offset* or *aligned* section plane.

Once the command is initiated, select the **Draw section** option. Pick the start point, using object snaps if necessary. See **Figure 15-17**. Continue picking points as needed. After picking the last point to define the section plane, press [Enter]. You are then prompted to specify a point in the direction of the section view. This point is on the opposite side of the section object from the viewer. Pick a point on the model using object snaps if necessary. The section plane is created.

Notice in **Figure 15-17** that the pick points created a section object that does not extend beyond the boundary of the model because the hole centers were selected. Also, the command "squares up" the section boundary to create a closed profile. Using section object grips, the section plane can be easily edited to include the entire solid. This is discussed in detail later in the chapter.

Create Orthographic Section Planes

The **Orthographic** option enables you to quickly place a section plane through the front, back, top, bottom, left, or right side of the object. See **Figure 15-18**. The origin is the center point of all objects in the model. Once the command is initiated, select the **Orthographic** option. Then, specify which orthographic plane you want to use as the section plane. The section object is created and all objects in the drawing are affected by it.

You may encounter a situation in which there is more than one solid on the screen and you want to use the **Orthographic** option to create a section object based on just one object. In this case, create a new layer, move objects you do not want to section to this layer, and then freeze the layer. The section object will be created based on the object that is visible. However, if the section plane passes through the objects on the frozen layers, those objects will be sectioned when the layers are thawed. A section plane affects all visible objects.

Exercise 15-3

Complete the exercise on the companion website.
www.g-wlearning.com/CAD

Figure 15-17.
Using the **Draw section** option of the **SECTIONPLANE** command to create a section object with multiple segments.

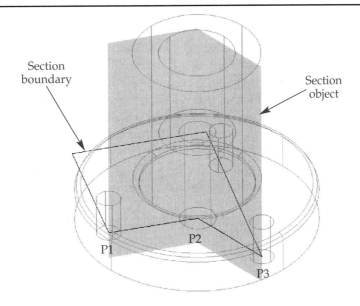

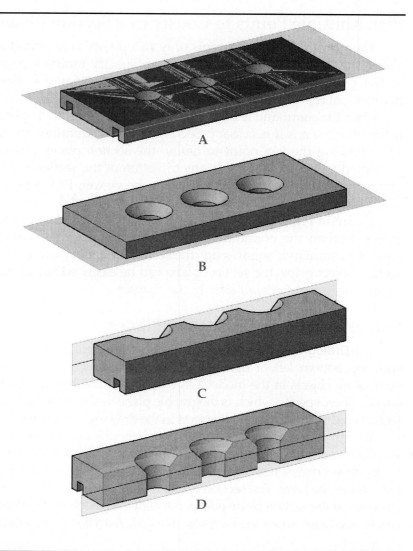

Figure 15-18.
Examples of
orthographic
SECTIONPLANE
options. A—Top.
B—Bottom. C—Left.
D—Right.

A

B

C

D

Editing and Using Section Planes

A wide range of section object editing and display options are available. To access these options, select the section object and then right-click to display the shortcut menu. From this menu, you can access all of the display and editing functions that apply to the section object.

Section Object States

There are three possible states for the section object created by the **SECTIONPLANE** command—section plane, section boundary, and section volume. See **Figure 15-19**. The section object can be changed from one state to another. Depending on which state is active, the section object produces different results on the solid(s).

When the section object is created by picking a face, picking two points, or using the **Orthographic** option, the object is in the *section plane state*. A transparent plane is displayed on each segment of the section object and a line connects the pick points (or the edges of the section object). See **Figure 15-19B**. The section plane extends infinitely in the section object's Z direction and along the direction of the object segment (unless connected to other segments).

When the **Draw section** option is used, the *section boundary state* is applied. A transparent plane is displayed on each segment of the section object. A 2D box extends to the XY-plane boundaries of the section object. See **Figure 15-19C**. The sectioned object fits inside of this footprint. The section plane extends infinitely in the section object's Z direction.

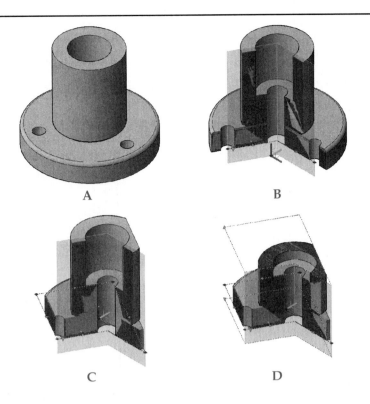

Figure 15-19.
Section object states. A—The original object. B—Section plane. C—Section boundary. D—Section volume.

A

B

C

D

The *section volume state* is not applied when the section object is created. The section object must be switched to this state once the object is created, as described in the next section. A transparent plane is displayed on each segment of the section object. In addition, a 3D box extends to the XYZ boundaries of the section object. The sectioned object fits inside of this box. See **Figure 15-19D**.

Section Object Properties

Once created, the properties of the section object can be changed. The **Section Object** category in the **Properties** palette contains properties specific to the section object. See **Figure 15-20**. These properties are described in the next sections.

Name

The default name of the first section object is Section Plane(1). Subsequent section planes are sequentially numbered, such as Section Plane(2), Section Plane(3), and so on. It may be beneficial to rename section objects so the name is representative of the section. For example, Front Half Section is much more descriptive than Section Plane(1). To rename a section object, select the Name property. Then, type a new name in the text box.

Type

As discussed earlier, the section object is in one of three states. The three states are section plane, section boundary, and section volume. To change the state of the section object, select the Type property. Then, pick the state in the drop-down list. The state can also be changed using the menu grip on the section object.

Live Section

Live sectioning is a tool that allows you to dynamically view the internal features of a solid, surface, or region as the section object is moved. This tool is discussed later in the chapter. To turn live sectioning on or off, select the Live Section property. Then, pick either Yes (on) or No (off) in the drop-down list. This is the same as turning live sectioning on or off using the shortcut menu or the ribbon.

Figure 15-20.
The properties of a section object can be changed in the **Properties** palette.

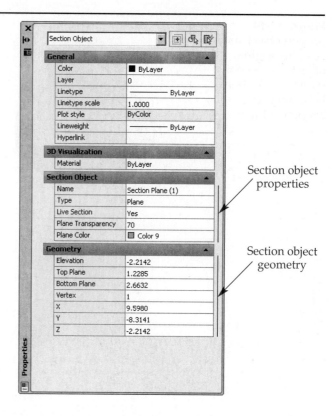

Section object properties

Section object geometry

Plane Transparency

The Plane Transparency property determines the opacity of the plane for the section object. The property value can range from 1 to 100. The lower the value, the more opaque the section plane object. See **Figure 15-21**.

Plane Color

The plane of the section object can be set to any color available in the **Select Color** dialog box. To change the color, select the Plane Color property and then select a color from the drop-down list. To choose a color in the **Select Color** dialog box, pick the Select Color... entry in the drop-down list. This property only affects the plane of the section object, not the lines defining the boundary, volume, or section line. The color of these lines is controlled by the Color property in the **General** category.

Editing the Section Object

When the translucent planes of a section object or the lines representing the section object state are picked, grips are displayed. The specific grips displayed are related to the current section object state. Refer to the grips shown in **Figure 15-19**. The types of grips are:
- Base grip.
- Menu grip.
- Direction grip.
- Second grip.
- Arrow grips.
- Segment end grips.

Base Grip

The *base grip* appears at the first point picked when defining the section object. See **Figure 15-22**. It is the grip about which the section object can be rotated and scaled. The section object can also be moved using this grip.

Figure 15-21.
A—The Plane Transparency property of the section plane object is set to 1 (or 1% transparent). B—The Plane Transparency property of the section plane object is set to 85 (or 85% transparent).

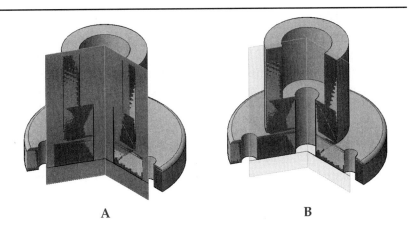

A   B

Figure 15-22.
The types of grips displayed on a section object.

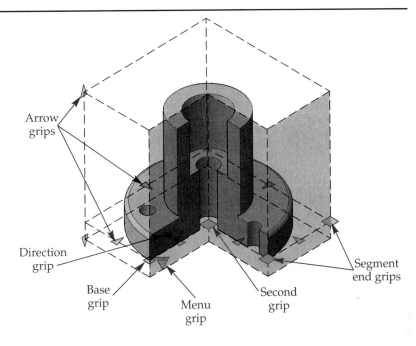

Menu Grip

The *menu grip* is always next to the base grip. Refer to **Figure 15-22**. Picking this grip displays the section state menu. See **Figure 15-23**. To switch the section object between states, pick the grip and then select the state from the menu. This is the same as changing the Type property in the **Properties** palette.

Direction Grip

The *direction grip* indicates the direction in which the section will be viewed. Refer to **Figure 15-22**. Pick the grip to flip the view 180°. The direction grip also shows the direction of the live section. Live sectioning is discussed later in this chapter.

Second Grip

The *second grip* appears at the second point picked when defining the section object. Refer to **Figure 15-22**. The section object can be rotated and stretched about the base grip using the second grip.

Figure 15-23.
Changing section object states.

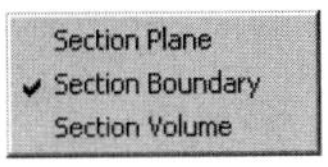

Section Plane
✔ Section Boundary
Section Volume

Arrow Grips

Arrow grips are located on all of the lines that represent the section plane, boundary, and volume. Refer to **Figure 15-22**. These grips are used to lengthen or shorten the section plane object segments or adjust the height of the section volume. The arrow grips at the top and bottom of the boundary box are used to change the height. Regardless of where the pointer is moved, the section object only extends in the segment's current plane. Changing the length of one segment of the section plane does not affect other segments.

In **Figure 15-17**, you saw an example of using the **Draw section** option to create a section object. The way in which the section object was created resulted in the section plane not extending beyond the solid object. This can quickly be corrected using the arrow grips. Notice in **Figure 15-24** that the arrow grip is being used to extend the left side of the section plane past the boundary of the solid model. This allows any subsequent section views to display the entire object rather than just a portion of it. The right side of the section plane can be extended in the same manner using the opposite arrow grip.

The arrow grips located on the line segments of the section plane move the position of the section plane. As a segment of the section plane is moved, it maintains its angular relationship and connection to any adjacent section plane segment.

Segment End Grips

The *segment end grips* are located at the end of each line segment defining the section object state. Refer to **Figure 15-22**. The number of displayed segment end grips depends on whether the section object is in the section plane, section boundary, or section volume state. These grips provide access to the standard grip editing options of stretch, move, copy, rotate, scale, and mirror. If the rotate option is used, the section plane is rotated about the selected segment end grip. Moving a segment end grip can change the angle between section plane segments.

Adding Jogs to a Section

Ribbon

Home
> Section
Solid
> Section

Add Jog

Type

SECTIONPLANEJOG
JOGSECTION

You can quickly add a jog, or offset, to an existing section object. First, select the section object. Then, right-click to display the shortcut menu and select **Add jog to section**. You can also enter the **SECTIONPLANEJOG** command. You are then prompted:

Specify a point on the section line to add jog:

Select a point directly on the section line. If any object snap is active, the **Nearest** object snap is temporarily turned on to ensure you pick the line. Once you pick, the jog is automatically added perpendicular to the line segment. See **Figure 15-25A**.

Figure 15-24.
A—The arrow grip is being used to extend the left side of the section plane past the boundary of the solid model. B—The edited section object. The right side can be corrected in the same manner.

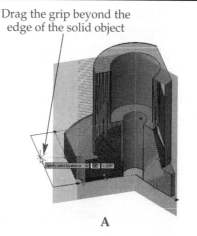

Drag the grip beyond the edge of the solid object

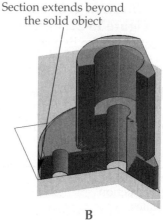

Section extends beyond the solid object

A

B

SECTIONPLANEJOG

Figure 15-25.
Adding a jog to a section object. A—Pick a point on the section line to add a jog. B—The jog is added, but it is not in the proper location. C—Using the arrow grip, the jog is moved to the proper location.

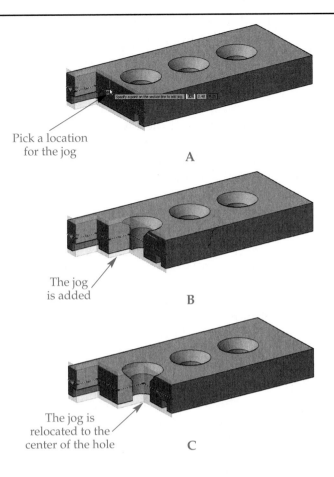

Pick a location for the jog

A

The jog is added

B

The jog is relocated to the center of the hole

C

It is not critical that you pick the exact location on the line where you want the jog to occur. Remember, grips allow you to easily adjust the section plane location. Notice in Figure 15-25B that the second jog barely cuts through the first hole. The intention is to run the section plane through the middle of the hole. To fix this, drag the arrow grip so the section plane segment is in the desired location, Figure 15-25C.

PROFESSIONAL TIP

If the section plane is not properly located, you can quickly change it. Simply pick the section object and right-click to display the shortcut menu. Then, select **Move**, **Scale**, or **Rotate** from the shortcut menu. Finally, adjust the section object location as needed.

Exercise 15-4

Complete the exercise on the companion website.
www.g-wlearning.com/CAD

Live Sectioning

Live sectioning is a tool that allows you to view the internal features of 3D solids, surfaces, and regions that are cut by the section plane of the section object. The view is dynamically updated as the section object is moved. This tool is used to visualize internal features and for establishing section locations from which 2D and 3D section views can be created. Live sectioning is either on or off.

LIVESECTION

Ribbon

Home
> **Section**
Solid
> **Section**

Live Section

Type

LIVESECTION

As you have seen, if the section plane is created by selecting a face, live sectioning is automatically turned on. However, when picking two points or using the **Draw** option of the **SECTIONPLANE** command, live sectioning is off. Live sectioning can be turned on and off for individual section objects, but only one section object can be "live" at any given time.

To turn live sectioning on or off, select the section object. Then, right-click to display the shortcut menu and pick **Activate live sectioning**. See **Figure 15-26**. A check mark appears next to the menu item when live sectioning is on. You can also use the ribbon or enter the **LIVESECTION** command and select the section object to toggle the on/off setting. When live sectioning is turned on, the material behind the viewing direction of the section plane is removed. The cross section of the 3D object is shown in gray (by default) and the internal shape of the 3D object is visible.

A wide variety of options allow you to change the appearance of not only the live sectioning display, but also of 2D and 3D section blocks that can be created from the sectioned display. These settings are found in the **Section Settings** dialog box. See **Figure 15-27**. To open this dialog box, select the section object, right-click, and pick **Live section settings...** in the shortcut menu. You can also pick the dialog box launcher button in the lower-right corner of the **Section** panel in the **Home** or **Solid** tab of the ribbon.

To change the settings for live sectioning, pick the **Live section settings** radio button at the top of the **Section Settings** dialog box. The categories displayed in the dialog box contain properties related to live sectioning. Settings for 2D and 3D sections and elevations are discussed later in this chapter.

The three categories for live sectioning are **Intersection Boundary**, **Intersection Fill**, and **Cut-Away Geometry**. To display a brief description of any property, hover the cursor over the property in the **Section Settings** dialog box. The description is displayed in a tooltip. A check box at the bottom of the **Section Settings** dialog box allows you to apply the properties to all section objects or to just the selected section object.

Figure 15-26.
Live sectioning can be turned on for any section state by selecting the section object, right-clicking to display the shortcut menu, and picking **Activate live sectioning**.

Turn live sectioning on and off

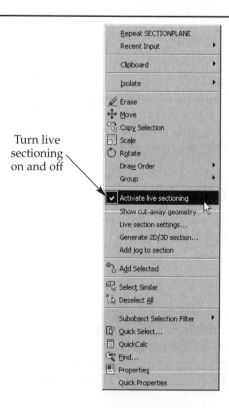

AutoCAD and Its Applications—Advanced

Figure 15-27.
Section settings. A—For a 2D block. B—For a 3D block. C—For live sectioning.

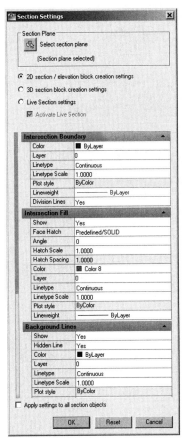

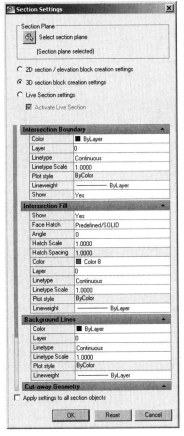

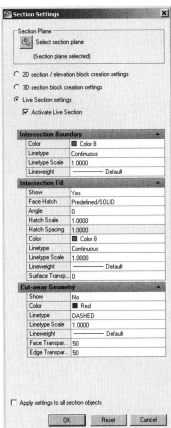

PROFESSIONAL TIP

Live sectioning can be quickly turned on or off by double-clicking on the section plane object.

Intersection Boundary

The intersection boundary is where the model is intersected by the section object. It is represented by line segments. You can set the color, linetype, linetype scale, and line weight of the intersection boundary lines.

Intersection Fill

The intersection fill is the material visible on the model surface where the section object cuts. It is displayed as a solid fill, by default. Any hatch pattern available in AutoCAD can be used as the intersection fill. See **Figure 15-28A**. The angle, hatch scale, hatch spacing, and color can be set. In addition, the linetype, linetype scale, and line weight can be changed. The fill pattern can even be set to be transparent.

Cutaway Geometry

The cutaway geometry is the part of the model removed by the live sectioning. By default, this geometry is not displayed. Changing the Show property to Yes displays the geometry. See **Figure 15-28B**. You can set the color, linetype, linetype scale, and line weight of the lines representing the cutaway geometry. In addition, the Face

Figure 15-28.
Live sectioning settings. A—The intersection fill can be displayed as a hatch pattern in any specified color. B—The cutaway geometry removed by the live sectioning is displayed.

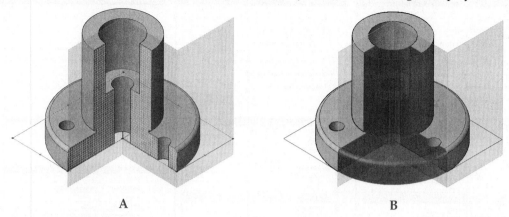

A   B

Transparency and Edge Transparency properties allow you to create a see-through effect, as seen in **Figure 15-29.** Each of these two properties is set to 50 by default.

PROFESSIONAL TIP

You can also display the cutaway geometry without using the **Section Settings** dialog box. Select the section object, right-click, and pick **Show cut-away geometry** in the shortcut menu. Live sectioning must be on.

Exercise 15-5

Complete the exercise on the companion website.
www.g-wlearning.com/CAD

Generating 2D and 3D Sections and Elevations

The **SECTIONPLANE** command provides a fast and efficient method of creating sections. The sections can be either 2D or 3D. Not only can the sections be displayed on the current drawing, they can also be exported as a file that can then be used in any other drawing or document for display, technical drawing, or manufacturing purposes.

Figure 15-29.
The cutaway geometry is displayed with 100% transparent faces and solid black lines.

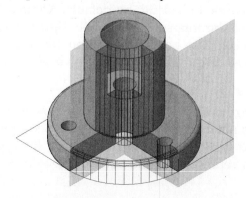

Creating Sections

To create a section, enter the **SECTIONPLANETOBLOCK** command. You can also select the section object, right-click, and pick **Generate 2D/3D section...** from the shortcut menu. The **Generate Section/Elevation** dialog box is displayed. See **Figure 15-30**. In this dialog box, you can specify whether the section will be 2D or 3D, select what is included in the section, and specify a destination for the section. To expand the dialog box, pick the **Show details** button, which looks like a down arrow.

To create a 2D section, pick the **2D Section/Elevation** radio button in the **2D/3D** area of the dialog box. A 2D section is projected onto the section plane, but is placed flat on the XY plane of the current UCS. To create a 3D section, pick the **3D Section** radio button. A 3D section is placed so its surfaces are parallel to the corresponding cut surfaces on the 3D object.

In the **Source Geometry** area of the dialog box, you can specify which geometry is included in the section. Picking the **Include all objects** radio button includes all 3D solids, surfaces, and regions in the section. To limit the section to certain objects, pick the **Select objects to include** radio button. Then, pick the **Select objects** button, select the objects on-screen, and press [Enter]. The number of selected objects is then displayed in the dialog box.

The **Destination** area of the dialog box is where you specify how the section will be placed. To place the section into the current drawing, pick the **Insert as new block** radio button. To update an existing section block, pick the **Replace existing block** radio button. Then, pick the **Select block** button, select the block on-screen, and press [Enter]. You will need to do this if the section object is changed. To save the section to a file for use in other drawings, pick the **Export to a file** radio button. Then, enter a path and file name in the text box.

Once all settings have been made, pick the **Create** button. The section is attached to the cursor and can be placed like a regular block. See **Figure 15-31**. Additionally, the options available are the same as if a regular block is being inserted. Once the block is inserted, it can be moved, rotated, and scaled as needed.

SECTIONPLANETOBLOCK

Ribbon

Home
> Section
Solid
> Section

Generate Section

Type

SECTIONPLANE-
TOBLOCK
GENERATESECTION

Figure 15-30.
The expanded
**Generate Section/
Elevation** dialog box.

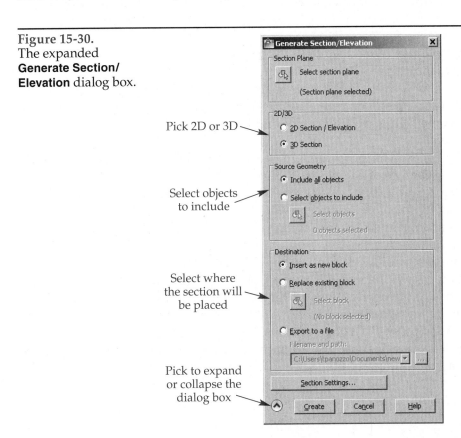

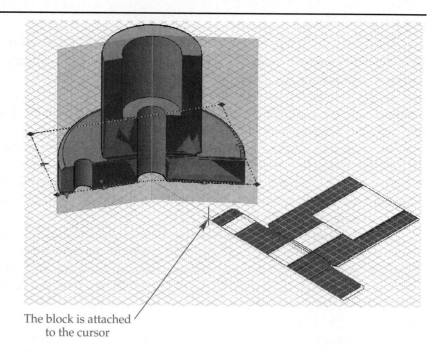

Figure 15-31.
Inserting a 2D
section block.

The block is attached
to the cursor

Section Settings

Picking the **Section Settings...** button at the bottom of the **Generate Section/ Elevation** dialog box opens the **Section Settings** dialog box discussed earlier. Using this dialog box, you can adjust all of the properties associated with the type of section being created. Depending on whether the **2D Section/Elevation** or **3D Section** radio button is selected in the **Generate Section/Elevation** dialog box, the appropriate categories and properties are displayed in the **Section Settings** dialog box. Refer to **Figure 15-27**.

The categories discussed earlier related to the **Live Section settings** radio button are available, although not all of the properties are displayed. Also, two additional categories are displayed for 2D and 3D sections:

- **Background Lines.** Available for 2D and 3D sections.
- **Curve Tangency Lines.** Only available for 2D sections.

These categories are discussed below. Examples of 2D and 3D sections inserted as blocks in the drawing are shown in **Figure 15-32**. Notice how properties can be set to show cutaway geometry in a different color and to change the section pattern, color, and linetype scale.

Background Lines. The properties in the **Background Lines** category provide control over the appearance of all lines that are not on the section plane. You can choose to have visible background lines, hidden background lines, or both displayed. They can be emphasized with color, linetype, or line weight. The layer, linetype scale, and plot style can also be changed. These settings are applied to both visible and hidden background lines.

Curve Tangency Lines. The properties in the **Curve Tangency Lines** category apply to lines of tangency behind the section plane. For example, the object shown in **Figure 15-32** has a round on the top of the base. This results in a line of tangency behind the section plane where the round meets the vertical edge. You can have these lines displayed or suppressed. In general, lines of tangency are not shown in a section view. If you choose to display these lines, you can set the color, layer, linetype, linetype scale, plot style, and line weight of the lines.

Figure 15-32.
A—The section object is created. B—A 2D section block is inserted into the drawing and the view is made plan to the block. C—A 3D section block is inserted into the drawing. Notice how the hatch pattern is displayed. D—The 3D section block is updated and now the cutaway geometry is displayed.

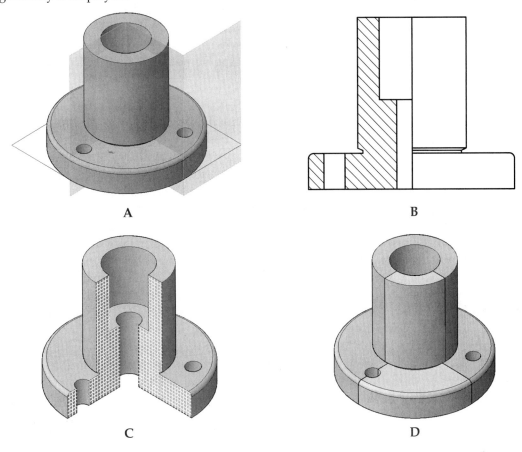

A

B

C

D

When a 3D section is created, you must turn off live sectioning to see the complete sectioned object in the block. With live sectioning on, only the cut surfaces appear in the block.

Updating the Section View

Once the section view is created, it is not automatically updated if the section object is changed. To update the section view, select the section object (not the block), right-click, and pick **Generate 2D/3D section...** from the shortcut menu. Then, in the **Destination** area of the **Generate Section/Elevation** dialog box, pick the **Replace existing block** radio button. If necessary, pick the **Select block** button and select the section block in the drawing. If you want to change the appearance of the section view, pick the **Section Settings...** button and adjust the properties as needed. Finally, pick the **Create** button in the **Generate Section/Elevation** dialog box to update the section block.

Exercise 15-6

Complete the exercise on the companion website.
www.g-wlearning.com/CAD

Once a solid model has been constructed, it is easy to create a multiview layout using the **SOLVIEW** command. This command allows you to create a layout containing orthographic, section, and auxiliary views. The **SOLDRAW** command can then be used to complete profile and section views. **SOLDRAW** must be used after **SOLVIEW**. The **SOLPROF** command can be used to create a profile of the solid in the current view. These commands may be typed or selected in the **Modeling** panel of the **Home** tab on the ribbon.

Creating Views with SOLVIEW

SOLVIEW

Ribbon
Home
> Modeling

Solid View

Type
SOLVIEW

The **SOLVIEW** command is used to create new floating viewports and to establish the display within those viewports. Therefore, you may want to delete the default viewport in the layout (paper space) tab before using the **SOLVIEW** command.

First restore the WCS. This will help avoid any confusion. Then, display a plan view. See Figure 15-33. It helps to have additional user coordinate systems created prior to using **SOLVIEW**. This allows you to construct orthographic views based on a specific named UCS.

Before using the **SOLVIEW** command, visualize which view is going to be the top view (or plan view) and how you would like the model rotated in relationship to the layout. With this in mind, look at the current UCS icon and make sure that the X axis is pointing to the "right" and the Y axis is pointing "up" in your imagined layout. If this is not the case, then you must restore the WCS, rotate the current UCS, or restore a saved UCS to correctly align the axes. Then, when you enter the **SOLVIEW** command in the layout, you can simply select the current UCS and you will be creating the top or plan view of your model.

Select a layout tab and delete the default viewport. Next, use the **SOLVIEW** command to create an initial view from which other views can project. This is normally the top or front. In the following example, the top view is constructed first by using the plan view of a UCS named Leftside.

Enter an option [Ucs/Ortho/Auxiliary/Section]: **U**↵
Enter an option [Named/World/?/Current] <Current>: **N**↵
Enter name of UCS to restore: **LEFTSIDE**↵
Enter view scale <1.0>: **.5**↵
Specify view center: *(pick a location in the layout for the center of the view)*
Specify view center <specify viewport>: ↵
Specify first corner of viewport: *(pick the first corner of a paper space viewport outside of the object)*
Specify opposite corner of viewport: *(pick the opposite corner of the viewport)*
Enter view name: **TOPVIEW**↵ *(the left of the object in AutoCAD is the top of the part)*
Enter an option [Ucs/Ortho/Auxiliary/Section]: *(leave the command active at this time)*

Figure 15-33.
Before using **SOLVIEW**, display a plan view of the WCS.

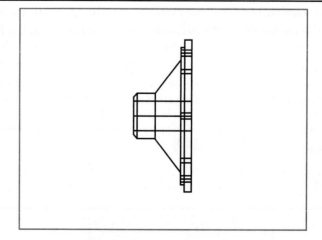

You must provide a name for the view. The result is shown in **Figure 15-34.**

The **SOLVIEW** command remains active until you press the [Enter] or [Esc] key. If you exit **SOLVIEW**, you can still return to the drawing and create additional orthographic viewports. With the command active, continue and create a section view to the right of the top view:

Enter an option [Ucs/Ortho/Auxiliary/Section]: **S**↵
Specify first point of cutting plane: *(pick the quadrant at Point 1 in* **Figure 15-35***)*
Specify second point of cutting plane: *(pick the quadrant at Point 2)*
Specify side to view from: *(pick Point 3)*
Enter view scale <0.5>: ↵
Specify view center: *(pick the center of the new section view)*
Specify view center <specify viewport>: ↵ *(this prompt remains active until [Enter] is pressed to allow you to adjust the view location if necessary)*
Specify first corner of viewport: *(pick one corner of the viewport)*
Specify opposite corner of viewport: *(pick the opposite corner of the viewport)*
Enter view name: **SECTION**↵
Enter an option [Ucs/Ortho/Auxiliary/Section]: ↵

Notice in **Figure 15-35** that the new view is shown in the current visual style and not as a section. This is normal. **SOLVIEW** is used to create the views. The **SOLDRAW** command draws the section lines. **SOLDRAW** is discussed later in this chapter.

A standard orthographic view can be created using the **Ortho** option of **SOLVIEW**. This is illustrated in the following example. The new orthographic view is shown in **Figure 15-36.**

Enter an option [Ucs/Ortho/Auxiliary/Section]: **O**↵
Specify side of viewport to project: *(pick the bottom edge of the left viewport)*
Specify view center: *(pick the center of the new view)*
Specify view center <specify viewport>: ↵
Specify first corner of viewport: *(pick one corner of the viewport)*
Specify opposite corner of viewport: *(pick the opposite corner of the viewport)*
Enter view name: **FRONTVIEW**↵

Figure 15-34.
The initial view created with the **Ucs** option of **SOLVIEW**.

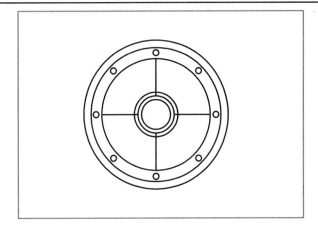

Figure 15-35.
The section view created with **SOLVIEW** (shown on the right) does not show projection lines. The pick points are shown on the left.

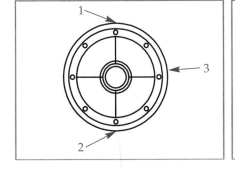
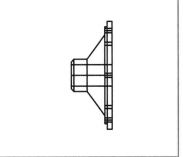

Figure 15-36.
An orthographic front view is created with the **Ortho** option of **SOLVIEW**. This is the view shown at the lower left.

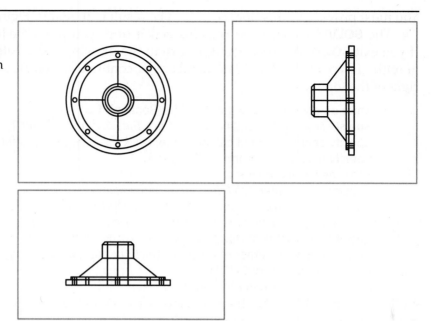

The **SOLVIEW** command creates new layers that are used by **SOLDRAW** when profiles and sections are created. The layers are used for the placement of visible, hidden, dimension, and section lines. Each layer is named as the name of the view with a three letter tag, as shown in the following table. The use of these layers is discussed later in this chapter.

Layer Name	Object
View name-VIS	Visible lines
View name-HID	Hidden lines
View name-DIM	Dimension lines
View name-HAT	Hatch patterns (sections)

Exercise 15-7

Complete the exercise on the companion website.
www.g-wlearning.com/CAD

Creating Auxiliary Views with SOLVIEW

Auxiliary views are used to display a surface of an object that is not parallel to any of the standard views. It may be an inclined or oblique surface. Refer to **Figure 15-37**. Sometimes these views are necessary to show or dimension a feature that is not being displayed in true size in any other view. The slot in the inclined surface in **Figure 15-37** is not shown in true size in any of the standard views.

The auxiliary view is taken from one of the other views where the inclined surface is shown as an edge. The auxiliary view will be projected perpendicular to this surface. The auxiliary view is created by picking two points on the surface in the front view and another point to indicate the line of sight.

> Enter an option [Ucs/Ortho/Auxiliary/Section]: **A**↵
> Specify first point of inclined plane: (*using object snaps, pick a point on one end of the inclined surface*)
> Specify second point of inclined plane: (*pick a point on the other end of the inclined surface*)

AutoCAD and Its Applications—Advanced

Figure 15-37.
An auxiliary view (shown in color) is created from the inclined plane in the front view.

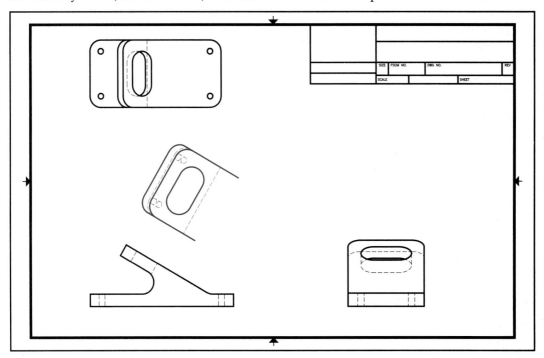

Specify side to view from: (*pick a point on the side of the surface from which you want to view it*)
Specify view center: (*pick the center*)
Specify view center: <specify viewport>↵
Specify first corner of viewport: (*pick one corner of the viewport*)
Specify opposite corner of viewport: (*pick the opposite corner of the viewport*)
Enter view name: **AUXILIARYVIEW**↵

Auxiliary views are often incomplete views, so it is acceptable to cut off portions of the view that are not necessary when you specify the corners of the viewport.

PROFESSIONAL
TIP

When creating an auxiliary view, you may want to move other viewports that may be in the way to make room for the view.

Creating Finished Views with SOLDRAW

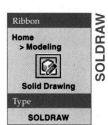

The **SOLVIEW** command saves information specific to each viewport when a new view is created. This information is used by the **SOLDRAW** command to construct a finished profile or section view. **SOLDRAW** first deletes any information currently on the *view name*-VIS, *view name*-HID, and *view name*-HAT layers for the selected view. Visible, hidden, and section lines are automatically placed on the appropriate layer. Therefore, you should avoid placing objects on any of the automatically generated layers other than the *view name*-DIM layer.

The **SOLDRAW** command automatically creates a profile or section in the selected viewport. If you select a viewport that was created using the **Section** option of **SOLVIEW**, the **SOLDRAW** command uses the current values of the **HPNAME**, **HPSCALE**, and **HPANG** system variables to construct the section. These three variables control the angle, scale factor, and name of the hatch pattern.

If a view is selected that was not created as a section in **SOLVIEW**, the **SOLDRAW** command constructs a profile view. All new visible and hidden lines are placed on the *view name*-VIS or *view name*-HID layer. All existing objects on those layers are deleted.

Once the command is initiated, you are prompted to select objects. Pick the border of the viewport(s) for which you want the profile or section generated. When all viewports are selected, press [Enter] and the profiles and sections are created.

After the profile construction is completed, lines that should be a hidden linetype are still visible (solid). This is because the linetype set for the *view name*-HID layer is Continuous. Change the linetype for the layer to Hidden and the drawing should appear as shown in **Figure 15-38**, depending on the current visual style and hatch settings. You may also want to change other layer properties such as color, line weight, and plot style.

Revising the 3D Model

If changes are needed after these views are created, the best practice is to modify the original 3D solid. However, the views created with **SOLVIEW** and **SOLDRAW** will not immediately reflect changes. To update the views, simply start the **SOLDRAW** command, select the viewports, and press [Enter]. The views are then updated with the changes.

PROFESSIONAL TIP

If you wish to add dimensions in model space for the views created with **SOLVIEW** and **SOLDRAW**, use the view-specific DIM layers. These layers are created for that purpose and are only visible in one view. **SOLDRAW** does not delete information on the DIM layers when it constructs a view. If you prefer to dimension in paper space, use a layer other than the DIM layers created by **SOLVIEW**.

Adding a 3D View in Paper Space to the Drawing Layout

If you want to add a paper space viewport that contains a 3D (pictorial) view of the solid, use the **MVIEW** command. Create a single viewport by picking the corners. The object will appear in the viewport. Next, activate the viewport and use any of the orbit commands or a preset isometric viewpoint to achieve the desired 3D view. Pan and

Figure 15-38.
The new front profile view shows hidden lines after the linetype is set to Hidden for the FRONTVIEW-HID layer.

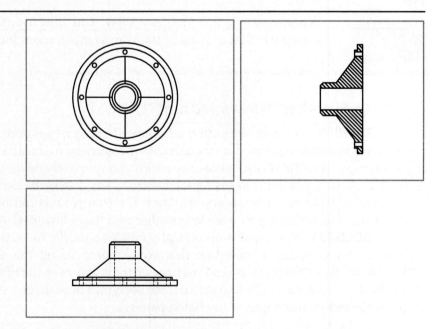

zoom as necessary. Change to the parallel or perspective projection if needed. You can also use the **Visual Style Controls** flyout in the viewport controls to adjust the display of the 3D viewport. The visual style set current for this viewport does not affect the displays in the other viewports. See **Figure 15-39**.

In order to have the hidden display correctly plotted, use the **MVIEW Shadeplot** option on the 3D viewport. Enter the command and select the **Shadeplot** option. Then, set the option to **Hidden**. If you have a hidden display shown in the viewport, you can also select **As displayed**. Then, pick the viewport when prompted to select objects.

Alternately, you can select the viewport and use the **Properties** palette to set the Shade plot property to As Displayed or Hidden. Any visual style display of the viewport can also be plotted in this manner by setting **MVIEW Shadeplot** to **As Displayed** (when the view is shaded) or **Rendered**.

Tips

Remember the following points when working with **SOLVIEW** and **SOLDRAW**.
- Use **SOLVIEW** first and then **SOLDRAW**.
- Do not draw on the *view name*-HID and *view name*-VIS layers.
- Place model space dimensions for each view on the *view name*-DIM layer for that specific view or simply dimension in paper space on a layer not created by **SOLVIEW**.
- After using **SOLVIEW**, use **SOLDRAW** on all viewports in order to create hidden lines or section views.
- Change the linetype on the *view name*-HID layer to Hidden and adjust other layer properties as needed.
- Create 3D viewports with the **MVIEW** or **VPORTS** command and an orbit command or a preset isometric view. Remove hidden lines when plotting with the **MVIEW Shadeplot** option set to **Hidden**.
- Plot the drawing in layout (paper) space at the scale of 1:1.

Exercise 15-8

Complete the exercise on the companion website.
www.g-wlearning.com/CAD

Figure 15-39.
Create a 3D viewport with the **MVIEW** command. You can hide the lines in the viewport, as shown at the lower right. To plot the viewport as a hidden display, use the **MVIEW Shadeplot** option and set it to Hidden.

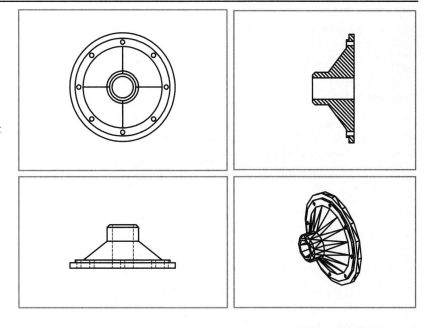

Creating a Profile with SOLPROF

SOLPROF

Ribbon

Home
> Modeling

Solid Profile

Type

SOLPROF

The **SOLPROF** command creates a profile view from a 3D solid model. This is similar to the **Profile** option of the **SOLVIEW** command. However, **SOLPROF** is limited to creating a profile view of the solid for the current view only.

SOLPROF creates a block of all lines forming the profile of the object. It also creates a block of the hidden lines of the object. The original 3D object is retained. Each of these blocks is placed on a new layer with the name of PH-*view handle* and PV-*view handle*. A *view handle* is a name composed of numbers and letters that is automatically given to a viewport by AutoCAD. For example, if the view handle for the current viewport is 2C9, the **SOLPROF** command creates the layers PH-2C9 and PV-2C9.

You must be in layout (paper) space and have a model space viewport active to use the command. Once the command is initiated, you are prompted to select objects:

> Select objects: *(pick the solid)*
> 1 found
> Select objects: ↵
> Display hidden profile lines on separate layer? [Yes/No] <Y>: ↵
> Project profile lines onto a plane? [Yes/No] <Y>: ↵

If you answer yes to this prompt, the 3D profile lines are projected to a 2D plane and converted to 2D objects. This produces a cleaner profile.

> Delete tangential edges? [Yes/No] <Y>: ↵

Answering yes to this prompt produces a proper 2D view by eliminating lines that would normally appear at tangent points of arcs and lines. If you wish to display the profile lines in a 3D view, do not delete the tangential edges. Once the profile is created, freeze the layer of the original object in the viewport. The original object and the profile created with **SOLPROF** are shown in Figure 15-40.

When plotting views created with **SOLPROF**, hidden lines may not be displayed unless you freeze the layer that contains the original 3D object.

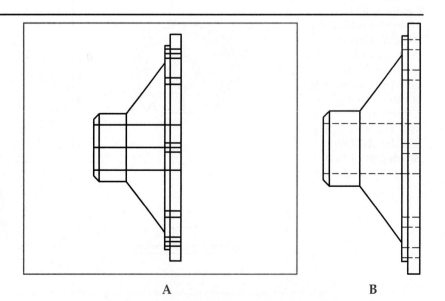

Figure 15-40.
A—The original solid. B—A profile created with **SOLPROF**.

A   B

Model Analysis

AutoCAD provides a number of tools for analyzing solid and surface models. Model analysis is conducted to evaluate various data and determine whether design changes are needed prior to manufacturing. Model analysis tools available in AutoCAD are discussed in the following sections.

Solid Model Analysis

Type

MASSPROP

The **MASSPROP** command allows you to analyze a solid model for its physical properties. The data obtained from **MASSPROP** can be retained for reference by saving the data to a file. The default file name is the drawing name. The file is an ASCII text file with a .mpr (mass properties) extension. The analysis can be used for third-party applications to produce finite element analysis, material lists, or other testing studies.

Once the command is initiated, you are prompted to select objects. Pick the objects for which you want the mass properties displayed and press [Enter]. AutoCAD analyzes the model and displays the results in the AutoCAD text window. See **Figure 15-41**. The following properties are listed.

- **Mass.** A measure of the inertia of a solid. In other words, the more mass an object has, the more inertia it has. Note: Mass is *not* a unit of measurement of inertia.
- **Volume.** The amount of 3D space the solid occupies.
- **Bounding box.** The dimensions of a 3D box that fully encloses the solid.
- **Centroid.** A point in 3D space that represents the geometric center of the mass.
- **Moments of inertia.** A solid's resistance when rotating about a given axis.
- **Products of inertia.** A solid's resistance when rotating about two axes at a time.
- **Radii of gyration.** Similar to moments of inertia. Specified as a radius about an axis.
- **Principal moments and X-Y-Z directions about a centroid.** The axes about which the moments of inertia are the highest and lowest.

PROFESSIONAL TIP

Advanced applications of solid model design and analysis are possible with Autodesk Inventor software. This product allows you to create parametric designs and assign a wide variety of physical materials to the solid model.

Figure 15-41.
The **MASSPROP** command displays a list of solid properties in the AutoCAD text window.

```
AutoCAD Text Window - Drawing1.dwg                      _|□|X|
Edit

Select objects:

---------------    SOLIDS    ---------------

Mass:                   3.3187
Volume:                 3.3187
Bounding box:      X: 5.4348  --  7.1848
                   Y: 11.5845 --  15.5845
                   Z: -0.5090 --  0.0000
Centroid:          X: 6.3098
                   Y: 13.5845
                   Z: -0.2569
Moments of inertia: X: 617.3021
                    Y: 133.3258
                    Z: 750.0485
Products of inertia: XY: 284.4649
                     YZ: -11.5807
                     ZX: -5.3791
Radii of gyration:  X: 13.6385
                    Y: 6.3383
                    Z: 15.0335
Principal moments and X-Y-Z directions about centroid:

Press ENTER to continue:
```

Surface Continuity Analysis

Surface continuity describes the type of transition formed between adjoining surfaces in a model. Surface continuity is discussed in Chapter 11. Depending on the type of model you are working with and manufacturing requirements, you may need to modify the curvature of the model to create a more smooth transition between surfaces. For example, if you are working with a surface model and continuity settings are available, you may need to adjust the settings to produce a more smooth contour.

In AutoCAD, *zebra analysis* is used to graphically check for surface continuity. Zebra analysis is also referred to as *zebra stripping*. The **ANALYSISZEBRA** command is used to conduct zebra analysis. When using this command, AutoCAD projects a map of zebra stripes onto the model. The flow of stripes from one surface to another is then checked to determine the surface continuity. Areas where the stripes line up indicate where the model has surface continuity.

The purpose of zebra stripping is to simply check for the quality of continuity or *flow* between surfaces. It does not tell you how accurate the surface is or the quality of the surface constructed. However, good surface continuity is important in many modeling applications. You will typically want the model to have a smooth surface shape and avoid sharp changes in curvature.

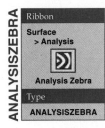

ANALYSISZEBRA

Ribbon

Surface
> Analysis

Analysis Zebra

Type

ANALYSISZEBRA

Before using zebra analysis, make sure a 3D visual style is set current. Then, initiate the **ANALYSISZEBRA** command and select the model surfaces to analyze. You can select one or more surfaces or solids. Press [Enter] after selecting the surfaces. In **Figure 15-42**, the three surfaces of the air duct are selected. The top and bottom surfaces of the model are loft surfaces. The center surface is a blend surface. In this part, smooth flow design is important. Notice the flow of the zebra stripes in **Figure 15-42B**. The stripes touch but do not align consistently. In this version of the model, the start and end continuity of the blend surface are both set to G0 (positional continuity). In **Figure 15-42C**, changing the continuity to G2 (curvature) creates a smoother flow between the surfaces.

To turn off the zebra stripping display, use the **ANALYSISZEBRA** command. Select the **Turn off** option. You can also select **Analysis Options** from the **Analysis** panel in the **Surface** tab of the ribbon to display the **Analysis Options** dialog box. In the **Zebra** tab, pick the **Clear Zebra Analysis** button to remove zebra stripping. The **Zebra** tab contains options for setting the stripe display, the stripe direction, and the colors and thickness used for the zebra stripping. This dialog box is also used to set options for surface curvature analysis and draft analysis, as discussed in the next sections.

Figure 15-42.
Zebra analysis. A—The original air duct model consists of two loft surfaces connected by a blend surface. B—The three surfaces are selected with the **ANALYSISZEBRA** command. Notice that the zebra stripes do not line up. In this version of the model, the start and end surface continuity of the blend surface are both set to G0. C—Adjusting the start and end surface continuity of the blend surface to G2 produces smoother continuity.

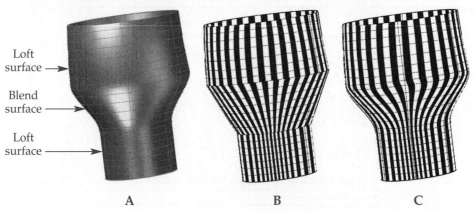

Surface Curvature Analysis

Once a model is designed, it is then analyzed to confirm the quality of the design before the actual part is manufactured. When working with models containing curved surfaces, you design with the intent that the model stays within specific curvature ranges for manufacturing purposes. If you design a model that falls outside specific design criteria, problems occur and the model cannot be manufactured. *Surface curvature analysis* assists in determining the overall smoothness of a 3D model design.

In AutoCAD, surface curvature analysis is performed with the **ANALYSISCURVATURE** command. When using this command, AutoCAD applies a color gradient to the surfaces of the model. This is known as *color mapping*. The model is then graphically analyzed. Different gradient colors indicate areas of high and low curvature and abrupt changes in curvature.

When you use the **ANALYSISCURVATURE** command, AutoCAD uses the settings specified in the **Curvature** tab of the **Analysis Options** dialog box, Figure 15-43. To display this dialog box, use the **ANALYSISOPTIONS** command. The options in the **Display style:** drop-down list in the **Color Mapping** area determine how the curvature is analyzed. The options are described as follows:

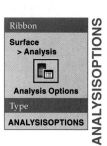

Ribbon

Surface
> Analysis

Analysis Options

Type

ANALYSISOPTIONS

ANALYSISOPTIONS

- **Gaussian.** This is the default option. AutoCAD analyzes areas of high and low curvature in the model. The color red is assigned to areas with positive curvature. A positive Gaussian value indicates that the surface has a bowl or spherical shape. The color green is assigned to areas with zero-value curvature (flat surface areas). Cylindrical and conical surfaces are examples of surfaces with zero-value Gaussian curvature. The color blue is assigned to areas with negative curvature. A negative Gaussian value indicates that the surface has a saddle or hyperbolic shape. The **Gaussian** option is suitable for general purposes and works well with swept objects.
- **Mean.** AutoCAD analyzes the mean curvature of values along the U and V directions of the surface. This option is useful when checking for quick or sharp changes in the shape of the surface.
- **Max radius.** AutoCAD analyzes the maximum curvature of values along the U and V directions of the surface. This option is useful for determining how flat or curved specific surfaces are.
- **Min radius.** AutoCAD analyzes the minimum curvature of values along the U and V directions of the surface. This option is useful for checking surfaces with small radius bends.

Figure 15-43.
The **Curvature** tab of the **Analysis Options** dialog box.

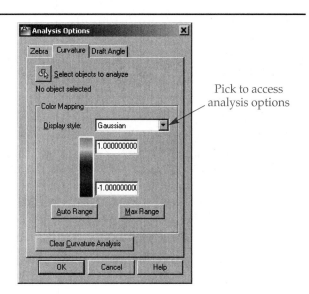

Pick to access analysis options

The values in the text boxes at the ends of the color gradient bar in the **Color Mapping** area define the curvature range from minimum to maximum curvature. The default values are 1.0 and –1.0. The values are used to define the acceptable curvature range. AutoCAD displays the corresponding color in the color gradient bar when a curvature value is reached. Maximum values are indicated in green and minimum values are indicated in blue. The resulting display can vary depending on the range of curvature values entered.

The **Auto Range** and **Max Range** buttons are used to calculate ranges of curvature values of selected objects. When the **Auto Range** button is picked, the calculation is based on 80% of the values in the curvature range. This can provide a starting point for setting the minimum and maximum curvature values. When the **Max Range** button is picked, the calculation is based on the maximum range of values.

To select surfaces to analyze, pick the **Select objects to analyze** button in the **Curvature** tab if the **Analysis Options** dialog box is open. Otherwise, use the **ANALYSISCURVATURE** command. Before using this command, make sure that a 3D visual style is set current. Once you initiate the command, select the surfaces and press [Enter]. You can select one or more surfaces or solids. If needed, access the **Analysis Options** dialog box to set the curvature range.

In **Figure 15-44**, a model created from a loft surface is analyzed using the **Min radius** option. In **Figure 15-44B**, the features at the end of the model are checked. In the resulting shaded display, the outer portion of the ring-shaped bend is shaded green. This indicates an area of maximum curvature. The recessed surface is shaded blue. This indicates an area of minimum curvature.

To turn off the curvature analysis display, use the **ANALYSISCURVATURE** command. Select the **Turn off** option. You can also pick the **Clear Curvature Analysis** button in the **Analysis Options** dialog box.

Draft Analysis

When designing a part that will be removed from a mold, a *draft angle* is designed on the part to ensure that the part can be pulled from the mold. *Draft analysis* is used to evaluate a solid or surface model for adequate draft. In AutoCAD, draft analysis is performed with the **ANALYSISDRAFT** command. The procedure is similar to that used with surface curvature analysis. When you select a model to

---

Figure 15-44.
Surface curvature analysis. A—The original loft surface. B—The **Min radius** option is used to analyze the features at the end of the model. The outer "ring" is shaded green. This indicates an area of maximum curvature. The inner recessed surface is shaded blue. This indicates an area of minimum curvature.

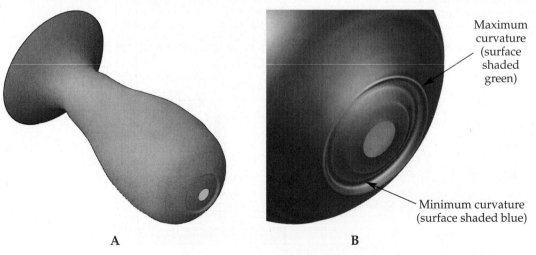

Maximum curvature (surface shaded green)

Minimum curvature (surface shaded blue)

A   B

analyze, AutoCAD maps a color gradient to indicate the angle variation of the model. Green colors are mapped to areas where the draft is at the highest angle specified, and blue colors are mapped to areas where the draft is at the lowest angle specified.

When using the **ANALYSISDRAFT** command, the high and low draft angles are specified in the **Draft Angle** tab of the **Analysis Options** dialog box. Use the **ANALYSISOPTIONS** command to open this dialog box, as previously discussed. Then, in the **Draft Angle** tab, use the **Angle:** text boxes to specify the highest and lowest draft angles allowed in the model. The default values are 3.0 and –3.0.

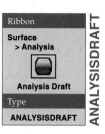

To select a model to analyze, pick the **Select objects to analyze** button in the **Draft Angle** tab if the **Analysis Options** dialog box is open. Otherwise, use the **ANALYSISDRAFT** command. Before using this command, make sure that a 3D visual style is set current. Once you initiate the command, select the model and press [Enter].

In **Figure 15-45**, a computer mouse is analyzed for adequate draft. The model was created as a mesh, converted to a surface, and then converted to a solid. Draft analysis is used to determine if a 4° draft angle can be used for the mold. The model after using the **ANALYSISDRAFT** command is shown in **Figure 15-45B**. The top of the model is shaded green and is within the parameters. However, there are two areas of concern along the side and bottom (where surfaces are shaded blue). This indicates that changes to the design are in order.

To turn off the draft analysis display, use the **ANALYSISDRAFT** command. Select the **Turn off** option. You can also pick the **Clear Draft Angle Analysis** button in the **Analysis Options** dialog box.

Solid Model File Exchange

AutoCAD drawing files can be converted to files that can be used for testing and analysis. Use the **ACISOUT** command or **Export Data** dialog box to create a file with a .sat extension. These files can be imported into AutoCAD with the **ACISIN** or **IMPORT** command.

Solids can also be exported for use with stereolithography software. These files have a .stl extension. Use the **3DPRINT** command to create STL files.

Figure 15-45.
Draft analysis. A—The original solid model. B—The resulting analysis based on angle values of 4.0 and –4.0. The top of the model is within the draft parameters. Areas along the side and bottom are shaded blue to indicate the draft is at the lowest angle specified.

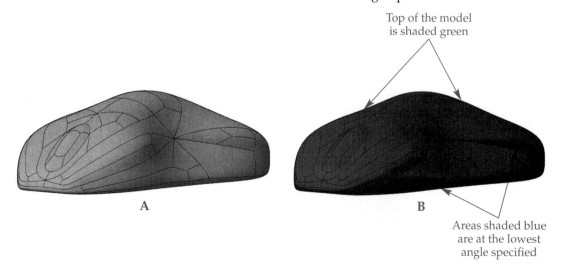

Importing and Exporting Solid Model Files

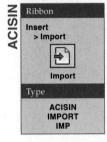

ACISOUT

Application Menu

Export
> Other Formats

Type

ACISOUT
EXPORT
EXP

A solid model is frequently used with analysis and testing software or in the manufacture of a part. The **ACISOUT** and **EXPORT** commands allow you to create a type of file that can be used for these purposes. Once the **ACISOUT** command is initiated, you are prompted to select objects. After selecting objects and pressing [Enter], a standard save dialog box is displayed. See **Figure 15-46**. When using the **EXPORT** command, the standard save dialog box appears first. After entering a file name and selecting a file type (SAT), you are then prompted to select objects.

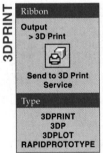

ACISIN

Ribbon

Insert
> Import

Import

Type

ACISIN
IMPORT
IMP

An SAT file can be imported into AutoCAD and automatically converted into a drawing file using the **ACISIN** or **IMPORT** command. Once either command is initiated, a standard open dialog box appears. Change the file type to SAT, locate the file, and pick the **Open** button.

Stereolithography Files

Stereolithography (SLA) is an additive manufacturing process that creates various plastic 3D model prototypes using a computer-generated solid model, a laser, and a vat of liquid polymer. This technology is also referred to as *rapid prototyping* or *3D printing*. Some additive manufacturing processes, such as fused-deposition modeling (FDM), add material in "layers" from a filament that is extruded through a heated nozzle. Other additive manufacturing processes are: selective laser sintering (SLS), 3D printing (3DP), multi-jet modeling (MJM), and electron beam melting (EBM). Using one of these processes, a prototype 3D model can be designed and formed in a short amount of time without using standard subtractive manufacturing processes.

Most CAD software today can create a stereolithograph file (STL file). AutoCAD can export a drawing file to the STL format, but *cannot* import STL files.

Using the 3DPRINT Command to Create STL Files

3DPRINT

Ribbon

Output
> 3D Print

Send to 3D Print Service

Type

3DPRINT
3DP
3DPLOT
RAPIDPROTOTYPE

The **3DPRINT** command is used to create an STL file. It can also be used to send a solid model to a 3D printer service provider. Solids and watertight meshes can be selected for use with the command. A *watertight mesh* is completely closed and contains no openings. Watertight meshes are converted to 3D solids when using the **3DPRINT** command.

Figure 15-46.
Exporting an ACIS file.

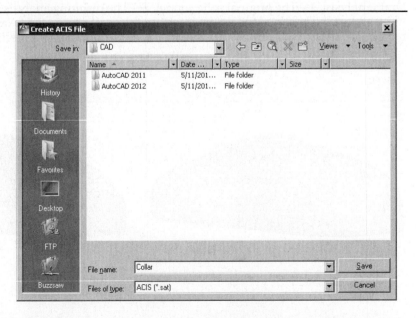

When the **3DPRINT** command is initiated, a dialog box appears with two options: **Learn about preparing a 3D model for printing** and **Continue**. Select the **Continue** option. Then, select the solids or watertight meshes and press [Enter]. The **Send to 3D Print Service** dialog box is displayed, **Figure 15-47**. This dialog box allows you to change the scale or select other objects. When done, pick the **OK** button to display a standard save dialog box. Type the file name in the **File name:** text box and pick the **Save** button. Next, a web page is displayed (if you are connected to the Internet). This page offers options for 3D printer service providers. Unless you are sending the file to a service provider, close this window. The STL file is ready to be sent to your additive manufacturing machine to create the model.

When using the **3DPRINT** command, the **FACETRES** system variable is automatically set to 10 for the creation of an STL file. The system variable is reset after the command is completed.

Exercise 15-9

Complete the exercise on the companion website.
www.g-wlearning.com/CAD

Figure 15-47.
This dialog box is displayed when creating an STL file with the **3DPRINT** command.

Pick to select other objects   Change the scale if needed   Use to navigate the preview

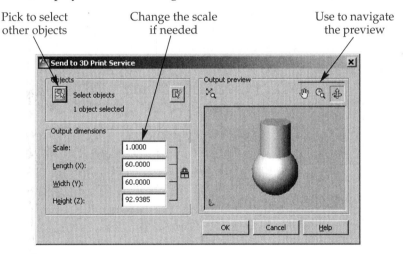

Chapter Review

Answer the following questions. Write your answers on a separate sheet of paper or complete the electronic chapter review on the companion website.
www.g-wlearning.com/CAD

1. Which AutoCAD system variable is used to control the display of a silhouette edge on a model?
2. Which option of the **VIEWBASE** command is used to change the default view when placing the base view?
3. Explain how to change the default projection used by AutoCAD when placing drawing views.
4. Explain two ways to place a projected view.
5. To what value should the **DIMASSOC** system variable be set in order to dimension drawing views with associative dimensions?
6. What does the **SECTIONPLANE** command create?
7. How is the **Face** option of the **SECTIONPLANE** command used?
8. Which option of the **SECTIONPLANE** command is used to create sections with jogs?
9. When a section object is created by picking a face or two points or using the **Orthographic** option of the **SECTIONPLANE** command, which section object state is established?
10. Which section object grips are used to accomplish the following tasks?
    A. Change the section object state.
    B. Lengthen or shorten the section object segment.
    C. Rotate the section view 180°.
11. How is live sectioning turned on or off?
12. Which category in the **Section Settings** dialog box provides control over the material that is removed by the section object?
13. What are the two types of section view blocks that can be created from a section object?
14. Which command should be used first, **SOLDRAW** or **SOLVIEW**?
15. Which option of the **SOLVIEW** command is used to create an orthographic view?
16. Name the layer(s) that the **SOLVIEW** command automatically create(s).
17. Which layer(s) in Question 16 should you avoid drawing on?
18. Which command can automatically complete a section view using the current settings of **HPNAME**, **HPSCALE**, and **HPANG**?
19. What is the function of the **MASSPROP** command?
20. What is the extension of the ASCII file that can be created by **MASSPROP**?
21. What command is used to graphically analyze surface continuity in a model?
22. Briefly explain how to set values defining the acceptable curvature range when using the **ANALYSISCURVATURE** command.
23. Which commands export and import solid models?
24. Which type of file has an .stl extension?
25. The **3DPRINT** command is used to accomplish two different operations. What are they?

Drawing Problems

1. In this problem, you will create a multiview layout of the base bracket by placing drawing views with the **VIEWBASE** command. Set up a page layout using the ANSI B (17.00 × 11.00 Inches) paper size. Select the appropriate scale for the orthographic views and the isometric view. Edit the views as needed to look like the layout shown.
   A. Construct the model shown using the dimensions provided.
   B. Create the front, top, right-side, and isometric views shown.
   C. The isometric view should be a shaded view.
   D. Dimension the drawing in the layout using the dimensions given. Make sure to use associative dimensions.
   E. Save the drawing as P15-01.

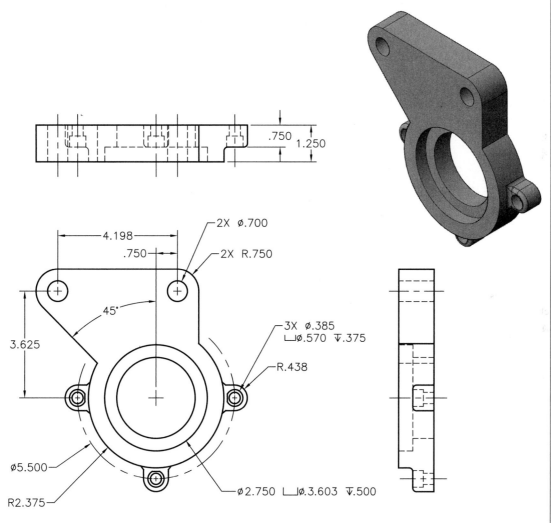

NOTE: ALL FILLETS AND ROUNDS R.125 UNLESS NOTED

Chapter 15   Model Display, Documentation, and Analysis

2. In this problem, you will create a multiview layout of the flanged coupler by placing drawing views with the **VIEWBASE** command. Set up a page layout using the ANSI B (17.00 × 11.00 Inches) paper size. Select the appropriate scale for the orthographic views and the isometric view. Edit the views as needed to look like the layout shown.

   A. Construct the model shown using the dimensions provided.
   B. Create the front, right-side, and isometric views shown.
   C. The isometric view should be a shaded view.
   D. Dimension the drawing in the layout using the dimensions given. Make sure to use associative dimensions.
   E. Save the drawing as P15-02.

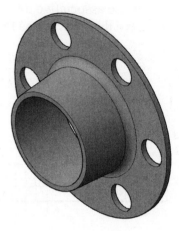

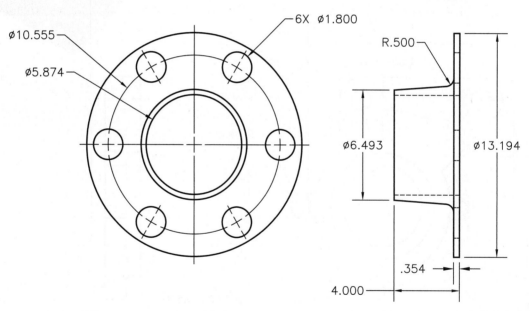

3. Open one of your solid model problems from a previous chapter and do the following.

   A. Create a multiview layout of the model. Use three orthographic views and one isometric view.
   B. Create a page layout and use an appropriate scale for the views.
   C. Use the **VIEWBASE** command to create the views.
   D. The isometric view should be shaded.
   E. Dimension the orthographic views.
   F. Save the drawing as P15_03.

4. Open one of your solid model problems from a previous chapter and do the following.
   A. Use the **Face** option of the **SECTIONPLANE** command to create a section object.
   B. Alter the section so that the section plane object cuts through features of the model.
   C. Change the section settings to display an ANSI hatch pattern.
   D. Save the drawing as P15_04.

5. Open one of your solid model problems from a previous chapter and do the following.
   A. Construct a section through the model using the **Draw section** option of the **SECTIONPLANE** command. Cut through as many features as possible.
   B. Display cutaway geometry with a 50% transparency.
   C. Display section lines using an appropriate hatch pattern.
   D. Generate a 3D section block that displays the cutaway geometry in a color of your choice.
   E. Create a layout with a viewport for the 3D block displayed at half the size of the original model.
   F. Save the drawing as P15_05.

6. Choose five solid model problems from previous chapters and copy them to a new folder. Then, do the following.
   A. Open the first drawing. Export it as an SAT file.
   B. Do the same for the remaining four files.
   C. Compare the sizes of the SAT files with the DWG files. Compare the combined sizes of both types of files.
   D. Begin a new drawing and import one of the SAT files.

7. Draw the object shown as a solid model. Do not dimension the object. Then, do the following.
   A. Construct a section object that creates a full section along the centerline of the hole.
   B. Generate a 2D section and display it on the drawing at half the size of the original. Specify section settings as desired.
   C. Generate a 3D section and display it on the drawing at half the size of the original. Do not display cutaway geometry. Specify section settings as desired.
   D. Activate live sectioning. Do not display the cutaway geometry.
   E. On the original solid model, display the intersection fill as an ANSI hatch pattern.
   F. Save the drawing as P15_07.

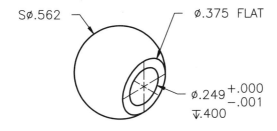

8. Draw the object shown as a solid model. Only half of the object is shown; draw the complete object. Do not dimension the object. Then, do the following.
    A. Construct a section plane that creates a full section, as shown.
    B. Display the intersection fill as an ANSI hatch pattern.
    C. Activate live sectioning and view the cutaway geometry with a high level of transparency.
    D. Save the drawing as P15_08.

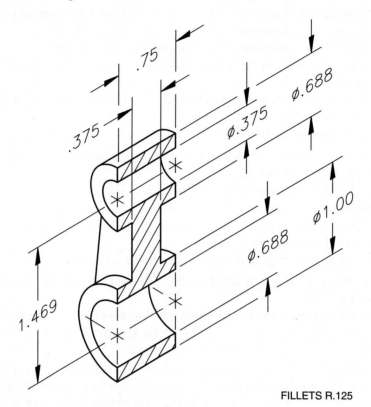

FILLETS R.125

9. Draw the object shown as a solid model. Only half of the object is shown; draw the complete object. Do not dimension the object. Then, do the following.
   A. Construct a section plane that creates a half section.
   B. Display the intersection fill as an ANSI hatch pattern.
   C. Activate live sectioning and view the cutaway geometry with a low level of transparency.
   D. Generate a 3D section and save it as a block.
   E. Use **SOLVIEW** to create a two-view orthographic layout. Use an appropriate scale to plot on a B-size sheet.
   F. Create a third floating viewport and insert the 3D section block scaled to half the size of the drawing.
   G. Save the drawing as P15_09.

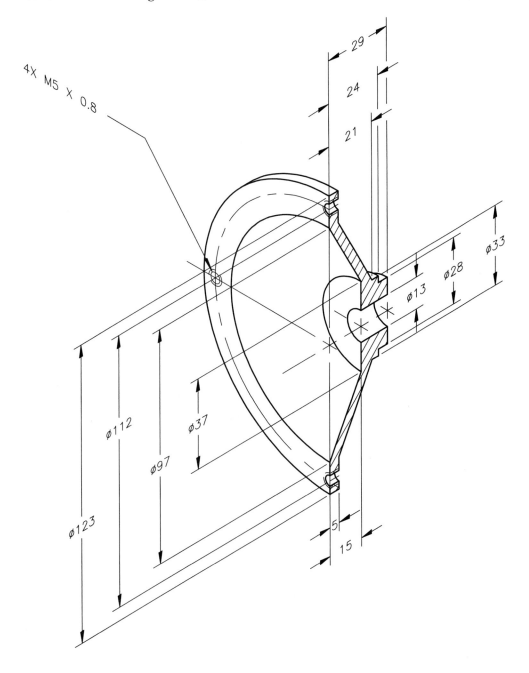

10. Draw the object shown as a solid model. Use your own dimensions. Then, do the following.
    A. Construct a section plane that creates an offset section. The section should pass through the center of two holes in the base and through the large central hole.
    B. Display the intersection fill as an ANSI hatch pattern.
    C. Activate live sectioning and view the cutaway geometry with a low level of transparency in the color red.
    D. Generate a 3D section of the sectioned solid model and save it as a block.
    E. Use **SOLVIEW** to create a two-view orthographic layout. One view should be a half section. Use an appropriate scale to plot on a B-size sheet.
    F. Create a third floating viewport and insert the 3D section block scaled to half the size of the drawing.
    G. Save the drawing as P15_10.

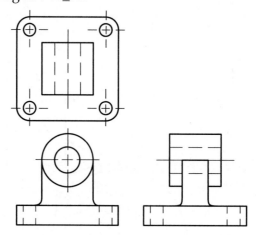

11. Draw the object shown as a solid model. Do not dimension the object. Then, do the following.
    A. Construct a section plane that creates a half section.
    B. Display the intersection fill as an ANSI hatch pattern.
    C. Activate live sectioning and view the cutaway geometry with a low level of transparency in the color red.
    D. Alter the section plane to create the section shown.
    E. Generate a 3D section and save it as a block.
    F. Use **SOLVIEW** to create a two-view orthographic layout. One view should be a full section. Use an appropriate scale to plot on an A-size sheet.
    G. Create a third floating viewport and insert the 3D section block scaled to half the size of the drawing.
    H. Save the drawing as P15_11.

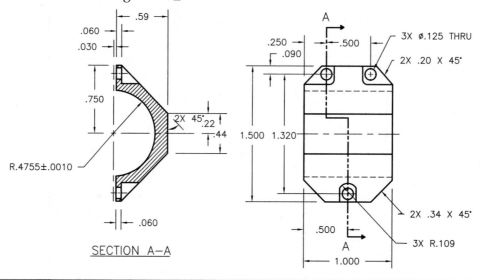

Chapter

Visual Style Settings and Basic Rendering

Learning Objectives

After completing this chapter, you will be able to:

✓ Describe the **Visual Styles Manager** palette.
✓ Change the settings for visual styles.
✓ Create custom visual styles.
✓ Export visual styles to a tool palette.
✓ Render a scene using sunlight.
✓ Save a rendered image from the **Render** window.

In Chapter 1, you were introduced to the default visual styles. In this chapter, you will learn about all visual style settings and how to redefine the visual style. You will also learn how to create your own visual style. Finally, this chapter provides an introduction to lights and rendering.

Overview of the Visual Styles Manager

The **Visual Styles Manager** palette provides access to all of the visual style settings. This palette is a floating window similar to the **Properties** palette. See **Figure 16-1**. Changes made in the **Visual Styles Manager** redefine the visual style in the current drawing.

At the top of the **Visual Styles Manager** are image tiles for the defined visual styles. See **Figure 16-2**. The default visual styles are 2D Wireframe, Conceptual, Hidden, Realistic, Shaded, Shaded with Edges, Shades of Gray, Sketchy, Wireframe, and X-Ray. User-defined visual styles also appear as image tiles. The image on the tile is a preview of the visual style settings. Selecting an image tile provides access to the properties of the visual style in the palette below. The name of the currently selected visual style appears below the image tiles and the corresponding image tile is surrounded by a yellow border.

To set a different visual style current using the **Visual Styles Manager**, double-click on the image tile. You can also select the image tile and pick the **Apply Selected Visual Style to Current Viewport** button immediately below the image tiles. An icon consisting of a small image of the button is displayed in the image tile of the visual style that is current in the active viewport, as shown in **Figure 16-2**. A drawing icon appears in the image tile if the visual style is current in a viewport that is not active. The AutoCAD icon appears in the image tiles of the default visual styles.

Figure 16-1.
The **Visual Styles Manager** palette.

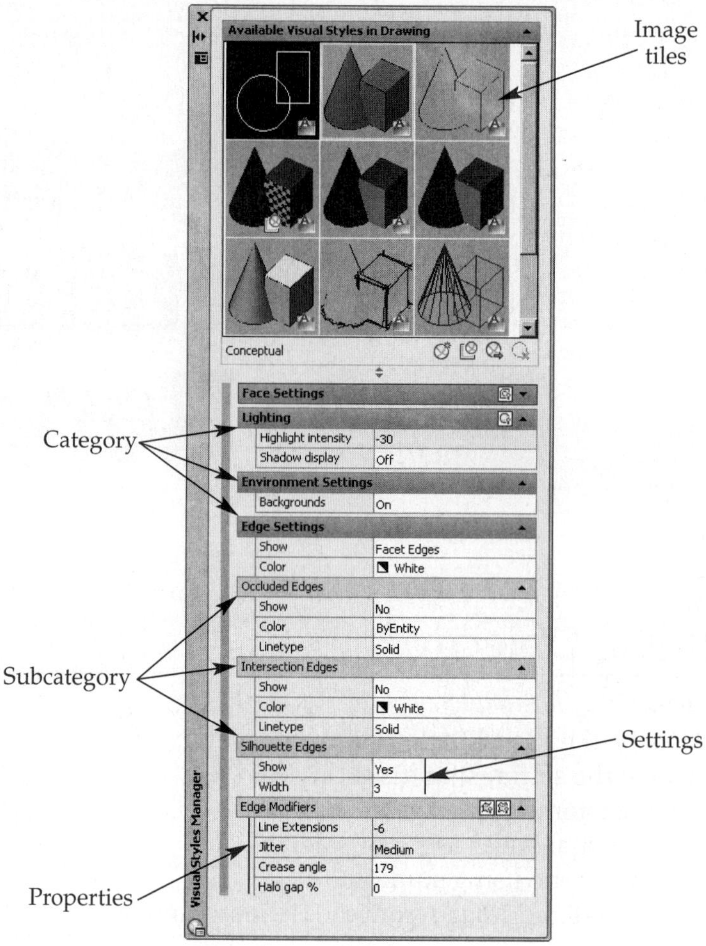

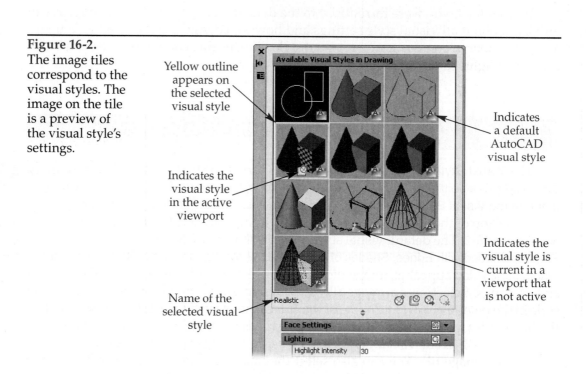

Image tiles

Category

Subcategory

Settings

Properties

Figure 16-2.
The image tiles correspond to the visual styles. The image on the tile is a preview of the visual style's settings.

Yellow outline appears on the selected visual style

Indicates the visual style in the active viewport

Name of the selected visual style

Indicates a default AutoCAD visual style

Indicates the visual style is current in a viewport that is not active

You can also set a different visual style current by selecting the style from the **Visual Style Controls** flyout in the viewport controls located in the upper-left corner of the drawing window. See **Figure 16-3**. Selecting **Visual Styles Manager...** in the **Visual Style Controls** flyout opens the **Visual Styles Manager**.

Exercise 16-1

Complete the exercise on the companion website.
www.g-wlearning.com/CAD

Visual Style Settings

The **Visual Styles** panel on the **View** tab of the ribbon provides several settings for altering the visual style. These settings are also available in the **Visual Styles Manager**. In addition, there are settings in the **Visual Styles Manager** that are not available on the ribbon. The next sections discuss settings available in the **Visual Styles Manager** for the default visual styles.

Changing any setting in the **Visual Styles Manager** *redefines* the visual style in the current drawing. Changes made using the ribbon are temporary and are not kept when a different visual style is set current. This is important to remember.

2D Wireframe

When the 2D Wireframe visual style is set current, lines and curves are used to show the edges of 3D objects. Assigned linetypes and lineweights are displayed. All edges are visible as if the object is constructed of pieces of wire soldered together at the intersections (thus, the name *wireframe*). Either the 2D or 3D wireframe UCS icon is displayed. OLE objects will display normally. In addition, the drawing window display changes to the 2D model space context and parallel projection. For the 2D Wireframe visual style, the **Visual Styles Manager** displays the following categories. See **Figure 16-4**.

- **2D Wireframe Options**
- **2D Hide—Occluded Lines**
- **2D Hide—Intersection Edges**
- **2D Hide—Miscellaneous**
- **Display Resolution**

Figure 16-3.
A visual style can be set current by selecting it from the **Visual Style Controls** flyout in the viewport controls.

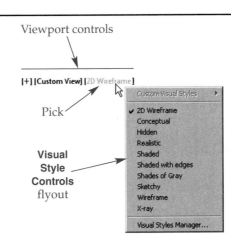

Figure 16-4.
The categories and
properties available
for the 2D Wireframe
visual style.

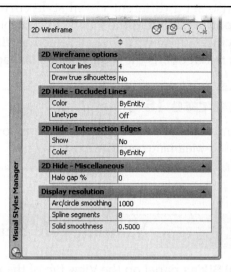

2D Wireframe Options

The Contour lines property controls the **ISOLINES** system variable. Isolines are the lines used to define curved surfaces on solid objects when displayed in a wireframe view. The setting is 4 by default and can range from 0 to 2047. Isolines are suppressed when the **HIDE** command is used with the 2D Wireframe visual style set current.

The Draw true silhouettes property controls the **DISPSILH** system variable. This determines whether or not silhouette edges are shown on curved surfaces. It is set to No by default, which is equivalent to a **DISPSILH** setting of 0.

2D Hide—Occluded Lines

The Color property in this category controls the **OBSCUREDCOLOR** system variable. This property determines the color of *occluded* lines (obscured lines hidden from the view) when the **HIDE** command is used. The default setting is ByEntity. This means that, when displayed, obscured lines are shown in the same color as the object.

The Linetype property controls the **OBSCUREDLTYPE** system variable. This property determines whether or not obscured lines are displayed and in which linetype they are displayed. The default setting is Off, which means that obscured lines are not displayed when the **HIDE** command is used. Selecting a linetype allows you to have obscured lines displayed instead of removed from the view. The available linetypes are: Solid, Dashed, Dotted, Short Dash, Medium Dash, Long Dash, Double Short Dash, Double Medium Dash, Double Long Dash, Medium Long Dash, and Sparse Dot. When a linetype is selected from the drop-down list, obscured lines are displayed in that linetype after the **HIDE** command is used.

The linetypes available in the Linetype property drop-down list are not the same as the linetypes loaded into the **Linetype Manager** dialog box. They are independent of zoom levels, which means the dash size will stay the same when zooming in and out.

2D Hide—Intersection Edges

This category is used to toggle the display of polylines at the intersection of 3D surfaces and set the color of the lines. The Show property controls the **INTERSECTIONDISPLAY** system variable. This property determines whether or not polylines are displayed at the intersection of non-unioned 3D surfaces. The default setting is No, which means that polylines are not displayed when the **HIDE** command is used.

The Color property in this category controls the **INTERSECTIONCOLOR** system variable. This property determines the color of the polylines displayed at intersection edges. By default, the setting is ByEntity. This means that, when displayed, the polylines at intersection edges are shown in the same color as the object.

2D Hide—Miscellaneous

The Halo Gap % property controls the **HALOGAP** system variable. This property determines the gap that is displayed where one object partially obscures another (between the foreground edge and where the background edge starts to show). The default setting is 0 and the value can range from 0 to 100. The value refers to a percentage of one unit. The gap is only displayed when the **HIDE** command is used. It is not affected by the zoom level.

Display Resolution

The Arc/circle smoothing property controls the zoom percentage set by the **VIEWRES** command. This determines the resolution of circles and arcs. The value can range from 1 to 20,000. The higher the value, the higher the resolution of circles and arcs.

The Spline segments property controls the **SPLINESEGS** system variable. This property determines the number of line segments in a spline-fit polyline. The value can range from −32,768 to 32,767.

The Solid smoothness property controls the **FACETRES** system variable. This property determines the number of polygonal faces applied to curved surfaces on solids. The default setting is .5 and the value can range from .01 to 10.0. A higher **FACETRES** value will create a smoother finish when 3D printing.

Polygon faces will not be visible if the Draw true silhouettes property is set to Yes. However, a higher setting for the Solid smoothness property will make curved edges smoother.

Exercise 16-2

Complete the exercise on the companion website.
www.g-wlearning.com/CAD

Conceptual

When the Conceptual visual style is set current, objects are smoothed and shaded. The shading is a transition from cool to warm colors. The transitional colors help highlight details. The previous projection is retained and that context is set current.

The categories and properties for the shaded visual styles (those other than 2D Wireframe) are the same. These settings are discussed in the Common Visual Style Settings section later in this chapter.

Hidden

The Hidden visual style removes obscured lines from your view and makes 3D objects appear solid. The previous projection is retained and that context is set current. The benefit of using Hidden is that you get sufficient 3D display, but it does not push the graphics system too hard. Objects are not shaded or colored. This is very useful when working on complex drawings and/or using a slow computer.

Realistic

As with the Conceptual visual style, the objects have smoothing and shading applied to them when the Realistic visual style is set current. In addition, if materials are applied to the objects, the materials are displayed. The previous projection is retained and that context is set current. This visual style is good for adjusting textures and patterns on materials and for a final look at the scene before rendering.

Shaded

This style is very similar to Realistic, except the lighting is smooth instead of smoothest. Also, textures are turned off.

Shaded with Edges

This style is the same as Shaded, except isolines are displayed with the default number of 4. Intersection edges are turned on and displayed as solid and white. Silhouette edges are also displayed with a width setting of 3.

Shades of Gray

This style is similar to the other shaded styles, except object colors are displayed as monochrome. Materials and textures are turned off. Facet edges are turned on, as are silhouette edges. This style is good for adjusting lights and shadows. It is much easier to see what is happening with lights when all objects are the same color.

Sketchy

The Sketchy visual style has most of the same settings as the Hidden visual style. However, objects look like they are hand-sketched. Line extensions and jitter are turned on, as are silhouette edges.

Wireframe

The Wireframe visual style is almost identical in appearance to the 2D Wireframe visual style. The main difference is that the 3D mode UCS icon is displayed. The previous projection is retained and that context is set current. When working in 3D, a wireframe view is sometimes necessary to select objects normally hidden from your view.

X-Ray

This visual style is very similar to the Shaded with Edges visual style, except the Opacity property is set to 50%. This makes objects appear somewhat transparent.

PROFESSIONAL TIP

Depending on the visual style chosen, drawing window colors for individual elements such as the background, grid lines, crosshairs, etc., may need to be adjusted to make it easier to see and work with the objects in the drawing. Access the **Options** dialog box and select the **Colors...** button in the **Display** tab to change these settings in the **Drawing Window Colors** dialog box. Colors used with 2D model space, 3D parallel projection, and 3D perspective projection are controlled separately.

Common Visual Style Settings

With the exception of 2D Wireframe, all of the visual styles share similar categories and settings in the **Visual Styles Manager**. The visual styles have **Face Settings**, **Lighting**, **Environment Settings**, and **Edge Settings** categories. See **Figure 16-5**. These categories and the properties available in them are discussed in the next sections.

Face Settings

The Face style property controls the **VSFACESTYLE** system variable. The options are the same as the **No Face Style**, **Realistic Face Style**, and **Warm-Cool Face Style** options available by selecting the buttons in the **Visual Styles** panel on the **View** tab of the ribbon. Remember, though, selecting a button on the ribbon does not redefine the visual style in the current drawing. Rather, it is a temporary change to the viewport display.

The Lighting quality property controls the **VSLIGHTINGQUALITY** system variable. This property determines whether curved surfaces are displayed smooth or as a series of flat faces. Lighting quality is unavailable if the Face style property is set to None.

Figure 16-5.
These categories and properties are similar for all of the visual styles except 2D Wireframe.

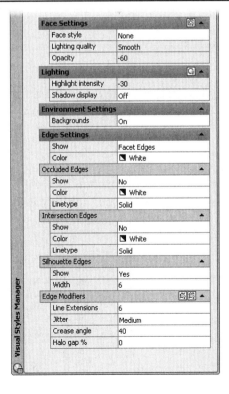

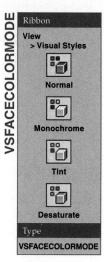

Ribbon

View
> Visual Styles

Normal

Monochrome

Tint

Desaturate

Type
VSFACECOLORMODE

The Color property controls the **VSFACECOLORMODE** system variable. This property determines how color is applied to the faces of an object. The effects can be temporarily applied by picking a button in the face colors flyout in the **Visual Styles** panel on the **View** tab of the ribbon. The choices are:

- Normal. The object color is applied to faces.
- Monochrome. One color is applied to all faces. This also displays and enables the Monochrome color property.
- Tint. A combination of the object color and a specified color is applied to faces. This also displays and enables the Tint Color property. The Tint property only works when the Material display property is set to Materials.
- Desaturate. The object color is applied to faces, but the saturation of the color is reduced by 30%.

The Monochrome color and Tint color properties control the **VSMONOCOLOR** system variable. This system variable determines the color that is applied when the face Color property is set to Monochrome or Tint.

The Opacity property controls the **VSFACEOPACITY** system variable. This property determines how transparent or opaque faces are in the viewport. The value can range from –100 to 100. When the setting is 0, the faces are completely transparent. When the setting is 100, the faces are completely opaque. Settings below 0 set the value, but turn off the effect. To quickly turn the effect on or off, pick the **Opacity** button on the **Face Settings** category title bar. See **Figure 16-6**. This changes the value from negative to positive, or vice versa. This property cannot be changed if the Face style property is set to None.

Opacity may also be controlled by picking the **X-Ray Effect** button in the **Visual Styles** panel on the **View** tab of the ribbon. The adjacent **Opacity** slider sets the value. However, remember, picking a button or making settings on the ribbon is only temporary. The change is not saved to the visual style.

The Material display property controls the **VSMATERIALMODE** system variable. When set to Off, objects display in their assigned color. When the setting is changed to Materials, the objects display the color of the material, but not the textures. When the setting is changed to Materials and textures, full materials are displayed.

CAUTION

Displaying materials and textures on 3D objects in a complex drawing will slow system performance. Set the Material display property to Materials and textures only when it is absolutely necessary.

Lighting

The Highlight intensity property controls the **VSFACEHIGHLIGHT** system variable. This property determines the size of the highlight on faces to which no material is assigned. A small highlight on an object makes it look smooth and hard. A large highlight on an

Figure 16-6.
Picking the **Opacity** button on the **Face Settings** category title bar turns the effect of opacity on or off.

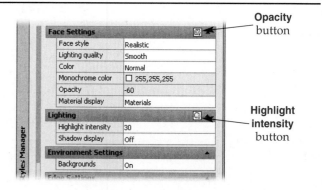

object makes it look rough or soft. The initial value is 30 or –30; the value can range from –100 to 100. The higher the setting is above 0, the larger the highlight. Settings below 0 set the value, but turn off the effect. To quickly turn the effect on or off, pick the **Highlight intensity** button on the **Lighting** category title bar. Refer to **Figure 16-6**. This changes the value from negative to positive or vice versa. This property cannot be changed if the Face style property is set to None.

The Shadow display property controls the **VSSHADOWS** system variable. This property controls if and how shadows are cast when the visual style is set current. The options are the same as those available by selecting a button in the shadows flyout in the **Visual Styles** panel on the **View** tab or in the **Lights** panel on the **Render** tab of the ribbon. Remember, however, a setting made using the ribbon is temporary and does not redefine the visual style.

If the property is set to Ground shadow, objects cast shadows on the ground, but not onto other objects. The "ground" is the XY plane of the WCS. The Mapped Object shadows setting, which corresponds to the **Full Shadows** button on the ribbon, only works if lights have been placed in the scene and hardware acceleration is enabled.

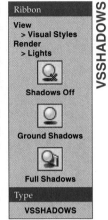

Use the **Performance Tuner** tool in the status bar or the **3DCONFIG** command to enable hardware acceleration.

Environment Settings

The Backgrounds property controls the **VSBACKGROUNDS** system variable. This property determines whether or not the preselected background is displayed in the viewport. Backgrounds can only be assigned to a view when a named view is created or edited. After the view is created, restore the view to display the background.

Edge Settings

Properties in the **Edge Settings** category determine the appearance of model edges. This category includes the **Occluded Edges, Intersection Edges, Silhouette Edges**, and **Edge Modifiers** subcategories. Refer to **Figure 16-1**.

The Show property controls the **VSEDGES** system variable. This property determines how edges on solid objects are represented when the visual style is set current. The options are the same as those available by selecting a button in the edge flyout in the **Visual Styles** panel on the **View** tab of the ribbon. The settings of Isolines and Facet Edges determine how edges and curved surfaces are displayed on a 3D model. This is discussed in detail in Chapter 15. Setting this property to None turns off isolines and facets and displays no edges. If the Face style property is set to None, the Show property cannot be set to None.

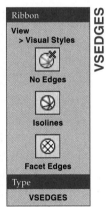

The Color property in the **Edge Settings** category controls the **VSEDGECOLOR** system variable. This property determines the color of all edges on objects in the drawing. It is disabled when the **Show** property is set to None.

The Number of lines and Always on top properties are displayed when the Show property is set to Isolines. The Number of lines property controls the **ISOLINES** system variable. The Always on top property controls the **VSISOONTOP** system variable. This property determines if isolines are displayed when the model is shaded or hidden. When set to Yes, edges are always displayed.

Additional edge settings are available in the **Edge Settings** category when the **Show** property is set to Isolines or Facet Edges. These settings are available in the **Occluded Edges, Intersection Edges**, and **Edge Modifiers** subcategories. The settings in the **Silhouette Edges** subcategory are available regardless of the Show property setting. The settings in the subcategories are discussed as follows.

Occluded Edges Subcategory. This subcategory is not available if the Show property in the **Edge Settings** category is set to None. The Show property in this subcategory controls the **VSOCCLUDEDEDGES** system variable. This property determines whether or not occluded edges are displayed in a hidden or shaded view. See **Figure 16-7.**

The Color property in this subcategory controls the **VSOCCLUDEDCOLOR** system variable. The Linetype property controls the **VSOCCLUDEDLTYPE** system variable. These properties are similar to the properties for occluded lines discussed earlier in this chapter in the 2D Hide—Occluded Lines section.

Intersection Edges Subcategory. This subcategory is not available if the Show property in the **Edge Settings** category is set to None. The Show property in this subcategory controls the **VSINTERSECTIONEDGES** system variable. This property determines whether or not lines are displayed where one 3D object intersects another 3D object. See **Figure 16-8.**

The Color property in this subcategory controls the **VSINTERSECTIONCOLOR** system variable. The Linetype property controls the **VSINTERSECTIONLTYPE** system variable. These properties are similar to the properties for intersection lines discussed earlier in this chapter in the 2D Hide—Intersection Edges section.

PROFESSIONAL TIP

Setting the intersection edges Color property to a color that contrasts with the objects in your model is a good way to quickly check for interference between 3D objects.

Figure 16-7.
A—Occluded lines are not shown. B—The Show property is set to Yes and occluded lines are shown.

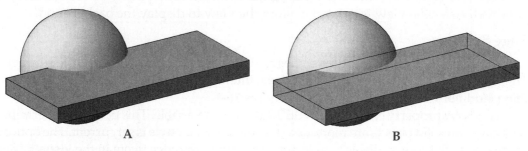

A   B

Figure 16-8.
A—A line does not appear where these two objects intersect. B—The Show property is set to Yes and a line appears at the intersection.

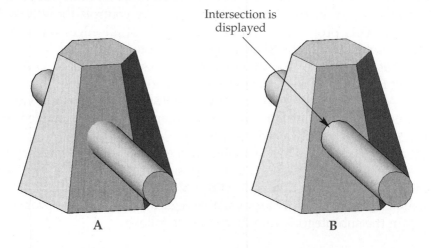

Intersection is displayed

A   B

Silhouette Edges Subcategory. This subcategory is available for each setting for the Show property in the **Edge Settings** category. The Show property in this subcategory controls the **VSSILHEDGES** system variable. It determines whether or not silhouette edges are displayed around the outside edges of all objects.

The Width property controls the **VSSILHWIDTH** system variable. This property determines the width of silhouette lines. It is measured in pixels and the value can range from 1 to 25.

Edge Modifiers Subcategory. This subcategory is not displayed if the Show property in the **Edge Settings** category is set to None. The Line Extensions property controls the **VSEDGELEX** system variable. This property can be used to create a hand-sketched appearance by extending the ends of edges. See **Figure 16-9A**. In order to make changes to this property, the **Line Extensions edges** button must be on in the **Edge Modifiers** subcategory title bar. The value for the Line Extensions property can range from –100 to 100, which is the number of pixels. The higher the setting, the longer the extension. A negative value sets the extension length, but turns off the property. Picking the button makes the value positive and applies the effect (or makes the value negative and turns off the effect).

The Jitter property controls the **VSEDGEJITTER** system variable. Jitter makes edges of objects look as if they were sketched with a pencil. See **Figure 16-9B**. In order to make changes to this property, the **Jitter edges** button must be on in the **Edge Modifiers** subcategory title bar. There are four settings from which to choose: Off, Low, Medium, and High. The number of sketched lines increases at each higher setting.

When the Show property in the **Edge Settings** category is set to Facet Edges, the Crease angle and Halo gap % properties are displayed in the **Edge Modifiers** subcategory. The Crease angle property controls the **VSEDGESMOOTH** system variable. This property determines how facet edges within a face are displayed based on the angle between adjacent faces. It does not affect edges between faces. See **Figure 16-10**. The value can range from 0 to 180. This is the number of degrees between edges below which a line is displayed. The Halo gap % property is similar to the setting discussed earlier in the 2D Hide—Miscellaneous section; however, this property controls the **VSHALOGAP** system variable.

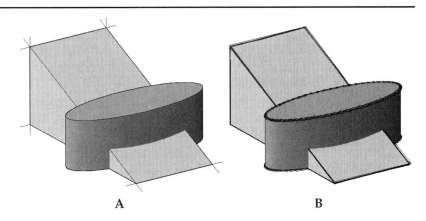

Figure 16-9.
A—Line extensions have been turned on for this visual style.
B—Jitter has been turned on for this visual style.

A   B

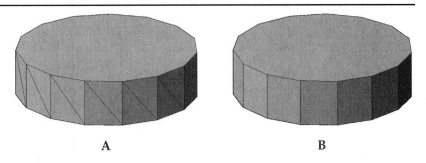

Figure 16-10.
A—The Crease angle property is set to 0. Notice the edges between facets within each face.
B—The Crease angle property is set to 10. The edges are no longer displayed.

A   B

Creating Your Own Visual Style

As you saw in the previous sections, you can customize the default AutoCAD visual styles. However, you may also want to create a number of different visual styles to quickly change the display of the scene. Custom visual styles are easy to create.

To create a custom visual style, open the **Visual Styles Manager**. Then, pick the **Create New Visual Style** button below the image tiles. You can also right-click in the image tile area and select **Create New Visual Style...** from the shortcut menu. In the **Create New Visual Style** dialog box that appears, type a name for the new style and give it a description. See **Figure 16-11**. Then, pick the **OK** button to create the new visual style.

An image tile is created for the new visual style. The name and description of the visual style appear as help text when the cursor is over the image tile. Select the image tile to display the default properties for the new visual style. Then, change the settings as needed to meet your requirements.

Custom visual styles are only saved in the current drawing. They are not automatically available in other drawings. To use the new visual styles in any drawing, they must be exported to a tool palette. This is discussed in the next section.

PROFESSIONAL TIP

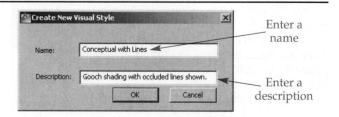

To return one of AutoCAD's visual styles to its default settings, right-click on the image tile in the **Visual Styles Manager** and select **Reset to default** from the shortcut menu.

Exercise 16-3

Complete the exercise on the companion website.
www.g-wlearning.com/CAD

Steps for Exporting Visual Styles to a Tool Palette

To have custom visual styles available in other drawings, export them to a tool palette. Use the following procedure.

1. Create and customize a visual style as described in the previous section.
2. Open the **Tool Palettes** window.
3. Right-click on the **Tool Palettes** title bar and pick **New Palette** from the shortcut menu.
4. Type the name of the new palette, such as My Visual Styles, in the text box that appears. See **Figure 16-12A**.
5. The new palette is added and active. You are ready to export your custom visual styles into it.
6. Select the image tile of the visual style in the **Visual Styles Manager**. Remember, a yellow border appears around the selected image tile.
7. Pick the **Export the Selected Visual Style to the Tool Palette** button below the image tiles in the **Visual Styles Manager**. You can also right-click on the image tile and select **Export to Active Tool Palette** from the shortcut menu.

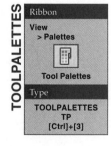

Ribbon

View
> Palettes

Tool Palettes

Type

**TOOLPALETTES
TP
[Ctrl]+[3]**

Figure 16-11.
Creating a new visual style.

Create New Visual Style

Name: Conceptual with Lines — Enter a name

Description: Gooch shading with occluded lines shown. — Enter a description

OK   Cancel

Figure 16-12.
A—Creating a new tool palette on which to place visual style tools. B—A visual style has been copied to the tool palette as a tool.

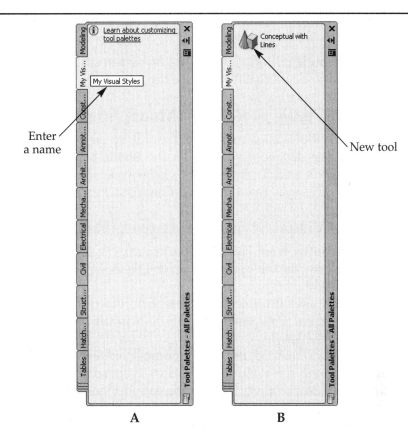

A new tool now appears in the palette with the same image, name, and description as the visual style in the **Visual Styles Manager**. See **Figure 16-12B**. Selecting the tool applies the visual style to the current viewport. You can also right-click on the tool to display a shortcut menu. Using this menu, you can apply the visual style to the current viewport, apply it to all viewports, or add the visual style to the current drawing. The shortcut menu also allows you to rename the tool, access the properties of the visual style, and delete the visual style from the palette.

PROFESSIONAL TIP

A visual style can be added as a tool on a tool palette by dragging its image tile from the **Visual Styles Manager** and dropping it onto the tool palette.

Exercise 16-4

Complete the exercise on the companion website.
www.g-wlearning.com/CAD

Deleting Visual Styles from the Visual Styles Manager

Custom visual styles can be deleted from the **Visual Styles Manager**. Pick the image tile of the visual style you want to delete. Then, pick the **Delete the Selected Visual Style** button below the image tiles. You can also right-click on the image tile and select **Delete** from the shortcut menu. You are *not* warned about the deletion. The default AutoCAD visual styles cannot be deleted, nor can a visual style that is currently in use.

Plotting Visual Styles

A visual style not only affects the on-screen display, it also affects plots. To plot objects with a specific visual style, use the following guidelines.

Plotting a Visual Style from Model Space

Open the **Plot** dialog box and expand it by picking the **More Options** (>) button. Then, select the desired display from the **Shade plot** drop-down list in the **Shaded viewport options** area. If the desired visual style is current in the viewport, you can also select As displayed from the drop-down list. Finally, plot the drawing.

Plotting a Visual Style from Layout (Paper) Space

When plotting from layout (paper) space, the shade plot properties of the viewports govern how the viewport is plotted. The viewports can be set to plot visual styles in three different ways.

In the first method, select the viewport in layout space and right-click to display the shortcut menu. Pick **Shade plot** to display the cascading menu. Then, select the appropriate visual style.

In the second method, use the **Properties** palette to set the Shade plot property of the viewport. To do this, select the viewport in layout space and open the **Properties** palette. Pick the Shade plot property in the **Misc** category and change the setting to the desired option. The Shade plot property is also available in the **Quick Properties** palette.

In the third method, use the **Visual Styles** suboption of the **Shadeplot** option of the **MVIEW** command. When prompted to select objects, pick the border of the viewport. Do not pick the objects in the viewport.

The visual style of the viewport may also be selected when you create a viewport configuration in the **Viewports** dialog box (**VPORTS** command). Select the viewport in the **Preview** area of the dialog box. Then, pick the visual style desired from the **Visual Style:** drop-down list at the bottom of the dialog box. The **VPORTS** command can be used in model space or layout (paper) space.

Introduction to Rendering

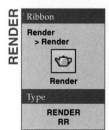

Visual styles provide a way to plot your 3D scene to paper or a file, but control over the appearance is limited to the visual style settings. In Chapter 1, you were briefly introduced to the **RENDER** command. The **RENDER** command offers complete control over the scene and, with its features, you can create photorealistic images. In this chapter, you will be introduced to AutoCAD's rendering and lighting tools. Materials, lights, and more advanced rendering features are discussed in later chapters.

When you render a scene, you are making a realistic image of your design that can be printed, displayed on a web page, or used in a presentation. To create an attractive rendering, you have to figure out what view you want to display, where the lights should be placed, what types of materials need to be applied to the 3D objects, and the kind of output that is needed. This section shows you how to create a quick rendering of your scene.

Introduction to Lights

Lights provide the illumination to a scene and are essential for rendering. There are three types of lighting in AutoCAD—default lighting, sunlight, and user-created lighting. AutoCAD automatically creates two default light sources in every scene. These lights ensure that all surfaces on the model are illuminated and visible. The types of lighting are discussed in more detail in Chapter 18.

A scene can be rendered with the default lights, but the results are usually not adequate to produce a photorealistic image. See **Figure 16-13**. The appearance is very artificial and no shadows are created. Shadows anchor objects to the scene and make them look real. See **Figure 16-14**. Without shadows, objects appear to float in space. Because the default lights do not cast shadows, other lights must be added to the scene and set to cast shadows. When a light is added to a scene, the default lights must be turned off. The first time you add a light, you receive a warning to this effect (unless the warning has been disabled).

In this section, you will learn how to add sunlight to the scene. Chapter 18 provides detailed information on lighting. Sunlight is produced by an automated distant light. Sunlight can be turned on by picking the **Sun Status** button in the **Sun & Location** panel on the **Render** tab of the ribbon. The button background is blue when sunlight is on. See **Figure 16-15**.

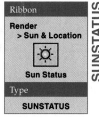

If the **Default Lighting** button is on in the **Lights** panel when the **Sun Status** button is turned on, a warning dialog box is displayed. This dialog box gives you choices to either turn off default lighting or keep it on. You cannot see the effects of sunlight with default lighting turned on, so it is recommended to turn it off.

If the current visual style is set to display full shadows, you should now see shadows in the scene, provided there are areas to receive shadows. Remember, hardware acceleration must be enabled to display full shadows.

The **Date** and **Time** sliders in the **Sun & Location** panel on the **Render** tab of the ribbon are active when sunlight is turned on. You can drag the sliders to adjust the date and time. The current date and time are displayed on the right-hand end of the sliders. As you drag the sliders, the shadows in the scene change to reflect the settings.

Figure 16-13.
A scene rendered with the default AutoCAD lighting.

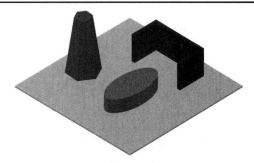

Figure 16-14.
A light has been added and set to cast shadows. Compare this rendering with Figure 16-13.

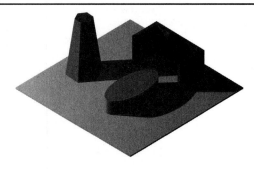

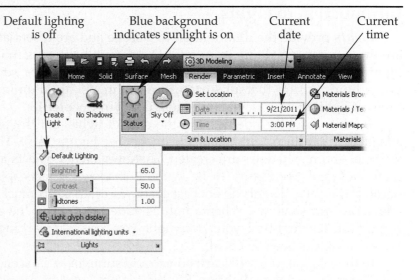

Default lighting is off

Blue background indicates sunlight is on

Current date

Current time

Rendering the Scene

The **Render** panel on the **Render** tab of the ribbon is shown in **Figure 16-16**. There are two buttons in the **Render** flyout to initiate a rendering. If you pick the **Render** button, the **Render** window appears (by default) and AutoCAD immediately starts rendering the viewport. You will see the rendered tiles appear in the image pane as they are calculated. The **Render** window is explained more in the next section.

If you pick the **Render Region** button in the **Render** flyout, you are prompted to pick two points in the viewport, similar to performing a window selection. The selected area is rendered in the viewport. Rendering a cropped area is often used to test areas of the scene for possible problems before performing the final rendering.

Also in the **Render** panel you will find the render preset drop-down list. This is located in the upper-right corner of the panel. This drop-down list gives you a selection of rendering presets based on image quality. The choices are:

- Draft
- Low
- Medium
- High
- Presentation

Figure 16-16.
The **Render** panel on the **Render** tab of the ribbon.

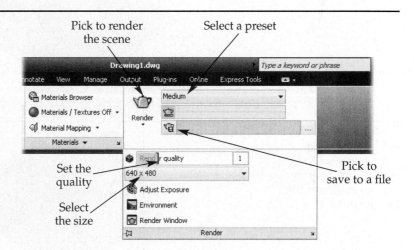

Pick to render the scene

Select a preset

Set the quality

Select the size

Pick to save to a file

The Draft preset produces the lowest-quality rendering. The Presentation preset produces the highest-quality rendering. The better the quality, the longer it takes to complete the rendering process.

PROFESSIONAL TIP

Rendering a complex drawing may take a very long time and you do not want to repeat it because of some small error. It is important to make sure that everything in the scene is perfect before the final rendering. By rendering a cropped region and using lower-quality renderings, you can verify the appearance of any questionable areas without performing a full rendering.

Introduction to the Render Window

The **Render** window is composed of three main areas. See **Figure 16-17**. The image pane is where the rendering appears. The statistics pane shows the current rendering settings. The history pane shows a list of all of the images rendered from the drawing, with the most recent at the top.

You can zoom into the image in the image pane for detailed inspection. Use the mouse scroll wheel or the **Zoom +** and **Zoom –** entries in the **Tools** pull-down menu of the **Render** window. In the **File** pull-down menu of the **Render** window, select **Save...** to save the image selected in the history pane to an image file. The symbol in front of the image in the history pane changes to a folder with a green check mark on it. The **Save Copy...** option in the **File** pull-down menu of the **Render** window creates a copy of the image without modifying the original in the history pane.

Figure 16-17.
The **Render** window.

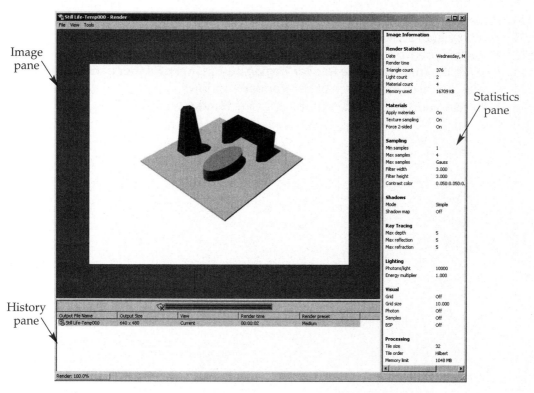

 Advanced rendering is discussed in Chapter 19.

 Exercise 16-5

Complete the exercise on the companion website.
www.g-wlearning.com/CAD

 # Chapter Review

Answer the following questions. Write your answers on a separate sheet of paper or complete the electronic chapter review on the companion website.
www.g-wlearning.com/CAD

1. What is the **Visual Styles Manager**?
2. Name the 10 default AutoCAD visual styles that can be edited in the **Visual Styles Manager**.
3. Describe the difference between setting the Lighting quality property to Smooth and Faceted.
4. What does the Desaturate setting of the Color property do?
5. What has to be added to a scene before full shadows are displayed?
6. If you want to make your scene look hand sketched, but the Line Extensions and Jitter properties are not available, what other setting(s) do you have to change?
7. How do you set a visual style to display silhouette edges?
8. List the four settings for the Jitter property.
9. What is an *intersection edge*?
10. How do you make your own visual styles available in other drawings?
11. Which visual styles cannot be deleted?
12. How can you turn on sunlight?
13. Explain the function of the **Render Region** button in the **Render** panel on the ribbon.
14. Name the three main areas of the **Render** window.
15. How can you save a rendered image in the **Render** window?

Drawing Problems

1. This problem demonstrates the differences in rendering time and image quality of the five different rendering presets.
   A. Open any 3D drawing from a previous chapter and display an appropriate isometric view. Change the projection to perspective, if it is not already current.
   B. Turn on sunlight and set the **Date** and **Time** sliders to place the shadows where you want them. Tip: Turning on full shadows allows you to locate the shadows without rendering.
   C. Render the scene once for each rendering preset: Draft, Low, Medium, High, and Presentation.
   D. In the history pane of the **Render** window, note the differences between the rendering time for each rendering.
   E. Save each image with a corresponding name: P16_01_Draft.jpg, P16_01_Low.jpg, P16_01_Medium.jpg, P16_01_High.jpg, and P16_01_Presentation.jpg.
   F. Save the drawing as P16_01.

2. In this problem, you will construct a living room scene using some simple shapes and blocks available through the Autodesk Seek website.
   A. Draw a 12′ × 12′ × 1′ box.
   B. Draw two boxes to represent walls, 12′ × 4′ × 8′. Position them as shown.

   C. Open the **Content Explorer** by selecting **Explore** from the **Content** panel on the **Plug-ins** tab on the ribbon. Pick on the drop-down list next to the home icon and select **Autodesk Seek**. Select the Autodesk Seek hyperlink (you must be connected to the Internet).
   D. Search for 3D drawing files for the following objects, download the files, open them, and then drag and drop them into the scene: sofa, table, end table, lamp, plant, entertainment center, and chair. The blocks do not have to be exactly the same as shown here and may need to be scaled up or down. Position them as shown.
   E. Apply each of the five default visual styles to the viewport and plot each. Use the **As displayed** option in the **Plot** dialog box. Note the differences in each one.
   F. Save the drawing as P16_02.

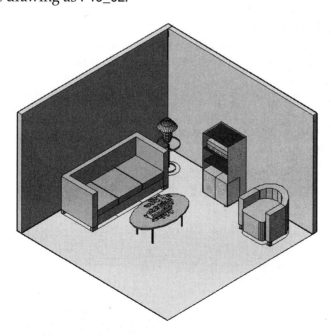

3. In this problem, you will create a new visual style to display the scene as if it is hand sketched.
   A. Open drawing P16_02.
   B. Create a new visual style named Hand Sketched with a description of Displays objects as sketched.
   C. Change the Line Extensions and Jitter settings to make the scene look as shown.
   D. Change any other settings you like.
   E. Plot the scene. Select the new visual style in the **Shade plot** drop-down list.
   F. Export the new visual style to a tool palette so that it can be used in other drawings.
   G. Save the drawing as P16_03.

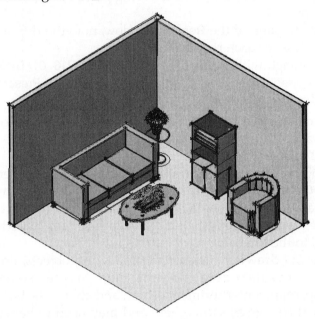

4. In this problem, you will create a realistic image of the car fender that you created in Chapter 9.
   A. Open drawing P09_03.
   B. Freeze any layers needed so that only the fender is displayed. Display the fender in the color you want it to be.
   C. Draw a planar surface to represent the ground.
   D. Set the Realistic visual style current. Then, turn on the highlight intensity and full shadows. Also, set the Show property in the **Edge Settings** category to None.
   E. Turn on sunlight. Adjust the **Date** and **Time** sliders to make the shadows look as shown.
   F. Render the scene and save it as a JPEG image. Name the file P16_04.jpg.
   G. Save the drawing as P16_04.

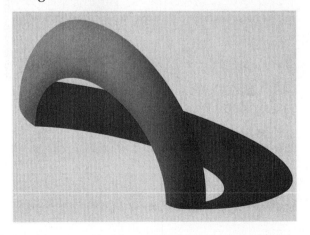

Chapter

Materials in AutoCAD

Learning Objectives

After completing this chapter, you will be able to:

✓ Attach materials to the objects in a drawing.
✓ Change the properties of existing materials.
✓ Create new materials.

A *material* is simply an image stretched over an object to make it appear as though the object is made out of wood, marble, glass, brick, or various other materials. AutoCAD provides an assortment of materials that can be used in your drawings to create a realistic scene. The materials are grouped into categories to make them easier to find.

Materials are easy to attach. They can be dragged and dropped onto the objects, attached to all selected objects, and even attached based on the object's layer. Once the material is attached, you can adjust how the material is *mapped* to the object. If the current visual style is set to display materials in the viewport, you can immediately see the effects on the object. The properties of a material can also be changed to make it look shinier, softer, smoother, rougher, and so on. When you finally render the scene, you will see the full effect of the materials.

The *materials library* is the location where all materials are stored. When you install AutoCAD, the Autodesk Material Library is installed. The images in the Base Resolution Image Library are low resolution (512 × 512) and are used with AutoCAD materials. The Medium Resolution Image Library contains medium-resolution images (1024 × 1024) that are good for close-up work or large-scale model rendering. The Medium Resolution Image Library is an additional software option that you must install after installing AutoCAD. When you attempt to render, you may be asked if you want to download the Medium Resolution Image Library. Just follow the prompts to accomplish this task.

Materials Browser

MATBROWSEROPEN

Ribbon

Render
> Materials

Materials Browser

Type

MATBROWSEROPEN
RMAT
MAT

In AutoCAD, the **Materials Browser** palette is used to manage the materials library. This is generically called the *materials browser*. It provides access to all materials that are available in the Autodesk libraries and from other sources. The **Materials** panel in the **Render** tab of the ribbon contains buttons for accessing the materials browser and the materials editor. See **Figure 17-1**.

There are two main areas in the materials browser, **Figure 17-2**. The **Document Materials** area contains materials that have been selected for use in the current drawing. The **Libraries** area shows all available libraries from which materials may be selected. These areas are discussed in more detail in the next sections.

Figure 17-1.
The **Materials Browser** button is located on the **Materials** panel in the **Render** tab of the ribbon. The materials editor is displayed by picking the dialog box launcher on the panel.

Pick to display the materials browser

Pick to display the materials editor

Figure 17-2.
The materials browser contains all of the available materials.

Click to set what is shown in area

Click to sort material swatches

Materials in drawing

Document Materials area

Library

Categories

Category with nested categories

Materials in library

Libraries area

Click to open or create a materials library

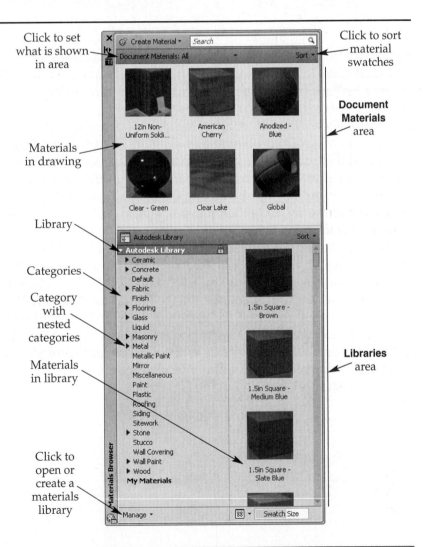

Document Materials Area

Materials are added to the **Document Materials** area by selecting them in the **Libraries** area. Single-clicking on a material swatch in the library adds it to the drawing. A swatch for the material then appears in the **Document Materials** area.

Picking the **Document Materials** title bar opens a drop-down menu. The options in this menu allow you to control which materials are displayed in the **Document Materials** area. You can show all materials, only the materials that are applied to objects in the drawing, selected materials, or unused materials. In addition, you can purge unused materials from the **Document Materials** area.

The options in the **Sort** drop-down menu control how the material swatches are arranged. You can sort materials by name, type, or material color.

Right-clicking on a material swatch displays a shortcut menu. There are several options available in the shortcut menu:

- **Edit.** This displays the materials editor for editing the selected material. The materials editor is discussed later in this chapter.
- **Duplicate.** A copy of the material is added to the drawing. A new material swatch is added with the same name as the original material, but with a sequential number added to the name.
- **Rename.** This option is used to change the name of the material. It is a good idea to rename all of your materials to meaningful names.
- **Delete.** This option removes the material from the drawing. If it is currently being used in the drawing, a warning alerts you to this. Continuing with the deletion removes the material from the objects in the drawing as well as from the **Document Materials** area. It is not, however, removed from the library.
- **Select Objects Applied To.** All objects in the drawing that have the material applied to them are selected in the drawing window.
- **Add to.** This option displays a cascading menu with two choices. The material may be added to any of your own libraries in the **Libraries** area. This is excellent for organizing the materials that you like to use and making them available to other drawings. You can also add the material to the active tool palette.
- **Purge All Unused.** This option removes from the drawing all materials not currently assigned to an object.

Libraries Area

The **Libraries** area of the materials browser contains the open library files. The libraries are recognizable by the name in bold. Pick the triangle to expand or collapse the library.

By default, two libraries exist when AutoCAD is installed: Autodesk Library and My Materials. The Autodesk Library is composed of the materials installed when AutoCAD was installed and cannot be edited. My Materials is empty by default, but you may add your own libraries to it. Custom materials are only available in the drawing in which they were created unless saved to a materials library located in My Materials. Custom materials you create in a drawing but do not save to a materials library are called *embedded materials*.

Libraries can have categories within them. Category names are not bold. If the category contains nested categories, a triangle appears next to the name. Refer to **Figure 17-2**. Pick the triangle to expand the item. Categories in the Autodesk Library are organized by material type. Categories in user libraries must be created by the user. If no category exists in the user library, AutoCAD assigns the library name as the category.

Materials are displayed in the right-hand column as a swatch and name. The materials in this column change when a different category is selected.

The **Manage** bar at the bottom provides options for managing materials libraries and categories:

- **Open Existing Library.** Allows you to select a library file (*.adsklib) and display it in the materials browser.
- **Create New Library.** Gives you the option of saving a library you assembled in the materials browser. Library files are saved with the .adsklib file extension.
- **Remove Library.** Used to delete libraries from the materials browser.
- **Create Category.** Allows you to add categories to your library.
- **Delete Category.** Used to remove categories.
- **Rename.** Allows you to rename libraries and categories.

To the right of the **Manage** bar, there is a flyout with three options for the display of material information. The **Grid View** option is active by default and shows the material swatch and its name. Picking the **List View** option shows the same information with columns to the right for type and category. Picking the **Text View** option turns off the swatches and just displays the material name, type, and category. The slider next to this button controls the swatch size in the materials browser. The right-hand button on the bottom of the materials browser opens the materials editor.

You cannot rename or remove any of the default Autodesk libraries or categories, nor can you add categories to or remove categories from them.

Exercise 17-1

Complete the exercise on the companion website.
www.g-wlearning.com/CAD

Applying and Removing Materials

To attach a material to an object in the drawing, you can drag the material from the materials browser and drop it onto an object. To apply a material only to a face on an object, hold the [Ctrl] key and pick the face. To apply a different material to an object, simply drag the new material to the object and pick.

You can use the **MATERIALATTACH** command to assign materials to the layers in your drawing. Once a material is assigned to a layer, any object on that layer is displayed in the material, as long as the object's material property is set to ByLayer. When objects are created in AutoCAD, the default "material" assigned to them is ByLayer. If your objects are organized on layers, this is the easiest way to attach materials. You can override the layer material by applying a material to individual objects.

Figure 17-3 shows the **Material Attachment Options** dialog box displayed by the **MATERIALATTACH** command. The list on the left side of the dialog box shows the materials loaded into the drawing. The right side of the dialog box shows the layers in the drawing and the material attached to each layer. When no material is attached to a layer, the material is listed as Global. The Global material is a "blank" material in every drawing. To attach a material to a layer, drag the material from the list on the left and drop it onto the layer name on the right. To remove a material from a layer, pick the **X** button next to the material name on the right side of the dialog box.

Figure 17-3.
Attaching materials to layers.

Layers in drawing

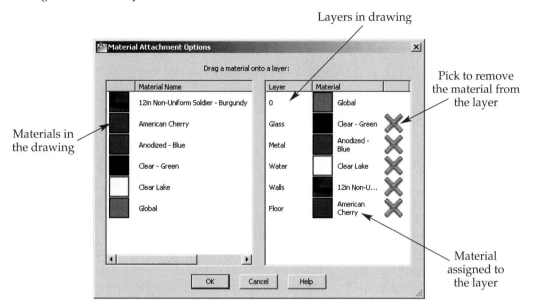

Materials in
the drawing

Pick to remove
the material from
the layer

Material
assigned to
the layer

The **Remove Materials** button on the **Materials** panel in the **Render** tab of the ribbon allows you to quickly set an object's material back to ByLayer. When the button is picked, a paintbrush selection cursor is displayed. If you select an object that has a material specifically attached to it, the material is removed. Selecting an object already set to ByLayer has no effect.

A material can also be removed using the **Properties** palette. To remove a material from an object or subobject, simply change its Material property in the **3D Visualization** category to Global. If a material has not been assigned to the object's layer, the property can also be set to ByLayer. See **Figure 17-4**.

Ribbon
Render
> Materials

Remove Materials

Remove Materials

**PROFESSIONAL
TIP**

A material can also be applied to an object by selecting the object in the drawing window, right-clicking on the material swatch in the materials browser, and selecting **Assign to Selection**. A material can be loaded into the drawing without attaching it to an object by picking the material swatch once in the **Libraries** area of the materials browser or by dragging and dropping it into a blank area of the drawing. This makes the material available in the drawing.

Figure 17-4.
Removing a material from an object. A—The material is assigned. B—The material is removed.

Material
assigned

Material
removed

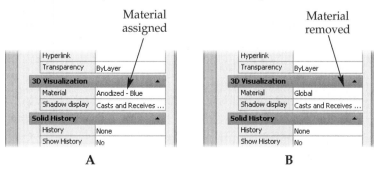

A

B

Exercise 17-2

Complete the exercise on the companion website.
www.g-wlearning.com/CAD

Material Display Options

VSMATERIALMODE

Ribbon

Render
> Materials

Materials/
Textures Off

Materials On/
Textures Off

Materials/Textures
On

Type

VSMATERIALMODE

As you learned in the previous chapter, visual styles control how materials are displayed in the viewport. The Material display property of a visual style can be set to display materials and textures, materials only, or neither materials nor textures. The **Materials** panel on the **Render** tab of the ribbon has three buttons in a flyout that correspond to, but override, this property setting:

- **Materials/Textures Off.** Objects are displayed in their assigned colors.
- **Materials On/Textures Off.** Objects are displayed in the basic color of the material, but no other material details are displayed.
- **Materials/Textures On.** Objects are displayed with the effects of all material properties visible.

Exercise 17-3

Complete the exercise on the companion website.
www.g-wlearning.com/CAD

Materials Editor

MATEDITOROPEN

Type

MATEDITOROPEN

The *materials editor* allows you to create new materials and edit existing materials to your liking. The next sections discuss creating and editing materials.

The **Materials Editor** palette is displayed by picking the dialog box launcher button at the bottom-right corner of the **Materials** panel in the **Render** tab of the ribbon. You can also type the **MATEDITOROPEN** command or double-click on a material swatch in the materials browser.

A preview of the material appears at the top of the materials editor. The geometry used for the preview can be changed by selecting a shape from the **Options** drop-down menu above the preview area. There are 12 shapes: sphere, cube, cylinder, canvas, plane, object, vase, draped fabric, glass curtain wall, walls, pool of liquid, and utility. The **Options** menu also allows you to choose the quality of the preview image. By default, this is set to **Fastest Renderer**. This option provides a good quality preview image. The **Mental Ray** options are used for previewing the material with a much higher quality, but the resulting display will also take longer to update. The updates occur whenever you change a material property. The name of the material can be changed by editing it in the **Name** text box.

Materials created or edited in the materials editor are then added to the **Document Materials** area of the materials browser. Once the material is in the materials browser, it is a good idea to add it to one of your material libraries. Right-click on the material swatch in the **Document Materials** area, select **Add to**, and then select the library from the shortcut menu. The materials browser may be accessed by picking the button to the right of the **Name** text box.

There are three different approaches to creating new materials. You can duplicate an existing material. You can also start with an existing material type as a template. Finally, you can start with a generic material, which is like starting from scratch. To access these options, click the **Create Material** drop-down menu at the top of the materials editor.

New Material from an Existing Material

By far the easiest way to create your own material is to start with an existing material, create a duplicate, and make any needed modifications. Look through the materials library in the materials browser to find a material that is close to what you want. When you select a material in the library, it is placed in the **Document Materials** area. Select the material swatch and launch the materials editor.

The material selected in the **Document Materials** area of the materials browser is displayed in the materials editor. Pick **Duplicate** from the **Create Material** drop-down menu at the top of the materials editor. See **Figure 17-5**. This creates a duplicate material with the same name and a sequential number. Change the name and the properties to your liking (discussed later) and close the materials editor. The new material is in the **Document Materials** area of the materials browser ready to use.

New Material Using an Existing Material Type

Another way to create a material is by using a material type as a template for your new material. Open the materials editor and select one of the material types from the **Create Material** drop-down menu under the **New using type:** option. Refer to **Figure 17-5**.

Figure 17-5.
This drop-down list can be used to select a material type or to copy the existing material.

Pick to display drop-down menu

Pick to copy the material

The material type provides certain default settings as a starting point. The following sections discuss the material types, their possible applications, and any special properties the material type may have.

Ceramic

The *ceramic* material type is designed for ceramic floors. However, this material type may serve well for other glossy surfaces, such as countertops, bathtubs, or dinnerware. The Type property can be Ceramic or Porcelain. The Finish property can be High Gloss/Glazed, Satin, or Matte. The **Finish Bumps** category contains a Type property that can be Wavy or Custom. The **Relief Pattern** category contains an Image property. The image determines the relief pattern.

Concrete

The *concrete* material type works well for concrete floors, walls, and sidewalks. This material type can be used for anything constructed of concrete, such as an in-ground swimming pool. The Sealant property can be None, Epoxy, or Acrylic. The **Finish Bumps** category contains a Type property that can be Broom Straight, Broom Curved, Smooth, Polished, or Stamped/Custom. The **Weathering** category contains a Type property that can be Automatic or Custom (based on a selected image).

Glazing

The glass in windows is called glazing. Therefore, the *glazing* material type is designed mostly for use on windows or thin glass objects. The solid glass material type is designed for thicker glass. The Color property can be Clear, Green, Gray, Blue, Blue-green, Bronze, or Custom. The Reflectance value determines how reflective the material is. The Sheets of glass property alters the material effect based on a virtual thickness of the material.

Masonry

Masonry includes brick walls, cobblestone, tile flooring, and so on. The *masonry* material type is designed for use on objects such as these. The Type property can be CMU (concrete masonry unit) or Masonry. The Finish property can be Glossy, Matte, or Unfinished. The **Relief Pattern** category contains an Image property. The image determines the relief pattern.

Metal

The *metal* material type works well for mechanical parts and other objects made from different types of metals. This material type primarily is for raw, or unfinished, metal. The Type property can be Aluminum, Anodized Aluminum, Chrome, Copper, Brass, Bronze, Stainless Steel, or Zinc. The Finish property can be Polished, Semi-polished, Satin, or Brushed. The **Relief Pattern** category contains a Type property that can be Knurl, Diamond Plate, Checker Plate, or Custom (based on a selected image). The **Cutouts** category contains a Type property that can be Staggered Circles, Straight Circles, Squares, Grecian, Cloverleaf, Hexagon, or Custom. Depending on which type of metal is selected, additional properties may be available for editing.

Metallic Paint

The *metallic paint* material type is for objects made of metal, but with a finish applied. This includes objects like car parts, lawn furniture, and kitchen appliances. The **Flecks** category contains Color and Size properties. The **Pearl** category includes a Type property that can be Chromatic or Second Color. The **Top Coat** category includes a Type property that can be Car Paint, Chrome, Matte, or Custom. The **Top Coat** category also includes a Finish property that can be Smooth or Orange Peel.

Mirror

The *mirror* material type is designed for very reflective objects, such as mirrors. It can also be used for water, glass, or any object that should have a high reflectivity. The only property for this material type is Tint Color.

Plastic

The *plastic* material type is designed for use on plastic objects. The plastic can be opaque, translucent, glossy, or textured. The Type property can be Solid, Transparent, or Vinyl. The Finish property can be Polished, Glossy, or Matte. The **Finish Bumps** category contains an Image property, as does the **Relief Pattern** category. The two images do not have to be the same.

Solid Glass

The *solid glass* material type is intended for thick glass objects. The glazing material type is designed for thin glass objects. The Color property can be Clear, Green, Gray, Blue, Blue-green, Bronze, or Custom. The Reflectance property determines the degree of reflectivity of the material. The Refraction property can be Air, Water, Alcohol, Quartz, Glass, Diamond, or Custom. The Roughness property determines the polish on the material. The **Relief Pattern** category contains a Type property that can be Rippled, Wavy, or Custom.

Stone

The *stone* material type works well for stone walls, stone walkways, and marble countertops. The **Stone** category contains an Image property, which is the image applied to the material. This can be a selected image or a specified texture. The **Stone** category also contains a Finish property that can be Polished, Glossy, Matte, or Unfinished. The **Finish Bumps** category contains a Type property that can be Polished Granite, Stone Wall, Glossy Marble, or Custom. The **Relief Pattern** category contains an Image property. This can be a selected image or a specified texture.

Wall Paint

The *wall paint* material type works well for interior or exterior painted walls and other objects. The Finish property can be Flat/Matte, Eggshell, Platinum, Pearl, Semi-gloss, or Gloss. The Application property can be Roller, Brush, or Spray.

Water

The *water* material type is designed for any liquid. Pools, reflecting ponds, rivers, lakes, and oceans are examples of where this material type may be used. The Type property can be Swimming Pool, Generic Reflecting Pool, Generic Stream/River, Generic Pond/Lake, or Generic Sea/Ocean. The Color property (when available) can be Tropical, Algae/Green, Murky/Brown, Generic Reflecting Pool, Generic Stream/River, Generic Pond/Lake, Generic Sea/Ocean, or Custom. The Wave Height property sets the amplitude of ripples in the liquid.

Wood

The *wood* material type works well for various finishes used for flooring, furniture, wood trim, and so on. The **Wood** category contains an Image property, which is the image applied to the material. This can be a selected image or a specified texture. When the **Stain** check box is checked, the color of stain can be set. The **Wood** category also contains a Finish property, which can be Unfinished, Glossy Varnish, Semi-gloss Varnish, or Satin Varnish. The Used For property can be Flooring or Furniture. The **Relief Pattern** category contains a Type property that can be Based on Wood Grain or Custom (which is based on a selected image).

Generic

The generic material type is the "blank canvas" material. All properties are available to create any material needed. This material type is used to create a material from scratch, as discussed in the next section.

PROFESSIONAL TIP

AutoCAD provides fantastic-looking materials to dress up the scene and make it look real. However, after you get comfortable with creating materials, start a library of your own materials. If your project is presented to a customer along with projects from competitors, and your competitors are using standard AutoCAD materials, your project will stand out from the crowd.

Exercise 17-4

Complete the exercise on the companion website.
www.g-wlearning.com/CAD

New Material from Scratch

Creating a material from scratch gives you complete control over the material. There are many options to consider, but it may be the only way to get exactly the material you are looking for to complete your scene. To start creating a material from scratch, select the generic material type by selecting **New Generic Material...** at the bottom of the **Create Material** drop-down menu in the materials editor. Refer to **Figure 17-5**. The properties and settings are discussed in the next sections.

Generic Category

The *diffuse color* is the color of the object in lighted areas, or the perceived color of the material. See **Figure 17-6**. It is the predominant color you see when you look at the object. The Color property sets the diffuse color of the material. In AutoCAD, the ambient and specular colors are determined by the diffuse color.

The *ambient color* is the color of the object where light does not directly provide illumination. It can be thought of as the color of an object in shadows. In nature, shadows cast by an object typically contain some of the ambient color.

The *specular color* is the color of the highlight (the shiny spot). It is typically white or a light color. The amount of specular color shown is determined by the glossiness and reflectivity of the material and the intensity of lighting in the scene.

You have two options for controlling the Color property of the material. Picking the button to the right of the color text box displays a drop-down list. Selecting **Color** in the drop-down list means that you can select whatever color you wish.

Figure 17-6.
The three colors of a material are illustrated here. In AutoCAD, the Color property sets the diffuse color. The ambient and specular colors are based on the diffuse color.

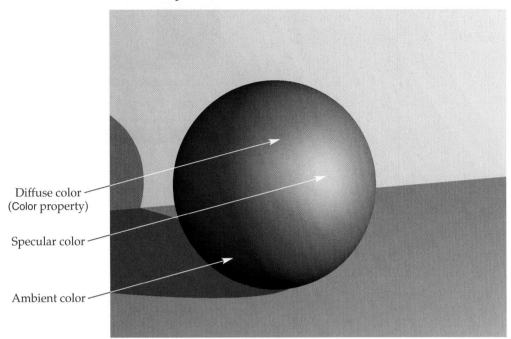

Diffuse color
(Color property)

Specular color

Ambient color

To set the color, pick in the text box; the **Select Color** dialog box is displayed. See **Figure 17-7.** If you select **Color by Object** in the drop-down list, the base color of the object is used as the diffuse color.

The Image property allows you to apply an image or texture to control the appearance of the material. Picking in the image preview area displays a standard open dialog box. When you select an image, it is applied to the material and is displayed in the material preview area, **Figure 17-8.** Also, the texture editor is displayed with the image loaded and ready for editing, if necessary. The button to the right of the image preview area is used to apply a texture instead of an image. Textures and texture editing are discussed later in this chapter.

Figure 17-7.
Setting a color for a material.

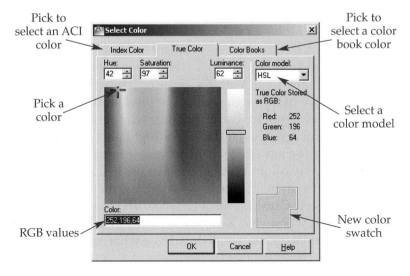

Pick to
select an ACI
color

Pick a
color

RGB values

Pick to
select a color
book color

Select a
color model

New color
swatch

Figure 17-8.
The property settings in the **Generic** category of a material created from scratch.

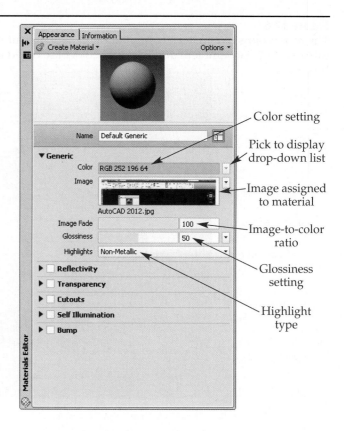

Color setting

Pick to display drop-down list

Image assigned to material

Image-to-color ratio

Glossiness setting

Highlight type

When an image, such as a bitmap image or digital photograph, is applied, the material color is replaced with the image. **Figure 17-9** shows an object with a computer screen image applied to the material attached to it. The Image Fade property controls the ratio between the image and the object color. It is only available when an image or texture is used. When set to 100, the image completely replaces the object color.

The Glossiness property controls how shiny the material appears, **Figure 17-10**. It is a measure of the surface roughness of a material. A setting of 100 specifies a very shiny material, such as a smooth surface. A setting of 0 specifies a matte (dull) material. The button to the right of the setting allows you to add a texture or image to the Glossiness property. The pattern is applied to the glossiness effect. Reflectivity must be turned on for the glossiness effect to be seen. Refer to the next section.

The Highlights property determines how the shiny areas are created. Choices for this property are Metallic or Non-metallic. Highlights are brighter when Metallic is selected.

PROFESSIONAL TIP

Highlights can be seen everywhere. Look around you right now at edges and inclined surfaces. The diffuse color of the surface typically has little to do with the color of the highlight. The color of the light source usually determines the predominant highlight color. The majority of highlights is white or near white because most light sources are white or nearly white. However, highlights in the interior of a home may have a yellow cast to them because incandescent light bulbs generally cast yellow light. Compact fluorescent lights cast a slightly different color, although many are designed to cast the same color as an incandescent bulb. Outside with a clear sky and bright sun, highlights may have a slight blue cast. Keep these points in mind when creating your own materials. These small details are what make a scene realistic.

Figure 17-9.
A material with an image applied instead of a color to simulate a tablet screen. A—The screen object's material does not have an image applied. B—A bitmap image has been applied to the image component of the screen object's material. C—The finished tablet.

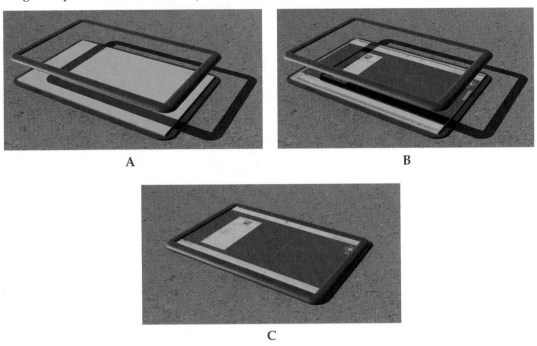

A

B

C

Figure 17-10.
Three different glossiness settings are illustrated here.

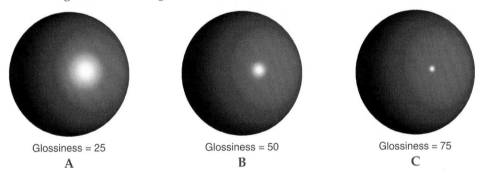

Glossiness = 25
A

Glossiness = 50
B

Glossiness = 75
C

Reflectivity Category

Reflectivity is a measure of how much light is bounced off of the surface. There are two basic property settings for reflectivity, Direct and Oblique. See **Figure 17-11.** The Direct property controls how much light is reflected back for surfaces that are more or less facing the camera. The Oblique property controls how much light is reflected back when the surface is at an angle to the camera.

Each property has a slider/text box that controls the amount of reflectivity. No reflections are created with a setting of 0. The maximum reflections are created with a setting of 100. Object color, lighting, surroundings, and other factors also determine just how reflective a material appears in the scene.

The buttons to the right of the text boxes allow you to add a texture or image to control the reflection. The white areas of the texture or image have a reflection. The black areas do not have a reflection. The degree of reflectivity varies for gray areas and the grayscale values of colors.

Figure 17-11.
Setting the reflectivity for a material.

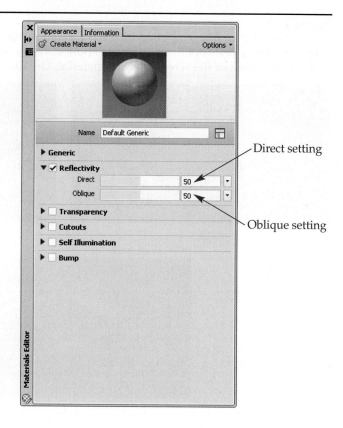

Direct setting

Oblique setting

Transparency Category

Transparency is a measure of how much light the material allows to pass through it. Glass, water, crystal, and some plastics, along with other materials, are nearly completely transparent. **Figure 17-12** shows an example of using a transparent material to show the internal workings of a mechanical assembly. The Transparency category contains a number of properties that combine to create any transparent or semitransparent material, **Figure 17-13**.

The amount of light that passes through the material is controlled by the Amount property. When this property is set to 0, the material is opaque. A setting of 100 creates a completely transparent material.

Figure 17-12.
The material used for the housing on this mechanism has a transparency setting of about 50.

Figure 17-13.
Setting transparency
for a material.

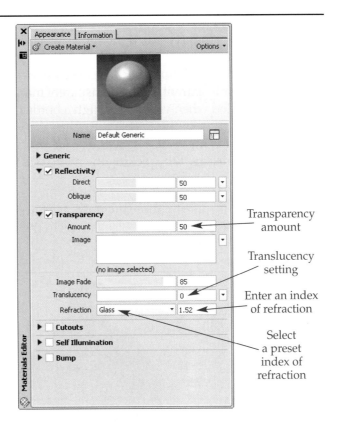

The Image property is used to add a transparency map to the material. White areas in the image or texture are transparent. Black areas are opaque. All other colors produce varying degrees of transparency based on their grayscale values. The Amount property is applied to the transparency map. Maps are discussed later in this chapter.

The Image Fade property determines how much of an impact the transparency map has on the transparency of the material. A setting of 100 means the transparency is completely see-through based on the transparency map. As the setting is decreased, a higher percentage of transparency is determined by the Amount property.

Translucency is a quality of transparent and semitransparent materials that causes light to be diffused (scattered) as it passes through the material. See **Figure 17-14.** This makes any object with the material applied to it appear as if it is being illuminated from

Figure 17-14.
The effect of translucency. A—The glass material has a translucency setting of zero. B—When the translucency setting is increased, light is diffused within the material. Notice how the glass appears slightly frosted.

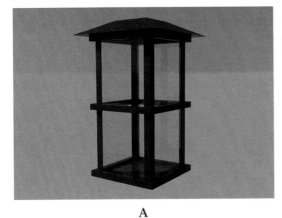

A

B

within, or glowing. The thicker the material, the more pronounced the effect. When the Translucency property is 0, light appears to travel through the material, lighting the opposite side. A setting of 100 creates a material similar in appearance to frosted glass.

The *index of refraction (IOR)* is a measure of how much light is bent (refracted) as it passes through transparent or semitransparent materials. Refraction is what causes objects to appear distorted when viewed through a bottle or glass of water. See **Figure 17-15**. The Refraction property sets the IOR. The higher the value, the more that light is bent as it passes through the material. The IOR of water is 1.3333. You can enter a value in the text box or select a preset IOR by picking the name that is displayed to the left of the text box.

Cutouts Category

The Cutouts property allows you to select an image or texture to use for a pattern of cutouts (holes), **Figure 17-16**. Black areas in the image will appear to be see-through,

Figure 17-15.
The effect of refraction. A—The transparent material on the sphere has a refraction setting of zero. B—When the refraction setting is increased, the cylinder behind the sphere is distorted as light is refracted by the material.

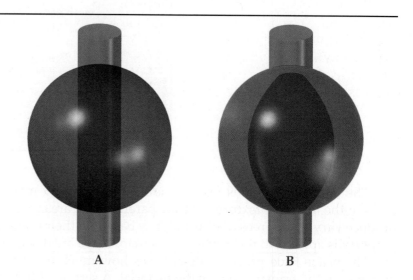

A   B

Figure 17-16.
Adding a cutout map to a material.

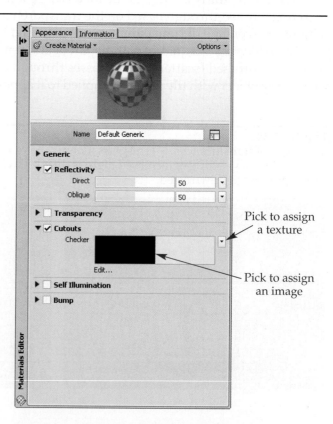

Pick to assign a texture

Pick to assign an image

as if there is no object in those areas. White areas in the image have the normal material colors. This is similar to using a transparency map, but without the other transparency settings. **Figure 17-17** shows an example of a cutout map applied to a material.

Self Illumination Category

Self illumination is an effect of a material producing illumination. See **Figure 17-18**. For example, the surface of a neon tube glows. However, in AutoCAD, a material with self illumination will not actually add illumination to a scene. This effect can be simulated with properly placed light sources. *Luminance* is defined as the value of light reflected off a surface. The Self Illumination category contains several properties related to self illumination and luminance, **Figure 17-19**.

The Filter Color property controls the color of the self illumination effect. Pick in the edit box to open the **Select Color** dialog box. An image may be selected instead of a color. Black areas of the image are not illuminated. White areas of the image are illuminated. Grayscale values control how much the rest of the image illuminates the material. To apply a texture to control the illumination, pick the button next to the edit box to display a drop-down list.

Luminance is expressed in candelas per square meter (cd/m^2). For example, 1 cd/m^2 is the equivalent of one candela of light radiating from a surface area that is one square meter. You can enter a specific value in the Luminance property text box. Or, you can pick the name displayed next to the text box to display a drop-down list with choices for typical materials. Some of these choices include Dim Glow, LED Panel, and Cell Phone Screen.

Figure 17-17.
The effect of applying an image to the Cutout property. A—This black and white image will be used as the cutout map. B—The material on the plane is completely opaque. C—When the cutout map is applied to the material, the dark areas of the map produce transparent areas on the object.

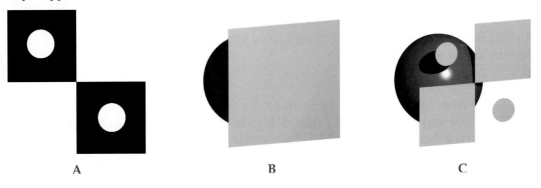

A   B   C

Figure 17-18.
The effect of self illumination/luminance. A—The globe of this light bulb does not have any self illumination. B—Self illumination is applied to the globe material.

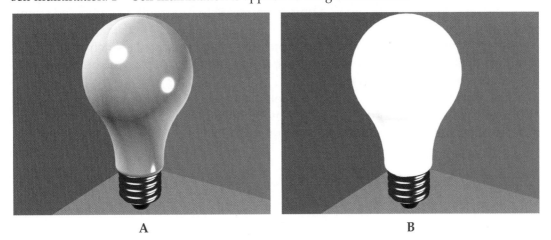

A   B

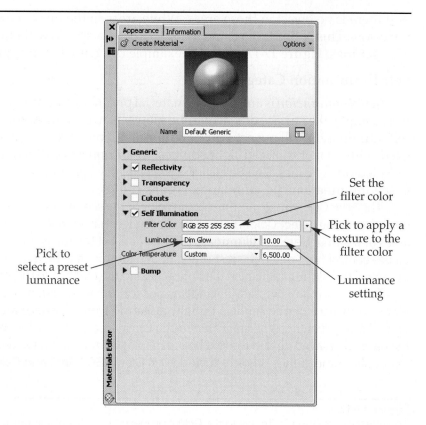

Figure 17-19.
Setting self illumination and luminance for a material.

Set the filter color

Pick to apply a texture to the filter color

Pick to select a preset luminance

Luminance setting

The Color Temperature property determines the warmth or coolness value of the color. The value is expressed as degrees Kelvin. Candles and incandescent bulbs are warm. Fluorescent lights and TV screens are cool. The drop-down list gives you choices for typical objects with different color temperatures, but you can enter a value directly in the text box.

Bump Category

The **Bump** category contains settings for making some areas of the material appear raised and other areas depressed, **Figure 17-20**. The image or texture used for this effect is called a *bump map*. The black, white, and grayscale values of the map are used to determine raised and depressed areas. Dark areas of the map appear raised and light areas appear depressed.

For example, to show the texture of a brick wall, you could physically model the grooves into the wall. This would take a lot of time to model and would immensely increase the rendering time because of the increased complexity of the geometry. Using the properties in the **Bump** category is an easier and more efficient way to accomplish the same task. **Figure 17-21** shows a bump map used to represent an embossed stamp on a metal case.

To apply an image as a bump map, pick the Image property swatch to display a standard open dialog box. To apply a texture as a bump map, pick the button next to the edit box to display a drop-down list.

The Amount property determines the relative height of the bump pattern. A setting of 0 results in a flat material, or no bumps. A setting of 1000 creates the maximum difference between low and high areas of the pattern.

Figure 17-20.
Making bump
settings for a
material.

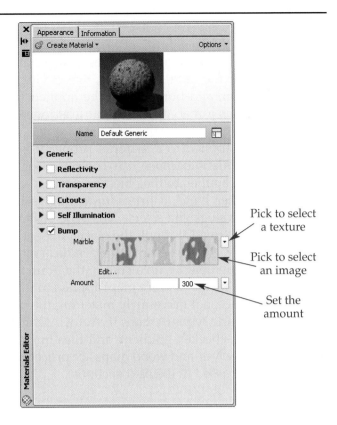

Pick to select
a texture

Pick to select
an image

Set the
amount

Figure 17-21.
The effect of a bump
map. A—This image
will be used as the
bump map.
B—When applied
to the material, the
bump map simulates
an embossed stamp
on the metal case.

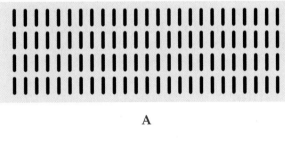

A

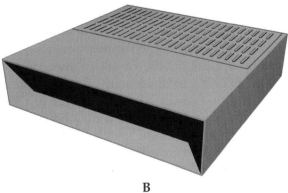

B

As discussed in previous sections, images and textures may be applied to materials to enhance the color, glossiness, reflectivity, transparency, translucency, cutouts, self illumination, and bumpiness. The applied image or texture is called a *map*. A material that has a map applied to at least one of its properties is called a *mapped material*.

An image applied to a material property is sometimes known as a *2D map* or *texture map*. This is because it is composed of a fixed set of pixels, kind of like a mosaic pattern, applied to the object's surfaces.

On the other hand, a *procedural map* is mathematically generated based on the colors and values you select. This type of map is sometimes known as a *3D map* because it may extend through the object, depending on its algorithm. For example, if you attach AutoCAD's wood map (procedural map) to a material property and assign the material to an object with a cutout, the grain in the cutout will match the grain on the exterior. An image (texture map) applied to this same object will "reapply" itself to the cutout surface, not necessarily matching the pattern on the adjacent surfaces.

There are nine types of maps in AutoCAD that can be applied to material properties. The image, checker, gradient, and tiles maps are texture maps (2D). The marble, noise, speckle, waves, and wood maps are procedural maps (3D). Each map has unique settings, as discussed in the next sections.

PROFESSIONAL TIP

The term *map* may be applied to the type of texture or the property to which the texture is applied. For example, a checker map may be used as a bump map. *Checker* is the type of map and *bump* is the property to which the texture is applied.

Texture Editor

The *texture editor* is used to adjust a map once it has been applied to a material property. It is automatically displayed once a map is assigned. **Figure 17-22** shows the texture editor with an image map displayed for editing. A preview of the map appears at the top of the texture editor. The triangle in the lower-right corner of the preview is used to resize the preview. Below the map preview, the type of map is indicated.

Maps rarely appear on the object in the correct position, scale, or angle. For example, if you are using an image map of a logo to be applied to a box for a packaging design, it may not be positioned in the center of the box by default. The map may be rotated in the wrong direction or it may be too large. The texture editor is where adjustments to the map are made.

The bottom of the texture editor contains the settings for the map. The properties that are available depend on the type of map being edited. As shown in **Figure 17-22**, the **Transforms**, **Position**, **Scale**, and **Repeat** categories are available and automatically expanded when the texture editor is opened for an image map. The next sections discuss the properties for each map type.

Image

Applying an *image map* is straightforward. There are several file types that can be applied as a map. When you specify Image as the map for a property, a standard open dialog box is displayed. Navigate to the image file and open it.

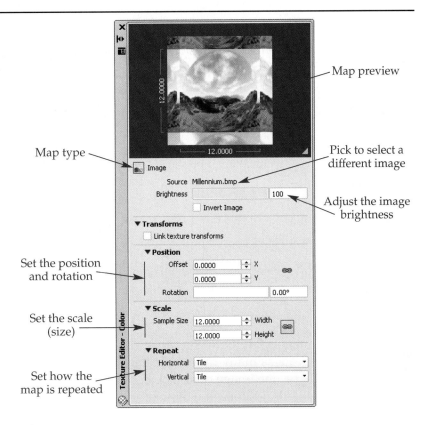

Figure 17-22.
The texture editor is used to adjust a map once it has been applied to a material property. Shown are the properties of an image map.

Map preview

Map type

Pick to select a different image

Adjust the image brightness

Set the position and rotation

Set the scale (size)

Set how the map is repeated

In the texture editor, the preview area shows the image with dimensions for the size of an individual tile. If the Sample Size property in the **Scale** category is changed, the preview dimensions reflect the change. Directly below the preview are the image file name and a slider/text box for adjusting the image brightness. Refer to **Figure 17-22**. To select a different image, pick the name to display a standard open dialog box. A brightness value of 100 means the image is at its full brightness. A value of 0 results in the image being all black. The check box below the slider is used to invert the image colors. This produces an effect similar to a photographic negative from an old film camera but is most useful for reversing the effects of a bump, cutout, or transparency map.

In the **Transforms** category is the **Link texture transforms** check box. It is very important to check this check box if you are using the same map for different properties in the material and need them to be synchronized in appearance. For example, to make a realistic tile floor, an image of tiles is applied to the color property. The same image map is also used as a bump map to make the grout look recessed. If the scaling is changed for the bump map, but the maps are not synchronized, the grout colors and indentations may not match. See **Figure 17-23**. By linking the texture transforms, all transform settings are the same. Transforms are not linked by default and if you turn on linking after you have made changes, the properties may not be synchronized. Linking must be turned on for *each* material map or it will not work.

The properties in the **Position** category control the location of the map on the object and its rotation. The Offset property moves the image in the X and Y directions. The link button to the right of these text boxes locks the X and Y values together. It is off by default. The Rotation property allows you to rotate the image on the material.

The properties in the **Scale** category control the size of the image. The Width and Height properties are locked together by default. This is important to maintain a proportional *aspect ratio* for the image. The properties can be unlocked, but be aware that entering different values for width and height will stretch and distort the image.

Figure 17-23.
A—This material has an image map assigned to the color and bump properties. B—If the bump map is not synchronized to transformations, it may not align with the grout lines if the object is transformed.

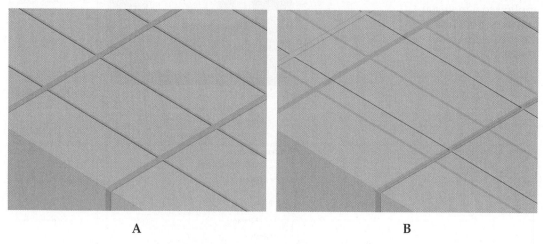

A   B

The **Repeat** category is where you can set the image to tile or not tile. *Tiling* means that the image repeats as many times as it takes to cover the object. If tiling is turned off, there will only be one image on the object. For example, if you are using an image for a label on a box, tiling should be turned off. Otherwise, the box will be completely covered with labels.

Checker

A *checker map* creates a two-color checkerboard pattern. By default, the colors are black and white, but different colors or images can be used as well. This map type can be used for checkerboard pattern floor materials. However, by changing various properties, you can simulate many different effects and materials.

When you specify Checker as the map for a property, the texture editor displays the properties for the map, **Figure 17-24**. In the **Appearance** category, the Color 1 and Color 2 properties set the color of the checkers. The button to the right of each property allows you to specify an image or texture for the checker. For example, you may add a texture map to the Color 1 property and a noise map to the Color 2 property. The possibilities are endless.

The Soften property is used to blur the edges between the checkers. To change the setting, enter a value in the text box or use the up and down arrows. A value of 0.00 creates sharp edges between the checkers. The maximum setting of 5.00 produces edges that are very blurred.

The properties in the **Transforms**, **Position**, **Scale**, and **Repeat** categories are the same as described for an image map. Refer to the Image section for details on these properties.

Gradient

The *gradient map* is a texture map that allows you to create a material blending colors in different patterns. This is similar to creating a gradient using the **HATCH** command's **Gradient Colors** option. When you specify Gradient as the map for a property, the texture editor displays the properties for the map, **Figure 17-25**.

The gradient is represented in the **Appearance** category with three nodes at the bottom edge of the ramp. Each node represents a different color in the ramp. By default, the left node (node 1) is black, the middle node (node 2) is gray, and the right node (node 3) is white. The middle node can be moved left or right, as discussed later in this section, to change where the color transitions.

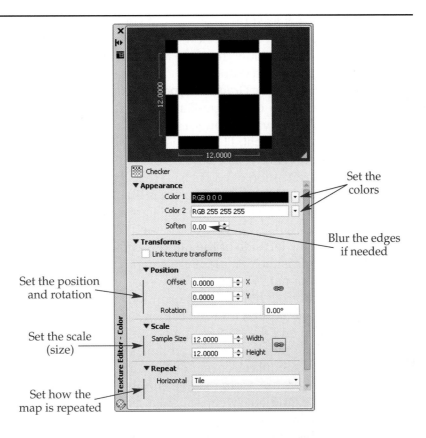

Figure 17-24.
Adjusting the properties of a checker map.

Set the colors

Blur the edges if needed

Set the position and rotation

Set the scale (size)

Set how the map is repeated

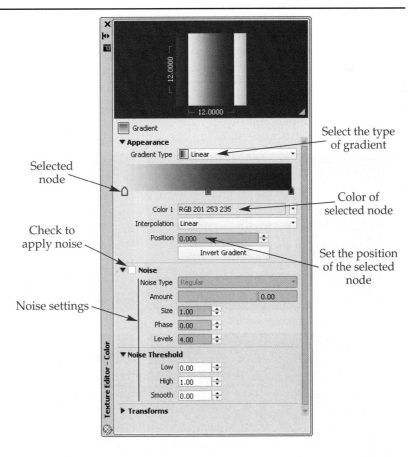

Figure 17-25.
Adjusting the properties of a gradient map.

Select the type of gradient

Selected node

Color of selected node

Check to apply noise

Set the position of the selected node

Noise settings

There must be at least three nodes, but you are not limited to three nodes. Picking anywhere in the ramp creates a new node. Selecting a node by picking on it changes the properties directly below the ramp to the settings for that node. The Color property sets the color of the selected node.

Above the ramp is the **Gradient Type** drop-down list. The setting in the drop-down list controls the pattern of the gradient ramp. See **Figure 17-26**. The default pattern is linear. This results in a typical pattern similar to the ramp display. The other options are:

- **Linear asymmetrical.** This gradient type is similar to the default linear type, but the transition between colors is not symmetrical.
- **Box.** The transition of colors is in the shape of a square.
- **Diagonal.** The transition of colors is linear, but rotated on the surface.
- **Light normal.** The intensity of the light source determines where the transition takes place. The right side of the ramp corresponds to the highest intensity of light and the left side of the ramp is equal to no light.
- **Linear.** This is the default. It is a smooth transition from one node to the next.
- **Camera normal.** The angle between the camera direction and the surface normal controls how the pattern is displayed. The left side of the ramp is 0° between the normal and the camera viewpoint. The right side is 90° between the normal and the camera viewpoint.
- **Pong.** This gradient is a rotated linear transition, similar to the diagonal type, but it pivots about the corner of a box and reverses in the middle of the pattern.
- **Radial.** Colors are arranged in a circular pattern similar to a target.
- **Spiral.** The gradient sweeps about a central point similar to the movement on a radar screen.
- **Sweep.** This gradient is similar to the spiral type, but the center of the sweep is at a corner instead of in the center. Also, the gradient does not repeat like the pong type.
- **Tartan.** This gradient resembles a plaid pattern. It is very similar to the box type.

Figure 17-26.
There are 12 different gradient types available for use in a gradient map. The four shown here are applied to the same object with the same lighting and mapping coordinates. A—Linear asymmetrical. B—Box. C—Diagonal. D—Light normal.

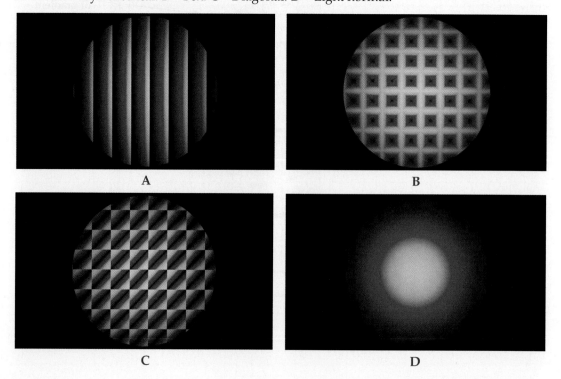

A

B

C

D

AutoCAD and Its Applications—Advanced

The Interpolation property controls the transition of colors from one node to the next. Transitions are applied to the nodes from left to right, regardless of the node number. The options are:

- **Ease in.** Shifts the transition closer to the node on the right.
- **Ease in out.** Shifts the transition toward the node, but it remains more or less centered on the node.
- **Ease out.** Shifts the transition closer to the node on the left.
- **Linear.** This is the default. The transition is constant from one node to the next.
- **Solid.** No transition between nodes. There is an abrupt change at each node.

The Position property is simply the position of the selected node in the ramp. The node on the left is at the 0 position and the node on the right is at the 1.000 position. Nodes in between will be at varying values between 0 and 1.000.

Picking the **Invert Gradient** button reverses all color values, inverting the gradient pattern. In effect, the ramp is flip-flopped from left to right.

Noise may be added to the gradient map to create an uneven appearance. The properties in the **Noise** and **Noise Threshold** categories are similar to those for the noise map. The noise map is described later in this chapter.

The properties in the **Transforms**, **Position**, **Scale**, and **Repeat** categories are the same as described for an image map. Refer to the Image section for details on these properties.

Tiles

A *tiles map* is a pattern of rectangular, colored blocks surrounded by colored grout lines. This may be the most versatile map in the whole collection. Tiles are used to simulate tile floors, ceiling grids, hardwood floors, and many different types of brick walls. When you specify Tiles as the map for a property, the texture editor displays the properties for the map, **Figure 17-27**.

You first need to define the pattern for the map. The **Pattern** category contains properties for defining the pattern. For the Type property, select one of the seven predefined tile patterns or Custom Pattern to create your own. The names of the predefined patterns bring to mind brick walls. For example, a mason may use a stack bond to build a brick wall. However, remember these are only *patterns*. You can also use a brick pattern to create tile floors and acoustic ceiling panels. Four of the tile patterns are shown in **Figure 17-28**. The Tile Count property sets how many tiles are in each row and column before the pattern repeats.

As the name implies, the properties in the **Tile Appearance** category determine what the tiles look like. You can choose any color you wish or apply a texture or image to the tiles. Ceramic tile floors look more realistic if each tile is slightly different in color. The Color Variance property can be used to alter the color of random tiles to create a more realistic appearance. The Fade Variance property is used to fade the color of random tiles. You will have to experiment with the color variance and fading to create the look you need. Start with very low values. The Randomize property is used to alter the random color variation in the tiles. This variation is automatically applied, but entering a different random seed changes the pattern. If your scene has more than one object with this material applied to it, duplicate the material and change the seed number of the new material, then apply it to the other object.

The properties in the **Grout Appearance** category control what the grout looks like. The grout is the line between the tiles. The Grout Color property is set to dark gray by default, but any color, image, or texture may be used. The Gap Width property determines how wide the grout lines are in relation to the tiles. There are horizontal and vertical settings. The settings are locked by default so the line widths are equal. In some cases, such as for a hardwood floor material, you will have to scale the pattern differently on the horizontal and vertical axes to make the gap thicker in one direction.

The properties in the **Stacking Layout** category are only available for a custom pattern. The Line shift property changes the location of the vertical grout lines in every other row

Figure 17-27.
Adjusting the
properties of a tiles
map.

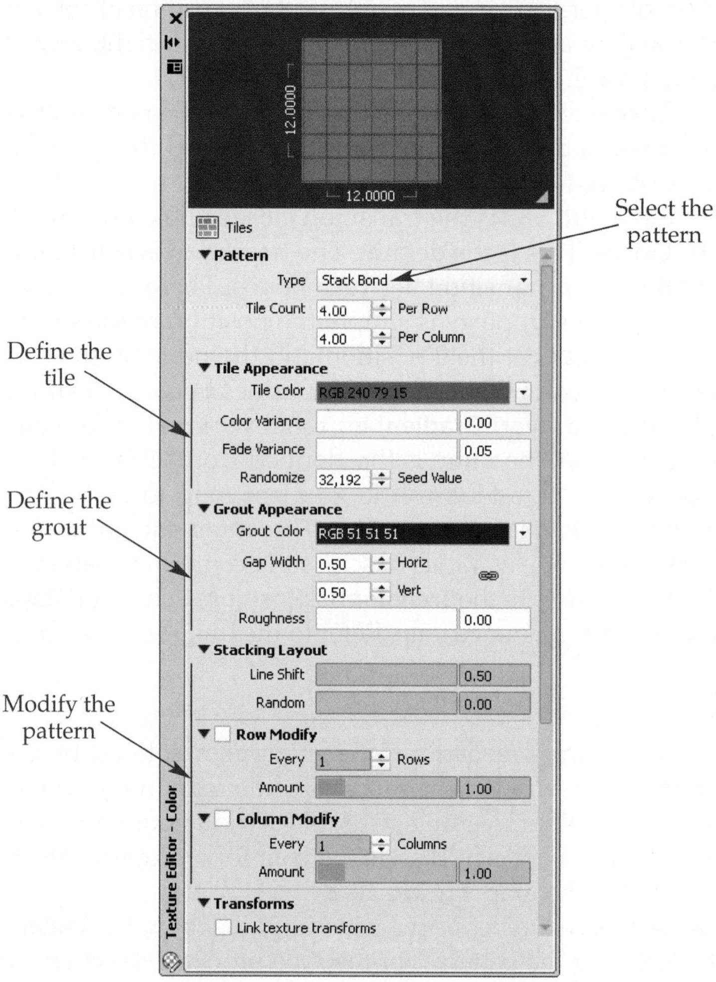

Select the
pattern

Define the
tile

Define the
grout

Modify the
pattern

Figure 17-28.
A tiles map
can have a
custom pattern
or predefined
pattern. Four of the
predefined patterns
are shown here.
A—Running bond.
B—English bond.
C—Stack bond.
D—Fine running
bond.

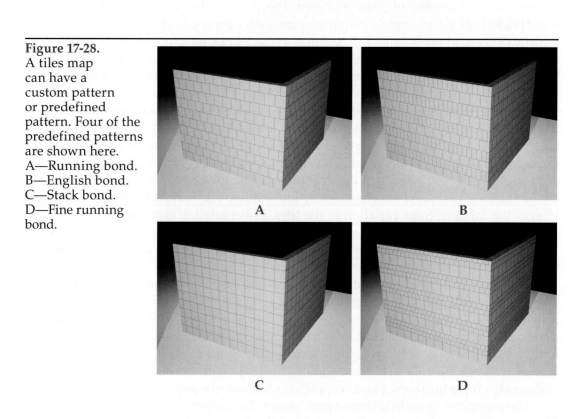

A

B

C

D

AutoCAD and Its Applications—Advanced

to create an alternate pattern of tiles. The default value is 0.50 and the range is from 0.00 to 100.00. The Random property randomly moves the same lines. This works nicely for hardwood floor materials. The default value is 0.00 and the range is 0.00 to 100.00.

The properties in the **Row Modify** and **Column Modify** categories are available with all tile pattern types, but may be disabled by default. To enable the settings, check the check box by the category name. The settings in these areas allow you to change the number of grout lines in the horizontal and vertical directions to create your own pattern. The two Every properties determine which rows and columns will be changed. When set to 0, no changes take place in the row or column. When set to 1, every row or column will be changed. When set to 2, every other row or column will be changed, and so on. The value must be a whole number. The Amount property controls the size of the tiles in the row or column. A setting of 1 means that the tiles remain their original size. A setting of 0.50 makes the tiles one-half of their original size, a setting of 2 makes the tiles twice their original size, and so on. A setting of 0.00, in effect, completely turns off the row or column and the underlying material color shows through.

The properties in the **Transforms**, **Position**, **Scale**, and **Repeat** categories are the same as described for an image map. Refer to the Image section for details on these properties.

Marble

A *marble map* is a procedural map based on the colors and values you set. It is used to simulate natural stone. When you specify Marble as the map for a property, the texture editor displays the properties for the map, **Figure 17-29**. The viewport may not reflect changes made in the texture editor, even if set to display materials and textures. You may have to render the scene to see the changes.

A marble map is based on two colors—stone and vein. The **Appearance** category contains the Stone Color and Vein Color properties. You can swap the vein and stone colors by picking the button to the right of the color definition and selecting **Swap Colors** from the drop-down list. The Vein Spacing property determines the relative

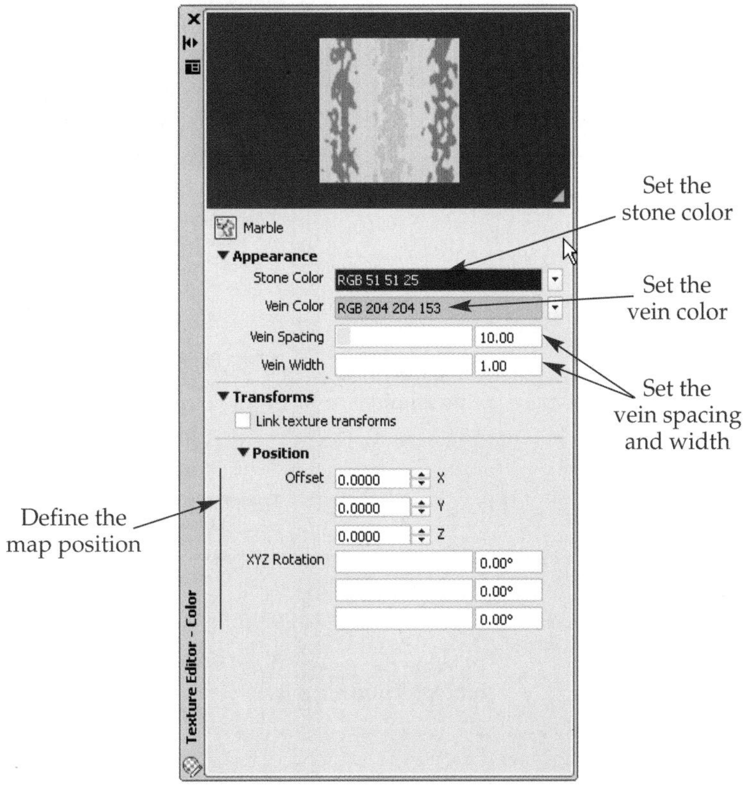

Figure 17-29.
Adjusting the properties of a marble map.

Set the stone color

Set the vein color

Set the vein spacing and width

Define the map position

distance between each vein in the marble. The Vein Width property determines the relative width of each vein. Each of these settings can range from 0.00 to 100.00.

The **Link texture transforms** check box in the **Transforms** category works as described earlier for an image map. Refer to the Image section for details on the **Transforms** properties.

Since this is a procedural (3D) map, the properties in the **Position** category are different from the maps previously discussed. The three Offset properties move the map in the X, Y, and Z directions on the object. Simply enter a value in the text boxes. The XYZ Rotation properties control the rotation of the map around the X, Y, or Z axis. You can move the sliders or enter an angle in the text boxes.

The mathematical calculations that create a procedural map are based on the world coordinate system. If you move or rotate the object, a different result is produced.

Noise

A *noise map* is a procedural map based on a random pattern of two colors used to create an uneven appearance on the material. It is most often used to simulate materials such as concrete, soil, asphalt, grass, and so on. When you specify Noise as the map for a property, the texture editor displays the properties for the map, **Figure 17-30**.

The properties in the **Appearance** area control how the noise looks. First, you need to select the type of noise. The options for the Noise Type property are:

- **Regular.** This is "plain" noise and useful for most applications.
- **Fractal.** This creates the noise pattern using a fractal algorithm. When this is selected, the Levels property in the **Noise Threshold** category is enabled.
- **Turbulence.** This is similar to fractal noise, except that it creates fault lines.

Figure 17-30.
Adjusting the properties of a noise map.

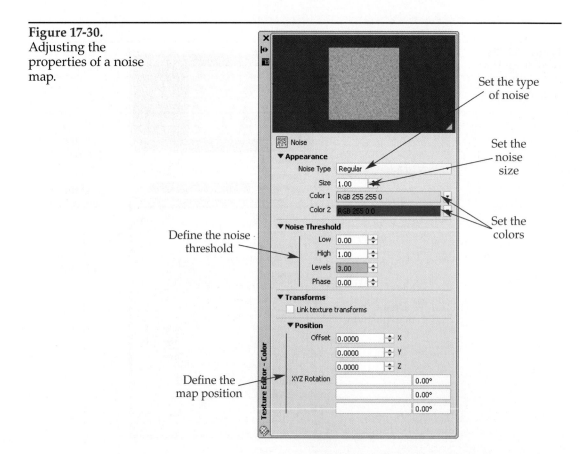

The Size property controls the size scale of the noise. The larger the value, the larger the size of the noise. The default value is 1.00 and the value can range from 0.00 to 1 billion. The Color 1 and Color 2 properties control the color of the pattern of noise. You can assign a color, image, or texture to the property. To swap the color definitions, pick the button next to the properties and select **Swap Colors** from the drop-down list.

The properties in the **Noise Threshold** category are used to fine-tune the noise effect. The properties in this category are:

- **Low.** The closer this setting is to 1.00, the more dominant color 1 is. The default setting is 0.00 and it can range from 0.00 to 1.00.
- **High.** The closer this setting is to 0.00, the more dominant color 2 is. The default setting is 1.00 and it can range from 0.00 to 1.00.
- **Levels.** Sets the energy amount for the fractal and turbulence types. Lower values make the fractal noise appear blurry and the turbulence lines more defined. The default setting is 3.00 and it can range from 1.00 upward.
- **Phase.** Randomly changes the noise pattern with each value. This allows you to have materials with the same noise map settings look slightly different. You should have different patterns on different materials. This adds a level of realism to your scene.

The properties in the **Transforms** and **Position** categories are the same as described for a marble map. Refer to the Marble section for details on these properties.

Speckle

A *speckle map* is a procedural map based on a random pattern of dots created from two colors. This map is very useful for textured walls, sand, granite, and so on. When you specify Speckle as the map for a property, the texture editor displays the properties for the map, **Figure 17-31.**

Figure 17-31.
Adjusting the properties of a speckle map.

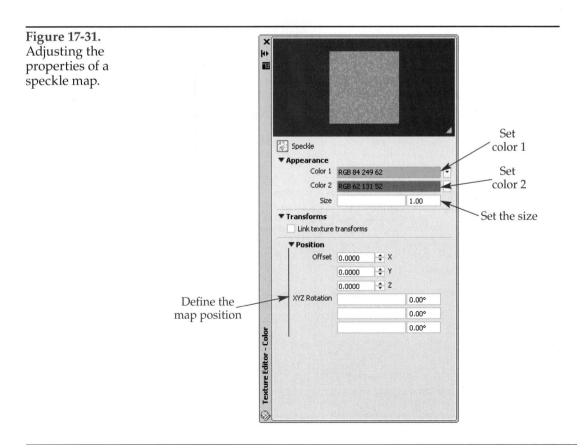

The settings for a speckle map are very simple. In the **Appearance** category, pick colors for the Color 1 and Color 2 properties. You cannot use maps, only colors. To swap the color definitions, pick the button next to the properties and select **Swap Colors** from the drop-down list. The Size property controls the size of the speckles.

The properties in the **Transforms** and **Position** categories are the same as described for a marble map. Refer to the Marble section for details on these properties.

Waves

A *waves map* is a procedural map in a pattern of concentric circles. Imagine dropping two or three stones into a pool of water and watching the ripples intersect with each other. A number of wave centers are randomly generated and a pattern created by the overlapping waves is the result. As the name implies, the waves map is usually used to simulate water. When you specify Waves as the map for a property, the texture editor displays the properties for the map, **Figure 17-32**.

In the **Appearance** category, pick colors for the Color 1 and Color 2 properties. You cannot use maps, only colors. To swap the color definitions, pick the button next to the properties and select **Swap Colors** from the drop-down list. The Distribution property can be set to 2D or 3D. This setting determines how the wave centers are distributed on the object. Selecting 3D means that the wave centers are randomly distributed over the surface of an imaginary sphere. This distribution affects all sides of an object. On the other hand, selecting 2D means that the wave centers are distributed on the XY plane. This is much better for nearly flat surfaces, such as the surface of a pond or lake.

The properties in the **Waves** category define the pattern of waves. The Number property sets the number of wave centers that are generating the waves. The Radius

Figure 17-32.
Adjusting the properties of a waves map.

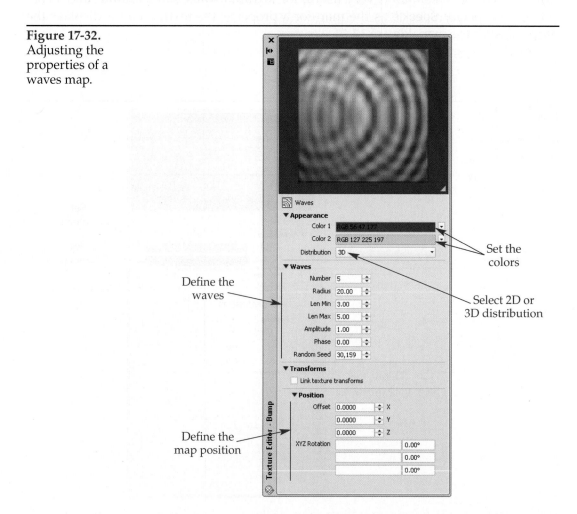

Set the colors

Select 2D or 3D distribution

Define the waves

Define the map position

property sets the radius of the circle or sphere from which the waves originate. The Len Min and Len Max properties define the minimum and maximum interval for each wave. The Amplitude property can be thought of as the "power" of the wave. The default value is 1.00, but the value can range from 0.00 to 10000.00. A value less than 1.00 makes color 1 more dominant. For a value greater than 1.00, color 2 is more dominant. The Phase property is used to shift the pattern and the Random Seed property is used to redistribute the wave centers.

The properties in the **Transforms** and **Position** categories are the same as described for a marble map. Refer to the Marble section for details on these properties.

PROFESSIONAL TIP

Remember, you can change the material swatch geometry in the materials editor, such as to a cube, sphere, or cylinder. Some maps, like a waves map, are easier to understand when displayed on a cube.

Wood

A *wood map* is a procedural map that generates a wood grain based on the colors and values you select. See **Figure 17-33**. When you specify Wood as the map for a property, the texture editor displays the properties for the map, **Figure 17-34**.

A wood map is based on two colors. The Color 1 and Color 2 properties in the **Appearance** category are used to specify these colors, usually one dark and one light color. To swap the color definitions, pick the button next to the properties and select **Swap Colors** from the drop-down list. The Radial Noise property determines the waviness of the wood's rings. The rings are found by cutting a tree crosswise. The Axial Noise property determines the waviness of the length of the tree trunk. The Grain Thickness property determines the relative width of the grain.

The properties in the **Transforms** and **Position** categories are the same as described for a marble map. Refer to the Marble section for details on these properties.

Exercise 17-5

Complete the exercise on the companion website.
www.g-wlearning.com/CAD

Figure 17-33.
A wood map is a procedural, or 3D, map. Note how the pattern matches on adjacent surfaces.

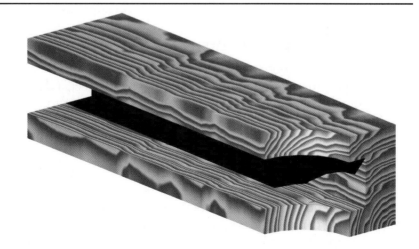

Figure 17-34.
Adjusting the properties of a wood map.

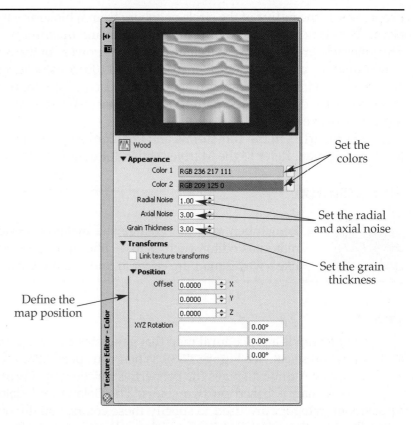

Set the colors

Set the radial and axial noise

Set the grain thickness

Define the map position

Adjusting Material Maps

Simply applying a map to a material property rarely results in a realistic scene when the scene is rendered. The maps usually need to be adjusted to produce the desired results. Maps can be adjusted at the material level or the object level. A combination of these two adjustments is usually required to produce a photorealistic rendering.

Material-Level Adjustments

Material-level adjustments involve changing the properties of the map in the material definition. Map properties are discussed in previous sections. Sometimes, these adjustments may be enough to get the materials looking the way you want them.

Other adjustments become necessary when the same material is applied to more than one object in the same scene. If you make changes at the material level, they affect all objects with that same material. If you need different objects to have different settings, then you will have to make object-level adjustments.

Object-Level Adjustments

Material mapping refers to specifying how a mapped material is applied to an object. When a mapped material is attached to an object, a default set of mapping coordinates, or simply *default mapping*, is used to apply the map to the object.

AutoCAD allows you to adjust mapping at the object level for texture-mapped (2D-mapped) materials. The **MATERIALMAP** command applies a grip tool, or *gizmo*, based on one of four mapping types: planar, box, spherical, or cylindrical. See **Figure 17-35.** The colored edge represents the start and end of the map. For best results, select the mapping type based on the general shape of the object to which mapping is applied. Do not be afraid to experiment with other mapping types, however. Any mapping type can be used on any object, regardless of the object's shape. However, only one mapping type can be applied to an object at any given time.

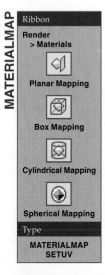

MATERIALMAP

Ribbon
Render
> Materials

Planar Mapping

Box Mapping

Cylindrical Mapping

Spherical Mapping

Type
MATERIALMAP
SETUV

Figure 17-35.
These are the four material map gizmos. From left to right: planar, box, spherical, and cylindrical. The colored edge represents the start and end of the map.

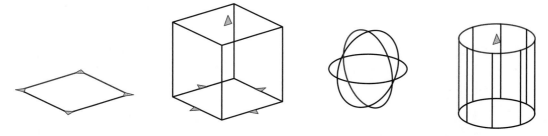

After one of the mapping types is selected, you are prompted to select the faces or objects. You can select multiple objects or faces. After making a selection, the gizmo is placed on the selection set. The command remains active for you to adjust the mapping or enter an option.

Drag the grips on the gizmo to stretch or scale the material. The effects of editing a color map are dynamically displayed if the current visual style is set to display materials and textures. Otherwise, exit the command and render the scene to see the effect of the edit. To readjust the mapping, select the same mapping type and pick the object again. The gizmo is displayed in the same location as before.

The **Move** and **Rotate** options of the command toggle between the move and rotate gizmos. Using the gizmos, you can move and rotate the map on the object. The **Reset** option of the command restores the default mapping to the object. The **Switch mapping mode** option allows you to change between the four types of mapping.

If the command is typed, there is an additional option. The **Copy mapping to** option provides a quick and easy way to apply the changes made to the current object to other objects in the scene. Enter this option, select the face or object to copy from, and then select the faces or objects to copy to. This option is also available if the **Switch mapping mode** option is entered.

For example, look at **Figure 17-36A**. The grain on the stair risers is running vertically when it should run horizontally. First, apply a planar map gizmo to the bottom riser. Next, rotate the mapping 90°, **Figure 17-36B**. Finally, use the **Copy mapping to** option to copy the mapping to the other risers, **Figure 17-36C**.

Mapping coordinates can be applied to procedural-mapped (3D-mapped) materials, but adjusting the coordinates has no effect. This is because the procedural map is generated from mathematical calculations based on the world coordinate system. Procedural-mapped materials must be adjusted at the material level.

PROFESSIONAL TIP

If you are using a reflectivity, self illumination, or bump map and need to adjust it at the object level, you cannot see the effects of the mapping change in the viewport. Apply the same map as a color map. Also, set the visual style to display materials and textures. Then, adjust the object mapping as needed. The edits are dynamically displayed in the viewport. When the image is in the correct location, remove the color map from the material.

Figure 17-36.
Correcting material mapping. A—The grain on the risers runs vertically instead of horizontally. B—Rotating the map with the rotate gizmo. C—The corrected rendering. (Model courtesy of Arcways, Inc., Neenah, WI)

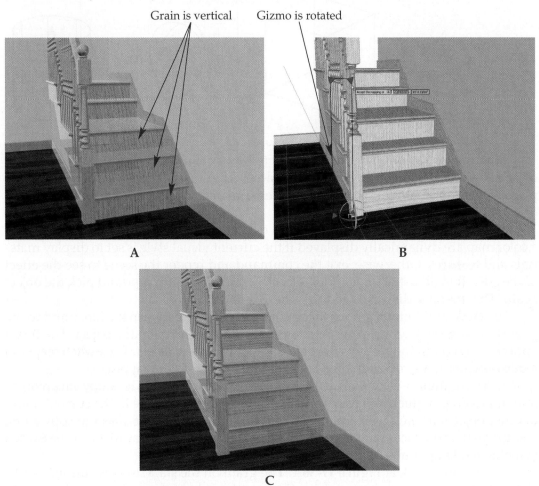

Exercise 17-6

Complete the exercise on the companion website.
www.g-wlearning.com/CAD

Chapter Review

Answer the following questions. Write your answers on a separate sheet of paper or complete the electronic chapter review on the companion website.
www.g-wlearning.com/CAD

1. Define *material*.
2. Define *materials library*.
3. What is the My Materials section of the **Libraries** area of the materials browser used for?
4. Describe how to attach a material using the materials browser.
5. How can materials be attached to layers?
6. By default, which material is attached to newly created objects?
7. Which material is used as the base material for creating new materials?
8. Name the 12 shapes that can be used to display the material in the preview in the materials editor.
9. How do you know if a material in the materials browser is being used in the drawing?
10. How can the name of an existing material be changed?
11. Name the 14 basic material types.
12. When creating a material to look like plastic, what is the benefit of using the plastic material type instead of starting from scratch using the generic material?
13. In the **Reflectivity** category of the generic material, there are Direct and Oblique properties. What are these used for?
14. An image mapped to a material property will normally repeat itself to cover the entire object. What is this called and how do you turn it off?
15. Describe the difference between a transparent material and a translucent material.
16. How much illumination does a self-illuminated material add to a scene?
17. How is a marble material created?
18. Explain how black and white areas of a map applied to the Transparency property affect the transparency of a material.
19. Explain what the nodes in the ramp of a gradient map are for.
20. Name the four types of mapping available for adjusting texture maps at the object level.

Drawing Problems

1. In this problem, you will create a scene with basic 3D objects, attach materials to the objects, and adjust the settings of the materials.
   A. Start a new drawing and set the units to architectural.
   B. Draw a 15' × 15' planar surface to represent the floor.
   C. Draw two boxes to represent two walls. Make the boxes 15' × 4' × 9'. Position them to form a 90° corner. Alternately, you can draw a polysolid of the same dimensions.
   D. Draw a R2' × 5'H cone in the center of the room.
   E. Open the materials browser and locate a material similar to the one shown on the floor. Attach this material to the floor.
   F. Locate an appropriate material for the wall and attach it.
   G. Locate an appropriate material for the cone and attach it.
   H. Turn on the sun and adjust the time to create good shadows. Refer to Chapter 16 for an introduction to sun settings.
   I. Render the scene. Save the rendering as P17_01.jpg.
   J. Save the drawing as P17_01.

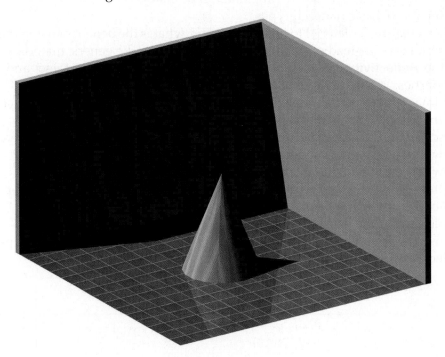

2. In this problem, you will attach materials to the objects in an existing drawing and render the scene.
   A. Open the drawing P16_02 from Chapter 16. If you did not complete this problem, do so now. Save the drawing as P17_02.
   B. Attach the materials of your choice to the objects in the scene. Do not be restricted by the names of the materials. For example, a concrete material may be suitable for foliage or even carpet with a simple color change. Be creative.
   C. If the items in the scene were inserted using the Autodesk Seek website, they may be blocks with nested layers. Instead of exploding the blocks, use the **MATERIALATTACH** command and attach materials to the layers on which the nested objects reside.
   D. Turn on the sun and adjust the time to create good shadows. Refer to Chapter 16 for an introduction to sun settings.
   E. Render the scene. Save the rendering as P17_02.jpg.
   F. Save the drawing.

3. In this problem, you will create custom wood and marble materials.
   A. Start a new drawing and save it as P17_03.
   B. Draw two 5 × 5 × 5 boxes and position them near each other. Using other primitives, cut notches and holes in the boxes. The boxes will be used to test the custom materials.
   C. In the materials editor, create two new materials. Name one Wood-*your initials* and the other Marble-*your initials*.
   D. Attach the wood material to one of the boxes and the marble material to the other box.
   E. Render the scene and make note of the wood grain and marble veins.
   F. Use the texture editor to change the properties of the materials.
   G. Render the scene again and make note of the changes. Using the **Render Region** button on the **Render** panel in the **Render** tab of the ribbon can save time when testing material changes.
   H. When you are satisfied with the materials, save the rendering as P17_03.
   I. Save the drawing.

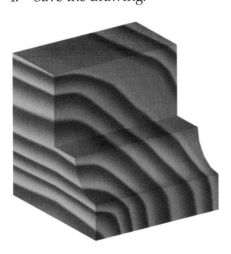

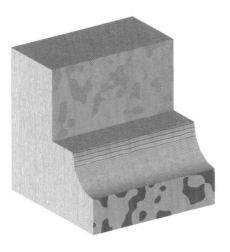

4. In this problem, you will create a bitmap and use it as a transparency map and a bump map.
   A. Draw a rectangle with an array of smaller rectangles inside of it as shown. Sizes are not important and the pattern can be varied if you like.
   B. Display a plan view of the rectangles. Then, copy all of the objects to the Windows Clipboard by pressing [Ctrl]+[C] and selecting the objects.
   C. Launch Windows Paint. Then, paste the objects into the blank file. Notice how the AutoCAD background outside of the large rectangle is also included.
   D. Use the select tool (rectangle) in Paint to draw a window around the large rectangle created in AutoCAD and the smaller rectangles within it. Copy this to the Windows Clipboard by pressing [Ctrl]+[C].
   E. Start a new Paint file without saving the current one and paste the image from the Clipboard into the new blank file. Now, the unwanted AutoCAD background is no longer displayed. If needed, change the small rectangles to black and the lattice to white using the tools in Paint. The colors should be the reverse of what is shown below. Then, save the image file as P17_04.bmp and close Paint.
   F. In AutoCAD, draw a solid box of any size.
   G. Using the materials editor, create a new material.
   H. Assign the P17_04.bmp image file you just created as a transparency map. Adjust the map so that it is scaled to fit to the object.
   I. Render the scene and note the effect.
   J. Turn off the transparency property by unchecking the check box in the category name.
   K. Assign the P17_04.bmp image file as a bump map. Adjust the map so that it is scaled to fit to the object.
   L. Render the scene and note the effect.
   M. Save the drawing as P17_04.

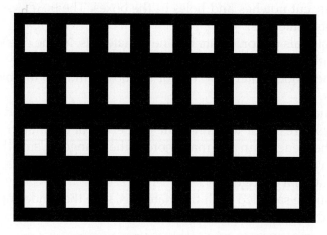

Lighting

Learning Objectives

After completing this chapter, you will be able to:

✓ Describe the types of lighting in AutoCAD.
✓ List the user-created lights available in AutoCAD.
✓ Change the properties of lights.
✓ Generate and modify shadows.
✓ Add a background to your scene and control its appearance.

In Chapter 15, you were introduced to lighting. You learned how to adjust lighting by turning off the default lights and adding sunlight. In this chapter, you will learn all about the lights available in AutoCAD. You will learn lighting tips and tricks to help make the scene look its best.

Types of Lights

Ambient light is like natural light just before sunrise. It is the same intensity everywhere. All faces of the object receive the same amount of ambient light. Ambient light cannot create highlights, nor can it be concentrated in one area. AutoCAD does not have an ambient light setting. Instead, it relies on indirect illumination, which is discussed in Chapter 19.

A *point light* is like a lightbulb. Light rays from a point light shine out in all directions. A point light can create highlights. The intensity of a point light falls off, or weakens, over distance. Other programs, such as Autodesk 3ds Max®, may call these lights *omni lights*. A *target point light* is the same as a standard point light except that a target is specified. The illumination of the target point light is directed toward the target.

A *distant light* is a directed light source with parallel light rays. This acts much like the Sun. Rays from a distant light strike all objects in your model on the same side and with the same intensity. The direction and intensity of a distant light can be changed.

A *spotlight* is like a distant light, but it projects in a cone shape. Its light rays are not parallel. A spotlight is placed closer to the object than a distant light. Spotlights have a hotspot and a falloff. The light from a standard spotlight is directed toward a target. A *free spotlight* is the same as a standard spotlight, but without a target.

A *weblight* is a directed light that represents real-world distribution of light. The illumination is based on photometric data that can be entered for each light. The light from a standard weblight is directed toward a target. A *free weblight* is the same as a standard weblight, but without a target point.

There are several factors that affect how a light illuminates an object. These include the angle of incidence, reflectivity of the object's surface, and the distance that the light is from the object. In addition, the ability to cast shadows is a property of light. Shadows are discussed in detail later in this chapter.

Angle of Incidence

AutoCAD renders the faces of a model based on the angle at which light rays strike the faces. This angle is called the *angle of incidence.* See **Figure 18-1**. A face that is perpendicular to light rays receives the most light. As the angle of incidence decreases, the amount of light striking the face also decreases.

Reflectivity

The angle at which light rays are reflected off of a surface is called the *angle of reflection.* The angle of reflection is always equal to the angle of incidence. Refer to **Figure 18-1**.

The "brightness" of light reflected from an object is actually the number of light rays that reach your eyes. A surface that reflects a bright light, such as a mirror, is reflecting most of the light rays that strike it. The amount of reflection you see is called the *highlight.* The highlight is determined by the angle from the viewpoint relative to the angle of incidence. Refer to **Figure 18-1**.

The surface quality of the object affects how light is reflected. A smooth surface has a high specular factor. The *specular factor* indicates the number of light rays that have the same angle of reflection. Surfaces that are not smooth have a low specular factor. These surfaces are called *matte.* Matte surfaces *diffuse,* or "spread out," the light as it strikes the surface. This means that few of the light rays have the same angle of reflection. **Figure 18-2** illustrates the difference between matte and high specular finishes.

Figure 18-1.
The amount of reflection, or highlight, you see depends on the angle from which you view the object.

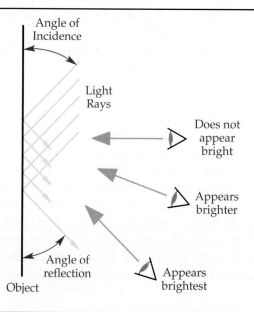

Figure 18-2.
Matte surfaces
produce diffuse
light. This is also
referred to as having
a low specular
factor. Shiny surfaces
evenly reflect light
and have a high
specular factor.

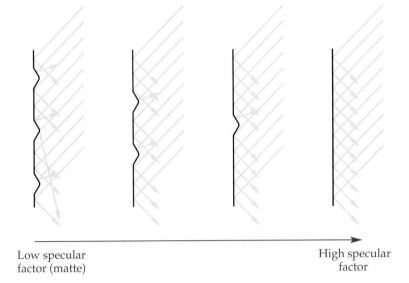

Low specular
factor (matte)

High specular
factor

Surfaces can also vary in *roughness*. Roughness is a measure of the polish on a surface. This also affects how diffused the reflected light is.

PROFESSIONAL TIP

When rendering metal and other shiny materials, it helps to have a bright object nearby to create some highlights on the object. Adding a simple box to the left and right of the object (out of the view, however) and applying a self-illuminating material to the box will create some nice highlights. Photographers use this trick in the studio.

Hotspot and Falloff

A spotlight produces a cone of light. The *hotspot* is the central portion of the cone, where the light is brightest. See **Figure 18-3**. The *falloff* is the outer portion of the cone,

Figure 18-3.
The hotspot of a
spotlight is the
area that receives
the most light. The
smaller cone is
the hotspot. The
falloff receives light,
but less than the
hotspot. The larger
cone is the falloff.

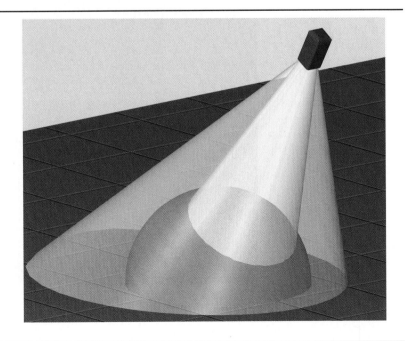

where the light begins to blend to shadow. The hotspot and falloff of a spotlight are not affected by the distance the light is from an object. Spotlights are the only lights with hotspot and falloff properties.

Attenuation

The farther an object is from a point light or spotlight, the less light that reaches the object. See **Figure 18-4**. The intensity of light decreases over distance. This decrease is called *attenuation*. All lights in AutoCAD, except distant lights, have some kind of attenuation. Often, attenuation is called *falloff* or *decay*. However, do not confuse this with the falloff of a spotlight, which is the outer edge of the cone of illumination. The following attenuation settings are available in AutoCAD.

- **None.** Applies the same light intensity regardless of distance. In other words, no attenuation is calculated.
- **Inverse Linear.** The illumination of an object decreases in inverse proportion to the distance. For example, if an object is two units from the light, it receives 1/2 of the full light. If the object is four units away, it receives 1/4 of the full light.
- **Inverse Squared.** The illumination of an object decreases in inverse proportion to the square of the distance. For example, if an object is two units from the light, it receives $(1/2)^2$, or 1/4, of the full light. If the object is four units away, it receives $(1/4)^2$, or 1/16, of the full light. As you can see, attenuation is greater for each unit of distance with the **Inverse Squared** option than with the **Inverse Linear** option.

PROFESSIONAL TIP

The intensity of the Sun's rays does not diminish from one point on Earth to another. They are weakened by the angle at which they strike Earth. Therefore, since distant lights are similar to the Sun, attenuation is not a factor with distant lights.

Figure 18-4.
Attenuation is the intensity of light decreasing over distance. Attenuation has been turned on in this scene.

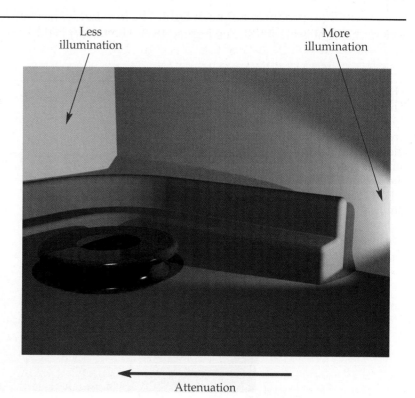

Less illumination

More illumination

Attenuation

AutoCAD and Its Applications—Advanced

AutoCAD Lights

AutoCAD has three types of lighting: default lighting, sunlight with or without sky illumination, and user-created lighting. ***Default lighting*** is the lighting automatically available in the scene. It is composed of two light sources that evenly illuminate all surfaces. As the viewpoint is changed, the light sources follow to maintain an even illumination of the scene. There is no control over default lighting and it must be shut off whenever one of the other types of lighting is used.

As you saw in Chapter 16, ***sunlight*** may be added to any scene. AutoCAD uses a distant light to simulate the parallel rays of the Sun. The date and time of day can be adjusted to create different sunlight illumination. ***Sky illumination*** may also be added with sunlight to simulate light bouncing off of objects in the scene and particles in the atmosphere. This helps create a more-natural feel.

User-created lighting results when you add AutoCAD light objects to the drawing. There are four types of user-created lights: distant light, weblight, point light, and spotlight. See **Figure 18-5**. A distant light is a directed light source with parallel light rays. A weblight is a directional point light containing light intensity (photometric) data. A point light is like a lightbulb with light rays shining out in all directions. A spotlight is like a distant light, but it projects light in a cone shape instead of having parallel light rays.

When created, point lights, weblights, and spotlights are represented by ***light glyphs***, or icons, in the drawing. To suppress the display of light glyphs, pick the **Light Glyph Display** button in the expanded area of the **Lights** panel on the **Render** tab of the ribbon. The button is blue when light glyphs are displayed. The default lights, sun, and distant lights are not represented by glyphs.

In this section, you will learn how to add lights. You will also learn how to adjust the various properties of sunlight and AutoCAD light objects. The tools for working with lights can be accessed using the command line, the tool palettes, and **Lights** panel on the **Render** tab of the ribbon. See **Figure 18-6**.

So that you will never work with a completely dark scene, default lighting is applied in the viewport and to the rendering if no other lights are added. In order for your lights to be applied, you must switch between default lighting and user lighting. To do this, pick the **Default Lighting** button in the expanded area of the **Lights** panel on the **Render** tab of the ribbon. This button toggles the lighting between default lighting and whatever lights are available in the scene. When default lighting is on, the button is blue. When off, the button is not highlighted. If you elected for AutoCAD to do so, the default lighting will be automatically shut off when sunlight is turned on or a user-created light is added to the scene.

Ribbon
Render > Lights
Light Glyph Display
Type
LIGHTGLYPHDISPLAY

LIGHTGLYPHDISPLAY

Figure 18-5.
AutoCAD has four types of user-created lights: distant, point, weblight, and spotlight. A weblight is really a targeted point light. It projects in all directions, but may be predominant in one direction.

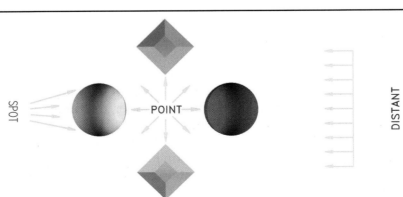

Figure 18-6.
The tools for adding and controlling lights.

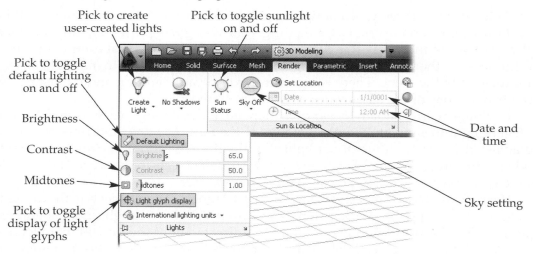

Figure 18-6.
The tools for adding and controlling lights.

Lighting Units

There are three types of *lighting units* available in AutoCAD: standard (generic), international (SI), and US customary (American). Generic lighting is the type of lighting that was used in AutoCAD prior to AutoCAD 2008. This lighting provides very nice results, but the settings are not based on any real measurements. International or US customary lighting is called photometric lighting. *Photometric lighting* is physically correct and attenuates at the square of the distance from the source. For more accuracy, photometric data files can be imported from lighting manufacturers.

The **LIGHTINGUNITS** system variable sets which type of lighting is used. A setting of 0 means that standard (generic) lighting is used. However, for more realistic lighting, it is recommended that photometric lighting be used. Entering a value of 1 specifies US customary (American) lighting units. A setting of 2 specifies international lighting units. This is the default setting. A setting of 1 or 2 results in photometric lighting. The only difference between a setting of 1 and 2 is that US customary (American) units are displayed as *foot-candles (fc)* and international units are displayed as *lux (lx)*.

Sunlight

To turn sunlight on or off, pick the **Sun Status** button in the **Sun & Location** panel on the **Render** tab of the ribbon. This button is blue when sunlight is on. Sunlight can also be turned on or off in the **Sun Properties** palette, which is discussed later in this chapter. Sunlight is not represented by a light glyph. The date, time, and geographic location can also be set in the **Sun & Location** panel on the **Render** tab of the ribbon. These properties determine how the scene is illuminated by the sun.

To change the current date, drag the **Date** slider left or right. Sunlight must be on for the slider to be enabled. As you drag the slider, the date is displayed on the right-hand end of the slider. The time is changed in the same manner as the date. Drag the **Time** slider left or right. As you drag the slider, the time is displayed on the right-hand end of the slider.

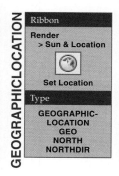

The location of the scene can be set to an actual geographic location. This is important if you want to replicate the lighting and shadows of an actual site. Picking the **Set Location** button on the **Sun & Location** panel on the **Render** tab of the ribbon opens a dialog box titled **Geographic Location – Define Geographic Location**. See **Figure 18-7**. The first option in this dialog box allows you to specify the location by importing a KML or KMZ file. *KML* stands for Keyhole Markup Language. A KML file contains

Figure 18-7.
This dialog box provides the options for defining the geographic location of the model.

Pick to import a KML file

Pick to import a location from Google Earth

Pick to set the location in the **Geographic Location** dialog box

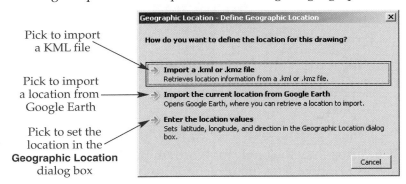

latitude, longitude, and, sometimes, other data to pinpoint a location on Earth. A KMZ file is a zipped KML file. The second option is to import the current location from Google Earth. In order for this to work, Google Earth must be installed and open with the location selected. The third option opens the **Geographic Location** dialog box. See **Figure 18-8**.

When a location is imported from Google Earth, you are asked to select a point in the drawing for the location. After picking a point in the drawing, you are asked for the North vector. Pick a point that should be in the direction of North. Once this is set, a *geographic marker* is added at the point selected for the location. This marker looks like a red and white thumbtack. Its visibility is controlled with the **GEOMARKERVISIBILITY** system variable. See **Figure 18-9**.

To set the location in the **Geographic Location** dialog box, enter the latitude and longitude. You can select to use either decimal or degrees/minutes/seconds values for these settings. Also, pick a time zone in the **Time Zone:** drop-down list and set the XYZ coordinates and elevation for the location of the geographic marker. The WCS origin

Figure 18-8.
The **Geographic Location** dialog box is used to input geographic location data.

Latitude setting

Longitude setting

Pick to select the location on a map

Direction of North in the drawing

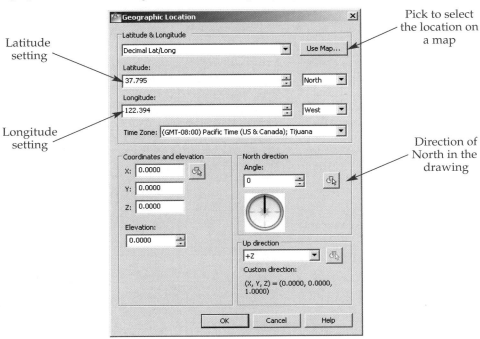

Figure 18-9.
The geographic marker is an icon that represents the point in the drawing selected for the geographic location of the scene.

is the default location. You can specify the angle for North in the **Angle:** text box or by picking points in the drawing window. The "up direction" is normally +Z, but you can change it to –Z. You can also set it to anything you want if you select **Custom Direction** in the drop-down list in the **Up direction** area.

The quickest way to set the location is to select it on a map. Pick the **Use Map...** button to open the **Location Picker** dialog box. See **Figure 18-10.** The crosshairs on the map indicate the current location. By default, picking a point on the map selects the nearest big city. Many other cities are available in the **Nearest City:** drop-down list. The **Region:** drop-down list is used to change the map to one of eight different areas of the world or the entire world. You can pick the time zone in the **Time Zone:** drop-down list, but AutoCAD attempts to match the time zone to the city you select when you pick the **OK** button to close the dialog box. An alert box is displayed that gives you the option to accept the new time zone or go back and pick a different one.

Once you have selected a location, close the **Geographic Location** dialog box. The geographic marker is automatically placed in the drawing at the specified coordinate location.

The properties of the sun are set in the **Sun Properties** palette, **Figure 18-11.** The **SUNPROPERTIES** command opens this palette. You can also pick the dialog box launcher button at the lower-right corner of the **Sun & Location** panel on the **Render** tab of the ribbon. The sun can be turned on or off using the **Sun Properties** palette. The date, time, and time zone can also be changed in the palette. These settings are

Figure 18-10.
The **Location Picker** dialog box provides a quick way to set the location. Pick a location on the map or select it in the drop-down lists.

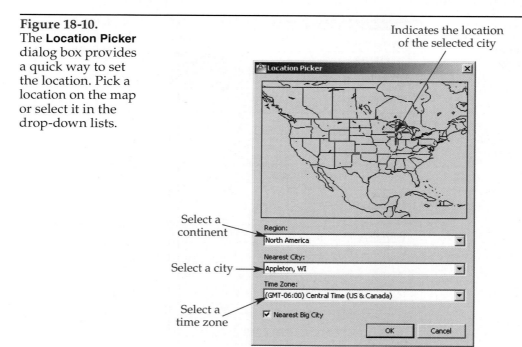

AutoCAD and Its Applications—Advanced

Figure 18-11.
The **Sun Properties** palette.

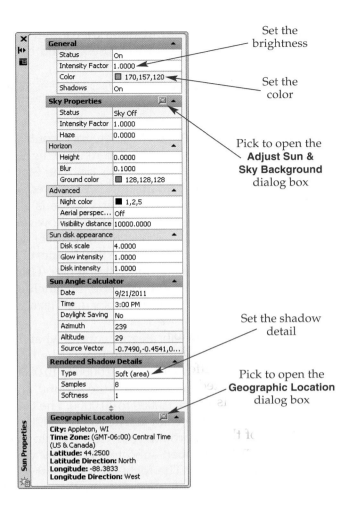

Set the
brightness

Set the
color

Pick to open the
**Adjust Sun &
Sky Background**
dialog box

Set the shadow
detail

Pick to open the
Geographic Location
dialog box

the same as previously discussed. There are other properties of the sun that are only available in the **Sun Properties** palette. The categories in this palette are discussed in the next sections.

General

The Intensity Factor property in the **General** category determines the brightness of the sun. Setting this property to zero, in effect, turns off sunlight. Increasing the property makes the sunlight brighter. The maximum value for the property is determined by the capabilities of your computer.

The Color property in the **General** category is used to set the color of the sun, if photometric lighting is off. By default, sunlight is white (true color 255, 255, 255) when photometric lighting is off. To change the color, pick the drop-down list and choose a new color. If you pick the Select Color... entry, the **Select Color** dialog box is displayed for selecting a color. Sunlight can be changed to any color, but be aware that changing the color of the light may drastically alter the appearance of a scene. This is especially true if materials are attached to objects in the drawing. The color of sunlight is often set to a very light blue for an outdoor scene to help convey a bright blue sky.

The Shadows property in the **General** category determines whether or not the sun casts shadows. The property is either on or off. Shadows are discussed in detail later in this chapter.

Photometric lighting must be off (**LIGHTINGUNITS** = 0) in order to set the sun color. If photometric lighting is on, the Color property is disabled and set to a preselected color based on the geographic location, date, and time.

Sky Properties

The settings in the **Sky Properties** category control the sky. The sky is used in conjunction with sunlight to generate more-realistic lighting in the scene. This category is only displayed if photometric lighting is on. The Status property determines if the sky effect is off, if just the sky background is displayed, or if both the background and illumination are active. The Intensity property value is a multiplier for the illumination provided by the sky. The Haze property controls how the sky illumination is diffused. The value can range from 0.0000 to 15.0000. The preset sun color is affected by this value.

Notice the button in the title bar of the **Sky Properties** category. Picking this button opens the **Adjust Sun & Sky Background** dialog box. This dialog box contains the same settings found in the **Sun Properties** palette, but includes a preview of the sun disk. Use this dialog box to preview Intensity Factor and Haze property changes, as discussed later in this chapter.

The settings in the **Horizon** subcategory control what the horizon looks like and where it is located. Changing the Height property moves the horizon up or down. The default value is 0.0000 and values can range from -10.0000 to 10.0000. The Blur property determines how much the horizon is blurred between the ground and the sky. The default value is 0.1000 and values can range from 0.0000 to 10.0000. The Ground color property controls the color of the ground. The default color is true color 128,128,128, which is a medium gray. To see how this affects your scene, rotate your viewpoint to make the horizon visible.

The settings in the **Advanced** subcategory allow you to control some of the more artistic settings of your scene. The Night color property sets the color of the night sky. This is only visible if sky illumination is turned on. In a city, the night sky may have an orange tint to it because of the streetlamps in the city. However, in the country, the night sky is nearly black due to the lack of artificial illumination. The Aerial perspective property determines whether aerial perspective is applied. This is a way of simulating distance between the camera and the sky/background. The setting is either on or off. Aerial perspective is controlled by the Visibility distance property. This sets the distance from the camera at which haze obscures 10% of the objects in the background. This is a very useful tool for creating the illusion of depth in a scene. The default value is 10000.0000, but the value can range from 0.0000 to whatever is needed. See **Figure 18-12**.

Normally, lights do not appear in the rendered scene at all, only the illumination provided by the lights. The settings in the **Sun disk appearance** subcategory control what the sun looks like in the sky, or on the background. The Disk scale property sets the size of the sun, or solar disk, as it appears on the background. The default value is 4.0000 and values can range from 0.0000 to 25.0000. The Glow intensity property determines the size of the glowing halo around the sun in the sky. The default value is 1.0000 and values can range from 0.0000 to 25.0000. The Disk intensity property controls the brightness of the sun on the background. The default value is 1.0000 and values can range from 0.0000 to 25.0000.

PROFESSIONAL TIP

The sky at night is always lighter than the Earth unless there are other light sources in the scene, such as streetlamps.

Figure 18-12.
In this scene, the red box is about 15 units from the viewer, the yellow cylinder is 200 units away, and the blue cone is 250 units away. Notice how the lower Visibility distance settings create the illusion of depth.

Visibility distance setting = 1000

A

Visibility distance setting = 100

B

Visibility distance setting = 10

C

Sun Angle Calculator

The settings in the **Sun Angle Calculator** category determine the angle of the sun in relationship to the XY plane. The Date and Time properties, discussed earlier, can be controlled from the **Sun & Location** panel on the **Render** tab of the ribbon. This category in the **Sun Properties** palette also includes the Daylight Saving property. This property is used to turn daylight saving on or off. The Azimuth, Altitude, and Source Vector properties display the current settings, but are read-only in the **Sun Properties** palette. These values are automatically calculated based on the settings in the **Geographic Location** dialog box.

Rendered Shadow Details

The Type property in the **Rendered Shadow Details** category determines the type of shadow cast by the sun, if shadows are cast. When the property is set to Sharp, raytraced shadows are cast. These shadows have sharp edges. Raytracing produces accurate shadows, but rendering may take longer. When the property is set to either Soft (mapped) or Soft (area), shadow-mapped shadows are cast. These shadows have soft edges. Shadow-mapped shadows may be calculated more quickly than raytraced shadows, but the resulting shadows are less precise. In addition, soft shadows do not work with transparent surfaces like windows. When the Type property is set to Soft (mapped), two additional settings are available in the category:

- **Map Size.** This property determines the number of subdivisions, or samples, used to create the shadow. By default, shadow maps are 256×256 pixels in size. If shadows look grainy, increasing this setting will make them look better.
- **Softness.** This property determines the sharpness of the shadow's edge. The value ranges from 1 to 10. The higher the value, the softer (less sharp) the edge of the shadow.

When the Type property is set to Soft (area), two additional settings are available:

- **Samples.** This property sets the number of samples used on the solar disk. The value can be from 0.0000 to 1000.0000.
- **Softness.** This property determines the sharpness of the shadow's edge, as described for the Soft (mapped) type.

When the Type property is set to Sharp, the other settings are read only and not applied.

 NOTE With photometric lighting on (**LIGHTINGUNITS** = 1 or 2), only the Soft (area) selection is available in the Type property drop-down list.

Geographic Location

The **Geographic Location** category at the bottom of the **Sun Properties** palette displays the current geographic location settings. Changes cannot be made here, but picking the **Launch Geographic Location** button in the category's title bar opens the **Geographic Location** dialog box for making changes, as described earlier.

 **Exercise 18-1**

Complete the exercise on the companion website.
www.g-wlearning.com/CAD

Distant Lights

Distant lights are user-created lights that have parallel light rays. See **Figure 18-13**. When a distant light is created, the location from where the light is originating must be specified along with the direction of the light rays. Distant lights are not represented in the drawing by light glyphs. Distant lights are often used to create even, uniform, overhead illumination, such as the illumination you would encounter in an office situation. The distance of the objects in the scene to the distant light has no effect on the intensity of the illumination. Distant lights do not attenuate. A distant light can also be used to simulate sunlight without having to set up a time and location. However, the light must be manually moved to change the illumination effect.

The **DISTANTLIGHT** or **LIGHT** command is used to create a distant light. You are first prompted to specify the direction from which the light is originating or to enter the **Vector** option. If you pick a point, it is the location of the light. Next, you are prompted to specify the point to which the light is pointing. This is simply the location where the light is aimed.

If you enter the **Vector** option instead of picking a "from" point, you must type the endpoint coordinates (in WCS units) of the direction vector. The light will point from the WCS origin to the entered endpoint. This option is not often used.

Ribbon

Render
> Lights

Distant

Type

DISTANTLIGHT
LIGHT

DISTANTLIGHT

Figure 18-13.
This example shows the use of a distant light to simulate sunlight shining through a window. Notice how the edges on the shadows of the grill are parallel.

After the light location and direction are determined, several other options are available. You can name the light, set the intensity, turn the light on or off, determine if and how shadows are cast, and set the light color.

AutoCAD provides a default name for new lights based on the type of light and a sequential number, such as Distantlight1, Distantlight2, and so on. It is a good idea to provide a meaningful name for a light. This is especially true if there are other lights in the scene. To rename a light, enter the **Name** option. Then, type the name of the light and press [Enter].

To set the brightness of the light, enter the **Intensity** option. Then, type a value and press [Enter]. The default value is 1.00. Setting the value to 0.00, in effect, turns off the light. The maximum value depends on the capabilities of your computer. This option is called **Intensity factor** if photometric lighting is enabled.

When a light is created, it is on. To turn the light off, enter the **Status** option. Then, change the setting to Off.

The **Photometry** option is available when photometric lighting is active. It controls the luminous qualities of visible light sources. Once you enter the **Photometry** option, you can select one of three options:

- **Intensity.** This is the power of the light source. The value is specified in candelas (cd).
- **Color.** This is the color of the light source. The value can be changed by typing in a name (to get a list of color names, enter ?) or by specifying the Kelvin temperature value (k).
- **Exit.** This exits the command option.

The **Shadow** option is used to determine if and how shadows are cast by the distant light. To turn off shadow casting, enter the option and select the Off setting. The Sharp setting creates raytraced shadows. The Softmapped setting casts shadow-mapped shadows. Shadows are discussed later in this chapter.

By default, new distant lights cast white light (true color 255, 255, 255). To change the color of the light, enter the **Color** option. This option is called **Filter Color** if photometric lighting is enabled. To specify a new true color, simply specify the RGB values and press [Enter]. To specify a color based on hue, saturation, and luminance (HSL), enter the **Hsl** option and specify the values. To enter an AutoCAD color index (ACI) number, enter the **Index color** option and specify the ACI number. To specify a color book color, enter the **Color Book** option and then specify the name of the color book followed by the name of the color.

Once all settings for the distant light have been made, use the **Exit** option to end the command and create the light. Do not press [Esc] to end the command. Doing so actually cancels the command and the light is not created.

Exercise 18-2

Complete the exercise on the companion website.
www.g-wlearning.com/CAD

Point Lights

Point lights are user-created lights that have light rays projecting in all directions. See **Figure 18-14**. When a point light is created, its location must be specified. Since point lights illuminate in all directions, there is no "to" location for a point light. A light glyph represents point lights in the drawing. See **Figure 18-15**. Point lights can be set to attenuate. In this case, the distance of the objects in the scene to the point light affects the intensity of the illumination.

Figure 18-14.
A—A point light is placed inside of the lamp fixture. Notice how the light projects in all directions. B—When shadow casting is turned on for the light, the lampshade blocks the light from illuminating objects below the shade.

A

B

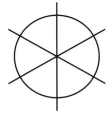

Figure 18-15.
This is the light glyph for a point light.

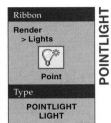

The **POINTLIGHT** or **LIGHT** command is used to create a point light. You are first prompted to specify the location of the point light. Once the location is established, several options for the light are available. You can name the light, set the intensity, turn the light on or off, adjust the photometry settings, determine if and how shadows are cast, set the attenuation, and set the light color. The **Name**, **Intensity** (or **Intensity Factor**), **Status**, **Photometry**, **Shadow**, and **Color** (or **Filter Color**) options work the same as the corresponding options for a distant light. However, additional photometry settings are available when photometric lighting is active. The **Intensity** option provides additional options for setting the power of the light source. The **Flux** option can be used to specify the power in lumens (lm). The **Illuminance** option can be used to specify the illuminance in lux (lx) or foot-candles (fc), depending on the current lighting units.

The **Attenuation** option is used to set attenuation for the point light when photometric lighting is off. When this option is selected, five more options are available:

- **Attenuation Type**
- **Use Limits**
- **Attenuation Start Limit**
- **Attenuation End Limit**
- **Exit**

The **Exit** option returns you to the previous prompt.

The **Attenuation Type** option is used to turn attenuation on and off and to set the type of attenuation. To turn attenuation off, select the option and then enter None. To turn attenuation on, select the option and then enter either **Inverse Linear** or **Inverse Squared**. Attenuation is discussed in detail in the Attenuation section in this chapter.

The **Use Limits** option determines if the attenuation of the light has a beginning and an end. When this option is set to Off, attenuation starts at the light and ends when the illumination reaches zero. When set to On, attenuation begins at the starting limit and ends at the ending limit.

To set the starting point for attenuation, enter the **Attenuation Start Limit** option. Then, specify the distance from the point light where attenuation will begin. The full intensity of the light provides illumination up to this point. From this point to the attenuation end limit, the light falls off.

To set the point where the illumination attenuates to zero, enter the **Attenuation End Limit** option. Then, specify the distance from the point light where the illumination is zero. Beyond this point, AutoCAD does not calculate the effect of the light.

Target Point Lights

The target point light is like a regular point light except that the command starts by asking for a *source* location and a *target* location. The rest of the options are the same. You can type TARGETPOINT to access this command or pick the **Targetpoint** option in the **LIGHT** command.

POINTLIGHT

Ribbon

Render > Lights

Point

Type

POINTLIGHT LIGHT

PROFESSIONAL TIP

A point light may be used as an incandescent lightbulb, such as the lightbulb in a table lamp. Most of these lightbulbs cast a yellow light. In these cases, you may want to change the color of the light to a light yellow. Compact fluorescent lightbulbs may cast white, light blue, or light yellow light. LED-based lights cast a color based on the color of the LED. Other colors can be used to give the impression of heat or colored lights.

Exercise 18-3

Complete the exercise on the companion website.
www.g-wlearning.com/CAD

Spotlights

Spotlights are user-created lights that have light rays projecting in a cone shape in one direction. See **Figure 18-16**. When a spotlight is created, the location from where the light is originating must be specified, along with the direction in which the light rays travel. A light glyph represents spotlights in the drawing. See **Figure 18-17**. Spotlights can be set to attenuate. In this case, the distance of the objects in the scene to the spotlight affects the intensity of the illumination.

The **SPOTLIGHT** or **LIGHT** command is used to create a spotlight. You are first prompted to specify the location of the light. This is from where the light rays will originate.

SPOTLIGHT

Ribbon
Render
> Lights

Spot

Type
SPOTLIGHT
LIGHT

Figure 18-16.
Three spotlights are used to simulate recessed ceiling lights. Notice how the light from each spotlight projects in a cone.

Figure 18-17.
This is the light glyph for a spotlight.

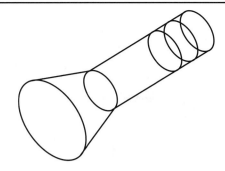

Next, you are prompted for the target location. This is simply the location where the light is aimed.

Once the location and target are established, several options for the light are available. The **Name**, **Intensity** (or **Intensity Factor**), **Status**, **Photometry**, **Shadow**, **Attenuation**, and **Color** (or **Filter Color**) options work the same as the corresponding options for a point light. However, a spotlight also has hotspot and falloff settings.

The *hotspot* is the inner cone of illumination for a spotlight. Refer to **Figure 18-18**. This is measured in degrees. To set the hotspot, enter the **Hotspot** option and then specify the number of degrees for the hotspot.

The *falloff*, not to be confused with attenuation, is the outer cone of illumination for a spotlight. Like the hotspot, it is measured in degrees. The falloff value must be greater than or equal to the hotspot value. It cannot be less than the hotspot value. In practice, the falloff value is often much greater than the hotspot value. To set the falloff, enter the **Falloff** option and then specify the number of degrees for the falloff.

Once the light is created and you select it in the viewport, grips are displayed. If you hover the cursor over a grip, a tooltip is displayed indicating what the grip will modify. You can use grips to change the location of the spotlight and its target, the hotspot, and the falloff. If you hover over a falloff or hotspot grip, the current angle is displayed in the wireframe cone (if dynamic input is on).

Free Spotlight

A free spotlight is like a standard spotlight except that you do not specify a target, only the light location. The rest of the options are the same. You can type FREESPOT to access this command or pick the **Freespot** option in the **LIGHT** command. When created, a free spotlight points down the Z axis (from positive to negative) of the current UCS. A free spotlight may be easier to control than a standard spotlight because you do not have to worry about the target point. If you want to change the angle or position of the light, use the **3DMOVE** and **ROTATE3D** commands.

Figure 18-18.
Hotspot and falloff for a spotlight are angular measurements.

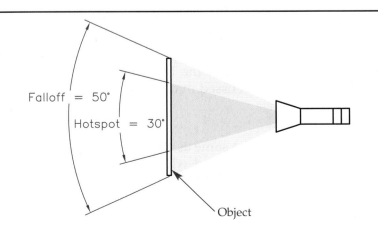

Falloff = 50°

Hotspot = 30°

Object

Exercise 18-4

Complete the exercise on the companion website.
www.g-wlearning.com/CAD

Weblight

A photometric weblight is really just a targeted point light. The difference is that a weblight provides a more precise representation of the light. Real-world lights appear to evenly illuminate from their source, but, in reality, the shape of the light, the material used in its manufacture, and other factors make all lights distribute their energy in different ways. These data are provided by light manufacturers in the form of light distribution data. Light distribution data can be loaded into the **Photometric Web** subcategory of the **General** category in the **Properties** palette when the light is selected. Select the Web file property, then pick the browse button (...) and select an IES file. IES stands for Illuminating Engineering Society. AutoCAD's online documentation has additional information on IES files.

Think of the web of a weblight as a spherical cage surrounding the light source. If the light is evenly distributed from its source, the cage is a true sphere. In actuality, a light may emit more light energy in the X direction than in the Z direction. In this case, the cage bulges out further in the X direction. The position of this bulge may be important to the illumination of the scene and you may need to rotate the web to apply more or less light in one direction or another.

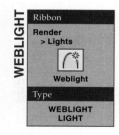

The **WEBLIGHT** or **LIGHT** command is used to create a weblight. You are prompted for source and target locations. The **Name**, **Intensity Factor**, **Status**, **Photometry**, **Shadow**, and **Filter Color** options work the same as the corresponding options for the previously discussed lights. However, weblights have an additional **Web** option. When this option is activated, these options are presented:

- **File.** Allows you to select an IES file.
- **X.** Rotates the web around the X axis.
- **Y.** Rotates the web around the Y axis.
- **Z.** Rotates the web around the Z axis.
- **Exit.** Exits the **Web** option.

Point and spotlights can be converted to weblights, and vice versa, using the **Properties** palette. Simply select an existing light and open the **Properties** palette. In the **General** category, the Type property determines whether the light is a point light, spotlight, or weblight. Select the type in the drop-down list. Using the **Properties** palette with lights is discussed in detail later in this chapter.

Free Weblight

A free weblight is the same as a standard weblight except there is no target. Only the source location is specified when placing the light. To change the location and direction of the light, use the **ROTATE3D** and **3DMOVE** commands.

 To create either standard or free weblights, photometric lighting must be enabled (**LIGHTINGUNITS** = 1 or 2).

Photometric Lights Tool Palette Group

Photometric lights may be easily added to the drawing using the tool palettes in the **Photometric Lights** tool palette group. This palette group contains four palettes: **Fluorescent, High Intensity Discharge, Incandescent,** and **Low Pressure Sodium.** Lights created with these tools have preset properties for **Intensity Factor, Shadow,** and **Filter Color.** The glyph for the long fluorescent lights has a yellow line passing through its center indicating the direction of the light. The high intensity discharge, low-pressure sodium, and regular incandescent lights are point lights. The incandescent halogen lights are free spotlights. The recessed incandescent lights are a special weblight designed for recessed light fixtures.

Lights in Model and Properties Palettes

The **Lights in Model** palette is extremely useful for controlling the lights in your scene, **Figure 18-19.** Using the **Lights in Model** palette in conjunction with the **Properties** palette, you can manage and edit all of the lights in a scene.

Lights in Model Palette

The **LIGHTLIST** command displays the **Lights in Model** palette. This palette can also be displayed by picking the dialog box launcher button at the lower-right corner of the **Lights** panel on the **Render** tab of the ribbon. All user-created lights in the scene are displayed in the list. To modify the properties of a light, either double-click on the light name or right-click on it and select **Properties** from the shortcut menu. This opens the **Properties** palette. See **Figure 18-20.** If the **Properties** palette is already open, you can simply select a light in the **Lights in Model** palette. You can select more than one light by pressing the [Ctrl] key and selecting the names in the **Lights in Model** palette, which allows you to change all of their settings at the same time. This is an excellent way to make the lighting in your scene uniform or to control a series of lights with a single edit.

A light can be deleted from the scene using the **Lights in Model** palette. To do so, simply right-click on the name of the light and select **Delete Light** from the shortcut menu. The light is removed from the drawing. Using the **UNDO** command restores the light.

Figure 18-19.
All user-created lights are listed in the **Lights in Model** palette.

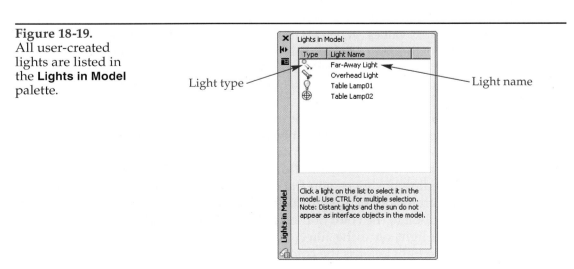

Light type

Light name

Figure 18-20.
The **Properties** palette with a weblight selected.

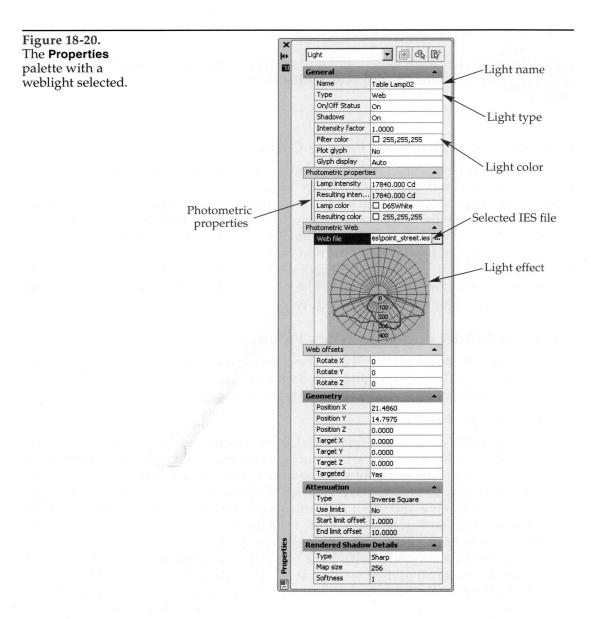

Light name

Light type

Light color

Photometric properties

Selected IES file

Light effect

Properties Palette

The **Photometric properties** subcategory of the **General** category in the **Properties** palette has special settings for photometric lights. The Lamp intensity property determines the brightness of the light. The value may be expressed in candelas (cd), lumens (lm), or illuminance (lux) values. When you select the Lamp intensity property, a button is displayed to the right of the value. Picking this button opens the **Lamp Intensity** dialog box, **Figure 18-21**. In this dialog box, you can change the illumination units and set the intensity (Lamp intensity property). You can also set an intensity scale factor. This is multiplied by the Lamp intensity property to obtain the actual illumination supplied by the light. The read-only Resulting intensity property in the **Properties** palette displays the result.

The **Lamp color** property in the **Photometric Properties** subcategory in the **General** category controls the color of the light. If you select the property, a button is displayed to the right of the value. Picking this button opens the **Lamp Color** dialog box. See **Figure 18-22**. This dialog box gives you the option to control the color of the light by either standard spectra colors or Kelvin colors. The color selected in the **Filter color:** drop-down list is applied to the color of the light. The **Resulting color:** swatch displays the color cast by the light once the filter color is applied. If the filter color is white (255,255,255), then the light color is the color cast by the light.

Figure 18-21.
The **Lamp Intensity** dialog box is used to set the intensity for a light.

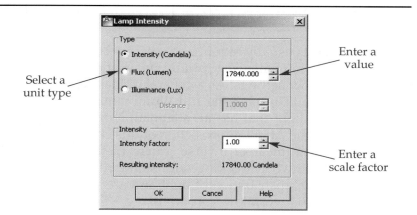

Select a unit type

Enter a value

Enter a scale factor

Figure 18-22.
The **Lamp Color** dialog box is used to set the color for the light and a filter color, if needed.

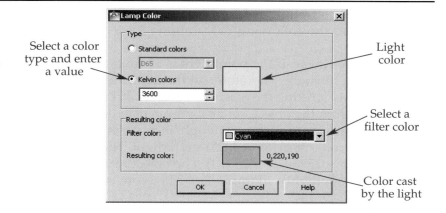

Select a color type and enter a value

Light color

Select a filter color

Color cast by the light

The **Photometric Web** subcategory in the **General** category is where you can specify an IES file for the light. Select the Web file property, then pick the browse button (**...**) and select the IES file. Once the file is selected, its location is displayed in the Web file property. The effect of the data is shown in a graph at the bottom of the **Photometric Web** subcategory. Refer to **Figure 18-20**.

The **Web offsets** subcategory in the **General** category allows you to rotate the web around the X, Y, and Z axes. This is discussed in the Weblight section of this chapter.

In the **Geometry** category, you can change the X, Y, and Z coordinates of the light. You can also change the X, Y, and Z coordinates of the light's target. If the light is not targeted, the Target X, Target Y, and Target Z properties are not displayed. To change the light from targeted to free, and vice versa, select Yes or No in the Targeted property drop-down list.

The properties in the **Attenuation** category are the same as those discussed earlier in this chapter in the Point Lights section. In order to change these properties in the **Properties** palette, photometric lighting must be off (**LIGHTINGUNITS** = 0).

The last category in the **Properties** palette is **Rendered Shadow Details**. The properties in this category are used to control shadows. Shadows are discussed later in this chapter.

PROFESSIONAL TIP

It is important to give your lights names that make them easy to identify in a list. If you accept the default names for lights, they will be called Pointlight1, Spotlight5, Distantlight7, Weblight2, etc., making them difficult to identify. Use the **Name** option when creating the light or, after the light is created, the **Properties** palette to change the name of the light.

Determining Proper Light Intensity

Placing lights usually requires adjustments to produce the results you are looking for. In addition, you will also typically spend some time determining the proper light intensity and other settings. As a general guideline, the object nearest to a point light or spotlight should receive the full illumination, or full intensity, of the light. Full intensity of any light that has an attenuation property is a value of one. Remember, as discussed in the Attenuation section of this chapter, attenuation is calculated using either the inverse linear or inverse square method. Both of these types of attenuation are available when working with standard lights. Photometric lights provide a more "real-world" distribution of light and are set to use inverse square attenuation. Depending on the workflow you are using, you can calculate the light intensity to establish an approximate starting point.

For example, suppose you have drawn an object and placed a point light and a spotlight. The point light is 55 units from the object. The spotlight is 43 units from the object. Use the following calculations to determine the intensity settings for the lights.

- **Inverse linear.** If the point light is 55 units from the object, the object receives 1/55 of the light. Therefore, set the intensity of the point light to 55 so the light intensity striking the object has a value of 1 (55/55 = 1). Since the spotlight is 43 units from the object, set its light intensity to 43 (43/43 = 1).
- **Inverse square.** If the point light is 55 units from the object, the object receives $(1/55)^2$, or 1/3025 ($55^2 = 3025$), of the light. Therefore, set the intensity of the point light to 3025 (3025/3025 = 1). The object receives $(1/43)^2$, or 1/1849 ($43^2 = 1849$), of the spotlight's illumination. Therefore, set the intensity of the spotlight to 1849 (1849/1849 = 1).

However, it should be noted that these settings are merely a starting point. You will likely spend some time adjusting lighting to produce the desired results. In some cases, it may take longer to light the scene than it did to model it.

PROFESSIONAL TIP

If you render a scene and the image appears black, all of the lights may have been turned off or have their intensity set to zero. A scene with no lights placed in it will be rendered with default lighting.

Shadows

Shadows are critical to the realism of a rendered 3D model. A model without shadows appears obviously fake. On the other hand, a model with realistic materials and shadows may be hard to recognize as computer generated. In AutoCAD, the sun, distant lights, point lights, spotlights, and weblights all can cast shadows. AutoCAD's default lighting does not cast shadows. There are two types of shadows that AutoCAD can create: shadow mapped and raytrace. The **Advanced Render Settings** palette provides settings for controlling the creation of shadows when rendering. This palette and its options are discussed in the next chapter.

The options for creating shadows are the same for all lights. The options can be set when the light is created or adjusted later using the **Properties** palette. In the case of sunlight, the **Sun Properties** palette is used to set the options.

AutoCAD and Its Applications—Advanced

Shadow-Mapped Shadow Settings

A *shadow-mapped shadow* is a bitmap generated by AutoCAD. A shadow map has soft edges that can be adjusted. Creating shadow-mapped shadows is the only way to produce a soft-edge shadow. However, shadow maps do not transmit object color from transparent objects onto the surfaces behind the object. **Figure 18-23** shows the difference between shadow-mapped shadows and raytraced shadows.

To specify shadow-mapped shadows and set the quality, or resolution, of the shadow, select the light and open the **Properties** palette (or **Sun Properties** palette). At the bottom of the **Properties** palette is the **Rendered Shadow Details** category, **Figure 18-24**. To specify shadow-mapped shadows, set the Type property to Soft (shadow map).

The Map size property determines the quality of the shadow. The value is the number of samples used to create the shadow. The higher the setting, the better the quality of the generated shadow. However, the higher the setting, the longer it will take to render.

The value of the Softness property determines how soft the edge of the shadow is. The higher the value, the softer or blurrier the edge of the shadow. A low value can produce a very hard edge. The value can range from 1 to 5.

A variation of shadow-mapped shadows is created when the Type property is set to Soft (sampled). In this case, different properties are displayed. The Samples property determines the number of "rays" used to generate the shadows. However, this is not considered raytracing. The Visible in rendering property determines whether the shape of the light is rendered. The Shape property sets the shape of the light. For spotlights, the shape can be either rectangular or circular (disk). For point lights and weblights, the shape can be linear, rectangular, circular (disk), cylindrical, or spherical. The remaining properties are based on the selected shape and are used to define the size of the shape.

Figure 18-23.
The shadow from the object in the foreground is a shadow-mapped shadow. The shadow from the object in the background is a raytraced shadow.

Figure 18-24.
The **Rendered Shadow Details** category in the **Properties** palette.

 For standard shadow-mapped shadows to be created, the Shadow Map property in the **Advanced Render Settings** palette must be set to On. Advanced render settings are covered in the next chapter.

Raytrace Shadow Settings

A *raytrace shadow* is created by beams, or rays, from the light source. These rays trace the path of light as they strike objects to create a shadow. Raytrace shadows have a well-defined edge. They cannot be adjusted to produce a soft edge. Raytrace shadows can be used with standard and photometric lighting.

All lights set to cast shadows, except those set for shadow-mapped shadows, cast raytraced shadows. To switch from shadow-mapped shadows to raytrace shadows, select the light object and open the **Properties** palette. In the **Rendered Shadow Details** category, set the Type property to Sharp. The other properties are disabled because they only apply to shadow-mapped shadows.

 Turning on shadow casting will increase rendering time because of the calculations that AutoCAD has to perform. It is difficult to determine if raytraced or shadow-mapped shadows will be quicker to render because every scene is different and there are many variables that come into play. You will have to experiment with your scene to determine the acceptable level of shadow detail versus rendering time.

Adding a Background

A *background* is the backdrop for your 3D model. The background can be a solid color, a gradient of colors, a bitmap file, the sun and sky, or the current AutoCAD drawing background color. By default, the background is the drawing background.

 To change the background for your drawing, you must first create a named view with the **VIEW** command. In the **View Manager** dialog box, pick the **New...** button to display the **New View/Shot Properties** dialog box. See **Figure 18-25**. The view name, category, and type are specified at the top of the dialog box. Near the bottom of the **View Properties** tab is the **Background** area. The drop-down list in this area is used to specify the type of background. The choices are: Default, Solid, Gradient, Image, and Sun & Sky. The Default setting uses the current AutoCAD viewport color. Photometric lighting must be on (**LIGHTINGUNITS** = 1 or 2) for Sun & Sky to appear in the drop-down list.

PROFESSIONAL TIP

 Before you create the named view, establish the viewpoint from which you want to see the final rendering. Set the perspective projection current, if desired. These settings are saved with the view. The **Views** drop-down list in the **Views** panel on the **View** tab of the ribbon makes it very easy to recall the view. You can also make a named view current by selecting it from the **Custom Model Views** menu in the **View Controls** flyout in the viewport controls.

AutoCAD and Its Applications—Advanced

Figure 18-25.
The **View Properties**
tab of the **New View/
Shot Properties**
dialog box. A
named view must
be created before
you can use a
background in your
scene.

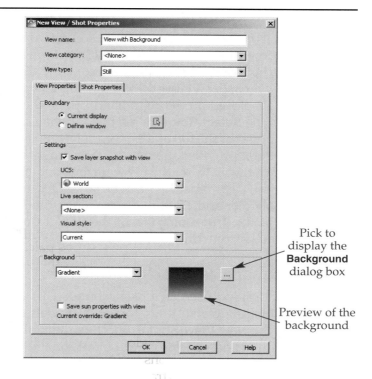

Pick to
display the
Background
dialog box

Preview of the
background

Solid Backgrounds

If you select Solid in the drop-down list, the **Background** dialog box is displayed. The **Type:** drop-down list in this dialog box is automatically set to Solid and the default color is displayed in the **Preview** area. See **Figure 18-26.** In the **Solid options** area of the dialog box, pick the horizontal **Color:** bar to open the **Select Color** dialog box. Then, select the background color that you desire. When the **Select Color** dialog box is closed, the color you picked is displayed in the **Preview** area of the **Background** dialog box. Close the **Background** dialog box, save the view, set the new view current, and close the **View Manager** dialog box.

Make sure to do a test rendering with the background color you selected. The final result may look quite different than your expectations.

Figure 18-26.
Creating a solid
background.

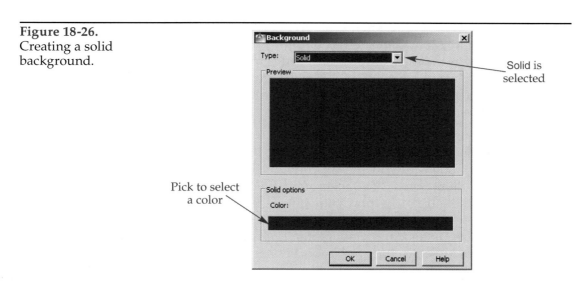

Solid is
selected

Pick to select
a color

Gradient Backgrounds

A gradient background can be composed of two or three colors. If you select Gradient in the drop-down list in the **New View/Shot Properties** dialog box, the **Background** dialog box is displayed with Gradient selected and the default gradient colors displayed in the **Preview** area. See **Figure 18-27**. The **Top color:**, **Middle color:**, and **Bottom color:** swatches are displayed on the right-hand side of the **Gradient options** area. Selecting a swatch opens the **Select Color** dialog box for changing the color. To create a two-color gradient composed of the top and bottom colors, uncheck the **Three color** check box. The **Rotation:** text box provides the option of rotating the gradient. Close the **Background** dialog box, save the view, set the view current, and close the **View Manager** dialog box.

Convincing, clear blue skies can be simulated using the **Gradient** option. Initially, set the **Top**, **Middle**, and **Bottom** color values the same. Then, change the lightness (luminance) in the **True Color** tab in the **Select Color** dialog box. Preview the background and make adjustments as needed.

Using an Image as a Background

An image can be used as a background. This technique can be used to produce realistic or imaginative settings for your models. If you select Image in the drop-down list in the **New View/Shot Properties** dialog box, the **Background** dialog box is displayed with Image selected and a blank image displayed in the **Preview** area. To locate the image file, pick the **Browse...** button to display a standard open dialog box. These image file types may be used for the background: TGA, BMP, PNG, JFIF (JPEG), TIFF, GIF, and PCX. Locate the image file and select **Open**. See **Figure 18-28**.

Once the image file is selected, it must be adjusted. The **Preview** area of the **Background** dialog box shows the image with a preview of a drawing sheet. This drawing sheet indicates how the image is going to be positioned in the view. Pick the **Adjust Image...** button to open the **Adjust Background Image** dialog box. See **Figure 18-29**.

In the **Image position:** drop-down list, pick how the image is applied to the viewport. The Center option centers the image in the view without changing its aspect ratio or scale. The Stretch option centers the image and stretches or shrinks it to fill the entire view. This is one way to plot an image file from AutoCAD. The Tile option keeps the image at its original size and shape, but moves it to the upper-left corner and duplicates it, if needed, to fill the view.

After the image is positioned, use the sliders to adjust it further. The sliders are disabled if Stretch is selected in the **Image position:** drop-down list. The slider function is based on which radio button is picked above the image:

- **Offset**. The sliders move the image in the X or Y direction.
- **Scale**. The sliders scale the image in the X or Y direction. This may distort the image if it is scaled too much in one direction. To prevent distortion, check the **Maintain aspect ratio when scaling** check box at the bottom of the dialog box.

The **Reset** button is located at the bottom-right corner of the preview pane. Picking this button returns the scale and offset settings to their original values.

Once the image is adjusted, pick the **OK** button to close the **Adjust Background Image** dialog box. Then, close the **Background** dialog box, save the view, set the view current, and close the **View Manager** dialog box.

Sun and Sky

If you select Sun & Sky in the drop-down list in the **New View/Shot Properties** dialog box, the **Adjust Sun & Sky Background** dialog box is displayed. See **Figure 18-30**. AutoCAD uses the settings in this dialog box to simulate the sun in the sky. This dialog box has a preview tile at the top and the **General**, **Sky Properties**, **Sun Angle Calculator**, **Rendered Shadow Details**, and **Geographic Location** categories. The settings in these categories are discussed in the Sky Properties section in this chapter.

Figure 18-27.
Creating a gradient
background.

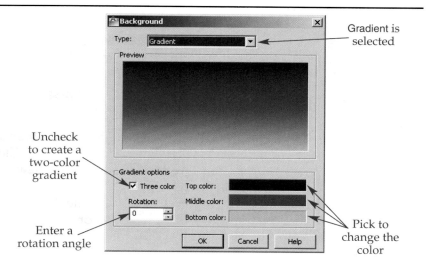

Gradient is
selected

Uncheck
to create a
two-color
gradient

Enter a
rotation angle

Pick to
change the
color

Figure 18-28.
Setting an image as
the background.

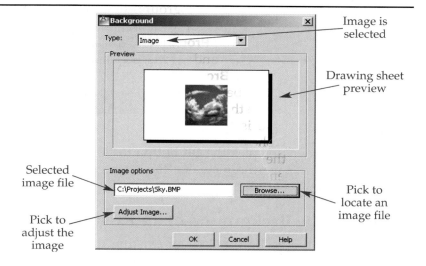

Image is
selected

Drawing sheet
preview

Selected
image file

Pick to
locate an
image file

Pick to
adjust the
image

Figure 18-29.
Adjusting the
background image.

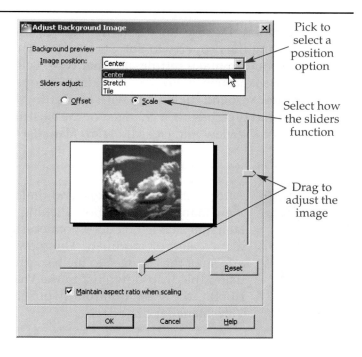

Pick to
select a
position
option

Select how
the sliders
function

Drag to
adjust the
image

Figure 18-30.
The **Adjust Sun & Sky Background** dialog box contains settings for the sun and sky illumination.

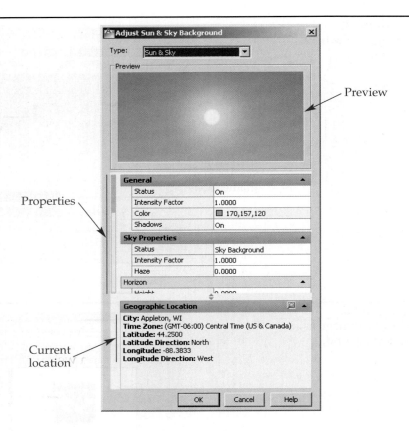

Preview

Properties

Current location

Once the sky is set, pick the **OK** button to close the **Adjust Sun & Sky Background** dialog box. Then, save the view, set the view current, and close the **View Manager** dialog box.

Changing the Background of an Existing View

To change the background of an existing named view, open the **View Manager** dialog box. Select the view name in the **Views** tree on the left-hand side of the dialog box. Then, in the **General** category in the middle of the dialog box, select the Background override property. See **Figure 18-31**. Next, pick the drop-down list for the property and make a selection. Picking None sets the background to the AutoCAD default background. Setting the property to Solid, Gradient, or Image opens the **Background** dialog box, where you can make settings for that type. Selecting Sun & Sky opens the **Adjust Sun & Sky Background** dialog box. Picking Edit opens the **Background** or the **Adjust Sun & Sky Background** dialog box with the settings of the current background. Once the background type has been changed or the existing background edited, pick the **OK** button to save the view and close the **View Manager** dialog box.

After exiting the **Background** dialog box, you are returned to the **View Manager** dialog box. Picking the **OK** button to exit the **View Manager** dialog box does not necessarily activate the view that you just created or modified. The view must be set current to see the effects of the changes to the background. A view can be set current using the drop-down list in the **Views** panel on the **View** tab of the ribbon, the **View Controls** flyout in the viewport controls, or the **View Manager** dialog box.

Figure 18-31.
Changing the background of an existing, named view.

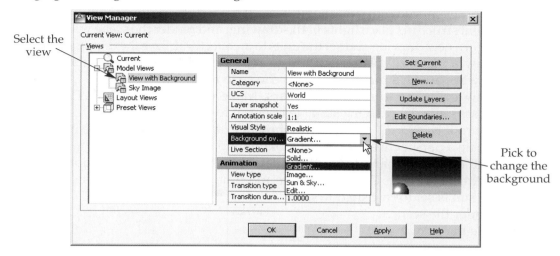

Select the view →

Pick to change the background →

Exercise 18-5

Complete the exercise on the companion website.
www.g-wlearning.com/CAD

Chapter Review

Answer the following questions. Write your answers on a separate sheet of paper or complete the electronic chapter review on the companion website.
www.g-wlearning.com/CAD

1. Compare and contrast *ambient light, distant lights, point lights, spotlights,* and *weblights.*
2. Define *angle of incidence.*
3. Define *angle of reflection.*
4. A smooth surface has a(n) _____ specular factor.
5. Describe *hotspot* and *falloff.* Which lights have these properties?
6. What is *attenuation*?
7. What are the three types of lighting in AutoCAD?
8. What are the four types of light objects in AutoCAD?
9. What are light glyphs and which lights have them?
10. List the types of shadows that can be created in AutoCAD. Which type(s) can have soft edges?
11. What must be created before a background can be added to a scene?
12. What are the four types of backgrounds in AutoCAD, other than the default background?
13. Why would you draw a line between the "from" point and "to" point of a distant light?
14. Describe how a gradient background can be used to represent a clear blue sky.

Drawing Problems

1. In this problem, you will draw some basic 3D shapes to create a building similar to an ancient structure, place lights in the drawing, and render the scene with shadows.
   A. Begin a new drawing and set the units to architectural. Save the drawing as P18_01.
   B. Draw a 32′ × 22′ planar surface to represent the floor. Using the materials browser, attach a material of your choice to the floor.
   C. Draw cylinders to represent pillars. Make each ∅2′ × 15′ tall. There are 10 pillars per side. Attach a suitable material to the pillars.
   D. The roof is 32′ × 22′ and 5′ tall at the ridge. Attach an appropriate material.
   E. Create a perspective viewpoint looking into the building.
   F. Turn on sunlight and turn off the default lighting. Set the geographic location to Athens, Greece. Change the date and time to whatever you wish. Make sure the sun is set to cast shadows.
   G. Render the scene.
   H. Save the drawing.

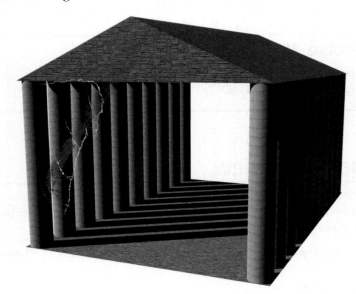

2. Using the drawing from Problem 1, you will experiment with different lighting types.
   A. Open drawing P18_01 and save it as P18_02.
   B. Turn off the sun.
   C. Place three point lights inside of the building. Evenly space the lights along the centerline of the ceiling. Adjust the light intensity so that the interior is not washed out. Set the color of the middle light to white. Set the color of the outside lights to red or blue. Render the scene.
   D. Turn off the point lights.
   E. Place two spotlights, one pointing from the front corner to the rear corner and the other pointing between the pillars on the left side of the building. Target them at the floor. Render the scene. Adjust the intensity, hotspot, and falloff as needed.
   F. Turn the point lights back on and render the scene with all six lights active. Adjust the light intensities again if the rendering is too washed out with light.
   G. Save the drawing.

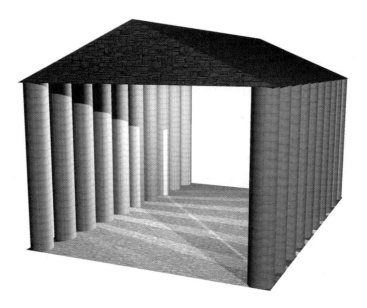

3. Using a previously created mechanical model, you will apply materials and lights to make it ready for presentation.
   A. Open drawing P08_05 created in Chapter 8. Save it as P18_03.
   B. Draw a planar surface below the flange to represent a tabletop.
   C. Attach an appropriate material, such as a wood or tile material, to the surface.
   D. Open the materials editor and create a new material based on the Metal material type. Experiment with different finish settings until you find one you like. Apply the material to the flange.
   E. Place two spotlights in the drawing and target them at the flange from different angles. Adjust their hotspot, falloff, and intensity to get the proper lighting.
   F. Create a perspective view of the scene. Then, render the scene.
   G. Save the drawing.

4. The building shown will be used to study passive solar heating at different times of the year. Model the building using the overall dimensions given. Use your own dimensions for everything else. The side with the windows should be facing South (–Y in AutoCAD, by default).

   A. Set the geographical location to a city in the northern hemisphere.
   B. Set the date to midsummer and the time to noon.
   C. Turn on sunlight and turn the default lighting off.
   D. Render the scene and note the location of the shadows inside the building.
   E. Change the date to late winter, render the scene again, and note the new location of the shadows. You can easily switch between the rendered images in the **Render** window by selecting each rendering in the **History** pane (this is discussed more in the next chapter).
   F. Observing the changes in the shadow locations, what design changes can be made to maximize Sun exposure in the cold winter months? What design changes can be made to minimize Sun exposure in the heat of summer?
   G. Change the geographical location to somewhere closer to the equator. Then, render the scene in summer and winter. How do the shadows compare to those in the previous location?
   H. Save the drawing as P18_04.

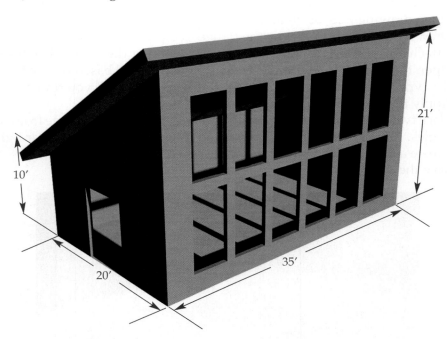

5. In this problem, you will be adding lights to a model and controlling the light properties to create a pleasing scene. Model the courtyard shown. The overall dimensions are 15′ × 16′ × 4′ (wall height). Use your own dimensions for everything else. Add lights as follows.

   A. Turn on the sun and turn off the default lighting. Set the time to late in the day so that the sun is close to the horizon.

   B. In the **Sky Properties** category of the **Sun Properties** palette, select Sky Background and Illumination for the Status property.

   C. Add a point light at the center of each sphere.

   D. Attach a glass material to the spheres that will make them look like they are glowing. There are materials in the Glass category of the Autodesk Library that can be used as a base for creating a new material.

   E. Create a fill light above to illuminate the scene. This can be a point light or spotlight.

   F. Render the scene using the low preset to see the lighting effects.

   G. Adjust the sun properties to create the look that you want.

   H. Adjust the properties of the point lights and any other lights in the scene. You may have to increase the intensity of the lights quite a bit to properly illuminate the scene.

   I. When the scene is illuminated the way you want it, render the scene using the medium or high preset.

   J. Save the drawing as P18_05.

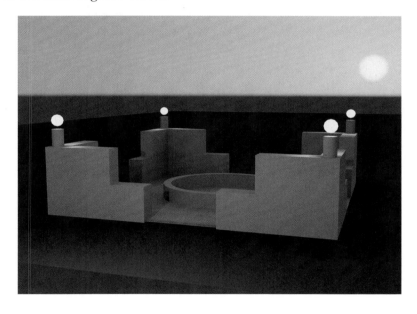

6. In Chapter 10, you completed the kitchen chair model that you started in Chapter 2. In this problem, you will be adding lights to the model and attaching materials to the various components in the chair.
   A. Open P10_07 from Chapter 10.
   B. Create a layer for each of the chair components: seat, legs/crossbars, seatback bow, and seatback spindles.
   C. Draw a planar surface to represent the floor. Place this on its own layer.
   D. Assign materials to each of the layers. You can use materials from the Autodesk Library or create your own materials.
   E. Set the Realistic visual style current.
   F. Adjust material mapping as needed.
   G. Add lighting to the scene.
   H. Render the scene. If you experience problems with the materials, try attaching by object instead of by layer.
   I. Save the drawing as P18_06.

Chapter 19

Advanced Rendering

Learning Objectives

After completing this chapter, you will be able to:
- ✓ Make advanced rendering settings.
- ✓ Set the resolution for a rendering.
- ✓ Save a rendering to an image file.
- ✓ Add fog/depth cueing to a scene.

In Chapter 16, you learned how to create a view of your scene that is more realistic than a visual style. In that chapter, you used AutoCAD's sunlight feature to create a simple rendering with mostly default settings. In this chapter, you will discover how to make a rendering look truly realistic, or *photorealistic*. Some of the advanced rendering features add significantly to the rendering time and you must learn how to balance the quality of the rendering with an acceptable time frame to get the job done.

Render Window

By default, a drawing is rendered in the **Render** window, unless you are rendering a cropped area. This window allows you to inspect the rendering, save it to a file, compare it with previous renderings, and take note of the statistics. See **Figure 19-1**. There are three main areas of the **Render** window—the image, history, and statistics panes.

Image Pane

As AutoCAD processes the scene, the image begins to appear in the image pane in its final form. There may be as many as four phases that the rendering goes through as it is being processed:

- **Translation.** This phase processes the drawing information and determines light intensity, shadow placement, colors, and so on. This phase is always completed.
- **Photon emission.** *Photon emission* is a technique for calculating indirect illumination that traces photons emitted by the light source until they come to rest on a diffuse surface. It determines which areas will be illuminated by indirect, or bounced, light. The photon emission phase may or may not be processed, depending on settings in the **Advanced Render Settings** palette.

Figure 19-1.
The **Render** window.

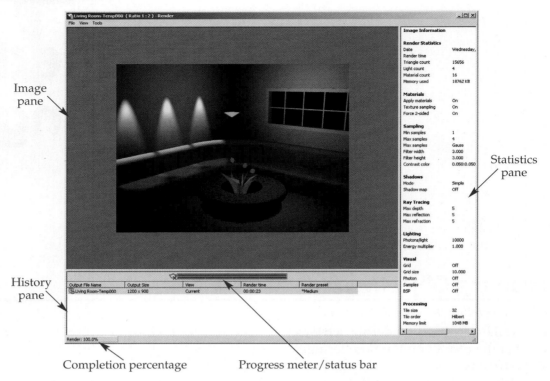

Image
pane

Statistics
pane

History
pane

Completion percentage   Progress meter/status bar

- **Final gather.** *Final gather* increases the number of rays used to calculate global illumination (GI). This phase will be processed if it is turned on in the **Advanced Render Settings** palette.
- **Render.** This phase converts the data into an image. This phase is always completed.

Immediately below the image pane is the progress meter/status bar. The top bar displays the progress of the current phase and the bottom bar indicates the progress of the entire rendering. Also, at the very bottom of the **Render** window, below the history pane, the status of the phase is shown with its percentage complete. The rendering can be canceled at any time by pressing the [Esc] key or picking the **X** button to the left of the progress meter.

As discussed in Chapter 16, you can zoom the rendering in and out to inspect it. You can also save it to an image file using the **File** pull-down menu in the **Render** window.

History Pane

The history pane contains a list of all of the renderings that were created in the drawing since it was created, not just in the current drawing session. The items in this list are called *history entries*. There are two types of history entries. These are indicated by icons, **Figure 19-2**. The entries are described as follows:

Figure 19-2.
The icon in front of
the name indicates if
the entry is normal
or temporary.

Temporary
entry

Normal
entry

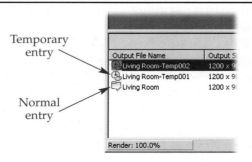

- **Normal.** The entry is saved to file. A link is maintained to that file. If the drawing is saved, closed, and reopened, you can pick the entry to view the rendering in the image pane.
- **Temporary.** The entry is available in the current drawing session, but is not saved to a file. If the drawing is closed, the image is lost. The name of the entry in the Output File Name column ends with -Temp*x*.

Right-clicking on an entry in the history pane displays a shortcut menu. The options in this menu can be used to save the image, render the image again, and manage the entry. The options in the shortcut menu are:

- **Render Again.** Renders the scene again using the same settings. A new entry is not added to the history pane.
- **Save.** Saves the rendered image to a file using a standard save dialog box. This changes the entry from a temporary entry into a normal entry.
- **Save Copy.** Saves the rendered image to a new file without changing the current entry.
- **Make Render Settings Current.** Makes all of the rendering settings of the entry the current rendering settings in the drawing. This allows you to render the current scene using the settings of the entry.
- **Remove From the List.** Deletes the entry from the history pane, but any image files saved from the entry remain.
- **Delete Output File.** Deletes the image file created by saving the entry. The entry remains in the history pane and any image files that were created as copies are retained.

Statistics Pane

The statistics pane shows the details of the rendering that is selected in the history pane. By selecting renderings in the history pane, you can see in the image pane which version provides the best result. Then, you can use the statistics pane to view the settings. The information under the Render Statistics heading (date, render time, etc.) is added when the rendering is completed. The rest of the information reflects the settings in the **Advanced Render Settings** palette and the **Render Presets Manager** dialog box at the time the rendering was created.

Exercise 19-1

Complete the exercise on the companion website.
www.g-wlearning.com/CAD

Advanced Render Settings

The quickest and easiest way to control the quality of a rendering is with render presets. AutoCAD provides five standard render presets: Draft, Low, Medium, High, and Presentation. The Draft preset produces the lowest-quality rendering. Each preset above Draft changes the advanced render settings to gradually improve the rendering quality, with the highest quality produced by the Presentation preset. However, as the quality is improved, the rendering time increases. The presets can be selected in the **Render** panel on the **Render** tab of the ribbon or from the drop-down list at the top of the **Advanced Render Settings** palette. Creating and using your own render presets is covered later in this chapter.

The **Advanced Render Settings** palette provides settings that give you complete control over how a rendering is created. The **RPREF** command opens the palette. The palette can also be displayed by picking the dialog box launcher button at the lower-right corner of the **Render** panel in the **Render** tab on the ribbon. There are five main categories in this palette: **General**, **Ray Tracing**, **Indirect Illumination**, **Diagnostic**, and **Processing**. These categories are explained in the next sections.

General

The **General** category provides properties for controlling the rendering destination, materials, sampling, and shadows, **Figure 19-3**. It contains four subcategories: **Render Context**, **Materials**, **Sampling**, and **Shadows**.

Render Context

The **Render Context** subcategory contains general properties that control the rendering. The Procedure property determines what will be rendered. The settings are View, Crop, and Selected. The View setting is the default and renders whatever you see in the drawing window. The Crop setting allows you to specify an area of the scene to render. This is very useful when you want to do a test rendering, but do not want to wait for the whole scene to render. The Selected setting allows you to pick which objects to render.

The Destination property determines where the rendered scene will be displayed. You can choose to have the rendering placed in the viewport or **Render** window.

If the **Determines if File is Written** button in the subcategory title bar is picked (depressed), the Output file name property is enabled. This property sets the name and location of the file to which the rendering will automatically be saved.

The Output size property sets the resolution, measured in pixels × pixels, for the rendered image. You can select standard resolutions or pick Specify Output Size... for a custom resolution. See **Figure 19-4**. If you want to prevent the image from stretching, make sure the **Lock image aspect** button is selected in the dialog box so that the height and width remain proportional. When you change the resolution, it is stored in the drawing.

Figure 19-3.
The **General** category of the **Advanced Render Settings** palette.

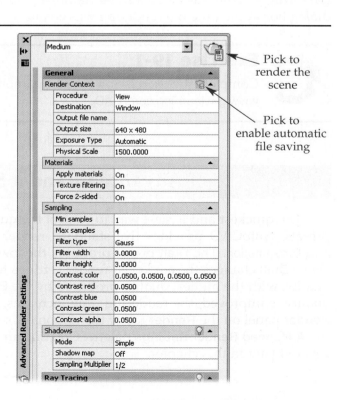

Pick to render the scene

Pick to enable automatic file saving

Figure 19-4.
Setting a custom
resolution.

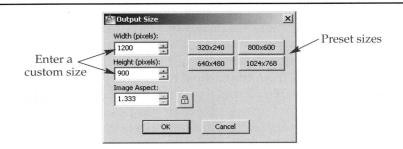

The Exposure Type property can be set to Automatic or Logarithmic. When set to Automatic, the entire image is sampled and some of the dim lighting effects are enhanced to make them more visible. When set to Logarithmic, the brightness and contrast are used to map physical values to RGB values. This is better for scenes with high dynamic ranges.

Exposure control needs a scale to work with and, if you are using non-physical lights (**LIGHTINGUNITS** set to 0), the Physical Scale property provides a scale. Standard lights have an Intensity Factor property that is multiplied by the Physical Scale value to determine the actual brightness of the light. The default value is 1500. In other words, a point light with an Intensity Factor value of 2 has an actual lamp intensity of 3000 candelas when the Physical Scale property is set to 1500.

Materials

The properties in the **Materials** subcategory determine how materials are handled in the rendering. The Apply materials property controls whether or not materials attached to objects are rendered. The property can be set to On or Off. If set to Off, objects are rendered in their own colors. The Texture filtering property determines whether or not antialiasing is applied to texture maps when rendered. Antialiasing is a way of reducing "jaggies" in the rendered image. The Force 2-sided property determines if AutoCAD renders both sides of all faces. This can fix problems where objects disappear in a rendering, but will increase rendering time.

Sampling

Sampling is a process that tests the scene color at each pixel and then determines what the final color should be. This is most important in transition areas, such as edges of objects or shadows. Increasing the sampling will smooth out the jagged edges and incorrect coloring, but increase rendering time. You may also notice thicker lines in the final rendering.

The Min samples and Max samples properties set the minimum and maximum number of samples computed per pixel. A value of 1 means one sample per pixel. A value of 1/4 means one sample for every four pixels. The Filter type property determines how the samples are brought together to determine the pixel value. The settings are described as follows:

- Box. Provides the quickest method; evenly combines samples and gives them equal weight.
- Triangle. Weights the samples based on a pyramid with samples in the center of the filter area receiving the most weight.
- Gauss. Weights the samples based on a bell curve with samples in the center of the filter area receiving the most weight.
- Mitchell. Provides the most accurate method; weights samples based on a curve centered on the filter area, like the Gauss method. However, the curve is steeper.
- Lanczos. Weights samples based on a curve centered on the filter area, like the Mitchell method, but diminishes the weight of samples at the edge of the filter area.

The Filter width and Filter height properties determine the size of the filter area. A larger filter area softens the image, but increases rendering time.

The Contrast color, Contrast red, Contrast blue, Contrast green, and Contrast alpha properties specify the threshold value of the colors and components involved in sampling. If a sample differs from the sample next to it by more than this value, AutoCAD takes more than one sample per pixel up to the Max samples property. Values for the Contrast red, Contrast blue, and Contrast green properties can be from 0.0 (black) to 1.0 (fully saturated). Values for the Contrast alpha property can be from 0.0 (fully transparent) to 1.0 (fully opaque). Increasing the value can reduce the amount of sampling and, therefore, speed up the rendering. However, it may also reduce the quality of the image.

Shadows

The properties in the **Shadows** subcategory control how the renderer handles shadows generated by the lights in the scene. For shadows to be applied, the button in the subcategory title bar must be on (yellow).

The Mode property controls a shader function that calculates light effects. There are three modes that determine how shading is calculated:
- Simple. Shaders are randomly created.
- Sorted. Shaders are called in order from the object to the light.
- Segment. Shaders are called in order from the volume shaders to the segments of the light rays between the object and the light.

The Shadow Map property determines whether shadow-mapped or raytraced shadows are created. When this property is set to On, shadow-mapped shadows are generated. When it is set to Off, raytraced shadows are created.

The Sampling Multiplier property limits shadow sampling for lights. The values are preset for the rendering presets: Draft = 0, Low = 1/4, Medium = 1/2, High = 1, and Presentation = 1. However, these values can be changed. Shadow sampling utilizes the same principles for sampling as described in the Sampling section, but instead of sampling pixels for object color, it is sampling for shadows.

Raytracing

The **Ray Tracing** category provides properties for controlling how the rendered image is shaded, **Figure 19-5**. *Raytracing* is a method of calculating reflections, refractions, and shadows by tracing the path of the light rays from the light sources. This is more accurate at producing shadows than shadow mapping, but it takes more time and the shadow edge is always sharp. To enable raytracing, pick the button in the category's title bar. If this is off (not yellow), there will be no raytracing and the properties are disabled.

The Max reflections property specifies the maximum number of times that a ray can be reflected. See **Figure 19-6**. The Max refractions property specifies the maximum number of times that a ray can be refracted. The Max depth property specifies the maximum number of reflections and refractions. For example, if this property is set to 5 and the Max reflections property is set to 3, then no more than two refractions will occur. A good way to figure out the required maximum depth is to imagine a light ray traveling through transparent objects or bouncing off of reflective objects in your scene. Count how many surfaces the object must contact and that is the maximum depth.

Figure 19-5.
The **Ray Tracing** category of the **Advanced Render Settings** palette.

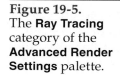

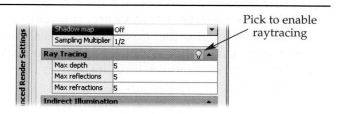

Pick to enable raytracing

Figure 19-6.
A—If the Max reflections property for raytracing is set to 3, a light ray will bounce three times and then stop at the next surface. B—If transparent objects exist in the scene and the Max refractions property is set to 2, the light ray will still make it to the back wall as long as the Max depth property is set to 5 or more. C—However, if the Max depth property is set to 4, the light ray will not make it to the back wall.

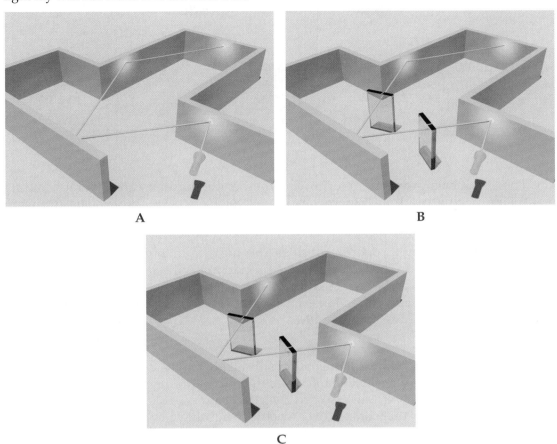

Indirect Illumination

Indirect illumination is an AutoCAD mechanism that simulates natural, bounced light. If indirect illumination is turned off and light does not directly strike an object, the object is black. Without indirect illumination enabled, other lights must be added to the scene to simulate indirect illumination. The properties in the **Indirect Illumination** category allow you to create a natural-looking scene. There are three subcategories in the **Indirect Illumination** category: **Global Illumination**, **Final Gather**, and **Light Properties**. See **Figure 19-7**.

Global Illumination

Global illumination (GI) is an indirect illumination process. Bounced light is simulated by generating photon maps on surfaces in the scene. These maps are created by tracing photons from the light source. Photons bounce around the scene from one object to the next until they finally strike a diffuse surface. When a photon strikes a surface, it is stored in the photon map. To enable global illumination, pick the button in the subcategory's title bar. If this button is off (not yellow), there will be no indirect illumination.

The Photons/sample property sets the number of photons used to generate the photon map. The higher the value, the less noise produced by global illumination. However, rendering time is longer and the image is blurrier.

Figure 19-7.
The **Indirect Illumination** category of the **Advanced Render Settings** palette.

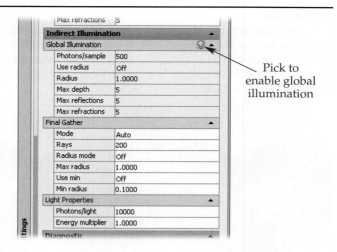

Pick to enable global illumination

The Use radius property determines whether the photons are a default radius or a user-specified radius. When the property is set to On, the Radius property sets the size of the photon. When set to Off, the radius of each photon is 1/10th the scene's radius.

The Max reflections property specifies the maximum number of times that a photon can be reflected. The Max refractions property specifies the maximum number of times that a photon can be refracted. The Max depth property specifies the maximum number of reflections and refractions. These properties function as explained in the Ray Tracing section. See also **Figure 19-6**.

Final Gathering

The settings in the **Global Illumination** subcategory may result in dark and light areas in the scene. *Final gathering* increases the number of rays in the rendering and cleans up these artifacts. It will also greatly increase rendering time. Final gathering works best with scenes that contain overall diffuse lighting. See **Figure 19-8**. The Mode property for final gathering can be set to:

- On. Turns on final gathering.
- Off. Turns off final gathering.
- Auto. Final gathering is turned on or off based on the sky light status. This is the default setting.

The Rays property sets the number of rays used to calculate indirect illumination. The higher the value, the better the result, but the longer it takes to render the scene.

The Radius mode property determines how the Max radius property is applied during final gathering. There are three possible settings:

- On. The Max radius value is used for final gathering and it is measured in world units.
- Off. The radius of each area processed by final gathering is 10% of the maximum model radius.
- View. The Max radius value is used for final gathering, but it is measured in pixels instead of world units.

The Max radius property determines the maximum radius of each area processed during final gathering. The lower the value, the higher the quality of the rendering because a larger number of smaller areas is processed. However, rendering time is higher.

The Use min property determines whether or not the Min radius property is applied for final gathering. The Min radius property sets the minimum radius of the processed areas. Increasing this setting improves quality, but increases rendering time.

Figure 19-8.
A—This scene has a single point light. B—Global illumination is turned on. Notice the unevenness of the lighting. This can be seen especially on the sofa and in the corner of the walls. C—Final gathering cleans up artifacts and provides a more even illumination.

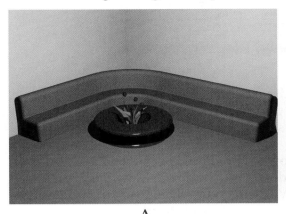
A

B

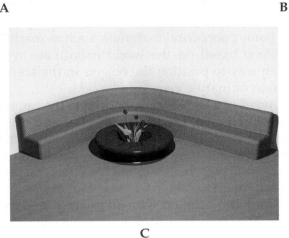
C

Light Properties

The properties in the **Light Properties** subcategory control how the lights in the scene are applied when calculating indirect illumination. The Photons/light property sets the number of photons emitted by each light. Increasing this number makes each light cast more photons and improves the rendering quality. The Energy multiplier property determines how much light energy is used in global illumination. The default value of 1.0000 does not increase or decrease the light energy. Values less than the default decrease the light energy. Values greater than the default increase the light energy.

PROFESSIONAL
TIP

If your scene looks washed out (flooded with light) with indirect illumination enabled, experiment with reducing the energy multiplier. This can have a dramatic effect on the scene.

Diagnostic

The properties in the **Diagnostic** category control tools to help you understand why the rendering produced the results it did, **Figure 19-9.** The scene can be rendered with photon maps, grids, and irradiance shown. These tools can help you diagnose and correct problems.

Indirect Illumination	▼
Diagnostic	▲
Visual	▲
Grid	Off
Grid size	10.0000
Photon	Off
Samples	Off
BSP	Off
Processing	▲

The Grid property determines if a coordinate grid is shown in the rendered image. The Grid size property sets the size of the grid. When the Grid property is set to Off, which is the default setting, the grid is not shown. There are three other settings:

- Object. A colored grid displays local coordinates (UVW coordinates), **Figure 19-10A**. Each object has its own set of local coordinates. UVW coordinates can be thought of as XYZ coordinates on the object. Materials are mapped to objects based on the object's UVW directions. This grid helps determine why a material may not be correctly positioned or rotated.

- World. World coordinates (XYZ coordinates) are displayed in a colored grid, **Figure 19-10B**. Some procedural materials, such as marble and wood, are positioned in the scene based on the world coordinate system. This grid helps determine which way to position the objects in the scene to control the direction of wood grain or marble vein.

- Camera. Coordinates of a UCS corresponding to the camera or current view are displayed in a colored grid, **Figure 19-10C**. Because this grid is based on the camera (viewer) position, it may be used to visualize the horizon location or clipping planes without having to completely render the scene.

The Photon property controls whether or not the effect of a photon map is shown in the rendering. When the property is set to Density or Irradiance and global illumination is on, the scene is rendered and overlaid with an image representing the photon map. See **Figure 19-11**. The Photon property settings are described as follows:

Figure 19-10.
Applying a grid to the rendering using the Grid property setting. A— The Grid property is set to Object. This grid helps to visualize the UVW directions on individual objects: red = U, green = V, blue = W. B— The Grid property is set to World. This helps to visualize the XYZ directions in the entire scene: red = X, green = Y, blue = W. C—The Grid property is set to Camera. This helps to visualize the XYZ directions in the entire scene based on the camera (viewer) direction: red = X, green = Y, blue = W.

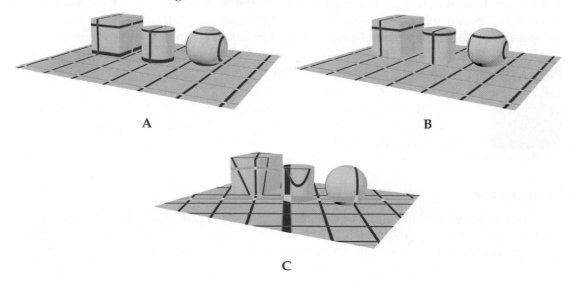

A   B

C

AutoCAD and Its Applications—Advanced

- Density. Shows the photon map projected onto the scene. Higher-density areas are red and lower-density areas are the cooler colors.
- Irradiance. The rendering is similar to one produced with the Density setting, but the photons are shaded based on their irradiance value. Maximum irradiance is red and lower irradiance values are shown in the cooler colors.

The Samples property can be set to On or Off. When set to On, a grid is rendered plan to the view and varying shades of gray and white are displayed in the scene. This tool is another way to evaluate the lighting in the scene.

The BSP property determines whether or not the effects of *binary space partitioning (BSP)* are shown. BSP is a raytrace-acceleration method. When rendering, if you receive a message about large depth or size values or the rendering is very slow, this tool may help you locate the problem. The BSP property settings are described as follows:
- Depth. The depth of the raytrace tree is displayed. Top faces are displayed in bright red. The deeper the faces are in the tree, the cooler the colors in which they are displayed, **Figure 19-12A**.
- Size. The size of the leaves in the raytrace tree is displayed. Different colors are used to identify different leaf sizes, **Figure 19-12B**.

Figure 19-11.
Applying a photon map to the rendering. In this example, the Photon property is set to Density.

Figure 19-12.
Showing the effects of binary space partitioning. A—The BSP property is set to Depth. B—The BSP property is set to Size.

A   B

Processing

The properties in the **Processing** category control how the final render processing takes place, **Figure 19-13.** The Tile size property controls the size of the tiles into which the total image is subdivided. The larger the tile size, the fewer tiles that have to be rendered and the fewer times the image has to update. Larger tiles usually mean a shorter rendering time. The Tile order property controls the order in which the tiles are rendered. The settings are described as follows:

- Hilbert. The "cost" of switching to the next tile determines which tile is rendered next.
- Spiral. The rendering begins with the tiles in the center of the image and then spirals outward.
- Left to Right. The tiles are rendered from bottom to top and left to right in columns.
- Right to Left. The tiles are rendered from bottom to top and right to left in columns.
- Top to Bottom. The tiles are rendered from right to left and top to bottom in rows.
- Bottom to Top. The tiles are rendered from right to left and bottom to top in rows.

The Memory limit property specifies the maximum memory allocated for the rendering process. When this limit is reached, some objects may be removed from rendering.

Exercise 19-2

Complete the exercise on the companion website.
www.g-wlearning.com/CAD

Render Presets

Ribbon

Render
> Render
> Manage
Render
Presets...

Type

RENDERPRESETS
RP

Once settings have been established that create a rendering with the desired results, the settings can be saved to a custom render preset. The **Render Presets Manager** dialog box is used to create custom render presets, **Figure 19-14.** The **RENDERPRESETS** command opens this dialog box.

The left side of the dialog box displays a tree that contains the standard render presets and any custom render presets. In the middle of the dialog box are all of the properties for the selected render preset. These are the same properties available in the **Advanced Render Settings** palette. On the right side of the dialog box are three buttons that allow you to make a preset current, make a copy of a preset, or delete a preset. You cannot delete one of the default presets.

The easiest way to create a custom preset is to start with a standard render preset and modify the properties until the desired result is produced. This preset will be indicated as the current preset in the **Render Presets Manager** dialog box, but there will be an asterisk (*) in front of its name. The asterisk indicates that the preset has been changed from its original settings. Next, pick the **Create Copy** button in the **Render Presets Manager** dialog box to make a copy. In the **Copy Render Preset** dialog box that appears, name the new render preset, provide a description, and pick the **OK** button. The new preset is saved in the Custom Render Presets branch of the tree.

Figure 19-13.
The **Processing** category of the **Advanced Render Settings** palette.

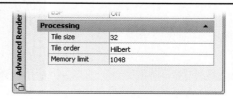

Figure 19-14.
The **Render Presets Manager** dialog box.

Saved custom presets

Pick to create a copy with the current settings

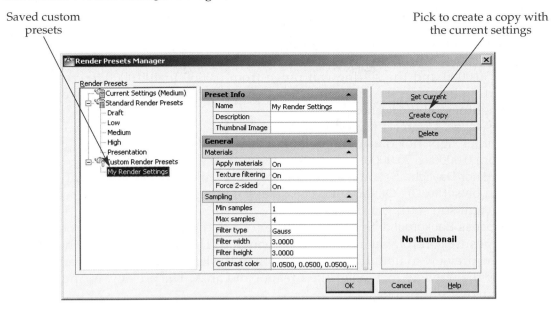

Render Exposure

The **RENDEREXPOSURE** command displays the **Adjust Rendered Exposure** dialog box. See **Figure 19-15**. In this dialog box, you can globally adjust the brightness, contrast, midtones, and exterior daylight of the scene. In order to use this command, photometric lighting must be on (**LIGHTINGUNITS** = 1 or 2) or the Exposure Type property in the **General** category of the **Advanced Render Settings** palette must be set to Logarithmic.

The **Preview** area in the **Adjust Rendered Exposure** dialog box displays the rendered scene with the changes you make in the dialog box so you can see how the scene will be altered. This saves the step of re-rendering the scene. Simply change the

Ribbon

Render
> Render

Adjust Exposure

Type

RENDEREXPOSURE

RENDEREXPOSURE

Figure 19-15.
Using the
RENDEREXPOSURE
command to adjust
the rendering.

Preview

Settings

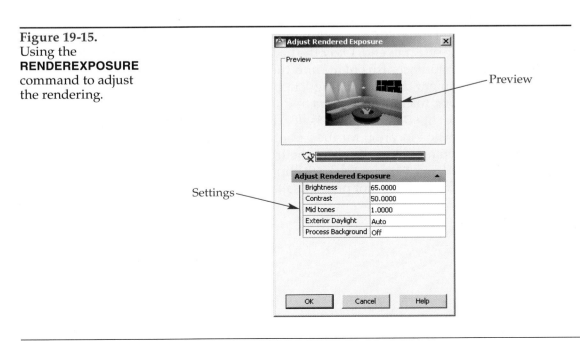

settings until the preview looks correct and then close the dialog box. The properties in this dialog box are described as follows:

- **Brightness.** Controls the brightness of the colors. The default value is 65.0000 and the setting can range from 0.0000 to 200.0000. Increasing the setting increases how light the colors in the scene appear.
- **Contrast.** Controls the contrast of the colors in the scene. The default value is 50.0000 and the setting can range from 0.0000 to 100.0000. Increasing the setting increases the difference between similar colors, in effect increasing the brightness of the scene.
- **Mid tones.** Controls the midtone values of the colors. The midtone colors are neither light nor dark. The default value is 1.0000 and the setting can range from 0.0000 to 20.0000.
- **Exterior Daylight.** Sets the exposure for scenes illuminated with sunlight. This can be set to On, Off, or Auto. The default setting is Auto. With this setting, the current sun settings are used.
- **Process Background.** Specifies whether or not the background is processed by exposure control when the scene is rendered. The setting is either on or off.

To force the preview to update, pick the button to the left of the rendering progress bars below the preview that looks like a teapot and an X. The preview is updated with the current settings.

Render Environment

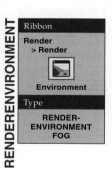

The render environment allows for the addition of fog or depth cueing to the scene. *Fog* and *depth cueing* in AutoCAD are actually ways of using color to visually represent the distance between the camera (viewer) and objects in the model. See **Figure 19-16.** This is similar to looking at an object from a distance and seeing that the object is a little obscured from haze in the air. The only difference between fog and depth cueing is the color. Fog is displayed as white or another light color and depth cueing is generally displayed as black. The **Render Environment** dialog box is used to add fog/depth cueing, **Figure 19-17.**

Creating a Camera

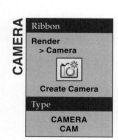

Before adding fog/depth cueing, a camera must be created that shows the view you want. Then, start the **3DCLIP** command and adjust the back clipping plane to where you want the effect to end. Only the back clipping plane needs to be active. The fog/depth cueing references this plane and the camera location.

Creating a camera and adjusting clipping planes is discussed in detail in Chapter 20. However, to create a camera and turn on the clipping plane(s), first select the command. Note: the **Camera** panel is not displayed by default in the **Render** tab of the ribbon. Then, pick a location for the camera followed by the location for its target. Next, enter the **Clipping** option. Turn on the front clipping plane, if desired, and enter the offset distance. Then, turn on the back clipping plane and enter the offset distance. Finally, end the command (do not press [Esc]).

Figure 19-16.
A—This scene has no fog/depth cueing applied. B—The scene has white fog applied (including the background). C—The scene has black depth cueing applied (including the background).

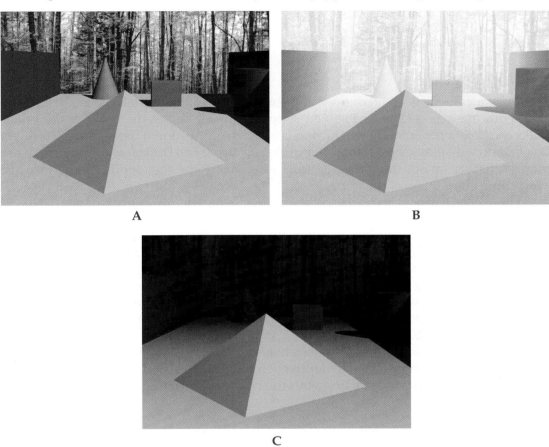

A   B

C

Figure 19-17.
The **Render Environment** dialog box is used to add fog/depth cueing to the scene.

Set to On to apply
fog/depth cueing

A camera is automatically created when a view is saved as a named view. This is another good reason to save your views.

Adding Fog

Once a camera is created, open the **Render Environment** dialog box. To turn on fog/depth cueing, set the Enable Fog property to On. To set the color of the effect, select the Color property. Then, choose a color in the drop-down list. Picking the Select Color... entry displays the **Select Color** dialog box. The Fog Background property determines whether or not the background is affected by the fog/depth cueing just like everything else.

The Near Distance property sets where the fog/depth cueing begins. This is a distance from the camera. The value can be from 0.0000 to 100.0000, which is a percentage of the total distance between the camera and the back clipping plane. The back clipping plane is where the target is located. The Far Distance property sets where the fog ends. This is also a distance from the camera. The value is also a percentage of the total distance from the camera to the back clipping plane and can be from 0.0000 to 100.0000. In other words, 100% ends at the back clipping plane.

The Near Fog Percentage property determines the opacity of the fog at its starting location. A value of 100 means the fog is 100% opaque. The near percentage is usually set to 0, or 0% opaque. The Far Fog Percentage property determines the opacity of the fog at its ending location. The fog/depth cueing will increase in opacity from the near distance to the far distance starting with the near fog percentage and ending with the far fog percentage.

Exercise 19-3

Complete the exercise on the companion website.
www.g-wlearning.com/CAD

Chapter Review

Answer the following questions. Write your answers on a separate sheet of paper or complete the electronic chapter review on the companion website.
www.g-wlearning.com/CAD

1. Describe the three panes of the **Render** window.
2. What are the three possible destinations for render output?
3. Once a rendering is completed and displayed in the **Render** window, how can it be saved to a file?
4. List the render presets AutoCAD provides.
5. What is *sampling* and what do the properties in the **Sampling** subcategory in the **Advanced Render Settings** palette control?
6. Raytracing calculates shadows, _____, and _____.
7. How does global illumination simulate bounced light?
8. What is the benefit of final gathering?
9. For what is the Energy multiplier property in the **Light Properties** subcategory in the **Advanced Render Settings** palette used?
10. For what are the properties in the **Diagnostic** category of the **Advanced Render Settings** palette used?
11. Describe how to create a custom render preset.
12. List the properties that can be changed in the **Adjust Render Exposure** dialog box.
13. What is *fog/depth cueing*?
14. Which color is normally used to display depth cueing?
15. What must be created before fog or depth cueing is added to the scene?

Drawing Problems

1. Using the drawing from Problem 2 in Chapter 18, you will experiment with advanced render settings.
   A. Open drawing P18_02 and save it as P19_01. If you did not complete this problem, do so now.
   B. Make sure all of the lights are active and render the scene to the **Render** window using the Medium or High render preset.
   C. In the **Advanced Render Settings** palette, enable global illumination. Then, render the scene again.
   D. Enable final gathering and render the scene again. This time it will probably take much longer to render.
   E. Which rendering has the best quality?
   F. Which setting impacted render time the most?
   G. Save the last image as P19_01.jpg.
   H. Save the drawing.

2. In this problem, you will set up fog/depth cueing.
   A. Start a new drawing and save it as P19_02.
   B. Draw a planar surface that is 50 units × 20 units.
   C. Randomly place various objects (cones, boxes, spheres, etc.) on the plane. Assign a different color or material to each object.
   D. Create a viewpoint that is almost at ground level looking down the length of the plane. Try to get as many objects in the view as possible. Save this as a named view and add a background of some type.
   E. Add a distant light source. Position it and adjust its intensity so that interesting shadows are created in the scene, but the objects are sufficiently illuminated.
   F. With the **3DCLIP** command, set up clipping planes with the back clipping plane at the far end of the plane. Make sure the back clipping plane is on.
   G. In the **Render Environment** dialog box, turn on fog and set the color to black. The far distance should be 100 and the percentage should be around 75.
   H. Render the scene.
   I. Change the fog color to white and render the scene again.
   J. Set the fog to affect the background and render the scene again.
   K. Save the image as P19_02.jpg.
   L. Save the drawing.

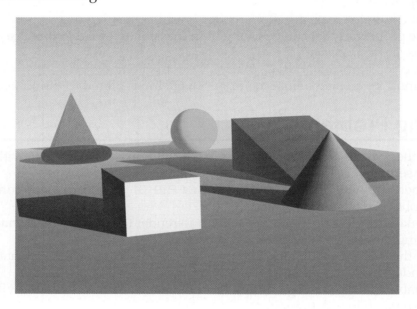

3. In this problem, you will experiment with the **RENDEREXPOSURE** command and final gathering. Open P18_05 created in Chapter 18 and save it as P19_03. If you did not complete this problem, do so now.

A. Open the **Advanced Render Settings** palette. In the **General** category, set the Exposure Type property to Logarithmic.

B. In the **Indirect Illumination** category, change the Mode property in the **Final Gather** subcategory to Off.

C. Render the scene and note the appearance.

D. Use the **RENDEREXPOSURE** command to display the **Adjust Rendered Exposure** dialog box. Note the appearance of the preview image.

E. Change the brightness setting to 80 and note how the preview changes.

F. Pick the **OK** button to close the **Adjust Rendered Exposure** dialog box and render the scene again. Does the rendered scene match the preview in the **Adjust Rendered Exposure** dialog box?

G. Turn on final gathering (Mode property = On) and open the **Adjust Rendered Exposure** dialog box. How does the preview look different? Why?

H. Adjust the brightness setting to get the exposure that you want in the preview. Then, close the dialog box and render the scene again.

I. Experiment with the other settings in the **Adjust Rendered Exposure** dialog box until you get the scene the way you want it.

J. Render the scene one last time and save the image as a file named P19_03.jpg.

K. Save the drawing.

Drawing Problems - Chapter 19

4. Using the same drawing from Problem 19-3, you will perform diagnostics to determine the effects of the lights on the final rendering.
   A. Open drawing P19_03 and save it as P19_04.
   B. In the **Advanced Render Settings** palette, select the Medium rendering preset. In the **Indirect Illumination** category, turn off final gathering (Mode property = Off).
   C. In the **Diagnostic** category of the **Advanced Render Settings** palette, set the Grid property to Object. Render the scene.
   D. In the **Diagnostic** category of the **Advanced Render Settings** palette, set the Grid property to World. Render the scene.
   E. In the **Diagnostic** category of the **Advanced Render Settings** palette, set the Grid property to Camera. Render the scene.
   F. Describe the differences and explain why this is helpful in analyzing a scene.
   G. In the **Indirect Illumination** category of the **Advanced Render Settings** palette, turn on global illumination. In the **Diagnostic** category, turn off the grid and set the Photon property to Density.
   H. Render the scene. Describe the effect and what can be learned from it.
   I. Set the Photon property to Irradiance and render the scene. What does this effect tell about the lighting in the scene?
   J. Which diagnostic tool worked the best and why?
   K. Save the drawing.

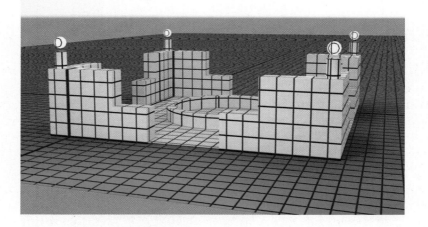

Cameras, Walkthroughs, and Flybys

Learning Objectives

After completing this chapter, you will be able to:

✓ Create a camera to define a static 3D view.
✓ Activate and adjust front and back clipping planes.
✓ Record a walkthrough of a 3D model to a movie file.
✓ Record a flyby of a 3D model to a movie file.
✓ Create walkthroughs and flybys by following a path.
✓ Control the viewpoint, speed, and quality of the animation.

Once you have a 3D design complete, or even while still in the conceptual phase of design, you may want to take a stroll through the model and have a look around. You may also want to strap on some wings and fly over and around the model to see it from above. A *walkthrough animation* shows a scene as a person would view it walking through the scene. Walkthroughs are typically used to show the interior of a building, but can be created for exterior scenes as well. A *flyby animation* is similar to a walkthrough, except that the person is not bound by gravity. In other words, the scene is viewed as a bird flying above would see it. Flybys often show the exterior of a building.

The **3DWALK** command is used to create a walkthrough by recording views as a camera "walks" through the scene. The **3DFLY** command is very similar, but the movement of the camera is not limited to a single Z value. A path can also be drawn and the camera linked to the path. This chapter discusses these commands and other methods used to create the animation you need. In addition, this chapter discusses creating and using cameras.

Creating Cameras

Cameras are used in AutoCAD to store a viewpoint and easily recall it later when needed for viewing or rendering the scene. After the camera is established, you can zoom, pan, and orbit as needed and then come back to the camera view. It is not necessary to create a camera before using the **3DWALK**, **3DFLY**, and **ANIPATH** commands (discussed later in this chapter) because these commands create their own cameras.

CAMERA

Ribbon
Render
> Camera

Create Camera

Type
CAMERA
CAM

The **CAMERA** command allows you to add a camera to the scene. Note: the **Camera** panel may not be displayed in the **Render** tab of the ribbon. You may have to toggle it on. Cameras are normally placed in the plan view of the scene to make it easy for you to pick where you want to "stand" and where you want to "look." Once the command is selected, you are first prompted to specify the camera location. A camera glyph is placed in the scene at the camera location, **Figure 20-1**. Next, you must specify the target location. As you move the cursor before picking the target location, a pyramid-shaped field of view indicates what will be seen in the view (if a 3D visual style is current). Once you select the target location, the command remains active for you to select an option:

Enter an option [?/Name/LOcation/Height/Target/LEns/Clipping/View/eXit]<eXit>:

The list, or ?, option allows you to list the cameras in the drawing. Select this option and type an asterisk (*) to show all of the cameras in the drawing. You can also enter a name or part of a name and an asterisk. For example, entering HOUSE* will list all of the cameras whose name begins with HOUSE, such as HOUSE_SW, HOUSE_SE, and HOUSE_PLAN.

The **Name** option allows you to change the name of the camera as you create it. If you do not rename the camera, it is given a default, sequential name, such as Camera1, Camera2, Camera3, and so on. It is always a good idea to provide meaningful names for cameras. Names such as Living Room_SW, Corner, or Hallway_Looking East leave no doubt as to what the camera shows. If you choose not to rename the camera at this point, it can be renamed later using the **Properties** palette.

The **Location** option allows you to change the placement of the camera. Enter the option and then specify the new location. You can enter coordinates or pick a location in the drawing.

The **Height** option allows you to change the vertical location of the camera. Enter the option and then enter the height of the camera. The value you enter is the number of units from the current XY plane. If you are placing the camera in a plan view, this option is used to tilt the view up or down from the current XY plane.

The **Target** option allows you to change the placement of the camera target. Enter the option and then specify the new location. You can enter coordinates or pick a location in the drawing.

The **Lens** option allows you to change the focal length of the camera lens. If you change the lens focal length, you are really changing the field of view, or the area of the drawing that the camera covers. The lower the lens focal length, the wider the field of view angle. The focal length is measured in millimeters.

The **Clipping** option is used to turn the front and back clipping planes on or off. These planes are used to limit what is shown in the camera view. Clipping planes are discussed later in this chapter.

Figure 20-1.
A camera is represented by a glyph. When the camera is selected, the field of view (shown in color) and grips are displayed.

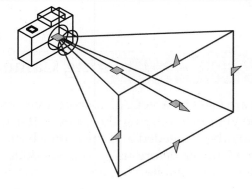

The **View** option is used to change the current view to that shown by the camera. This option has two choices—**Yes** or **No**. If you select **Yes**, the active viewport switches to the camera view and the **CAMERA** command ends. If you select **No**, the previous prompt returns.

Once you have made all settings, press [Enter] or select the **Exit** option to end the command. The view (camera) is listed with the other saved views in the drop-down list in the **Views** panel on the **View** tab of the ribbon and in the **View Controls** flyout of the viewport controls. Selecting the view makes it the current view in the active viewport. The view is also listed under the Model Views branch in the **View Manager** dialog box. It can be made current by selecting the view, picking the **Set Current** button, and then picking the **OK** button.

PROFESSIONAL TIP

In addition to the camera name, many other camera properties can be changed in the **Properties** palette. The camera and target locations can be changed, the lens focal length and field of view can be adjusted, and the clipping planes can be modified. Also, you can change the roll angle, which is the rotation about a line from the camera to the target, and set the camera glyph to plot.

Camera System Variables

The **CAMERADISPLAY** system variable controls the visibility of camera glyphs. When set to 1, which is the default, camera glyphs are displayed. When set to 0, camera glyphs are not displayed. Creating a camera automatically sets the variable to 1. The **Show Cameras** button in the **Camera** panel on the **Render** tab of the ribbon toggles the display of camera glyphs off and on. Remember, this panel may not be displayed by default.

When creating a camera, if you pick the camera and target locations without using object snaps, you may assume that the camera and target are located on the XY plane (Z coordinate of 0) of the current UCS. This may or may not be true. The **CAMERAHEIGHT** system variable determines the default height of the camera if a Z coordinate is not provided. It is a good idea to set this variable to a typical eye height before placing cameras. There is no corresponding system variable for the target because the target is usually placed by snapping to an object of interest. If X and Y coordinates are entered for the target location, but a Z coordinate is not provided, the Z value is automatically 0. If a camera was previously created in the drawing session and the **Height** option was used, that height value becomes the default camera height.

Cameras Tool Palette

The **Cameras** tool palette provides a quick way to add a camera, but the default tools do not allow for the options described earlier. The **Normal Camera** tool creates a camera with a 50 mm focal length. This camera simulates normal human vision. The **Wide-angle Camera** tool creates a camera with a 35 mm focal length. This type of view is commonly used for scenery or interior views where it is important to show as much as possible with minimal distortion. The **Extreme Wide-angle Camera** tool creates a camera with a 6 mm focal length. This camera produces a fish-eye view, which is very distorted and mainly useful for special effects.

Changing the Camera View

Once the camera is placed, it is easy to manipulate. If you select a camera, the **Camera Preview** window is displayed by default. This window shows the view through the camera, **Figure 20-2**. The view in the window can be displayed in any of the 3D visual styles or any named visual style. Select the visual style in the drop-down list in the window. If the **Display this window when editing a camera** check box at the bottom of the window is unchecked, the window is not displayed the next time a camera is selected. The next time the drawing is opened, this setting is restored (checked).

When a camera is selected, grips are displayed. Refer to **Figure 20-1**. If you hover the cursor over a grip, a tooltip appears indicating what the grip will alter. Picking the base grip on the camera allows you to reposition the camera in the scene. If the **Camera Preview** window is open, watch the preview as you move the camera to help guide you. Selecting the grip on the target allows the target to be repositioned. Again, use the preview in the **Camera Preview** window as a guide. The grip at the midpoint between the camera and target can be used to reposition the camera and target at the same time. If you pick and move one of the arrow grips on the end of the field of view, the lens focal length and field of view are changed.

Camera Clipping Planes

Clipping planes allow you to suppress objects in the foreground or background of your scene. Picture these clipping planes as flat, 2D objects perpendicular to the line of sight that can be moved closer to or farther from the viewer. Only the objects between the front and back clipping planes and within the field of view are seen in the camera view. This is helpful for eliminating walls, roofs, or any other clutter that may take away from the focus of the scene. Also, as mentioned in Chapter 19, the back clipping plane must be enabled when applying fog/depth cueing using the **Render Environment** dialog box. Clipping planes can be set while creating the camera or later using the **Properties** palette. Clipping planes can be adjusted using grips.

To set the clipping planes while creating the camera, enter the **Clipping** option. You are prompted:

Enable front clipping plane? [Yes/No] <No>:

To enable the front clipping plane, enter **YES**. You are then asked to specify the offset from the target plane. The target plane is described next. Once you enter the offset, or if you answer **No**, you are prompted:

Enable back clipping plane? [Yes/No] <No>:

Figure 20-2.
The **Camera Preview** window is displayed, by default, when a camera is selected.

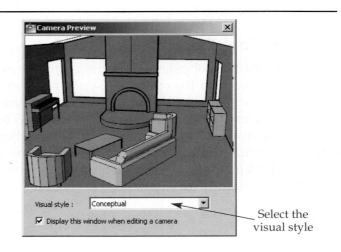

Select the visual style

AutoCAD and Its Applications—Advanced

To enable the back clipping plane, enter **YES** and then specify the offset from the target plane.

The *target plane* is the 2D plane that is perpendicular to the line of sight and passing through the camera's target point. Offsets for both front and back clipping planes are from this plane. Positive values place the clipping planes between the camera and the target plane. Negative values place the planes on the opposite side of the target plane from the camera. You can place the clipping planes anywhere in the scene from the camera location to infinity. You cannot, however, place the back clipping plane between the front clipping plane and the camera.

The best way to adjust clipping planes is using the **Properties** palette or grips. Create the camera and then display a plan view of the camera and target (an approximate plan view is okay). Select the camera and open the **Properties** palette. In the **Clipping** category, select the Clipping property. In the property drop-down list, select Front on, Back on, or Front and back on to turn on the appropriate clipping plane(s). Notice that the clipping planes are visible in the viewport, **Figure 20-3**. Next, enter offset values for the Front plane and Back plane properties as appropriate or use grips to set the locations of the clipping planes. By displaying a plan view of the camera and target, you can see where the clipping planes are located and visualize their effect on the scene. If the **Camera Preview** window is open, the clipping is displayed in the preview.

Exercise 20-1

Complete the exercise on the companion website.
www.g-wlearning.com/CAD

Figure 20-3.
Adjusting the clipping planes for a camera.

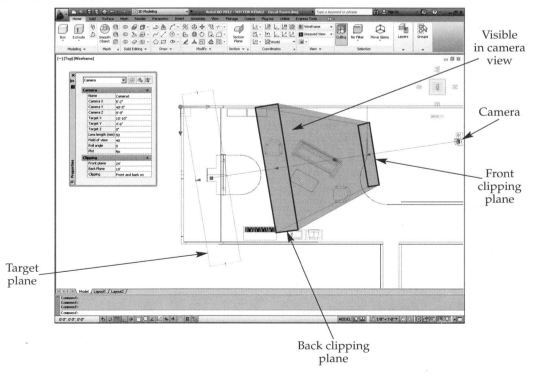

Animation Preparation

The tools presented in this chapter make it easy to lay out a path, plan camera angles, and record the movement of the camera. The resulting animation can be directly output to a number of movie file types that can be shared with others. However, there are some decisions to make first.

It is important to plan out exactly what you want to see in the animation. Think like a movie director and plan the "shots." Ask these questions:

- What will be visible from each camera angle?
- Is there a background in place?
- Is the lighting appropriate?
- Will a simple walkthrough suffice or will a flyby be necessary?
- How close is the viewer (camera) going to be to the objects in the scene?

The answers to these questions will help determine the modeling detail required. Do not model anything that will not be seen. Also, do not place detailed materials on objects that are not the focus of the animation. Processing the animation may take a long time. Unnecessary detail may bog down the computer. In addition, walkthroughs and flybys must be created in views with perspective, not parallel, projection.

The "visual quality" of the scene has the biggest impact on the time involved in rendering the animation. An animation can be rendered in any visual style or using any render preset that is available in the drawing. It is a natural tendency to render at the highest level to make the animation look the best. However, a computer animation has a playback rate of 30 frames per second (fps). If a single frame (view) takes three minutes to render using the Presentation render preset, how long will it take to render a 30 second animation? An animation 30 seconds in length has 900 frames (30 fps × 30 seconds). If each frame takes three minutes to render, the entire animation will take 2700 minutes, or 45 hours, to render.

Are you willing to wait two or three days for a 30 second movie? How about your boss or your client? There are trade-offs and concessions to be made. Perform test renderings on static views and note the rendering time. Then, decide on the acceptable level of quality versus rendering time and move ahead with it.

Walking and Flying

The process for creating a walkthrough or a flyby is the same. First, the command is initiated. Then, the movement is defined and recorded. Finally, the recorded movement is saved to an animation file. Note: the **Animations** panel in the **Render** tab on the ribbon may not be displayed by default.

When using the **3DWALK** and **3DFLY** commands, you can move through the scene using the arrow keys or the [W], [A], [S], and [D] keys on the keyboard to control your movements. Once either command is initiated, a message appears from the **Help** flyout in the **InfoCenter**, if balloon notifications are turned on. If you expand this message, the key movements are explained. See **Figure 20-4**. To redisplay this message while the command is active, press the [Tab] key.

- **Move forward.** Up arrow or [W].
- **Move left.** Left arrow or [A].
- **Move right.** Right arrow or [D].
- **Back up.** Down arrow or [S].

You can also navigate through the scene using the mouse. Press and hold the left mouse button and then drag the mouse in the active viewport to "steer" through the scene. With the **3DWALK** command, the camera remains at the same Z value. With the **3DFLY** command, the Z position of the camera can change. The steps for creating a walkthrough or flyby are provided at the end of this section.

Figure 20-4.
This message from
the **InfoCenter**
shows the keys
that can be used to
navigate through an
animation.

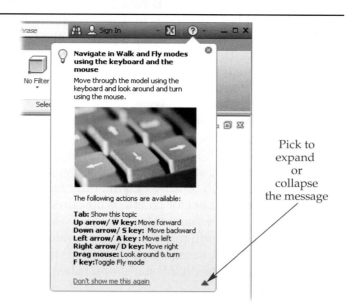

Pick to
expand
or
collapse
the message

Position Locator

When the **3DWALK** or **3DFLY** command is initiated, the **Position Locator** palette
appears. See **Figure 20-5**. The preview in this palette shows a plan view of the scene.
The purpose of this window is to provide an overview of the scene, in plan, while you
develop the animation. It does not need to be displayed to create an animation and can
be closed if it takes up too much space or slows down the rendering.

Position and target indicators appear in the plan view to show the location of
the camera and its target. The green triangular shape displays the field of view. The
field of view (FOV) is the area within the camera's "vision." The field of view indicator
is only displayed when the target indicator is displayed. By default, the position indicator is red. The target indicator is green by default. These properties can be changed
in the **General** category at the bottom of the **Position Locator** palette.

You can reposition the camera and the target in the plan view simply by picking
and dragging either indicator. The effect of the change is visible in the active viewport.
Moving the position and target indicators closer together reduces the field of view.

Figure 20-5.
The **Position Locator**
palette.

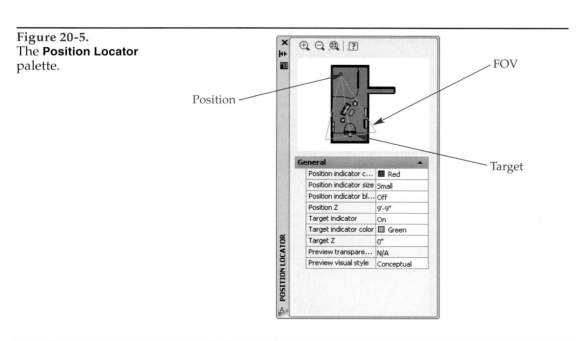

Position

FOV

Target

Picking the field of view lines and dragging moves the position and target indicators at the same time. The Position Z property in the **General** category sets the Z coordinate value for the position indicator. The Target Z property in the **General** category sets the Z coordinate value for the target indicator. The Z coordinate value determines eye level.

In addition to changing the color of the position and target indicators, you can use the properties in the **General** category to modify the display in the **Position Locator** palette. The Position indicator size property determines if both indicators are displayed small, medium, or large. If the Position indicator blink property is set to On, both indicators flash on and off in the preview. The Preview visual style property sets the visual style for the preview. This setting does not affect the current viewport or the animation. The Preview transparency property is set to 50% by default, but can be changed to whatever you want. If the view in the **Position Locator** palette is obscured by something (a roof, perhaps), you may want to set the Preview visual style property to Hidden and the Preview transparency property to 80% or 90%. This will make the objects under the roof visible. If hardware acceleration is on, then the Preview transparency property is disabled. The **Performance Tuner** tool in the status bar indicates the on/off status of hardware acceleration.

Exercise 20-2

Complete the exercise on the companion website.
www.g-wlearning.com/CAD

Walk and Fly Settings

WALKFLYSETTINGS

Ribbon
**Render
> Animations**

**Walk and Fly
Settings**

Type
WALKFLYSETTINGS

General settings for walkthroughs and flybys are made in the **Walk and Fly Settings** dialog box. See **Figure 20-6**. Open this dialog box by picking the **Walk and Fly Settings** button in the **Animations** panel on the **Render** tab of the ribbon (in the **Walk** flyout). This panel may not be displayed by default. The dialog box can also be displayed by picking the **Walk and Fly...** button in the **3D Modeling** tab of the **Options** dialog box.

The three radio buttons at the top of the dialog box are used to determine when the message shown in **Figure 20-4** is displayed. The check box determines if the **Position Locator** palette is automatically displayed when the **3DWALK** or **3DFLY** command is entered.

The text boxes in the **Current drawing settings** area determine the size of each step and the number of steps per second. The **Walk/fly step size:** setting controls the **STEPSIZE** system variable. This is the number of units that the camera moves in one step. The **Steps per second:** setting controls the **STEPSPERSEC** system variable. This is the number of steps the camera takes each second. Together, these two settings determine how fast the camera moves in the animation.

Figure 20-6.
General settings for the walkthrough or flyby are made in the **Walk and Fly Settings** dialog box.

Check to automatically display the **Position Locator** palette

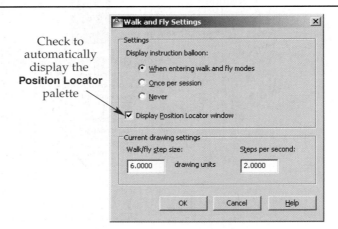

You will have to experiment with step size and steps per second values to make an animation that is easy to watch. Start with low numbers (for slow movement) and work your way up. Fast movement is disorienting and makes the viewer feel as if something was not seen, or missed. The viewer should be able to watch at a comfortable pace and get a good look at your design.

To get a feel for the proper speed for a walkthrough, pay attention to the next movie or TV show that you watch. When the director wants you to get a good look at the setting for the scene, the camera very slowly pans around the room. To emphasize distance, the camera slowly zooms in to a target object or person.

Camera Tools

The expanded **View** panel on the **Home** tab of the ribbon contains some tools for quickly adjusting the camera before starting the animation. See **Figure 20-7**. The **Lens length** slider controls how much of the scene is seen by the camera. The *lens length* refers to the focal length of the camera lens. The higher the number, the closer you are to the subject. The range is from about 1 to 100,000; 50 is a good starting point. There are stops on the slider for standard lens lengths. You can enter a specific value for the lens length or field of view by selecting the text in the slider, typing the value, and pressing [Enter]. The current view must be a camera view for the slider to be enabled.

Below the **Lens length** slider are text boxes for the camera and target positions. These can be used to change the X, Y, and Z coordinates of the camera or target before starting the animation.

Animation Tools

The **Animations** panel on the **Render** tab of the ribbon contains the tools for controlling the recording and playback of the animation. See **Figure 20-8**. Remember, this panel may not be displayed by default. The **Record Animation** button is used to

Figure 20-7.
Camera tools are located in the expanded **View** panel on the **Home** tab of the ribbon.

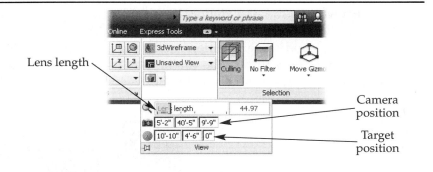

Figure 20-8.
The **Animations** panel on the **Render** tab of the ribbon is where you can record and play back the walkthrough or flyby animation. This panel is not displayed by default.

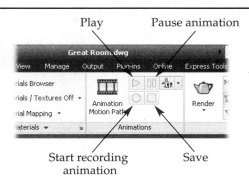

initiate recording of camera movement. After the **3DWALK** or **3DFLY** command is activated, pick the button to start recording. Make sure that you are ready to start moving when you pick the button because recording starts as soon as it is picked.

Picking the **Pause Animation** button temporarily stops recording. This allows you to make adjustments to the view without recording the adjustment. When you are ready to begin recording again, pick the record button to resume.

Picking the **Play Animation** button stops the recording and opens the **Animation Preview** dialog box, where the animation is played. See **Figure 20-9**. The controls in this dialog box can be used to rewind, pause, and play the animation. The slider can be dragged to preview part of the animation or move to a specific frame. The visual style can also be set using the drop-down list. If the animation is created using a render preset, the file must be played in Windows Media Player or another media player to view the rendered detail. Render presets are available in the **Animation Settings** dialog box, as discussed in the next section.

Picking the **Save Animation** button in the **Animation** panel stops recording and opens the **Save As** dialog box. Name the animation file, navigate to a location, and pick the **Save** button.

CAUTION

While the **3DWALK** or **3DFLY** command is active and the record button is on, you are creating an animation. If you move the camera in the **Position Locator** palette and start re-recording the animation to correct a problem, but do not first exit the current **3DWALK** or **3DFLY** command session, you are adding another segment to the animation you just previewed. To start over, first exit the current command session.

Exercise 20-3

Complete the exercise on the companion website.
www.g-wlearning.com/CAD

Animation Settings

The **Animation Settings** dialog box may contain the most important settings pertaining to walkthroughs and flybys. See **Figure 20-10**. These animation settings determine how good the animation looks, how long it is going to take to complete, and how big the file will be. The dialog box is displayed by picking the **Animation settings...** button in the **Save As** dialog box displayed when saving an animation.

Figure 20-9.
The animation is played in the **Animation Preview** dialog box.

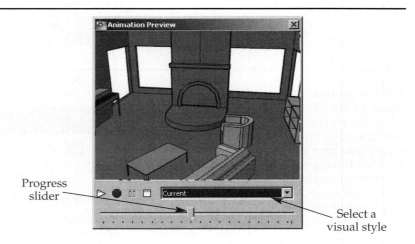

Progress slider

Select a visual style

Figure 20-10.
The **Animation Settings** dialog box contains important settings pertaining to walkthroughs and flybys.

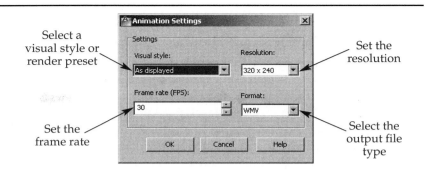

Select a visual style or render preset

Set the frame rate

Set the resolution

Select the output file type

The **Visual style:** drop-down list is used to set the shading level in the animation. The name of this drop-down list is a little misleading because visual styles and render presets are available. The higher the shading or rendering level selected in this drop-down list, the longer the rendering time and the bigger the file size. If you have numerous lights casting shadows, detailed materials, and global illumination and final gathering enabled, settle in for a long wait. A simple, straight-ahead walkthrough of 10 or 15 feet can easily result in 300 frames of animation. If each frame takes about five seconds to render, that equals 1500 seconds, or 25 minutes, to create an animation file that is only 10 seconds long.

The **Frame rate (FPS):** text box sets the number of frames per second for the playback. In other words, this sets the speed of the animation playback. The default is 30 fps, which is a common playback rate for computers.

The **Resolution:** drop-down list offers standard choices of resolution, from 160 × 120 to 1024 × 768. These are measured in pixels × pixels. Remember, higher resolutions mean longer processing times and larger file sizes.

The **Format:** drop-down list is used to select the output file type. The file type must be set in this dialog box. It cannot be changed in the **Save As** dialog box. The choices of output file type are:

- **WMV.** The standard movie file format for Windows Media Player.
- **AVI.** Audio-Video Interleaved is the Windows standard for movie files.
- **MOV.** QuickTime® Movie is the standard file format for Apple® movie files.
- **MPG.** Moving Picture Experts Group (MPEG) is another very common movie file format.

Depending on the configuration of your computer, you may not have all of these file type options or you may have additional options not listed here.

PROFESSIONAL TIP

Other AutoCAD navigation modes may be used to create a walkthrough or flyby. Any time that you enter constrained, free, or continuous orbit, you can pick the **Record Animation** button to record the movements. After you activate the **3DWALK** or **3DFLY** command, right-click and experiment with some of the other options as you record your animation. You can even combine some of these navigation modes with walking or flying. For example, use **3DFLY** to zoom into a scene, pause the animation, switch to constrained orbit, restart the recording, and slowly circle around your model. Animation settings are also available in this shortcut menu.

Exercise 20-4

Complete the exercise on the companion website.
www.g-wlearning.com/CAD

Steps to Create a Walkthrough or Flyby

1. Plan your animation. Determine where you are moving from and to, what you are going to be looking at, and what will be the focal point of the scene.
2. Set up a multiple-viewport configuration of three or four viewports.
3. In one of the viewports, create or restore a named view with the appropriate starting viewpoint. Make sure a background is set up, if desired.
4. Start the **3DWALK** or **3DFLY** command. Note in the **Position Locator** palette the camera location, target location, and field of view. Adjust these in the expanded **View** panel on the ribbon, if necessary.
5. Position your fingers over the navigation keys on the keyboard.
6. Pick the **Record Animation** button.
7. Practice navigating through the view and then pick the **Play Animation** button to preview your animation. When you are done practicing, make sure to cancel the command and reposition the camera at the starting point.
8. Pick the **Record Animation** button to start over.
9. Start navigating through the view. Try to keep the movements as smooth as possible. Any jerks and shakes will be visible in the animation.
10. When you are done, stop moving forward and then pick the stop (**Save Animation**) button.
11. In the **Save As** dialog box, pick the **Animation Settings...** button. In the **Animation Settings** dialog box, select the desired visual style, frame rate, resolution, and output file format. Pick the **OK** button to close the **Animation Settings** dialog box. In the **Save As** dialog box, name and save the file.
12. The **Creating Video** dialog box is displayed as AutoCAD processes the frames, **Figure 20-11**.
13. When the **Creating Video** dialog box is automatically closed, the animation file is saved and you can take a look at it. Pick the **Play Animation** button and watch the animation in the **Animation Preview** window. You can also locate the file using Windows Explorer. Then, double-click on the file to play the animation in Windows Media Player (or whichever program is associated with the file type).
14. Exit the command. If you are not satisfied with the results and want to try it again, make sure to exit the command before you make another attempt at the walkthrough or flyby.

Figure 20-11.
The **Creating Video** dialog box is displayed as AutoCAD generates the animation.

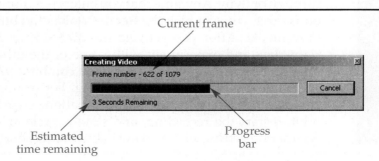

Current frame

Estimated time remaining

Progress bar

Motion Path Animation

You may have found it difficult to create smooth motion using the keyboard and mouse. Fortunately, AutoCAD provides an easy way to create a nice, smooth animated walkthrough or flyby. This is done through the use of a motion path. A *motion path* is simply a straight or curved path along which the camera, target, or both travel during the animation.

One method of using a motion path is to link the camera and target to a single path. The camera and its line of sight then follow the path much like a train follows tracks. See **Figure 20-12.**

Another option when using a motion path is to link the camera to a single point in the scene and the target to a path. For example, the target can be set to follow a circle or arc. The camera swivels on the point and "looks at" the path as if it is being rotated on a tripod. See **Figure 20-13.**

Figure 20-12.
A—The camera and target are linked to the same path (shown in color). B—The camera looks straight ahead as it moves along the path.

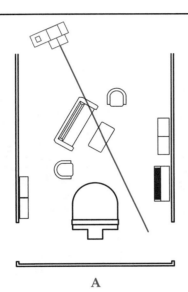

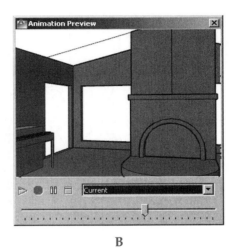

A   B

Figure 20-13.
A—The camera is linked to a point so it remains stationary. The target is linked to the circle. B—The camera view rotates around the room as if the camera is on a swivel tripod.

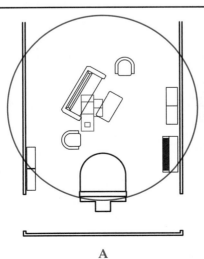

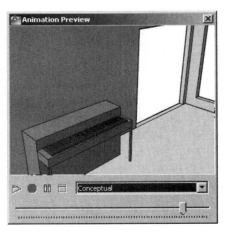

A   B

A third way to use a motion path is to have the camera follow a path, but have the target locked onto a stationary point. This is similar to riding in a vehicle and watching an object of interest on the side of the road. As the vehicle moves, your gaze remains fixed on the object. See **Figure 20-14**.

The fourth method of using a motion path is to have both the camera and target follow separate paths. Picture yourself walking into an unfamiliar room. As you walk into the center of the room, your gaze sweeps left and right across the room. In this case, the camera (you) follows a straight line path and the target (your gaze) follows an arc from one side of the room to the other.

The **ANIPATH** command is used to assign motion paths. The command opens the **Motion Path Animation** dialog box. See **Figure 20-15**. This dialog box has three main areas: **Camera**, **Target**, and **Animation settings**. These areas are described in detail in the next sections. The steps for creating a motion path animation are provided at the end of this section.

<table>
<tr><td>ANIPATH</td></tr>
<tr><td>Ribbon</td></tr>
<tr><td>Render
> Animations</td></tr>
<tr><td>Animation Motion
Path</td></tr>
<tr><td>Type</td></tr>
<tr><td>ANIPATH</td></tr>
</table>

Figure 20-14.
A—The camera is linked to the spline path and the target is linked to the point (shown in color). B—As the camera moves along the path, it always looks at the point.

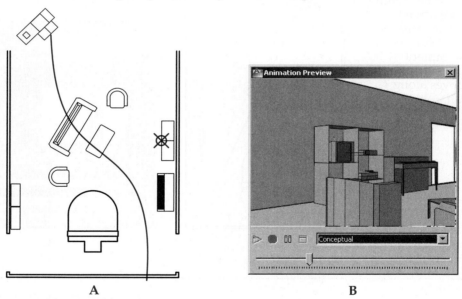

A	B

Figure 20-15.
The **Motion Path Animation** dialog box is used to create an animation that follows a path.

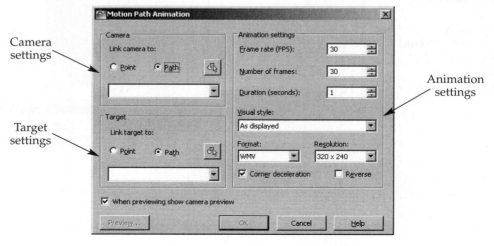

Selecting a motion path automatically creates a camera. You cannot add a motion path to an existing camera.

Camera Area

The camera can be linked to a path or a point. To select a path, pick the **Path** radio button and then pick the "select" button next to the radio buttons. The dialog box is temporarily closed for you to select the path in the drawing. The path may be a line, arc, circle, ellipse, elliptical arc, polyline, 3D polyline, spline, or helix, but it must be drawn before the **ANIPATH** command is used. Splines are nice for motion paths because they are smooth and have gradual curves. The camera moves from the first point on the path to the last point on the path, so create paths with this in mind.

To select a stationary point, pick the **Point** radio button. Then, pick the "select" button next to the radio buttons. When the dialog box is hidden, specify the location in the drawing. You can use object snaps or enter coordinates. It may be a good idea to have a point drawn and use object snaps to select the point.

The camera must be linked to either a path or a point. If neither is selected, the command cannot be completed. If you want the camera to remain stationary as the target moves, select the **Point** radio button and then pick the stationary point in the drawing.

Once a point or path has been selected, it is added to the drop-down list. All named motion paths and selected motion points in the drawing appear in this list. Instead of using the "select" button, you can select the path or point in this drop-down list.

Target Area

The target is the location where the camera points. Like the camera, the target can be linked to a point or a path. To link the target to a path, select the **Path** radio button. Then, pick the "select" button and select the path in the drawing. If the camera is linked to a point, the target must be linked to a path. If the camera is set to follow a path, then you actually have three choices for the target. It can be linked to a path, point, or nothing. To link the target to a point, pick the **Point** radio button. Then, pick the "select" button to select the point in the drawing. The None option, which is selected in the drop-down list, means that the camera will look straight ahead down the path as it moves.

Animation Settings Area

Most of the settings in the **Animation Settings** area have the same effect as the corresponding settings in the **Animation Settings** dialog box discussed earlier. However, there are four settings unique to the **Motion Path Animation** dialog box.

The **Number of frames:** text box is used to set the total number of frames in the animation. Remember, a computer has a playback rate of 30 fps. Therefore, if the frame rate is set to 30, set the number of frames to 450 to create an animation that is 15 seconds long ($30 \times 15 = 450$).

The value in the **Duration (seconds):** text box is the total time of the animation. This value is automatically calculated based on the frame rate and number of frames. However, you can enter a duration value. Doing so will automatically change the number of frames based on the frame rate.

By default, the **Corner deceleration** check box is checked. This slows down the movement of the camera and target as they reach corners and curves on the path. If this is unchecked, the camera and target move at the same speed along the entire path, creating very jerky motion on curves and at corners. It is natural to decelerate on curves.

The **Reverse** check box simply switches the starting and ending points of the animation. If the camera (or target) travels from the first endpoint to the second endpoint, checking this check box makes the camera (or target) travel from the second endpoint to the first.

Previewing and Completing the Animation

To preview the animation, pick the **Preview...** button at the bottom of the **Motion Path Animation** dialog box. The camera glyph moves along the path in all viewports. If the **When previewing show camera preview** check box is checked, the **Animation Preview** window is also displayed and shows the animation.

To finish the animation, pick the **OK** button in the **Motion Path Animation** dialog box. The **Save As** dialog box is displayed. Name the file and specify the location. If you need to change the file type, pick the **Animation settings...** button to open the **Animation Settings** dialog box. Change the file type, close the dialog box, and continue with the save.

Steps to Create a Motion Path Animation

1. Plan your animation. Determine where you are moving from and to, what you are going to be looking at, and what will be the focal point of the scene.
2. Draw the paths and points to which the camera and target will be linked. Draw the path in the direction the camera should travel (first point to last point). Do not draw any sharp corners on the paths and make sure that the Z value (height) is correct.
3. Start the **ANIPATH** command.
4. Pick the camera path or point.
5. Pick the target path or point (or None).
6. Adjust the frames per second, number of frames, and duration to set the length and speed of the animation.
7. Select a visual style, the file format, and the resolution.
8. Preview the animation. Adjust settings, if needed.
9. Save the animation to a file.

Exercise 20-5

Complete the exercise on the companion website.
www.g-wlearning.com/CAD

Chapter Review

Answer the following questions. Write your answers on a separate sheet of paper or complete the electronic chapter review on the companion website.
www.g-wlearning.com/CAD

1. Which system variable controls the display of camera glyphs?
2. Name the three camera tools available on the **Cameras** tool palette and explain the differences between them.
3. When is the **Camera Preview** window displayed, by default?
4. From where is the offset distance for the camera clipping planes measured?
5. What is the difference between the **3DWALK** and **3DFLY** commands?
6. How do you "steer" your movement when creating a walkthrough or flyby animation?
7. What is the *field of view*?
8. What is the purpose of the **Position Indicator** palette?

9. In the **Walk and Fly Settings** dialog box, which settings combine to control the speed of the animation?
10. How do you start recording a walkthrough or flyby?
11. What must be done before correcting a motion error in a walkthrough or flyby?
12. Motion path animation involves linking a camera or target to _____ or _____.
13. Which types of objects may be used as a motion path?
14. If None is selected as the target "path," what does the camera do in the animation?
15. Explain *corner deceleration*.

Drawing Problems

1. In this problem, you will create and manipulate a camera in a drawing from a previous chapter.
   A. Open drawing P19_02 from Chapter 19 and save it as P20_01.
   B. Create at least two viewports and display a plan view in one of them.
   C. Use the **CAMERA** command to create a camera looking at the objects from the southwest quadrant. Change the camera settings as needed to display a pleasing view of the scene.
   D. Name the camera SW View.
   E. Turn on both the front and back clipping planes. Adjust them to eliminate one object in the front and one object in the back.
   F. Open the **Cameras** tool palette and, using the tools in the palette, create three more cameras looking at the scene from various locations. Change their names to Normal, Wide-angle, and Fish-eye to match the type of camera.
   G. Save the drawing.

2. In this problem, you will draw some basic 3D shapes to represent equipment in a small workshop. Then, you will create an animated walkthrough.
   A. Start a new drawing and set the units to architectural. Save it as P20_02.
   B. Draw a planar surface that is 15′ × 30′.
   C. Draw three 9′ tall walls enclosing the two long sides and one short side.
   D. Use boxes and a cylinder to represent equipment and shelves. Refer to the illustration shown. Use your own dimensions.
   E. Use the **3DWALK** command to create an animation of walking into the workshop. Turn and look at the shelves at the end of the animation.
   F. Set the visual style to Conceptual and the resolution to 640 × 480.
   G. Save the animation as P20_02.avi (or the file format of your choice).
   H. Save the drawing.

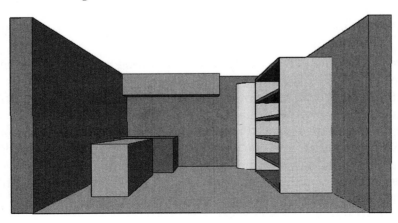

Chapter 20   Cameras, Walkthroughs, and Flybys

3. In this problem, you will create a motion path animation for the workshop drawn in Problem 20-2.
   A. Open drawing P20_02 and save it as P20_03.
   B. Draw a line and an arc similar to those shown (in color). The dimensions are not important.
   C. Move both objects so they are 4' off of the floor.
   D. Using the **ANIPATH** command, link the camera to the line and the target to the arc. Set the resolution to 640 × 480.
   E. Preview the animation. Adjust the animation settings as necessary. You may need to slow down the animation quite a bit. How do you do this?
   F. Save the animation as P20_03.wmv (or the file format of your choice).
   G. Save the drawing.

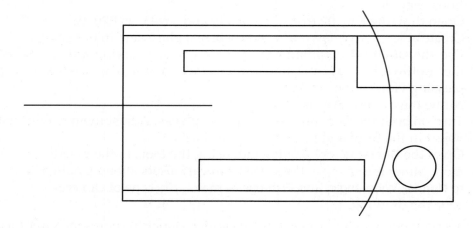

4. In this problem, you will create a motion path animation for the presentation of a mechanical drawing from a previous chapter.
   A. Open drawing P18_03 and save it as P20_04.
   B. Draw a circle centered on the flange with a radius of 300.
   C. Move the circle 200 units in the Z direction.
   D. Using the **ANIPATH** command, link the camera to the circle and the target to the center of the flange.
   E. Preview the animation and adjust the animation settings as necessary. Due to the materials in the scene, the preview may play slowly, depending on the capabilities of your computer.
   F. Select a visual style that your computer can handle. Set the resolution to 320 × 240.
   G. Save the animation as P20_04.avi (or the file format of your choice).
   H. Save the drawing.

5. In this problem, you will create a flyby of the building that you created in Chapter 18.
   A. Open drawing P18_01 and save it as P20_05.
   B. Create a perspective view of the scene that shows the building from slightly above it. Save the view.
   C. Start the **3DFLY** command. Practice with the movement keys to make sure you know how to fly around the building. Then, cancel the command.
   D. Restore the starting view and select the **3DFLY** command.
   E. Record the flyby and save the animation as P20_05.wmv (or the file format of your choice).
   F. Save the drawing.

Customizing the AutoCAD Environment

Learning Objectives

After completing this chapter, you will be able to:

✓ Identify and manage AutoCAD file and system settings.
✓ Assign colors and fonts to the text and graphics windows.
✓ Control AutoCAD system variables.
✓ Set options that control display quality and AutoCAD performance.
✓ Control shortcut menus.

AutoCAD provides a variety of options for customizing the user interface and working environment. These options permit users to configure the software to suit personal preferences. You can define colors for the individual window elements, assign preferred fonts to the command line window, control shortcut menus, and assign properties to program icons.

The options for customizing the AutoCAD user interface and working environment are found in the **Options** dialog box, **Figure 21-1**. This dialog box is commonly accessed by right-clicking in the drawing area or command line with nothing selected and no command active and picking **Options...** from the shortcut menu. It can also be displayed by picking the **Options** button at the bottom of the application menu or by entering the **OPTIONS** command.

OPTIONS	
Application Menu	
Options	
Type	
OPTIONS OP	

Changes made in the **Options** dialog box do not take effect until either the **Apply** or **OK** button is picked. If you pick the **Cancel** button or the close button (**X**) before picking **Apply**, all changes are discarded. Each time you change the options settings, the system registry is updated and the changes are used in this and subsequent drawing sessions. Settings that are stored within the drawing file have the AutoCAD drawing icon next to them. These settings do not apply to other drawings. Settings without the icon affect all AutoCAD drawing sessions.

Many of the options and settings specified in AutoCAD are made through the use of system variables. *System variables* store values that control AutoCAD functions. A system variable can be changed by typing the variable name and specifying a value. When setting options with AutoCAD commands, system variables are often changed dynamically. While different options exist for managing the settings of system variables in AutoCAD, the simplest method is to use the **Options** dialog box.

Figure 21-1.
The **Options** dialog
box is used to
customize the
AutoCAD working
environment. Each
tab contains a
variety of options
and settings.

Select the
appropriate tab

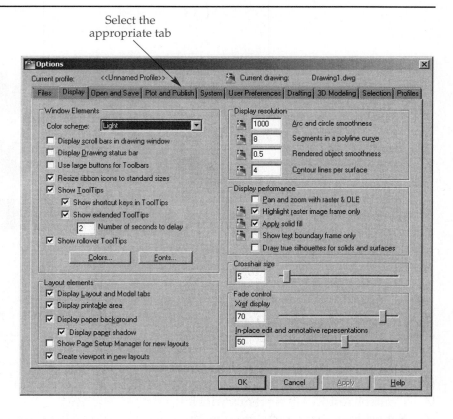

Managing AutoCAD File and Folder Paths

There are numerous settings that control the manner in which AutoCAD behaves in the Windows environment. AutoCAD stores settings to specify such items as which folders to search for driver and menu files and the location of temporary and support files. The default settings created during installation are often adequate, but changing the settings may result in better performance. The following sections discuss the settings available in the **Files** tab of the **Options** dialog box.

File Locations

When AutoCAD is used in a network environment, some files pertaining to AutoCAD may reside on a network drive so all users can access them. Other files may reside in folders specifically created for a particular AutoCAD user. These files may include external reference files, custom menu files, and drawings containing blocks.

The **Files** tab of the **Options** dialog box is used to specify the path AutoCAD searches to find support files and driver files. It also contains the paths where certain types of files are saved and where AutoCAD looks for specific types of files. Support files include text fonts, menus, AutoLISP files, ObjectARX files, blocks, linetypes, and hatch patterns.

The folder names shown under the Support File Search Path heading in the **Search paths, file names, and file locations:** list are automatically created by AutoCAD during the installation. For example, **Figure 21-2** shows that the support files are stored in seven different folders. Folders are searched in the order in which they are listed under Support File Search Path. As previously mentioned, some of these paths are created for a specific user. The first path listed consists of several subfolders and ultimately ends with the user-specific \Support folder. In the example shown, this path starts on the C: drive in a folder named \Users. This path may be different depending on the version of

Figure 21-2.
Folder paths can be customized in the **Files** tab of the **Options** dialog box.

Folder for
specific user

Add a new folder
to the selected path

Folders in
support
search
path

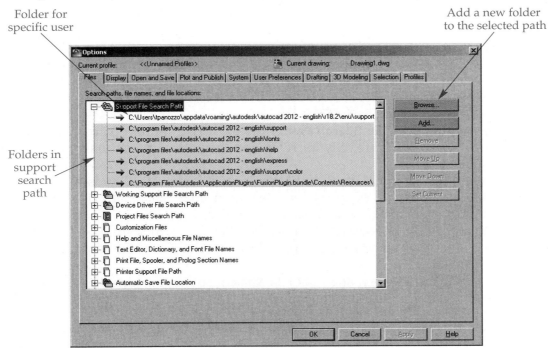

Windows you are using. The next folders to be searched are, in order, \Support, \Fonts, \Help, \Express, \Support\Color, and a \Resources folder for Inventor Fusion (if Inventor Fusion is installed). These paths are not user specific; they are located in the AutoCAD installation path.

You can add the path of any new folders you create that contain support files. As an example, suppose all of the blocks you typically use are stored in a separate folder named \Blocks. By placing this folder name in the support files search path, it is not necessary to include the entire path to the blocks in any command macros you create to automatically insert your blocks (creating command macros is discussed in Chapter 22).

A folder can be added to the existing search path in two ways. The first method is to highlight the Support File Search Path heading and pick the **Add...** button. This places a new, empty listing under the heading. You can now type C:\Blocks to complete the entry. Alternately, instead of typing the path name, after picking **Add...** you can pick the **Browse...** button to display the **Browse for Folder** dialog box. You can then use this dialog box to select the desired folder. The new setting takes effect as soon as you pick **Apply** or the **OK** button and close the **Options** dialog box.

PROFESSIONAL TIP

In addition to the paths listed under Support File Search Path, AutoCAD will search two other folders. The folder that contains the AutoCAD executable file, acad.exe, (typically, C:\Program Files\ Autodesk\AutoCAD 2012) is searched, as is the folder that contains the current drawing file. These folders are searched only if the desired file name is not found in any of the listed folders. It is not advisable to store files such as block files in the AutoCAD folder. However, if you store block files in the same folder as the current drawing, you do not need to add that folder to the search path.

Other File Settings

Another setting that can be specified in the **Files** tab is the location of device driver files. *Device drivers* are specifications for peripherals that work with AutoCAD and other Autodesk products. By default, the drivers supplied with AutoCAD are placed in the \Drv folder. If you purchase a third-party driver to use with AutoCAD, be sure the driver is loaded into this folder. If the third-party driver must reside in a different folder, you should specify that folder using the Device Driver File Search Path setting. Otherwise, the search for the correct driver is widespread and likely to take longer. Some other file paths listed in the **Files** tab include:

- **Working Support File Search Path.** Lists the active support paths AutoCAD is using. These paths are only for reference; the paths are read-only and cannot be amended.
- **Project Files Search Path.** Specifies project path names and folders.
- **Customization Files.** Specifies the names of the main and enterprise customization files. Also, the location of custom icon files is specified.
- **Help and Miscellaneous File Names.** Specifies which files are used for the help file and the default Internet location. Also, the location of the configuration file is specified.
- **Text Editor, Dictionary, and Font File Names.** Specifies which files are used for the text editor application, the main and custom dictionaries, the alternate font file, and the font mapping file.
- **Print File, Spooler, and Prolog Section Names.** Sets the file names for the plot file for legacy plotting scripts, the print spool executable file, and the PostScript prolog section name.
- **Printer Support File Path.** Specifies the print spool file path, the printer configuration search path, the printer description file search path, and the plot style table search path.
- **Automatic Save File Location.** Sets the path where the autosave (.sv$) file is stored. An autosave file is only created if the **Automatic save** option is checked in the **Open and Save** tab of the **Options** dialog box.
- **Color Book Locations.** Specifies paths for color book files that can be used when specifying colors in the **Select Color** dialog box.
- **Data Sources Location.** Specifies the path for database source (.udl) files.
- **Template Settings.** Specifies the default location for drawing and sheet set template files and the file names for the defaults.
- **Tool Palettes File Locations.** Specifies the path for tool palette support files.
- **Authoring Palette File Locations.** Specifies the location of authoring palette files.
- **Log File Location.** Specifies the path for the AutoCAD log file. A log file is only created if the **Maintain a log file** option is checked in the **Open and Save** tab of the **Options** dialog box.
- **Action Recorder Settings.** Specifies search and storage paths for the action recorder macro (.actm) files.
- **Plot and Publish Log File Location.** Specifies the path for the log file for "plot and publish" operations. A log file is only created if the **Automatically save plot and publish log** check box in the **Plot and Publish** tab of the **Options** dialog box is checked.
- **Temporary Drawing File Location.** Sets the folder where AutoCAD stores temporary drawing files.
- **Temporary External Reference File Location.** Indicates where temporary external reference files are placed.
- **Texture Maps Search Path.** Specifies the location of texture map files for rendering.
- **Web File Search Path.** Specifies the folders to search for files associated with photometric weblight lighting.

- **i-drop Associated File Location.** Specifies the folder used by default to store downloaded i-drop content.
- **DGN Mapping Setups Location.** Specifies the location of the mapping files for translation to and from MicroStation drawing files (.dgn files). This folder must have read/write file permissions enabled in order for the MicroStation commands to properly function.

Customizing the Graphics Window

Numerous options are available to customize the graphics window to your personal liking. Select the **Display** tab in the **Options** dialog box to view the display-control options, **Figure 21-3**.

The **Window Elements** area has a setting for the color scheme. The color scheme controls the outline color of the ribbon and status bar. Using the drop-down list, you can select between light and dark color schemes.

There are also check boxes for turning the scroll bars and drawing status bar on or off; using large buttons for toolbars; resizing ribbon icons to standard sizes; showing tooltips; showing shortcut keys and extended commands in tooltips; and showing rollover tooltips. The element colors and font settings can be changed using the buttons at the bottom of the **Window Elements** area.

On the right-hand side of the tab near the bottom, the **Crosshair size** setting is available for controlling the size of the crosshairs. The setting is a percentage of the drawing screen area. The higher the value, the further the crosshairs extend. The settings in the **Fade control** area determine the display intensity of the unselected objects in reference edit mode and in-place edit and annotative representations. A higher value means the unselected objects are less visible. Other options are discussed in the next sections.

Figure 21-3.
Use the **Display** tab to set up many of the visual elements of the AutoCAD environment.

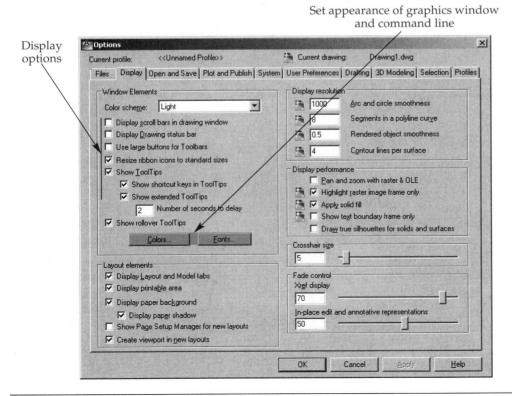

Changing Colors

By customizing colors, you can add your personal touch and make AutoCAD stand out among other active Windows applications. AutoCAD provides this capability with the **Drawing Window Colors** dialog box, **Figure 21-4**. This dialog box is accessed by picking the **Colors...** button in the **Window Elements** area of the **Display** tab in the **Options** dialog box.

To change a color, first select a context. A *context* is one of the environments, or modes, in AutoCAD, such as the 3D perspective projection mode that is set current when a new drawing is started based on the acad3D.dwt template. The **Context:** list box contains the names of all contexts. The context that was current when the **Options** dialog box is opened is initially selected. A preview of the context and its settings is displayed in the **Preview:** area at the bottom of the dialog box.

Each context contains several interface elements. An *interface element* is an item that is visible, or can be made visible, in a given context, such as the grid axis, autosnap marker, or light glyphs. Once a context is selected, pick the element to change in the **Interface element:** list.

With a context and element selected, the color of the element can be changed. Use the **Color:** drop-down list to change the color. If you pick the Select Color... entry, the **Select Color** dialog box is displayed. Below the **Color:** drop-down list is the **Tint for X, Y, Z** check box. This check box is available when certain elements are selected. When checked, a tint is applied along the X, Y, and Z axes. The elements to which a tint can be applied are: crosshairs, autotrack vector, drafting tooltip background, grid major lines, grid minor lines, and grid axis lines.

Along the right side of the dialog box are buttons for restoring the default settings. Picking the **Restore current element** button resets the currently selected element to its default color. Picking the **Restore current context** button resets *all* of the elements of

Figure 21-4.
Change AutoCAD color settings using the **Drawing Window Colors** dialog box.

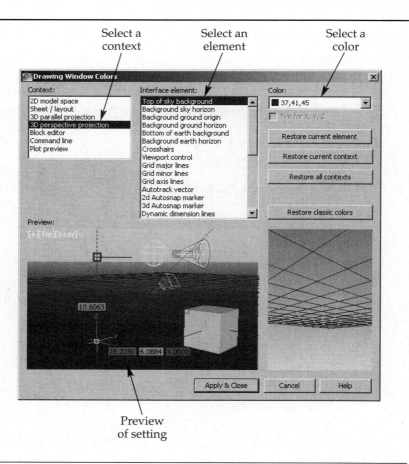

Select a context

Select an element

Select a color

Preview of setting

the currently selected context to their default colors. Picking the **Restore all contexts** button resets *all* of the elements in *all* of the contexts to their default colors. Lastly, picking the **Restore classic colors** button sets the AutoCAD graphics screen to the traditional black background in 2D model space and the grid and glyph colors to appropriate colors for the black background. These are the settings found in older releases of AutoCAD.

Once the colors are changed as needed, pick the **Apply & Close** button. Then, pick the **OK** button in the **Options** dialog box. The graphics window regenerates and displays the color changes you made.

Changing Fonts

You can change the font used in the command line window. The font you select has no effect on the text in your drawings, nor is the font used in the AutoCAD dialog boxes, the ribbon, or the pull-down menus.

To change the font used in the command line window, pick the **Fonts...** button in the **Display** tab of the **Options** dialog box. The **Command Line Window Font** dialog box appears, **Figure 21-5**.

The default font used by AutoCAD for the command line window is Courier New. The font style for Courier New is regular (not bold or italic) and the default size is 10 points. Select a new font from the **Font:** list. This list displays the system fonts available for use. Also, set a style and size. The **Sample Command Line Font** area displays a sample of the selected font. Once you have selected the desired font, font style, and font size for the command line window, pick the **Apply & Close** button to assign the new font.

Figure 21-5.
The command line window font can be changed to suit your preference.

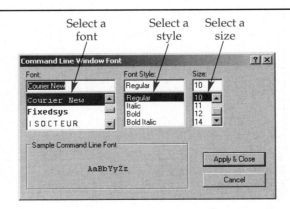

Exercise 21-1

Complete the exercise on the companion website.
www.g-wlearning.com/CAD

Layout Display Settings

The appearance of a layout in paper space is different from the appearance of model space. A layout is made active by selecting a layout tab. The theory behind the default layout tab settings is to provide a picture of what the drawing will look like when plotted. You can see if the objects will fit on the paper or if some of the objects are outside of the margins. The following options, which are found in the **Layout elements** area of the **Display** tab in the **Options** dialog box, are illustrated in **Figure 21-6**.

- **Display Layout and Model tabs.** Displays the **Model** and layout tabs at the bottom of the drawing screen area. This is checked by default.
- **Display printable area.** The margins of the printable area are shown as dashed lines on the layout paper. Any portion of an object outside of the margins is not plotted.
- **Display paper background.** Displays the paper size specified in the page setup.
- **Display paper shadow.** Displays a shadow to the right and bottom of the paper. This option is only available if **Display paper background** is checked.
- **Show Page Setup Manager for new layouts.** Determines if the **Page Setup** dialog box is displayed when a new layout is selected or created. By default, this option is unchecked.
- **Create viewport in new layouts.** Determines whether a viewport is automatically created when a new layout is selected or created. Many users uncheck this option since they will be creating their own floating viewports.

Figure 21-6.
Customizing the display of layouts. These options are set in the **Options** dialog box.

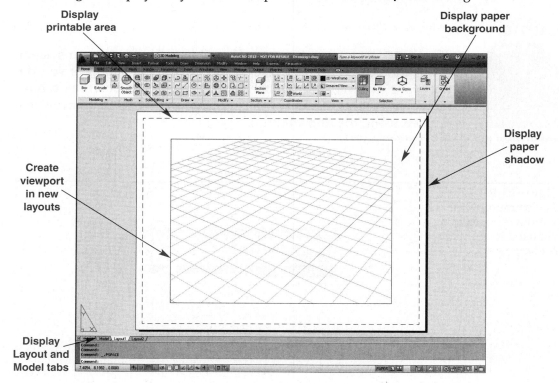

Display Resolution and Performance Settings

The settings in the **Display resolution** and **Display performance** areas of the **Display** tab in the **Options** dialog box affect the performance of AutoCAD. The settings can affect regeneration time and realtime panning and zooming. The following options are available in the **Display resolution** area. If the AutoCAD drawing icon is shown next to the setting, the value is saved in the current drawing, not in the AutoCAD system registry.

- **Arc and circle smoothness.** This setting controls the smoothness of circles, arcs, and ellipses. The default value is 1000; the range is from 1 to 20000. The value is also controlled with the **VIEWRES** command.
- **Segments in a polyline curve.** This value determines how many line segments will be generated for each spline-fit polyline. The default value is 8; the range is a nonzero value from –32768 to 32767. The system variable equivalent is **SPLINESEGS**.
- **Rendered object smoothness.** This setting controls the smoothness of curved solids when they are hidden, shaded, or rendered. This value is multiplied by the **Arc and circle smoothness** value. The default value is 0.5; the range is from 0.01 to 10. The system variable equivalent is **FACETRES**.
- **Contour lines per surface.** This value controls the number of contour lines per surface on solid objects. The default value is 4; the range is from 0 to 2047. The system variable equivalent is **ISOLINES**.

The following options are available in the **Display performance** area.

- **Pan and zoom with raster & OLE.** If this is checked, raster images are displayed when panning and zooming. If it is unchecked, only the frame is displayed during the operation. The system variable equivalent is **RTDISPLAY**.
- **Highlight raster image frame only.** If this is checked, only the frame around a raster image is highlighted when the image is selected. If this option is unchecked, the image displays a diagonal checkered pattern to indicate selection. The system variable equivalent is **IMAGEHLT**.
- **Apply solid fill.** Controls the display of solid fills in objects. Affected objects include hatches, wide polylines, solids, and multilines. The system variable equivalent is **FILLMODE**.
- **Show text boundary frame only.** This setting controls the Quick Text mode. When checked, text is replaced by a rectangular frame. The system variable equivalent is **QTEXTMODE**.
- **Draw true silhouettes for solids and surfaces.** Controls whether or not the silhouette curves are displayed for solid objects. The system variable equivalent is **DISPSILH**.

NOTE

After changing display settings, use the **REGEN** or **REGENALL** command to make the settings take effect on the objects in the drawing.

PROFESSIONAL TIP

If you notice performance slowing down, you may want to adjust display settings. For example, if there is a lot of text in the drawing, you can activate Quick Text mode to improve performance. When the drawing is ready for plotting, deactivate Quick Text mode.

The settings specified in the **Open and Save** tab of the **Options** dialog box control how drawing files are saved, safety precautions for files, how file names display in the application menu, the behavior of xrefs, and the loading of ObjectARX applications and proxy objects. This tab is shown in **Figure 21-7**. The options in this tab are discussed in the next sections.

Default Settings for Saving Files

The settings in the **File Save** area determine the defaults for saving files. The setting in the **Save as:** drop-down list determines the default file type. Drawing files created with AutoCAD 2010, 2011, and 2012 are all compatible. You may want to change this setting if you are saving drawing files using the file format of a previous AutoCAD release or saving drawings as DXF files.

The **Maintain visual fidelity for annotative objects** check box controls how annotative objects are displayed when the drawing is opened in AutoCAD 2007 or earlier versions. If you work primarily in model space, this can be left unchecked. If you use layouts and expect the drawing files to be saved for an older version of AutoCAD, this should be checked. When the check box is checked, and the drawing is saved and then opened in an older version of AutoCAD, the scaled representations of annotative objects are divided into separate objects. These objects are stored in anonymous blocks saved on separate layers. The block names are based on the layer's original name appended with a number. When the drawing is opened once again in AutoCAD 2008 or later, the annotative objects are restored to normal. The system variable equivalent for this toggle is **SAVEFIDELITY**. Checking the check box sets this variable to 1 (on). Unchecking it sets the variable to 0 (off).

Figure 21-7.
The **Open and Save** tab settings control default save options, file safety features, how file names are displayed in the application menu, xref options, and ObjectARX application options.

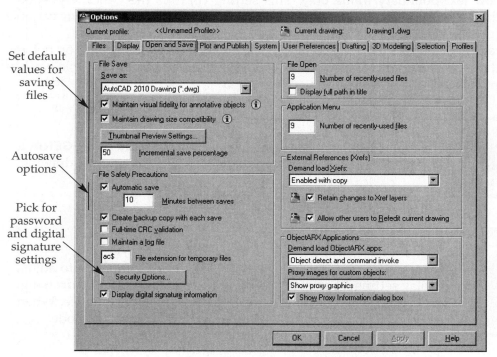

The **Maintain drawing size compatibility** check box determines how drawings with individual objects greater than 256 MB are handled. Files created with versions previous to AutoCAD 2010 are not allowed to have objects larger than 256 MB. When saving a drawing to a legacy format (2009 or earlier), problems may be encountered when attempting to open that drawing due to compatibility issues with these large objects. Checking this option maintains compatibility with earlier versions of AutoCAD by telling AutoCAD to check for objects larger than 256 MB when attempting to save the drawing. An alert box will be displayed noting that the issue needs to be resolved before the file can be saved. If you do not plan on working on your drawings in earlier versions of AutoCAD, this option box can be unchecked. The system variable equivalent for this toggle is **LARGEOBJECTSUPPORT**. Unchecking the **Maintain drawing size compatibility** check box sets the system variable to 1 (on) and allows the ability to create large objects. When the check box is checked, the system variable is set to 0 (off).

The **Incremental save percentage** value determines how much of the drawing is saved when the **SAVE** or **QSAVE** command is used. If the quantity of new data in a drawing file reaches the specified percentage, a full save is performed. To force a full save to be performed each time, set the value to 0.

If you pick the **Thumbnail Preview Settings...** button, the **Thumbnail Preview Settings** dialog box is displayed, **Figure 21-8**. If the **Save a thumbnail preview image** check box in this dialog box is checked, a preview image of the drawing will be displayed in the **Select File** dialog box when the drawing is selected for opening. The system variable equivalent for this setting is **RASTERPREVIEW**; a setting of 1 creates a preview.

The two radio buttons below the **Save a thumbnail preview image** check box are used to set what is used as the basis for the thumbnail. Selecting the **Use view when drawing last saved** radio button bases the thumbnail image on the last zoom location of the drawing. Selecting the **Use Home View** radio button bases the thumbnail on the view defined as the home view.

The home view can be set to the current view by picking the **Set current View as Home** button in the **Home view** area. This can also be done in the drawing area by using the view cube. See Chapter 4 for a detailed explanation of the view cube. To change the home view to the default setting, pick the **Reset Home to default** button. A preview of the current home view is shown to the left of the buttons.

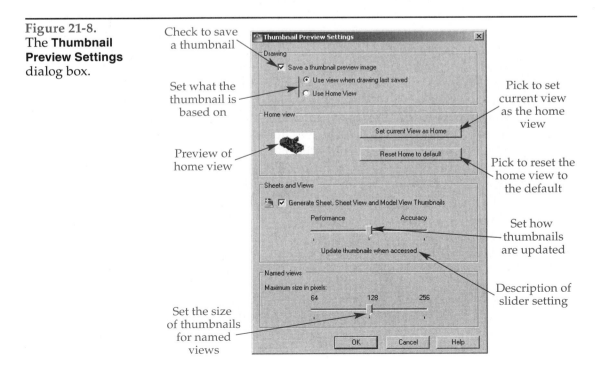

Figure 21-8.
The **Thumbnail Preview Settings** dialog box.

Check to save a thumbnail

Set what the thumbnail is based on

Preview of home view

Pick to set current view as the home view

Pick to reset the home view to the default

Set how thumbnails are updated

Description of slider setting

Set the size of thumbnails for named views

When the **Generate Sheet, Sheet View, and Model View Thumbnails** check box is checked in the **Thumbnail Preview Settings** dialog box, the thumbnails in the **Sheet Set Manager** are updated based on the position of the slider below this check box. The slider can be set to one of three positions. A description of the current setting appears below the slider. When the slider is in the middle position (the default setting), thumbnails are updated when they are accessed. When the slider is in the left-hand position, thumbnails must be manually updated. When the slider is in the right-hand position, the thumbnails are updated when the drawing is saved. The system variable equivalent is **UPDATETHUMBNAIL**. The settings are:

- **0.** The **Generate Sheet, Sheet View, and Model View Thumbnails** check box is unchecked.
- **7.** The check box is checked and the slider is in the left-hand position.
- **15.** The check box is checked and the slider is in the middle position.
- **23.** The check box is checked and the slider is in the right-hand position.

The **Maximum size in pixels:** slider in the **Thumbnail Preview Settings** dialog box controls size of thumbnails for named views. The slider can be set to 64, 128, or 256 pixels. This is the square size of the thumbnail. The system variable equivalent is **THUMBSIZE**, where a setting of 0 is 64 pixels, 1 is 128 pixels, and 2 is 256 pixels.

CAUTION

To maintain forward compatibility of drawings, annotative objects should not be edited in older versions of AutoCAD. Doing so may compromise the annotative properties. For example, exploding an annotative block in an older version of AutoCAD and then opening that drawing in AutoCAD 2012 results in each of the scaled representations becoming a separate annotative object.

Autosave Settings

When working in AutoCAD, data loss can occur due to a sudden power outage or an unforeseen system error. AutoCAD provides several safety precautions to help minimize data loss when these types of events occur. The settings for the precautions are found in the **File Safety Precautions** area in the **Open and Save** tab of the **Options** dialog box.

When the **Automatic save** check box is checked, AutoCAD automatically creates a backup file at a specified time interval. The **Minutes between saves** edit box sets this interval. This is the value of the **SAVETIME** system variable. Removing the check sets **SAVETIME** to 0.

The automatic save feature does not overwrite the source drawing file with its incremental saves. Rather, AutoCAD saves temporary files. The path for autosave files is specified in the **Files** tab in the **Options** dialog box, as discussed earlier in this chapter. The autosave file is stored in the specified location until the drawing is closed. When the drawing is closed, the autosave file is deleted. Autosave files have a .sv$ extension with the drawing name and some random numbers generated by AutoCAD. If AutoCAD unexpectedly quits, the autosave file is not deleted and can be renamed with a .dwg extension so it can be opened in AutoCAD.

The interval setting should be based on working conditions and file size. It is possible to adversely affect your productivity by setting your **SAVETIME** value too small. For example, in larger drawings, a save can take a significant amount of time. Ideally, it is best to set your **SAVETIME** value to the greatest amount of time you can afford to repeat. While it may be acceptable to redo the last 15 minutes or less of work, it is unlikely that you would feel the same about having to redo the last hour of work.

Backup Files

AutoCAD can create a backup of the current drawing file whenever the current drawing is saved. The backup file uses the same name as the drawing, but has a .bak file extension. The backup is not overwritten when a different drawing is opened or saved. When the **Create backup copy with each save** check box in the **Open and Save** tab of the **Options** dialog box is checked, the backup file feature is enabled. If the check box is not checked, the file is not backed up when you save. Unless you prefer to take unnecessary risks, it is usually best to have this feature enabled.

CRC Validation

A *cyclic redundancy check*, or CRC, verifies that the number of data bits sent is the same as the number received. Checking the **Full-time CRC validation** check box in the **Open and Save** tab of the **Options** dialog box enables this feature. This is a feature you can use when drawing files are being corrupted and you suspect a hardware or software problem. When using full-time CRC validation, the CRC is done every time data are read into the drawing. This ensures that all data are correctly received.

Log Files

The log file can serve a variety of purposes. The source of drawing errors can be determined by reviewing the commands that produced the incorrect results. Additionally, log files can be reviewed by a CAD manager to determine the need for staff training or customization of the system.

When the **Maintain a log file** check box is activated in the **Open and Save** tab of the **Options** dialog box, AutoCAD creates a file named with the drawing name, a code, and the .log file extension. The name and location of the log file can be specified using the Log File Location listing in the **Files** tab of the **Options** dialog box. When activated, all prompts, messages, and responses that appear in the command line window are saved to this file. The log file status can also be set using the **LOGFILEON** and **LOGFILEOFF** commands.

Security Settings

Picking the **Security Options...** button in the **Open and Save** tab of the **Options** dialog box opens the **Security Options** dialog box. This dialog box can be used to apply a password and digital signature to the drawing. Checking the **Display digital signature information** check box allows the information to be shown when opening a drawing that has a digital signature attached to it.

File Opening Settings

The **File Open** area of the **Open and Save** tab in the **Options** dialog box contains two settings. The value in the **Number of recently-used files** text box controls the number of drawing files listed in the **File** pull-down menu. The pull-down menus are not displayed, by default, unless the AutoCAD Classic workspace is set current. The value can be from 0 to 9. The **Display full path in title** check box controls whether the entire drawing file path or just the file name is displayed in the title bar of the AutoCAD window. If the check box is checked, the full file path is displayed.

Application Menu Settings

The **Application Menu** area of the **Open and Save** tab in the **Options** dialog box controls the number of drawing files displayed in the **Recent Documents** entry in the application menu. The setting in the **Number of recently-used files** text box is the number of files displayed and can range from 0 to 50.

External Reference Settings

The external reference options in the **Open and Save** tab of the **Options** dialog box are important if you are working with xrefs. These options are found in the **External Reference (Xrefs)** area of the tab. The **Demand load Xrefs:** setting can affect system performance and the ability for another user to edit a drawing currently referenced into another drawing. You can select Enabled, Disabled, or Enabled with copy from the drop-down list. This setting is also controlled by the **XLOADCTL** system variable.

If the **Retain changes to Xref layers** option is checked, xref layer settings are saved with the drawing file. The **VISRETAIN** system variable also controls this setting.

The **Allow other users to Refedit current drawing** setting controls whether or not the drawing can be edited in-place when it is referenced by another drawing. This setting is also controlled by the **XEDIT** system variable.

ObjectARX Options

The **ObjectARX Applications** area of the **Open and Save** tab of the **Options** dialog box controls the loading of ObjectARX applications and the displaying of proxy objects. The **Demand load ObjectARX apps:** setting specifies if and when AutoCAD loads third-party applications associated with objects in the drawing. The **Proxy images for custom objects:** setting controls how objects created by a third-party application are displayed. When a drawing with proxy objects is opened, the **Proxy Information** dialog box is displayed. To disable the dialog box, uncheck the **Show Proxy Information dialog box** check box. This is checked by default.

The settings in the **Plot and Publish** tab of the **Options** dialog box control how plots are generated. Plotting is covered in detail in *AutoCAD and Its Applications—Basics*.

Options for the pointing device, graphic settings, general system options, dbConnect, and web-based help can be found in the **System** tab of the **Options** dialog box, **Figure 21-9**. These settings affect the interaction between AutoCAD and your operating system.

In the **3D Performance** area is the **Performance Settings** button. Selecting this button displays the **Adaptive Degradation and Performance Tuning** dialog box. The options available in this dialog box are discussed in the next section.

The **Current Pointing Device** area determines the pointing device used with AutoCAD. The default is the current system pointing device (usually your mouse). If you have a digitizer tablet, you will want to select the **Wintab Compatible Digitizer** option. You must configure your tablet before it can be used.

The **Layout Regen Options** setting determines what is regenerated and when it is regenerated when working with layout tabs.

The options in the **dbConnect Options** area control how AutoCAD works with drawings containing database information. The following options are available.

- **Store Links index in drawing file.** When this option is checked, the database index is saved within the drawing file. This makes the link selection operation quicker, but increases the drawing file size.
- **Open tables in read-only mode.** This option determines whether tables are opened in read-only mode.

The settings in the **General Options** area control general system functions. When the **Hidden Messages Settings** button is picked, the **Hidden Message Settings** dialog box is displayed, **Figure 21-10**. All messages hidden by the user picking the "do not show again" option in a message box are available in this dialog box. For example, if

---

Figure 21-9.
General AutoCAD system options and hardware settings can be controlled in the **System** tab.

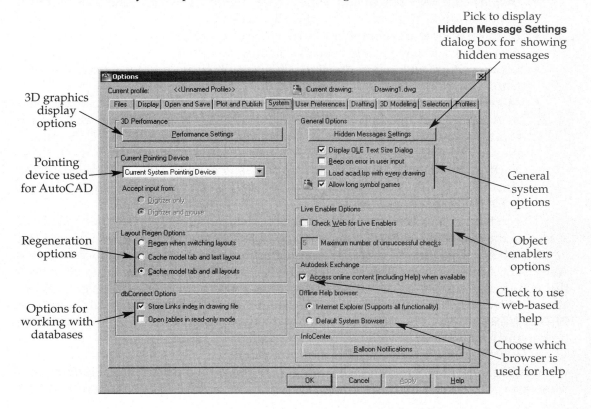

Figure 21-10.
Turning on the display of previously hidden messages.

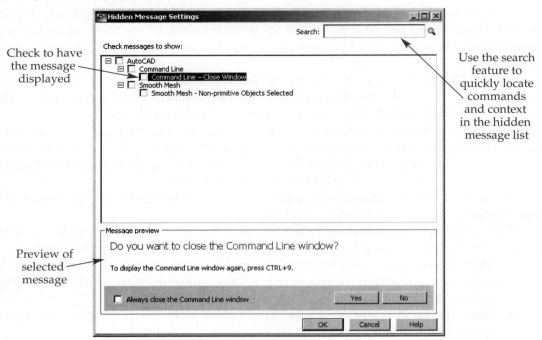

Check to have the message displayed

Use the search feature to quickly locate commands and context in the hidden message list

Preview of selected message

you close the command line window, a message is displayed with a warning. If you choose to always hide this message, it appears in the **Hidden Message Settings** dialog box. Check the entry for this message in the dialog box to have it once again displayed when you are presented with the related option or message.

The following additional options are available in the **General Options** area of the **System** tab.

- **Display OLE Text Size Dialog**. When inserting an OLE object, the **OLE Text Size** dialog box is displayed if this option is checked.
- **Beep on error in user input**. Specifies whether AutoCAD alerts you of incorrect user input with an audible beep. By default, this is off.
- **Load acad.lsp with every drawing**. This setting turns the persistent AutoLISP feature on or off. By default, it is off.
- **Allow long symbol names**. Determines if long symbol names can be used in AutoCAD. If this option is checked, up to 255 characters can be used for layers, dimension styles, blocks, linetypes, text styles, layouts, UCS names, views, and viewport configurations. If unchecked, symbol names are limited to 31 characters. By default, this option is checked. Also, it is saved in the drawing, not the AutoCAD system registry. The system variable equivalent is **EXTNAMES**.

The setting in the **Live Enabler Options** area determines whether or not AutoCAD searches the Autodesk website for object enablers. An object enabler allows your version of AutoCAD to open and manipulate drawings created in applications like AutoCAD Architecture without the display of proxy object errors. It also allows the use of the custom-made objects created in those applications. Object enablers are provided for free by Autodesk. The system variable that controls this setting is **PROXYWEBSEARCH**. The **Maximum number of unsuccessful checks** edit box indicates how many times AutoCAD will search the website.

The settings in the **AutoCAD Exchange** area determine how assistance is obtained when using the **HELP** command. Checking the **Access online content (including Help) when available** check box means that a web browser will be opened when the **HELP** command is accessed and online help from Autodesk's website will be displayed. Unchecking the check box allows help to be accessed from the file listed in the Help Location entry under the **Help and Miscellaneous File Names** listing in the **Files** tab of the **Options** dialog box. The radio buttons below the check box allow you to choose Internet Explorer or a default browser that you may be using (i.e., Firefox or Google Chrome). Using Internet Explorer ensures all features of Autodesk's online help pages are supported.

The **InfoCenter** area is used to access settings for balloon notifications issued from the **InfoCenter**. Selecting the **Balloon Notifications** button opens the **InfoCenter Settings** dialog box, Figure 21-11. The **Enable balloon notification for these sources:** check box determines whether any balloon notifications appear. If this check box is checked, balloon notifications are enabled, and the two check boxes below become available. The **Live Update channel (new software updates)** check box determines whether balloon notifications related to product updates and announcements appear. The **Did You Know messages** check box determines whether **Did You Know** messages and tips appear. The two radio buttons below this check box determine how the balloon notifications are dismissed. If the **Use balloon notification display time** radio button is selected, notifications "time out" based on the value in the **Number of seconds balloon notification displays** edit box. If the **Display until closed** radio button is selected, the notification displays until it is manually closed. The value in the **%Transparency of balloon notification** edit box controls how transparent balloon notifications appear when they are displayed. The value can be changed using the keyboard or by dragging the slider.

All help is provided via a web browser, whether you are using online help from Autodesk or a file on a local computer.

3D Performance Settings

With all of the powerful 3D and solid modeling features that are built into AutoCAD, there are many display-related tasks being handled by AutoCAD, the graphics card, and the computer itself. Materials, lights, shadows, shading, and rendering require a lot of computing power in order to project a quality representation

Figure 21-11. The **InfoCenter Settings** dialog box is used to manage settings for balloon notifications issued from the **InfoCenter**.

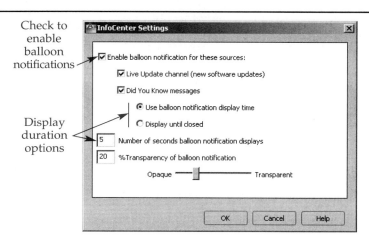

of the model onto the monitor screen. Often, there is no reduction in quality to any of the desired effects if the materials are not too complex, few lights are used, or if you have shadows turned off. Sometimes, in order to make one effect look good, fewer resources have to be assigned to other effects. The software and hardware, working together, usually do an adequate job assigning these resources. However, it may be necessary for you to assist in this decision-making process. The settings for this process are made in the **Adaptive Degradation and Performance Tuning** dialog box, **Figure 21-12**. To display this dialog box, pick the **Performance Tuner** tool in the status bar and select **Performance Tuner...** from the shortcut menu or pick the **Performance Settings** button in the **Systems** tab of the **Options** dialog box. You can also enter the **3DCONFIG** command.

The left-hand side of the **Adaptive Degradation and Performance Tuning** dialog box contains settings for controlling adaptive degradation. *Adaptive degradation* controls system performance by turning off features or preventing them from using resources. The check box at the top of the area controls whether or not adaptive degradation is being used. When unchecked, adaptive degradation is turned off and all effects are using all resources. This may result in graphics lagging or becoming slow and "choppy" as you zoom and pan around your drawing. The orbiting commands are even more affected by this being turned off. By checking the check box, adaptive degradation is activated.

AutoCAD tracks its graphics performance in terms of frames per second (fps). Just below the **Adaptive degradation** check box is a text box for setting this value. You may enter a new value in the text box or use the arrows to increase or decrease the value. The higher the number, the sooner resources start being reassigned. If performance dips below this level, resources are taken away from the various effects that create the displayed graphics.

The effects that can be controlled while adaptive degradation is turned on are shown in the **Degradation order:** list box. Certain effects that you deem important can be unchecked so they are not degraded and operate using maximum resources. The top-to-bottom order in which the effects are listed determines the priority in which resources are removed. This order can be changed by selecting an effect and picking the **Move Up** or **Move Down** button on the right side of the list.

Figure 21-12.
The **Adaptive Degradation and Performance Tuning** dialog box is used to turn on and prioritize adaptive degradation.

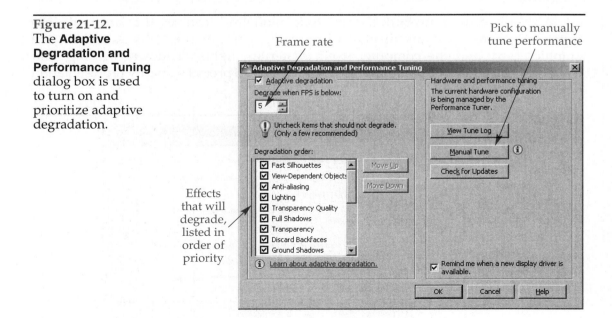

On the right-hand side of the **Adaptive Degradation and Performance Tuning** dialog box is the **Hardware and performance tuning** area. Picking the **View Tune Log** button displays a log of any features or effects that have been turned off. Information and details regarding your computer, amount of RAM, and 3D graphics card are also shown. The log can be saved as a file. Picking the **Manual Tune** button displays the **Manual Performance Tuning** dialog box. This dialog box allows control over hardware settings (*hardware acceleration* and hardware effects), general settings (discard back faces, smooth faces, and quality of transparency), and dynamic tessellation settings (surface and curve tessellation settings and number of tessellations to cache). At the bottom of the **Adaptive Degradation and Performance Tuning** dialog box is the **Remind me when a new display driver is available** check box. When checked, AutoCAD displays a pop-up message in the graphics area whenever a new graphics card driver is available.

When the settings have been adjusted as desired in the **Adaptive Degradation and Performance Tuning** dialog box, pick the **OK** button to return to the **Options** dialog box. Then, close the **Options** dialog box.

User Preferences

A variety of settings are found in the **User Preferences** tab of the **Options** dialog box. See **Figure 21-13**. AutoCAD allows users to optimize the way they work in AutoCAD by providing options for double-click editing, shortcut menu functions, insertion units, hyperlink icon display, working with fields, coordinate data entry, associative dimensions, undo/redo control, block editor settings, default lineweight settings, and scale list settings. All of these items are controlled in this tab.

Figure 21-13.
The **User Preferences** tab allows you to set up AutoCAD in a manner that works best for you.

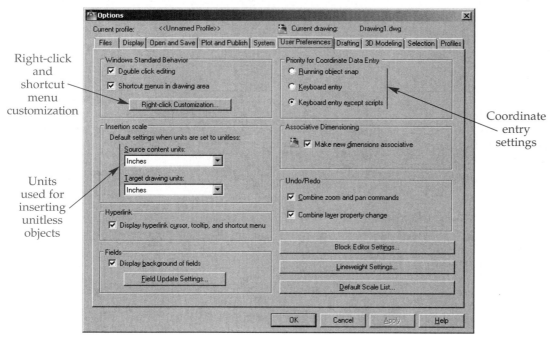

Shortcut Menus and Double-Click Editing

AutoCAD has tools that provide easy access to commonly used editing commands and options. Two of these tools are double-click editing and shortcut menus. Double-clicking on an object calls the most appropriate editing tool for that object type, often the **Properties** palette. Shortcut menus are displayed by right-clicking and are *context sensitive*. This means the options available in the shortcut menu are determined by the active command, cursor location, or selected object.

To enable double-click editing, check the **Double click editing** check box in the **Windows Standard Behavior** area in the **User Preferences** tab of the **Options** dialog box. To enable shortcut menus, check the **Shortcut menus in drawing area** check box in the same area. Disabling the shortcut menus makes a right mouse click the equivalent of pressing the [Enter] key. In general, this is not recommended.

You can also customize the setting for the right mouse button. Pick the **Right-click Customization...** button to access the **Right-Click Customization** dialog box. See Figure 21-14. The **Turn on time-sensitive right-click:** check box controls the right-click behavior. A quick click is the same as pressing [Enter]. A longer click displays a shortcut menu. You can set the duration of the longer click in milliseconds. If the check box is checked, the **Default Mode** and **Command Mode** areas of the dialog box are disabled.

Different settings can be used for the three different shortcut menu modes. Each of the three menu modes has a separate area with options in the **Right-Click Customization** dialog box.

- **Default Mode**. In this mode, no objects are selected and no command is active. The **Repeat Last Command** option activates the last command issued. The **Shortcut Menu** option displays the shortcut menu.
- **Edit Mode**. In this mode, an object is selected, but no command is active. The **Repeat Last Command** option activates the last command issued. The **Shortcut Menu** option displays the shortcut menu.
- **Command Mode**. In this mode, a command is active. The **ENTER** option makes a right-click the same as pressing [Enter]. The **Shortcut Menu: always enabled** option means that the shortcut menu is always displayed in command mode. The **Shortcut Menu: enabled when command options are present** option means the shortcut menu is only displayed when command options are available on the command line. When there are no command options, a right-click is the same as pressing [Enter]. This is the default option.

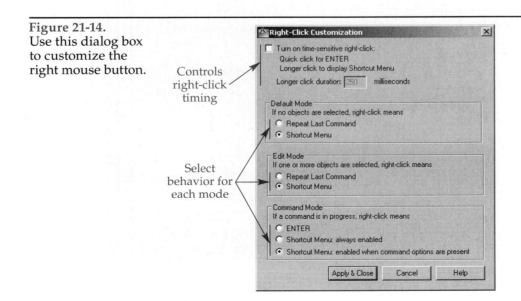

Figure 21-14.
Use this dialog box to customize the right mouse button.

Controls right-click timing

Select behavior for each mode

Insertion Scale

In the **Insertion scale** area of the **User Preferences** tab, unit values can be set for objects when they are inserted into a drawing. This applies to "unitless" objects dragged from **DesignCenter** or inserted using the i-drop method. The **Source content units:** setting specifies the units for objects being inserted into the current drawing. The **Target drawing units:** setting determines the units in the current drawing. These settings are used when there are no units set with the **INSUNITS** system variable.

Hyperlinks

In the **Hyperlink** area of the **User Preferences** tab, you can set whether or not the hyperlink cursor and tooltip are displayed when the cursor is over a hyperlink. If the **Display hyperlink cursor, tooltip, and shortcut menu** check box is checked, the hyperlink icon appears next to the crosshairs when the cursor is over an object containing a hyperlink. The tooltip is also displayed. Additional hyperlink options are available from the shortcut menu when an object with a hyperlink is selected.

Fields

A *field* is a special type of text object that displays a specific property value, setting, or characteristic. Fields can display information related to a specific object, general drawing properties, or information related to the current user or computer system. The text displayed in the field can change if the value being displayed changes. Refer to *AutoCAD and Its Applications—Basics* for more information on using fields.

In the **Fields** area of the **User Preferences** tab, you can set whether or not a field is displayed with a nonplotting background. When the **Display background of fields** check box is checked, the field background is displayed in light gray.

Picking the **Field Update Settings...** button in the **Fields** area opens the **Field Update Settings** dialog box, **Figure 21-15.** In this dialog box, you can set when fields are automatically updated. The five options are **Open, Save, Plot, eTransmit**, and **Regen**. Check as many of the options as appropriate.

Coordinate Data Priority

The **Priority for Coordinate Data Entry** area of the **User Preferences** tab controls how AutoCAD responds to input of coordinate data. The system variable equivalent for this is **OSNAPCOORD**. The three options are:

- **Running object snap.** When this option is selected, object snaps always override coordinate entry. This is equivalent to an **OSNAPCOORD** setting of 0.
- **Keyboard entry.** When this option is selected, coordinate entry always overrides object snaps. This is equivalent to an **OSNAPCOORD** setting of 1.
- **Keyboard entry except scripts.** When this option is selected, coordinate entry will override object snaps except object snaps contained within scripts. This is the default and equivalent to an **OSNAPCOORD** setting of 2.

Figure 21-15.
Setting when fields
are automatically
updated.

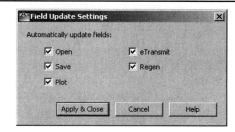

Associative Dimensions

By default, all new dimensions are associative. This means that the dimension value automatically changes when a dimension's defpoints are moved. However, you can turn this option off in the **User Preferences** tab of the **Options** dialog box. When the **Make new dimensions associative** check box in the **Associative Dimensioning** area is unchecked, any dimensions drawn do *not* have associativity. This is equivalent to a **DIMASSOC** setting of 1.

Undo/Redo

The **Undo/Redo** area of the **User Preferences** tab allows you to control how multiple, consecutive zooms and pans are handled within the **UNDO** and **REDO** commands. By checking the **Combine zoom and pan commands** check box, back-to-back zooms and pans are considered a single operation for undo and redo purposes. In other words, performing an undo or redo undoes or redoes the entire zoom/pan sequence. Unchecking the check box allows each zoom or pan to be considered a separate operation.

When the **Combine layer property change** check box is checked, all changes made in the **Layer Properties Manager** palette are considered one operation. If this is unchecked, each change is considered a separate operation when using **UNDO** and **REDO**.

Block Editor Settings, Lineweight Settings, and Edit Scale List

At the bottom of the **User Preferences** tab are the **Block Editor Settings...**, **Lineweight Settings...**, and **Default Scale List...** buttons. Picking one of these buttons opens a dialog box with corresponding settings.

Picking the **Block Editor Settings...** button opens a dialog box that contains settings for the colors, fonts, and sizes of the various elements that appear when working in the **Block Editor**, Figure 21-16. The **Block Editor** is discussed in detail in *AutoCAD and Its Applications—Basics*. At the bottom of the dialog box, there are three check boxes that control the highlight of dependent objects during selection, display of tick marks for parameters with value sets, and display of action bars. These check boxes control the **BDEPENDENCYHIGHLIGHT**, **BTMARKDISPLAY**, and **BACTIONBARMODE** system variables.

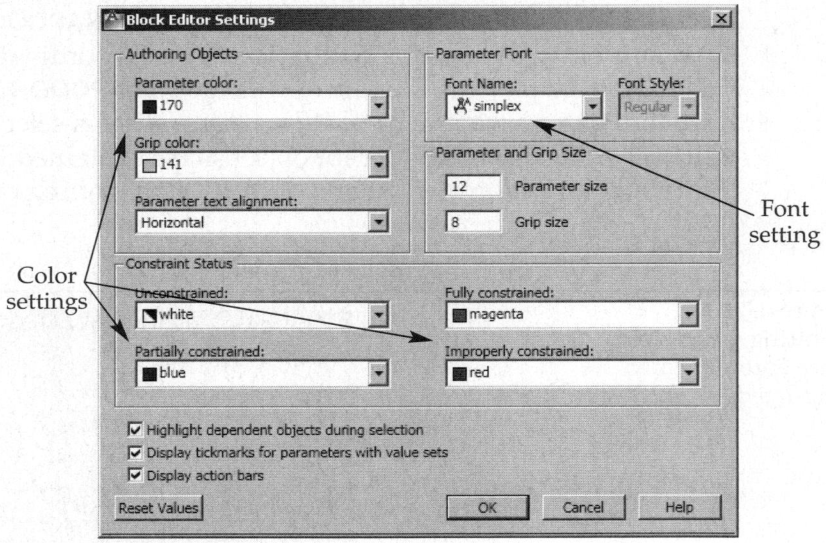

Figure 21-16.
Use the **Block Editor Settings** dialog box to adjust the appearance of objects when working in the **Block Editor**.

AutoCAD and Its Applications—Advanced

Picking the **Lineweight Settings...** button opens the **Lineweight Settings** dialog box, which you can use to change default lineweight settings. This is discussed in detail in *AutoCAD and Its Applications—Basics*.

A default list of scales appears in various dialog boxes for use with viewports, page setups, and plot scaling. You can add custom scales to or remove scales from this list so that it is more appropriate for your application. Picking the **Default Scale List...** button at the bottom of the **User Preferences** tab displays the **Default Scale List** dialog box. See **Figure 21-17**. The dialog box displays one of the two available lists of scales. The radio buttons at the top of the dialog box determine whether a metric scale list or a US customary (Imperial) scale list is displayed. The buttons on the right side of the dialog box allow you to add a new scale, edit an existing scale, move a scale up or down within the list, delete a scale, or reset the list to the default set of scales.

To add a scale, pick the **Add...** button. In the **Add Scale** dialog box that appears, enter a name for the scale in the **Name appearing in scale list:** text box. Then, in the **Scale properties** area of the dialog box, enter values to indicate how many paper units equal how many drawing units. Finally, pick the **OK** button to return to the **Default Scale List** dialog box. The new scale appears in the current list, under metric or US customary (Imperial) units, and is available wherever the scale list is displayed. All scales from both lists are available whenever the scale list is displayed.

3D Display Properties

There are many ways to customize your system specifically for working in a 3D environment. The **3D Modeling** tab of the **Options** dialog box allows you to control the various settings having to do with working in 3D, **Figure 21-18**.

The **3D Crosshairs** area of the tab contains check boxes for displaying the Z axis on the crosshairs, labeling the axes of the standard crosshairs, and labeling the axes of the dynamic UCS icon. There are three labeling possibilities from which to choose:

- X, Y, and Z.
- N (north), E (east), and z.
- Custom label for each axis.

Figure 21-17.
The **Default Scale List** dialog box allows you to change the scale list that appears when using various viewport, page setup, and plot scaling dialog boxes.

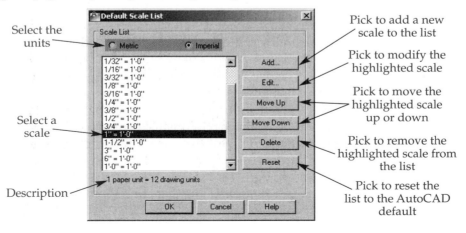

Select the units

Select a scale

Description

Pick to add a new scale to the list

Pick to modify the highlighted scale

Pick to move the highlighted scale up or down

Pick to remove the highlighted scale from the list

Pick to reset the list to the AutoCAD default

Figure 21-18.
Use the **3D Modeling** tab to customize settings related to working in 3D.

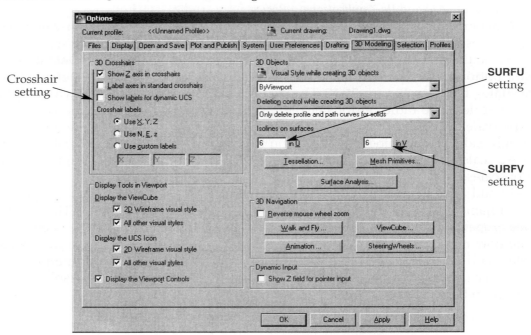

The **Display Tools in Viewport** area determines the display nature of the view cube, the UCS icon, and the viewport controls. The view cube can be displayed when the 2D Wireframe visual style is active or when any other visual style is active by checking the appropriate check box under the **Display the ViewCube** heading. There are identical check boxes controlling the display of the UCS icon. If the **Display the Viewport Controls** check box is checked, the viewport controls appear in the upper-left corner of the viewport.

The setting in the **Visual Style while creating 3D objects** drop-down list in the **3D Objects** area determines which visual style is set current when objects are created. The setting in the **Deletion control while creating 3D objects** drop-down list determines how geometry is handled when creating 3D objects. For example, when **Delete profile curves** is selected, the profile and path curves are deleted after a sweep is created. The two edit boxes in the **3D Objects** area set the **SURFU** and **SURFV** system variables for defining the isoline display on surface objects. Picking the **Tessellation...** and **Mesh Primitives...** buttons open dialog boxes that contain settings for controlling the options associated with mesh tessellation and mesh primitives. These settings are discussed thoroughly in Chapter 3. Picking the **Surface Analysis...** button opens a dialog box for changing the display of zebra, curvature, or draft angle analyses.

The **3D Navigation** area has a check box for reversing the zoom direction of the mouse wheel. There are also four buttons in this area that allow access to settings for walkthroughs and flybys, animations, the view cube, and steering wheels. Selecting the **Walk and Fly...** button opens the **Walk and Fly Settings** dialog box. This dialog box contains settings used when creating walkthroughs and flybys. Selecting the **Animation...** button opens the **Animation Settings** dialog box. This dialog box contains settings that control the actual animation of a walkthrough or flyby. The **Walk and Fly Settings** and **Animation Settings** dialog boxes are discussed in detail in Chapter 20. Selecting the **View Cube...** button opens the **View Cube Settings** dialog box. Selecting the **Steering Wheels...** button opens the **Steering Wheels Settings** dialog box. These dialog boxes are discussed in Chapter 4. The **Dynamic Input** area has a check box for showing the Z field for dynamic input.

Chapter Review

1. List three methods used to open the **Options** dialog box.
2. List the tabs found in the **Options** dialog box.
3. You have created two folders under the C: drive named \Projects and \Symbols. You want to store your drawings in the \Projects folder and your blocks in the \Symbols folder. What should you enter in the Support File Search Path area so these folders are added to the search path?
4. How do you open the **Drawing Window Colors** dialog box to change the color of AutoCAD screen elements?
5. For which AutoCAD features can you customize the font (not the font within a drawing)?
6. In which tab of the **Options** dialog box can you change settings for layout tabs?
7. Briefly describe how to turn on the automatic save feature and specify the save interval.
8. How do you select the folder in which the autosave file is saved?
9. What are the advantages of toggling the log file open?
10. Name the two commands that toggle the log file on and off.
11. How would you set the right mouse button to perform an [Enter], rather than displaying shortcut menus?
12. How do you open the **Default Scale List** dialog box from within the **Options** dialog box?
13. On which tab of the **Options** dialog box is the **Block Editor Settings...** button located?
14. How do you access the view cube display setting from within the **Options** dialog box?

Drawing Problems

1. Using the methods described in this chapter, change the AutoCAD screen colors to your liking. Customize the 3D parallel and 3D perspective contexts.

2. Set up AutoCAD so that profile curves are retained when creating 3D objects. Also, increase the isolines for meshes. To determine the best default setting, draw various meshes, change the setting, and draw additional meshes.

Customizing Interface Elements and Commands

Learning Objectives

After completing this chapter, you will be able to:

✓ Explain the features of the **Customize User Interface** dialog box.
✓ Describe partial CUIx files.
✓ Create custom commands.
✓ Create new toolbars, ribbon tabs, and ribbon panels.
✓ Customize ribbon tabs and panels with submenus and drop-down lists.
✓ Explain how to customize menus.

One of the easiest ways to alter the AutoCAD environment is by customizing interface elements. *Interface elements* are graphic command-entry components of AutoCAD, such as the ribbon, toolbars, and menus. They can be quickly modified by removing and adding commands. New commands can also be created and assigned to an existing ribbon panel, toolbar, or menu or to new interface elements. The most powerful aspect of customizing ribbon panels, toolbars, and menus is the ability to quickly create entirely new functions to help you in your work. The key to good customization can be broken down into four simple rules:

- Always make a backup of the original files, such as the acad.cuix file, *before* customizing.
- Do not over-customize your work. Plan your customization in steps to minimize confusion and maximize productivity. Anticipate workflow and where needs exist.
- It is best to locate customized files in folders other than the default AutoCAD folders. This makes it easier to upgrade AutoCAD in the future.
- Thoroughly test your customizations before implementing them. This will save many headaches for you and the end user.

The **Customize User Interface** dialog box funnels all major graphical user interface elements of AutoCAD into one central area where they can be tailored for productivity, **Figure 22-1**. All AutoCAD commands are linked for customization. The interface elements that will be discussed in this chapter are the ribbon tabs and panels, toolbars, and menus located in the menu bar.

The **CUI** command displays the **Customize User Interface** dialog box. This dialog box can also be displayed by right-clicking on a toolbar (when displayed) and picking **Customize...** from the shortcut menu. The changes made in the **Customize User Interface** dialog box are saved in a customization (CUIx) file. By default, this is the acad.cuix file.

The upper-left pane of the **Customize User Interface** dialog box is initially labeled **Customizations in All Files**. The drop-down list located below the pane name contains the name of the main CUIx file and any other currently loaded partial CUIx files. Partial CUIx files are discussed later in this chapter. By default, the main file acad.cuix and partial CUIx files custom.cuix, acautocadws.cuix, autodesk-seek.cuix, modeldoc.cuix, contentexplorer.cuix, acfusion.cuix, and acetmain.cuix (express tools) files are installed. If you select a different entry from the drop-down list, the name of the pane changes to reflect the selection, either **Customizations in All Files** or **Customizations in Main File**.

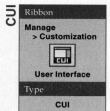

Figure 22-1.
The **Customize User Interface** dialog box is used to edit existing interface elements, create new interface elements, create custom commands, and create command icons.

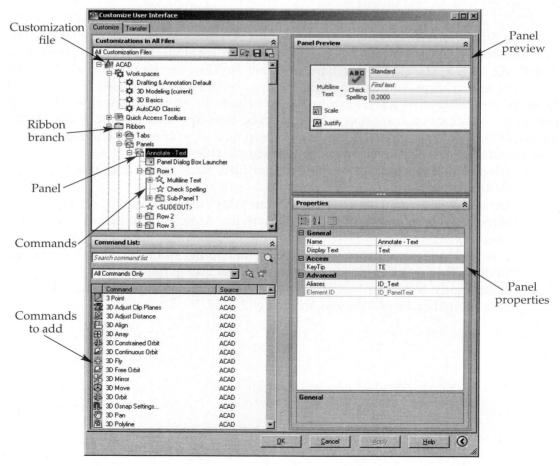

By default, the customization file acad.cuix is the main CUIx file. In the box located below the drop-down list, the selected CUIx file is displayed in a tree. The top level of the tree is the ACAD branch, which is the name of the selected CUIx file, and the AutoCAD logo icon is shown next to it. The tree under the ACAD branch lists the various customizable interface elements. For example, to see the list of available toolbars, expand the Toolbars branch (node) by picking the plus sign located just to its left. Any other partially loaded CUIx files that have toolbars in them will be listed under the Partial Customization Files branch in the tree and can have their toolbar list similarly expanded.

The shortcut menus accessed by right-clicking in the **Customize User Interface** dialog box provide editing options based on the branch or item selected. Options are available for creating new items; renaming, removing/deleting, copying/pasting, and duplicating items; adding menus or submenus; and inserting separators. For example, to delete a ribbon panel from the interface, expand the Panels branch in the Ribbon branch, locate the desired panel, right-click, and pick **Delete** from the shortcut menu. When prompted, pick the **Yes** button to delete the item. Then, pick the **Apply** or **OK** button in the **Customize User Interface** dialog box to make the deletion permanent. However, a better method is to remove a ribbon panel from the workspace. This way, the ribbon panel is still available to other workspaces. Refer to Chapter 25 for complete details on workspaces.

Content can be renamed. For example, to rename a toolbar, select the toolbar in the Toolbars branch, right-click, and pick **Rename** from the shortcut menu. The existing name of the toolbar in the tree turns into an edit box with the current name highlighted. Type a new name in the edit box and press [Enter]. The new toolbar name is displayed in the tree. You can also rename content by editing the Name property in the **Properties** pane on the right-hand side of the dialog box. Pick the **Apply** or **OK** button in the **Customize User Interface** dialog box to make the change permanent.

You can modify all existing content by deleting and adding commands. The **Command List:** pane of the **Customize User Interface** dialog box contains all commands, including those that are not available by default on an interface element. Custom commands can be assigned to all interface elements, such as ribbon panels, toolbars, and menus.

The right-hand side of the **Customize User Interface** dialog box displays specific information of the highlighted content. Panes that will appear are **Information**, **Preview**, and **Properties**, depending on the selected content. It is easy to read and navigate the **Customize User Interface** dialog box by remembering that general information is stored in the upper-left pane, and proceeding down to the **Command List:** pane and over to the panes on the right-hand side of the dialog box for more specific information.

All changes made in the **Customize User Interface** dialog box are saved in the CUIx file, including workspace, ribbon, toolbar, menu, and shortcut key customizations. This chapter discusses customizing the ribbon, toolbars, and menus. Customizing shortcut keys is discussed in Chapter 23. Customizing workspaces and customizing the **Quick Access** toolbar are covered in Chapter 25.

Adding a Command to an Interface Element

All commands are available in the **Command List:** pane of the **Customize User Interface** dialog box. Included are many commands not found on the default ribbon, toolbars, or menus. Any custom commands you have created are also available in this pane.

To add an existing command to any interface element, first expand the tree for the element in the **Customizations in All Files** pane. For example, to add a command to a toolbar, expand the Toolbars branch so the branch is visible for the toolbar to which you want the command added. Then, select a command from the **Command List:** pane. The list is alphabetized. If you hover the cursor over a command, the macro or command is displayed in a tooltip. See **Figure 22-2**.

You can search the command list by picking the **Find command or text** button at the top of the **Command List:** pane. The drop-down list at the top of the pane can be used to filter the list so that only commands in a certain category appear in the list. You can also filter the list by typing in the text box at the top of the pane. Only those commands containing the characters in this text box are displayed in the list.

Once the command is located, pick and hold on the command in the **Command List:** pane and drag it into the **Customizations in All Files** pane. A horizontal "I-bar" appears in the pane as you drag the command. This represents the location where the command will be inserted. Position the new command between the commands where you would like it to appear and release the left mouse button. The new command is added to the branch for the interface element. Pick the **Apply** or **OK** button in the **Customize User Interface** dialog box to make the addition permanent.

Selecting the **Cancel** button in the **Customize User Interface** dialog box after selecting the **Apply** button does *not* cancel the changes made before the **Apply** button was selected. Also, picking the Windows close button (the X) before picking the **Apply** button *cancels* the changes.

Figure 22-2.
The **Command List:** pane of the **Customize User Interface** dialog box displays all predefined and custom commands. These commands can be added to interface elements.

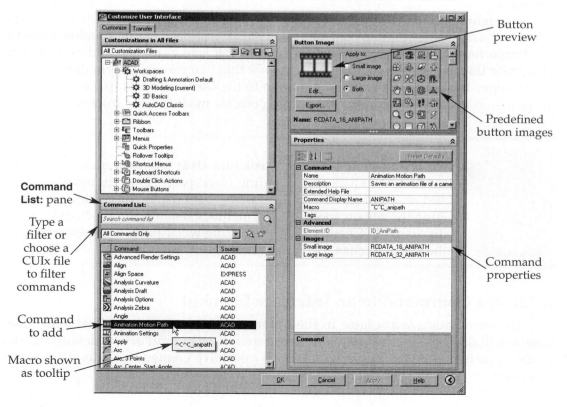

Deleting a Command from an Interface Element

To delete a command from an interface element, expand the tree for the element in the **Customizations in All Files** pane. You may need to expand more than one level to see the command you wish to delete. All of the commands currently on the interface element are displayed as branches below the element name.

Select the command you wish to delete, right-click, and pick **Remove** from the shortcut menu. You can also select the command and press the [Delete] key. Pick the **Apply** or **OK** button in the **Customize User Interface** dialog box to make the deletion permanent. The command is, however, still available in the **Command List:** pane of the **Customize User Interface** dialog box. It can be referenced to another interface element.

Moving and Copying Commands

You can move and copy commands between any interface element. First, in the upper-left pane of the **Customize User Interface** dialog box, expand the tree for both elements that you wish to edit. To move a command from one element to another, pick and hold on the command and drag it to the other interface element. The horizontal "I-bar" cursor appears as you drag. Position the cursor between the commands where you want the new command to appear and release the left mouse button. The command is moved from the first element to the second.

Use this same process to copy a command between elements, but hold the [Ctrl] key before you release the left mouse button. The command remains on the first element and a copy is placed on the second element.

You can also drag commands from the **Customize User Interface** dialog box and drop them onto toolbars and tool palettes that are currently displayed in the AutoCAD window. However, this method cannot be used to add a command tool to the ribbon.

A command can be removed from a displayed toolbar while the **Customize User Interface** dialog box is open by dragging it from the toolbar into the drawing area and releasing. A message appears asking if you want to remove the button. Pick **OK** to remove the button.

Buttons can also be rearranged on displayed toolbars while the **Customize User Interface** dialog box is open by simply dragging a button to a new position. However, it is recommended that you use the **Customize User Interface** dialog box to make edits to toolbars until you are completely comfortable with the drag-and-drop method.

PROFESSIONAL TIP

When dragging a command to a tool palette, if the desired palette is not current (on top), simply pause the cursor over the palette name until the palette is made current. Tool palette customization is discussed in detail in Chapter 24.

Adding a Separator to an Interface Element

A *separator* is a vertical or horizontal line that can be used in toolbars and menus to create visual groupings of related commands. On ribbon panels, a separator is a gap between tools. For example, look at the **Draw** panel in the **Home** tab of the ribbon with the 3D Modeling workspace set current. Between the **3D Polyline** button and the **Arc** drop-down list, there is a gap between the two indicating a separator. There is also a separator between the **Line** button and the **Circle** drop-down list and between the **Rectangle** button and the **Ellipse** drop-down list. These separators are vertical gaps because the ribbon is docked along the top edge. If the ribbon is docked along the left or right side of the screen, the separators will be horizontal gaps. Likewise, if a toolbar is horizontal, the separator is a vertical line. If a toolbar is vertical, the separator is a horizontal line.

To add separators to an interface element, first open the **Customize User Interface** dialog box. Then, in the **Customizations in All Files** pane, expand the branch for the element to which you want separators added.

For a toolbar or menu, right-click on the command in the tree below which you want the separator added. Select **Insert Separator** from the shortcut menu. A separator, represented by two dashes, appears in the tree below the selected command.

For a ribbon panel, right-click on the row in the panel to which you want a separator added. Then, select **Add Separator** from the shortcut menu. A separator is added to the bottom of the row's branch.

Once a separator is added, it can be dragged to a new location in the tree. When done adding and moving separators, pick the **OK** button to close the **Customize User Interface** dialog box.

A ribbon panel can also have a panel separator, called a *slideout*. The commands in the tree below the slideout appear in the expanded panel. For example, the **Draw** panel in the **Home** tab of the ribbon (with the 3D Modeling workspace current) has a slideout below the row of buttons containing the **Polygon**, **Rectangle** and **Ellipse** buttons. The slideout is automatically added to all panels.

A separator can be removed from an interface element in the same manner as removing a command.

Partial CUIx Files

A *partial CUIx file* is any CUIx file that is not the main CUIx file (acad.cuix). To load a partial CUIx file, pick the Open... entry in the drop-down list in the **Customizations in All Files** pane. Remember, the name of this pane may be different, depending on what is currently selected in the drop-down list. You can also pick the **Load partial customization file** button to the right of the drop-down list. Next, in the **Open** dialog box that is displayed, navigate to the folder where the CUIx file is located, select the file, and pick the **Open** button. If the partial CUIx file that has been opened contains any workspaces, the AutoCAD alert shown in **Figure 22-3** is displayed. Any workspace information contained in the CUIx file is not automatically available. Workspaces are covered in Chapter 25.

Once you open the CUIx file, it is automatically selected in the drop-down list. The name of the pane changes to **Customizations in Main File**. Now, you can manage the items contained within the partial CUIx.

If you select either the main CUIx file or All Customization Files in the drop-down list, the Partial Customization Files branch appears in the tree. Expanding this branch, you can see the partial CUIx files that are loaded. Expanding the branch for a partial

Figure 22-3.
This warning appears when loading a partial CUIx file that contains workspaces. Workspaces are covered in Chapter 25.

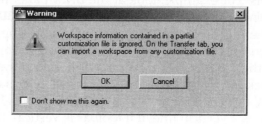

CUIx file, you can see the items contained within the CUIx file. These items can be copied from the partial CUIx file to the main CUIx file as needed.

To unload a partial CUIx file, select All Customization Files in the drop-down list in the "customizations" pane. Then, expand the Partial Customization Files branch, right-click on the name of the CUIx file, and select **Unload** *file_name*.cuix from the shortcut menu.

PROFESSIONAL TIP

You can also unload a partial CUIx file by typing **MENULOAD** or **MENUUNLOAD** at the Command: prompt. Then, in the **Load/Unload Customizations** dialog box, select the CUIx file to unload and pick the **Unload** button.

Creating New Commands

You are not limited to AutoCAD's predefined commands. *Custom commands* can be created and then added to ribbon panels, menus, tool palettes, and toolbars. First, however, you must create the new command. To create a custom command, first pick the **Create a new command** button in the **Command List:** pane of the **Customize User Interface** dialog box. This button is to the right of the drop-down list. A new command is added to the list in the **Command List:** pane. Also, the **Button Image** and **Properties** panes are displayed for the new command. See **Figure 22-4**.

Figure 22-4.
The first step in adding a custom command to an interface element is to create the custom command.

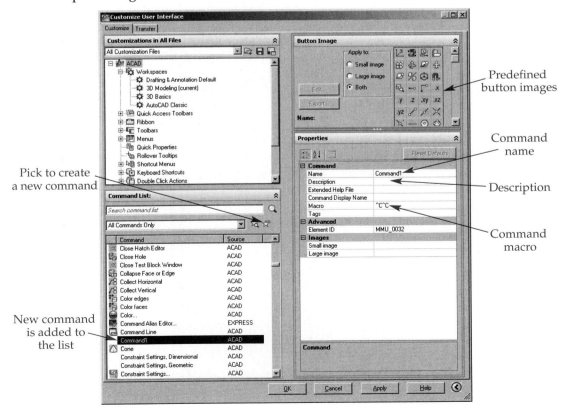

General Command Properties

By default, the new command name is **Command***n*, where *n* is a sequential number. To give the command a descriptive name, highlight the command in the **Command List:** pane. Next, pick in the Name property edit box in the **Command** category of the **Properties** pane. Then, type the new name and press [Enter]. This property is displayed as the command name in the tooltip. The name should be logical and short, such as **Draw Box**. The entry in the Command Display Name property is what appears in the command-line section of the tooltip.

The text that appears in the Description property text box in the **Command** category of the **Properties** pane appears in the tooltip when the cursor is over the button. This text, called the *help string*, should also be logical, but can be longer and more descriptive than the command name.

The Extended Help File property is used to specify an Extensible Application Markup Language (XAML) file to use as extended help. The *extended help* is displayed in the tooltip when the cursor is paused over a tool for a longer period of time. By default, if you pause the cursor for two seconds, the extended help is displayed (if the tool contains extended help). To assign an XAML file, select the property and pick the ellipsis button (...) on the right-hand side of the text box. Then, in the standard open dialog box that appears, locate and open the file. For information on creating XAML files, search the Internet for resources. Many resources can be found on the Microsoft website.

As an example, you will create a command that draws a rectangular border for an ANSI E-size sheet (44″ × 34″) using a wide polyline, sets the drawing limits, and finishes with **ZOOM Extents**. To start, create a new command and enter **E-Border** as the name. Also, enter Draws E-size border, sets limits, and zooms extents. for the Description property. In the Command Display Name property, enter Draw Border. See **Figure 22-5**. In the next sections, you will complete the command and its associated image.

Figure 22-5.
The custom command is named and a help string and tag are assigned to it.

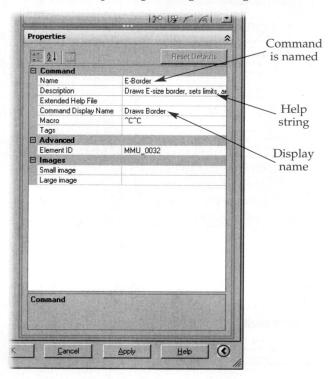

Button Image

The **Button Image** pane in the **Customize User Interface** dialog box is used to define the image that appears on the command button. The image should graphically represent the function of the command. AutoCAD provides several predefined images. One of these can be selected as the button image. You can also right-click on the list of images and select **Import Image...** from the shortcut menu to import an image.

The **CUI Editor—Image Manager** dialog box can be used to control and store custom images in a CUIx file. See **Figure 22-6**. To display this dialog box, pick the **Image Manager...** button to the right of the drop-down list at the top of the **Customizations in All Files** pane. Any image stored in a loaded CUIx file is available in the list of predefined images.

The large button image can be different from the small button image or you can use the same image for both button sizes. It may be a good idea for a button to have a separate image for each of the two button sizes. Pick the appropriate radio button in the **Button Image** pane of the **Customize User Interface** dialog box and select an image. The name of the image appears in the **Images** category in the **Properties** pane. The small image also appears next to the command name in the **Command List:** pane.

However, confusion may arise if your custom command has the same button image as an existing AutoCAD command. It is best to create custom button images for use with your custom commands. The **CUI Editor—Image Manager** dialog box makes it easy to manage the button images. You can either modify an existing button image or create a new image from scratch. In either case, a predefined image must be selected from the list of existing images. Then, pick the **Edit...** button in the **Button Image** pane to open the **Button Editor** dialog box. This is described in the next section.

 If you use the same image for both small and large buttons, the image is appropriately scaled as needed.

Figure 22-6.
The **CUI Editor—Image Manager** dialog box is used to store button images in a CUIx file.

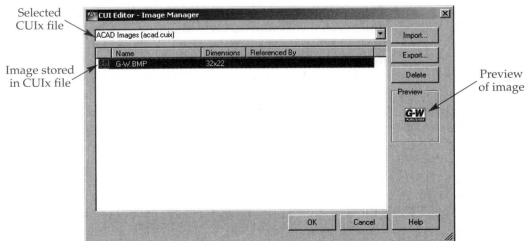

Selected CUIx file

Image stored in CUIx file

Preview of image

Creating a Custom Button Image

The **Button Editor** dialog box has basic "pixel-painting" tools and several features to simplify the editing process. The four tools are shown as buttons at the top of the dialog box. The pencil paints individual pixels. The line tool allows you to draw a line between two points. The circle tool allows you to draw center/radius style ellipses and circles. The erase tool clears the color from individual pixels. The current color is selected from the color palette on the left-hand side of the dialog box and indicated by a depressed color button. Anything you draw appears in the current color. A preview of the button image appears to the right of the tools.

Drawing a button image is usually much easier with the grid turned on. The grid provides outlines for each pixel in the graphic. Each square represents one pixel. Picking the **Grid** check box toggles the state of the grid.

When the toolbar buttons are set to their default, small size, the button editor provides a drawing area of 16 pixels × 16 pixels. If **Use large buttons for Toolbars** is turned on in the **Display** tab of the **Options** dialog box, then the button image drawing area is 32 pixels × 32 pixels. The images in the **Customize User Interface** dialog box are displayed at the current size setting (small or large).

There are several other tools available in the **Button Editor** dialog box. These include the following.

- **Clear.** If you want to erase everything and start over, pick the **Clear** button to clear the drawing area. This is the button you will use to clear the existing image and start a button image from scratch.
- **Undo.** You can undo the last operation by picking this button. Only the last operation can be undone. An operation that has been undone cannot be redone.
- **Save.** Names the current button image and saves it to the current CUIx file.
- **Import.** Use this button to open an existing bitmap (BMP) file that does not appear in the **Button Image** pane of the **Customize User Interface** dialog box. The image is automatically resized to fit the current button size.
- **Export.** This button saves a file using a standard save dialog box. Use this when you do not want to alter the original button image.
- **Close.** Ends the **Button Editor** session. A message is displayed if you have unsaved changes.
- **Help.** Provides context-sensitive help.
- **More.** Opens the standard **Select Color** dialog box. This allows you to use colors in the button other than those in the default color palette.

Once a button image is saved, it appears in the list of predefined images in the **Button Image** pane of the **Customize User Interface** dialog box. All images saved for use as button images must be stored where AutoCAD will find them. AutoCAD provides the \Icons folder within the user's support file search path. This is the default folder when using the **Export...** button in the **Button Editor** dialog box. If you choose to use a different folder, it must be added to the support file search path, which is specified in the **Files** tab of the **Options** dialog box.

Rather than using an existing button image for the **E-Border** command, an entirely new button image will be created. With **E-Border** highlighted in the **Command List:** pane, select any one of the images in the **Button Image** pane and pick the **Edit...** button. The **Button Editor** dialog box is displayed. Now, select the **Clear** button to completely remove the existing image.

Figure 22-7A shows a 16 × 16 pixel image created for the **E-Border** button with the **Grid** option activated. Use the pencil and line tools to create this or a similar image. Using the **Save...** button, save your image with a name of E-border; it will be stored in the current CUIx. Pick the **Close** button to return to the **Customize User Interface** dialog box. Your newly created image now appears in the list of existing images in the **Button Image** pane, as shown in **Figure 22-7B**. It is automatically associated with the command.

Figure 22-7.
A—A custom button image is created in the **Button Editor** dialog box. B—The new button image has been saved and appears in the list.

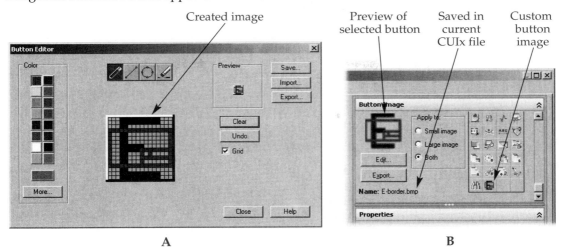

A

B

Consider the needs of the persons who will be using your custom commands when you design button images. Simple, abstract designs may be recognizable to you because you created them. However, someone else may not recognize the purpose of the command from the image. For example, the standard buttons in AutoCAD show a graphic that implies something about the command the button executes. A custom command will be most effective if its button image graphically represents the actions the command will perform.

Associating a Custom Image with a Command

There are two ways to associate a new, custom button image with a command. You can use the **Button Image** pane or the **Properties** pane in the **Customize User Interface** dialog box. Once a button image is associated with a command, the image is used for that command on *all* ribbon panels, menus, and toolbars where the command is inserted.

To use the **Button Image** pane to assign an image to a command, first make sure the command is selected in the **Command List:** pane. Then, select the **Large image**, **Small image**, or **Both** radio button in the **Button Image** pane to determine for which size of button the image will be used. Next, pick the button image in the list of predefined button images. Finally, pick the **Apply** button at the bottom of the **Customize User Interface** dialog box to assign the image to the button.

You can also use the **Properties** pane to associate a saved button image file with the command. Make sure the command is selected in the **Command List:** pane. Then, in the **Properties** pane, expand the **Images** category to display the Small image and Large image properties. If there is an image currently associated with the property, the path to the image is displayed in the text box, **Figure 22-8**. If there is no path displayed, the image is saved in a CUIx file. Pick in each property text box and type the path and file names of the saved image files. Alternately, you can pick the ellipsis button (...) to display a standard open dialog box and locate the file. This button appears when the property is selected. Finally, pick the **Apply** button at the bottom of the **Customize User Interface** dialog box to assign the image(s) to the button.

Figure 22-8.
The custom button image has been assigned to the custom command.

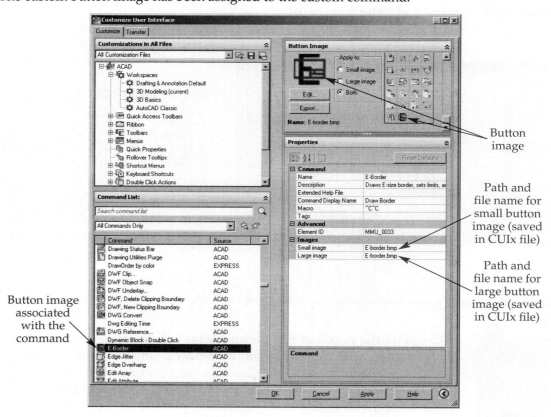

If you only designate an image file for small buttons, the button for the command will be blank when you switch to large buttons. This is because no image has been designated for that size. Be sure to specify an image for both small and large buttons. When using custom icons, the system variable that controls the resizing of ribbon images to standard sizes of 16 × 16 (small) and 32 × 32 (large) icons is **RIBBONICONRESIZE**. The system variable has an integer value of either 0 (off) or 1 (on). By default, this system variable is set to 1, which means that icons are resized. This system variable is also controlled by the **Resize ribbon icons to standard sizes** option in the **Display** tab of the **Options** dialog box.

PROFESSIONAL TIP

To open and edit a button image that is not shown in the button image list, select any button image and then pick the **Edit...** button to display the **Button Editor** dialog box. Then, use the **Import...** button to open the button image you want to edit. To assign an image that does not appear in the image list, use the **Properties** pane as previously described. Additionally, you may use the **CUI Editor—Image Manager** dialog box to import images into a loaded CUIx file. The images are then available in the image list.

Defining a Custom Command

Now, you need to define the action that the custom command will perform. A text string called a *macro* defines the action performed by the command. This text string appears in the Macro property text box in the **Command** category in the **Properties** pane of the **Customize User Interface** dialog box. In many cases, this "command" is actually a macro that invokes more than one command. By default, the text ^C^C appears in the text box. The text ^C is a cancel command. This is the same as pressing the [Esc] key. The default text, then, represents two cancels.

Whenever a command is not required to operate transparently, it is best to begin the macro with two cancel keystrokes (^C^C) to fully exit any current command and return to the Command: prompt. Typically, two cancels are required to be sure you begin at the Command: prompt. One cancel may not completely exit some commands or functions. For example, when grips are active, one cancel deactivates grips, but a second cancel is required to fully exit the command and return to the Command: prompt.

The macro must perfectly match the requirements of the activated commands. For example, if the **LINE** command is issued, the subsequent prompt expects a coordinate point to be entered. Any other data input is inappropriate and will cause an error in the macro. It is best to manually "walk through" the desired macro, writing down each step and the data required by each prompt. The following command sequence creates the rectangular polyline border with a .015 line width. Absolute coordinates are used, not relative coordinates.

> Command: **PLINE**↵
> Specify start point: **1,1**↵
> Current line-width is 0.0000
> Specify next point or [Arc/Halfwidth/Length/Undo/Width]: **W**↵
> Specify starting width <0.0000>: **.015**↵
> Specify ending width <0.0150>: ↵
> Specify next point or [Arc/Halfwidth/Length/Undo/Width]: **42,1**↵
> Specify next point or [Arc/Close/Halfwidth/Length/Undo/Width]: **42,32**↵
> Specify next point or [Arc/Close/Halfwidth/Length/Undo/Width]: **1,32**↵
> Specify next point or [Arc/Close/Halfwidth/Length/Undo/Width]: **C**↵
> Command:

Creating the macro for your custom **E-Border** command involves duplicating these keystrokes, with a couple of differences. Some symbols are used in menu macros to represent keystrokes. For example, a cancel (^C) is not entered by pressing [Esc]. Instead, the [Shift]+[6] key combination is used to place the *caret* symbol, which is used to represent the [Ctrl] key in combination with the subsequent character (a C in this case). Another keystroke represented by a symbol is the [Enter] key. An [Enter] is placed in a macro as a semicolon (;). A space can also be used to designate [Enter]. However, the semicolon is more commonly used because it is very easy to count to make sure that the correct number of "enters" is supplied.

AutoCAD system variables and control characters can be used in menus. They can be included to increase the speed and usefulness of your menu commands. Become familiar with these variables so you can make use of them in your menus.

- **^B.** Snap mode toggle.
- **^C.** Cancel.
- **^D.** Dynamic UCS toggle.
- **^E.** Cycles to next isoplane.
- **^G.** Grid mode toggle.
- **^H.** Issues a backspace.
- **^I.** Issues a tab.
- **^M.** Issues a return.
- **^O.** Ortho mode toggle.
- **^P. MENUECHO** system variable toggle.
- **^Q.** Toggles echoing of prompts, status listings, and input to the printer.
- **^R.** Toggles command versioning. Allows macros written in previous releases to properly function.
- **^T.** Tablet toggle.
- **^V.** Switches current viewport.
- **^Z.** Suppresses the addition of the automatic [Enter] at the end of a command macro.
- **\.** Pauses for user input.

Keeping the discussed guidelines in mind, the following macro draws the polyline border.

^C^CPLINE;1,1;W;.015;;42,1;42,32;1,32;C;

Compare this with the command line entry example to identify each part of the macro.

The next steps that the command will perform are to set the limits and zoom to display the entire border. To do this at the command line requires the following entries.

Command: **LIMITS**↵
Reset Model space limits:
Specify lower left corner or [ON/OFF] <0.0000,0.0000>: **0,0**↵
Specify upper right corner <12.0000,9.0000>: **44,34**↵
Command: **ZOOM**↵
Specify corner of window, enter a scale factor (nX or nXP), or
[All/Center/Dynamic/Extents/Previous/Scale/Window/Object] <real time>: **E**↵ *(this prompt will differ if the current view is perspective, but the entry is the same)*
Command:

Continue to develop the macro by adding the following text string (shown in color) immediately after the previous one.

^C^CPLINE;1,1;W;.015;;42,1;42,32;1,32;C;**LIMITS;0,0;44,34;ZOOM;E**

An "enter" is automatically issued at the end of the macro, so it is not necessary to place a semicolon at the end. The macro for the custom command is now complete.

To assign the macro to your custom **E-Border** command, first make sure the command is selected in the **Command List:** pane of the **Customize User Interface** dialog box. Then, pick in the Macro property text box in the **Properties** pane and enter the complete macro shown above. For a long macro such as this one, you can pick the ellipsis button (...) at the end of the text box to display the **Long String Editor** dialog box. See **Figure 22-9**. Enter the macro in this dialog box and pick the **OK** button to return to the **Customize User Interface** dialog box. Finally, pick the **Apply** button to associate the macro with the custom command.

Exercise 22-1

Complete the exercise on the companion website.
www.g-wlearning.com/CAD

Placing a Custom Command on an Interface Element

The custom command is now fully defined. The macro has been written and associated with the command. A custom button image has also been created and associated with the command. Now, you can add the custom command to a ribbon panel, menu, or toolbar just as you would one of the predefined AutoCAD commands. This was introduced earlier and is covered in detail later in this chapter in the specific sections on the ribbon, toolbars, and menus.

Figure 22-9.
The **Long String Editor** dialog box can be used to write longer macros. The text string will automatically "wrap" in this dialog box, which does not affect the macro.

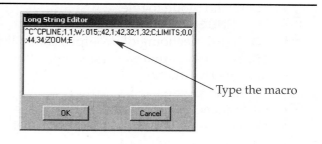

Type the macro

AutoCAD and Its Applications—Advanced

After adding the custom command to an interface element, it should be fully functional when you exit the **Customize User Interface** dialog box. Once you close the dialog box, test the command to make sure it works. If it does not, edit the macro in the **Customize User Interface** dialog box as needed.

Overview of the Ribbon

By default, the ribbon is docked at the top of the drawing area. The ribbon can also be docked to the left or right of the drawing area, or it can be floating. The ribbon contains commands and tools on *panels*. The panels are grouped on *tabs* that can be individually displayed. Think of the tabs as the containers that hold the ribbon panels. Together, the panels and tabs make up the ribbon.

You may quickly change the display from the full ribbon view to one of the minimized options by double-clicking any ribbon tab title. See **Figure 22-10**. By default, the ribbon is set to cycle through all views. To the right of the last tab name is a button with an arrow icon. This button is used to set the appearance of the docked ribbon and its cycling. The three options to set the view of tabs are minimize to tabs, minimize to panel titles, and minimize to panel buttons. When minimized to panel titles or buttons, hover the cursor over the minimized icon to display the corresponding panel. When minimized to tabs, pick the tab name to display the tab.

A ribbon panel may contain rows of command buttons, drop-down lists, or sliders. You can choose which panels are visible by right-clicking on the ribbon to display a shortcut menu. See **Figure 22-11**. Select either **Show Tabs** or **Show Panels** and choose which content to display. The currently displayed items are checked in the submenus.

Figure 22-10.
The four display states of the ribbon when it is docked. Double-click on a tab title to cycle through the views. A—Full. B—Minimized to panel buttons. C—Minimized to panel titles. D—Minimized to tabs.

A

B

C

D

Figure 22-11.
A—The **Show Tabs** submenu is used to choose which tabs are displayed in the ribbon.
B—The **Show Panels** submenu is used to choose which panels are displayed in the current tab.

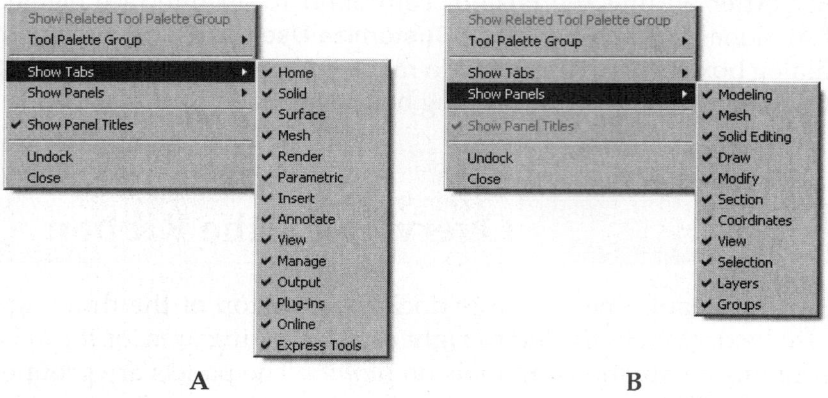

A   B

The items displayed in the **Show Panels** submenu are based on which tab is current (on top). There are separate panels for each tab.

Workspaces are typically used to set which ribbon components are displayed. Chapter 25 discusses customizing workspaces. Also, the ribbon can dynamically change when commands are accessed. *Contextual tabs* may be displayed on the ribbon when a command is active and then hidden when the command is finished. Contextual tabs are discussed further in the next section.

Tabs and Panels

The ribbon has three customization branches in the **Customize User Interface** dialog box: the Tabs branch, Panels branch, and Contextual Tab States branch. These branches are located below the Ribbon branch. Since the tabs contain the panels, they are displayed first in the tree. In this section, you will examine each area to gain a better understanding of the composition of the panels and how panels relate to tabs.

In the **Customizations in All Files** pane, expand the Tabs branch. There are 18 default tabs associated with the main user interface. These appear at the top of the branch. The remaining tabs have "contextual" in their name. These are used to reference the contextual tab states. Some of the main ribbon tab names have the suffix 2D or 3D. These help identify the types of commands contained on the tab. They are then included in the appropriate workspace, either 2D Drafting & Annotation, 3D Modeling, or 3D Basics. For example, select the Home - 3D branch. Notice that the name of this branch does not match the name displayed on the AutoCAD screen. With the branch selected, look at the **Properties** pane, **Figure 22-12**. The name displayed on the AutoCAD screen is the value in the Display Text property.

Expand the Home - 3D branch. It contains 11 branches: eight with a prefix of Home 3D, two with a prefix of Home, and one with a prefix of View. These branches correspond to the panels associated with the **Home** tab in the 3D Modeling workspace. Notice that there are no branches below these. The Tabs branch contains branches for tabs and, below those, branches for panels. The Panels branch contains branches for panels and, below those, branches for the commands on each panel.

Now, expand the Panels branch. All of the available panels are shown as branches below the Panels branch. Expand the Home 3D - Modeling branch. Notice that it consists of the Panel Dialog Box Launcher branch, two rows, and a Slideout branch. See **Figure 22-13**. Any row listed below the Slideout branch is only visible when the ribbon panel is expanded. For the **Modeling** panel in the ribbon, row 2 is located in the expanded portion of the panel.

You can expand the branches for the rows in a tab. Notice row 1 contains two drop-down lists (Solid Primitives Drop-down and Solid Creation Drop-down) and a subpanel. If you expand the Solid Primitives Drop-down branch, you can see the commands associated with the drop-down list. In this case, one drop-down list contains the solid primitive commands and the other one contains commands such as **LOFT** and **EXTRUDE**.

Figure 22-12.
The properties of the Home - 3D tab.

Tab name

Name displayed on
AutoCAD screen

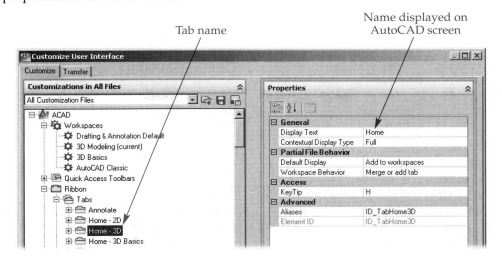

Figure 22-13.
Notice how a ribbon panel is composed in the **Customize User Interface** dialog box.

Drop-down lists

Subpanel
row 1

Subpanel
row 2

Row 1

Drop-down
list in row 1

Subpanel
in row 1

Rows in
subpanel

Slideout (panel
separator)

Row 2

Panel
properties

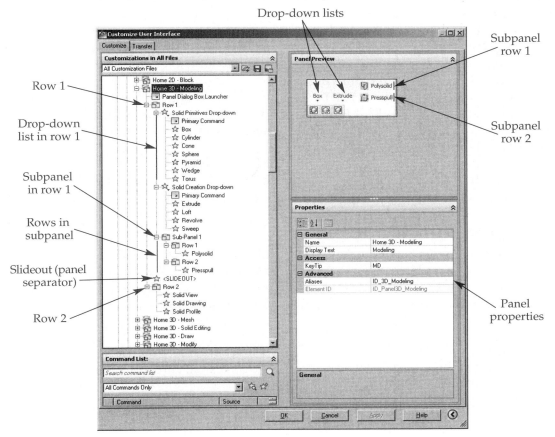

Figure 22-14.
Notice how a ribbon panel is composed on the ribbon.

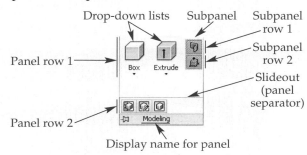

Next, expand the branch for the subpanel (Sub-Panel 1). This subpanel contains two rows. Expanding the branch for each row displays the commands contained in it. Notice how the drop-down list and subpanel contained in row 1 of the Home 3D - Modeling branch are fitted together in the **Modeling** panel in the ribbon. See **Figure 22-14.**

Expand the branch for row 2 in the Home 3D - Modeling branch. Notice this branch contains commands, but no drop-down lists or subpanels. Also, notice that row 2 is below the Slideout branch, which is the panel separator. This means it is displayed in the expanded panel. Refer to **Figure 22-14.**

The final branch in the Ribbon branch is Contextual Tab States. When you expand this branch, several branches of AutoCAD commands and features are displayed. Some branches have contextual tab panels assigned to them and others do not. For example, expand the Text Editor in progress branch. Below it is the Text Editor Contextual Tab branch. This indicates the **Text Editor** contextual tab will be displayed when text is being created or edited in the drawing.

For example, launch the **MTEXT** command to create sample multiline text. While the command is active, the **Text Editor** tab is displayed in the ribbon and made active. See **Figure 22-15.** When the command is complete, the tab is automatically removed from the ribbon.

You can associate tabs for any command or feature listed in the Contextual Tab States branch in the **Customize User Interface** dialog box. To do so, drag the tab from the Tabs branch and drop it in the desired branch in the Contextual Tab States branch.

Button Properties

In the ribbon, command buttons have additional properties from the basic command properties. These are located in the **Appearance** section of the **Properties** pane.

Buttons may be displayed in one of five default sizes: large with text (vertical), large with text (horizontal), large without text, small with text, and small without text.

Figure 22-15.
When the **MTEXT** command is active, the **Text Editor** contextual tab is displayed in the ribbon. Contextual tabs are added in the Contextual Tab States branch in the **Customize User Interface** dialog box.

The size is set in the **Customize User Interface** dialog box. The size setting is actually the maximum display size for the button. When the subpanel is set to do so, AutoCAD adjusts the button size smaller as needed based on the space available for the ribbon.

To set the size of a command button, expand the Ribbon branch and then the Panels branch in the **Customizations in All Files** pane. Then, expand the branches for the panel and row that contain the command. Next, select the command in the **Customizations in All Files** pane. If the panel branch is currently selected, you can also pick the button in the **Panel Preview** pane to select the command. Finally, set the Button Style property in the **Appearance** section of the **Properties** pane. See **Figure 22-16**. Generally, text labels are not shown when small buttons are specified. When large buttons are specified, the vertical orientation is usually selected so the text is below the button. This helps reduce the width of the panel to preserve space on the ribbon.

Customizing a Panel

To add a command to a panel, open the **Customize User Interface** dialog box. In the **Customizations in All Files** pane, expand the Ribbon branch and then the Panels branch. Next, expand the branches for the row to which the command will be added. In the **Command List:** pane, locate the command to add to the panel. Drag the command from the **Command List:** pane and drop it into position in the tree in the **Customizations in All Files** pane.

To remove a command from a panel, right-click on the command in the panel's branch in the **Customizations in All Files** pane and select **Remove** from the shortcut menu. You can also select the command and press the [Delete] key.

Row 1 is the top of the panel. In addition, the top of a row branch is the left-hand side of the panel. Commands are displayed in this order on the panel. The commands in a panel can be rearranged. In the **Customizations in All Files** pane, select the command to move and drag it to a new location, either within its current row or in a different row. Rows can also be rearranged by dragging them within the tree in the **Customizations in All Files** pane. After you drag a row to a new location, all rows are automatically renumbered. The row at the top of the panel branch is always row 1 and all other rows are sequentially numbered. The panel separator (Slideout branch) can also be dragged to a new location. Remember, rows listed after the panel separator are not displayed until the panel is expanded.

Figure 22-16.
A—Buttons in a ribbon panel can be displayed in one of five sizes. The small size can be displayed either with or without a label. B—Setting the appearance of a button in a panel.

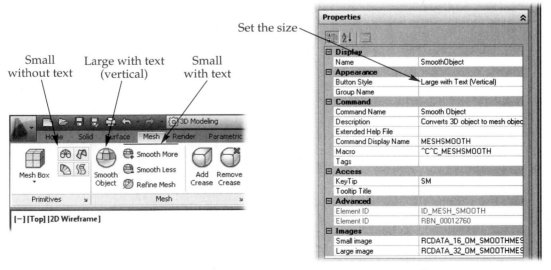

A new row can be added to a panel. In the **Customizations in All Files** pane, right-click on the row *after* which you would like the new row added. To add a new first row, right-click on the panel branch name. Then, select **New Row** from the shortcut menu. The new row is added and all other rows are renumbered. Once a row is added, commands can be added to it.

 There are certain conditions in which rows in a panel cannot be rearranged. If you attempt to drag a row to a different location and you cannot, just realize you have encountered one of these situations.

Adding a Drop-Down List to a Ribbon Panel

A drop-down list is added to a row in a ribbon panel using the **Customize User Interface** dialog box. To add a drop-down list, right-click on the row branch in the **Customizations in All Files** pane. Then, select **New Drop-down** from the shortcut menu. A branch for the new drop-down list is added to the bottom of the row's branch. Now you can drag commands from the **Command List:** pane into the drop-down list branch.

Notice the Primary Command branch below the drop-down list branch. The command directly below this is displayed as the button for the drop-down list on the ribbon. For example, the **Modeling** panel on the **Home** tab of the ribbon has the **Box** button displayed for the solid primitives drop-down list. In the **Customize User Interface** dialog box, the **Box** command is listed directly below the Primary Command branch.

With the branch for the drop-down list selected in the **Customizations in All Files** pane, look at the **Properties** pane. See **Figure 22-17**. Drop-down lists have appearance properties in addition to the Button Style property: Behavior and Split Button List Style. The Button Style properties are the same as discussed earlier. The other settings determine how the drop-down list on the ribbon is controlled. The Behavior property sets whether the top button in the drop-down list executes a command or displays the Name property. The Split Button List Style property determines how the buttons and names appear when the list drops down from the panel.

Figure 22-17.
Properties for a drop-down list in a ribbon panel.

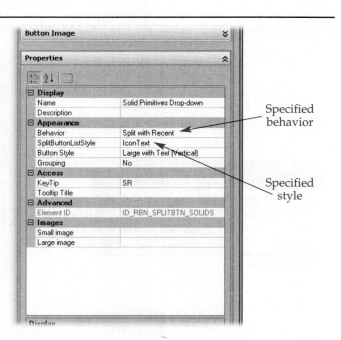

Specified behavior

Specified style

The default Behavior setting is Split with Recent and the default Split Button List Style setting is Icon Text. These settings are typically most desirable for standard AutoCAD workflow. This behavior means that the drop-down list on the ribbon will display in two parts (split). The upper part is the most recent command and the lower part, displayed when the button is selected, shows additional command icons in the drop-down list.

Creating a New Tab or Panel

To create a new panel, open the **Customize User Interface** dialog box. Then, in the **Customizations in All Files** pane, right-click on the Panels branch and select **New Panel** from the shortcut menu. A new panel with the default name of Panel*x* is added to the bottom of the Panels branch. The name appears in a text box in the tree. Enter a name for the new panel, either in the tree or in the **Properties** pane. Expand the branch for the new panel and notice that the Dialog Box Launcher, Row 1, and Slideout branches are automatically added when the panel is created. Add commands and rows to the new panel as needed. A new tab is similarly created by right-clicking on the Tabs branch.

You will now create a new ribbon panel containing the custom command you created earlier. First, create a new panel and name it My Panel. Then, in the **Command List:** pane, locate the **E-Border** command. Drag the command into the tree in the **Customizations in All Files** panel. The I-bar cursor appears as you drag through the tree. When the I-bar is below the Row 1 branch, drop the command. See **Figure 22-18**.

Now, create a new tab to hold the new panel. Name the tab My Stuff. In the **Properties** pane for the new tab, change the Display Text property to Border Tools. Next, locate the My Panel branch and drag it into the My Stuff branch. As you drag, the same I-bar appears in the tree. When the I-bar is below the My Stuff branch, release the mouse button.

Figure 22-18.
Adding a command to a custom ribbon panel.

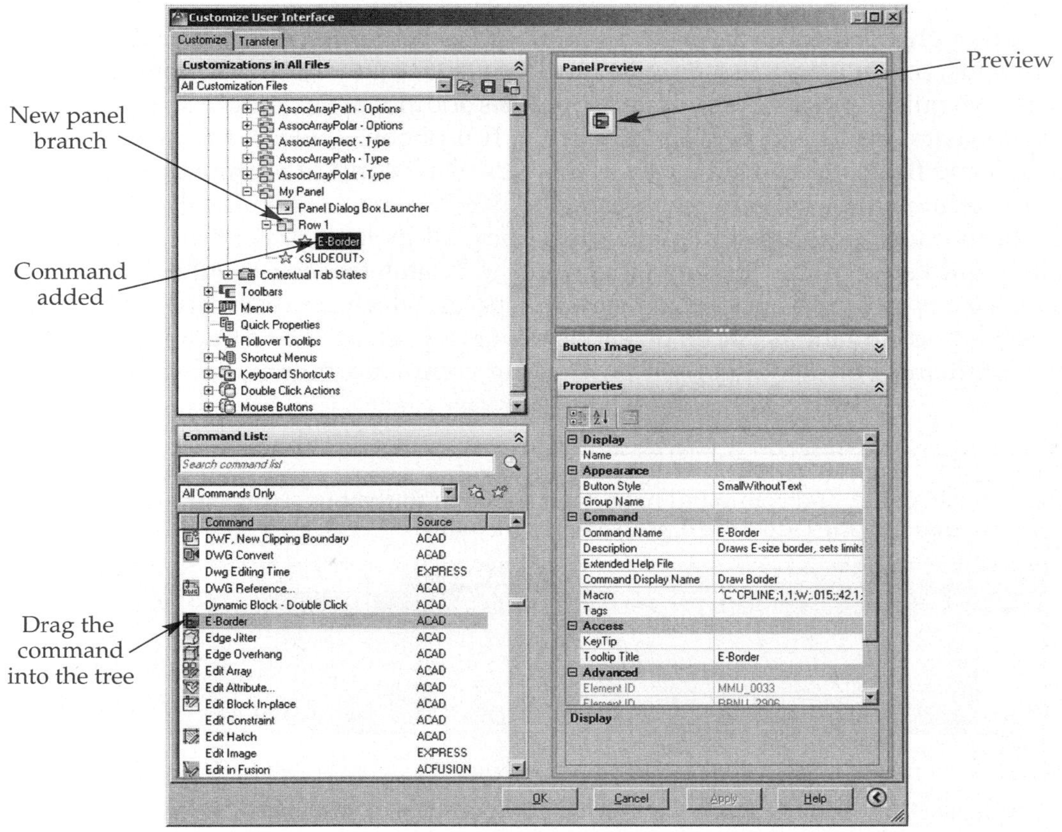

For the tab to be displayed, it must be added to the current workspace. Workspaces are covered in detail in Chapter 25. To add the tab, select the workspace in the **Customizations in All Files** pane. Then, drag the tab from the **Customizations in All Files** pane and drop it into the Ribbon Tabs branch in the **Workspace Contents** pane.

Close the **Customize User Interface** dialog box. After the menu compiles, the new tab is displayed in the ribbon, **Figure 22-19**. It is located on the right-hand side of the ribbon. Test the **E-Border** button on the panel to make sure the command properly functions.

Associating a Tool Palette Group with a Ribbon Tab

A tool palette group can be associated with a tab in the ribbon. Then, the associated tool palette group is displayed in the **Tool Palettes** window when you right-click on the tab and select **Show Related Tool Palette Group** from the shortcut menu.

To associate a tool palette group with a tab, right-click on the tab in the ribbon (not on a panel). This is done with the **Customize User Interface** dialog box closed. Next, select **Tool Palette Group** in the shortcut menu and then the name of the group in the submenu. See **Figure 22-20**. You can now right-click on the tab and select **Show Related Tool Palette Group** from the shortcut menu. The tool palette group you associated with the tab is displayed in the **Tool Palettes** window.

Exercise 22-2

Complete the exercise on the companion website.
www.g-wlearning.com/CAD

Overview of Toolbars

Although the ribbon is the primary graphic interface for accessing the main AutoCAD commands, *toolbars* can also provide quick access to many AutoCAD commands with one or two quick "picks." This interface provides additional flexibility, especially considering toolbars are a fraction of the size of the full ribbon. Toolbars are moved, resized, docked, and floated in the same way as toolbars in all Windows-compatible software.

In addition to positioning and sizing toolbars, you can customize the toolbar interface. In the **Customize User Interface** dialog box, all toolbars, except the **Quick Access** toolbar, are listed in the Toolbars branch in the **Customizations in All Files** pane. The **Quick Access** toolbar is customized via workspaces, which are discussed in Chapter 25.

When a command is placed on a toolbar, it is represented by a button. You can add new command buttons or reposition existing command buttons for quicker access.

Figure 22-19.
A custom tab, panel, and command have been added to the ribbon. Notice the display name for the tab matches the Display Text property for the tab.

Custom panel   Tool on panel   Custom tab

Figure 22-20.
Associating a tool palette group with a ribbon tab.

Right-click
on tab

Select tool
palette group

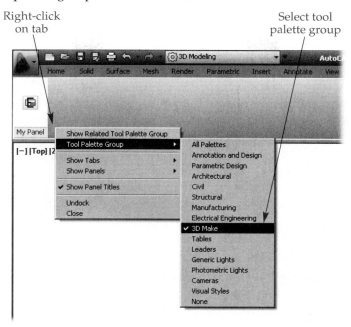

Infrequently used commands can be removed from the toolbar or repositioned to a less prominent location. Entirely new toolbars can be created and filled with predefined or custom commands.

Toolbar Visibility

By default, toolbars other than the **Quick Access** toolbar are not displayed except in the AutoCAD Classic workspace. To display toolbars, you can use the **Toolbars** button on the **Windows** panel of the **View** tab in the ribbon or use the **-TOOLBAR** command. Picking the button displays a drop-down list. Select the menu group and then the toolbar name.

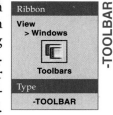

When the **-TOOLBAR** command is used, you are first prompted for the toolbar name. The complete toolbar name consists of the menu group and toolbar name, separated by a period. For example, the toolbar name for the **Draw** toolbar is ACAD.DRAW. The menu group name can be omitted when only one menu is currently loaded or if the toolbar name is not duplicated in another menu group. After specifying the toolbar name (or selecting **ALL** for all toolbars), you can select an option.

> Command: **-TOOLBAR**↵
> Enter toolbar name or [ALL]: **ACAD.DRAW**↵
> Enter an option [Show/Hide/Left/Right/Top/Bottom/Float] <Show>:

These options are used to hide, show, or specify a location for the toolbar.
- **Show.** Makes the toolbar visible.
- **Hide.** Causes the toolbar to be invisible.
- **Left.** Places the toolbar in a docked position at the left side of the AutoCAD window.
- **Right.** Places the toolbar in a docked position at the right side of the AutoCAD window.
- **Top.** Places the toolbar in a docked position at the top of the AutoCAD window.
- **Bottom**. Places the toolbar in a docked position at the bottom of the AutoCAD window.
- **Float.** Places the toolbar as a floating toolbar.

For example, to dock the **Zoom** toolbar on the left side of the AutoCAD window, use the following command sequence.

Command: **-TOOLBAR.**↵
Enter toolbar name or [ALL]: **ACAD.ZOOM.**↵
Enter an option [Show/Hide/Left/Right/Top/Bottom/Float] <Show>: **LEFT.**↵
Enter new position (horizontal, vertical) <0,0>: ↵
Command:

Another way to hide a floating toolbar is to pick its menu control button. This is the X in the corner of the toolbar. If you wish to hide a docked toolbar, you can first move it away from the edge to make it a floating toolbar. Then, pick the menu control button. When you hide a previously docked toolbar in this manner, it will appear in the floating position when you again make it visible.

When using floating toolbars, it is also possible to overlap the toolbars to save screen space. To bring a toolbar to the front, simply pick on it. Be sure to leave part of each toolbar showing.

Toolbar Display Options

Located in the **Window Elements** area of the **Display** tab of the **Options** dialog box are four check boxes and a text box relating to toolbars. See **Figure 22-21**.

When the **Use large buttons for Toolbars** check box is checked, the size of toolbar buttons is increased from 16×16 pixels to 32×32 pixels. At higher screen resolutions, such as 1280×1024, the small buttons may be difficult to see. At lower screen resolutions, such as 800×600, the large buttons take up too much of the display area.

When the **Show ToolTips** check box is checked, the name of the button to which you are pointing is displayed next to the cursor. Below this check box is the **Show shortcut keys in ToolTips** check box. When this option is checked, the shortcut key combination for the command is displayed in the tooltip. The **Show extended ToolTips** check box determines whether extended tooltips are displayed. When checked, additional tooltips are displayed when the cursor is hovered over a button for the number of seconds entered in the **Number of seconds to delay** text box. When tooltips are turned off, these two check boxes and the text box are grayed out.

Figure 22-21.
The **Display** tab of the **Options** dialog box contains settings for toolbars.

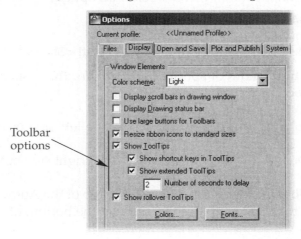

Toolbar options

Creating a New Toolbar

To create a new toolbar, open the **Customize User Interface** dialog box. Then, right-click on the Toolbars branch in the upper-left pane and pick **New Toolbar** in the shortcut menu. A new toolbar is added at the bottom of the Toolbars branch. An edit box is displayed in place of the toolbar name with a default name highlighted. Type a descriptive name for the toolbar and press [Enter].

After the new toolbar is named, it is highlighted in the Toolbars branch. The properties for the toolbar are displayed in the **Properties** pane of the **Customize User Interface** dialog box. See **Figure 22-22**. A preview of the toolbar also appears in the **Toolbar Preview** pane, but since the new toolbar is empty, there is not currently a preview. You can change the name of the toolbar and add a description in the General category of the **Properties** pane. The description appears as a tooltip when the cursor is over the docked toolbar. In the **Appearance** category, you can specify the default settings for the toolbar, including whether it is included in the current workspace, whether it is floating or docked, the location of the toolbar's upper-left corner, and the number of rows for the toolbar. The settings in the **Advanced** category are used for programming applications.

Adding a Command to a Toolbar

To add a command to a toolbar, first expand the Toolbars branch in the **Customizations in All Files** pane in the **Customize User Interface** dialog box. Next, expand the branch for the toolbar to which the command will be added. Then, select a command from the **Command List:** pane and drag it into position in the tree in the **Customizations in All Files** pane. As you drag the command in the tree, an I-bar cursor is displayed. When the I-bar is below the command where you want the new command placed, release the mouse button.

Figure 22-22.
The properties of a toolbar can be changed in the **Properties** pane of the **Customize User Interface** dialog box.

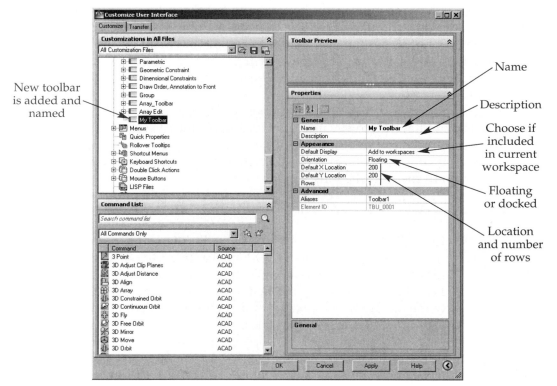

PROFESSIONAL TIP

If the toolbar branch is selected, you can also add a command by dragging and dropping the command into the **Toolbar Preview** pane, rather than into the tree in the **Customizations in All Files** pane. However, the toolbar must contain at least one command in order for the preview to appear.

Exercise 22-3

Complete the exercise on the companion website.
www.g-wlearning.com/CAD

Overview of Menus

When the menu bar is displayed, the names of the standard (classic) *menus* appear at the top of the AutoCAD drawing window. The menu bar is displayed in the AutoCAD Classic workspace. To display the menu bar in other workspaces, pick the arrow icon at the right-hand end of the **Quick Access** toolbar and select **Show Menu Bar** in the shortcut menu or type the **MENUBAR** command. Menus are selected by placing the cursor over the menu name and picking. You can also use the access (mnemonic) keys to select menus.

Once you understand how menus are designed, you can customize existing menus and create your own. Some basic information about menus includes:

- By default, AutoCAD has 13 menus displayed on the menu bar. If the Express tools are not installed, there are 12 menus.
- If no menus are defined in the current CUIx file or workspace, AutoCAD inserts default **File**, **Window**, and **Help** menus. This is similar to how AutoCAD is displayed without a drawing open (with the menu bar displayed).
- The name of the menu should be as concise as possible. On low-resolution displays, long menu names may cause the menu bar to be displayed on two lines, which reduces the drawing area.
- Menu item names can be any length. The menu is displayed as wide as its longest menu item name.
- Each menu can have multiple submenus (cascading menus).
- A menu can have up to 999 items, including submenus.
- To create an access (mnemonic) key for a menu or menu item, place an ampersand (&) before the desired access key character. Access and shortcut keys are discussed in the next section.

In earlier releases of AutoCAD, pull-down and context shortcut menus were referred to as POP menus. In addition, a series of menu files (MNU, MNC, MNR, and MNS) were used to define the menus. Pull-down menus were defined in the POP1 through POP499 sections of the menu file. Context shortcut menus were defined in the **POP500** through **POP999** sections. These POP designations are still used by AutoCAD for the sake of compatibility with older menus being used in the current version of AutoCAD. In the **Customize User Interface** dialog box, the POP names appear as aliases for these menus.

Shortcut and Access Keys

Before getting started with menu customization, it is important to understand the difference between shortcut keys and access keys. *Shortcut keys*, also called *accelerator keys*, are key combinations used to initiate a command. For example, [Ctrl]+[1] displays or closes the **Properties** palette. Custom shortcut keys can be created to initiate specific AutoCAD commands or macros. Creating custom shortcut keys is covered in Chapter 23.

Access keys, also called *mnemonic keys*, are keys used to access a menu or menu item via the keyboard. Pressing the [Alt] key activates the access keys for the menus. The access keys are shown as underlined (underscored) letters. Most access keys (underscores) are not displayed in the menu bar until the [Alt] key is depressed. For example, notice that the letter M is underlined in the **Modify** menu name. Pressing the [M] key accesses the **Modify** menu.

Any letter in the menu or menu item name can be defined as the access key, but an access key must be unique for a menu or submenu. Notice within the **Modify** pull-down menu that the M is used for **Match Properties**, so **Mirror** and **Move** use the i and v, respectively. The letter T can be used for both **Trim** and **Text** because **Text** is in the **Object** submenu, while **Trim** is in the "main" **Modify** menu. When creating custom menus, you can add custom access keys to the menu.

PROFESSIONAL TIP

Once the access (mnemonic) keys are activated, you can use the arrow keys to navigate through the pull-down menu structure.

Creating a New Menu

A new menu is created within the **Customize User Interface** dialog box. First, a menu is added to the Menus branch. Then, commands are added to the new menu. The process is basically the same as creating a new ribbon panel or toolbar, as described earlier in this chapter. The basic procedure is:

1. Open the **Customize User Interface** dialog box.
2. In the **Customizations in All Files** pane, expand the Menus branch. All of the existing menus are displayed.
3. Right-click on the Menus branch to display the shortcut menu. Pick **New Menu** from the shortcut menu. A new menu is added to the bottom of the list of existing menus. See **Figure 22-23A**. The name is highlighted in an edit box so the default name can be changed.
4. Give the menu an appropriate name.
5. Drag the desired commands from the **Command List:** pane and drop them into the new menu. See **Figure 22-23B**.
6. To add a separator, right-click on the command below which it should be inserted and select **Insert Separator** from the shortcut menu.

When adding a new menu, it is automatically assigned an alias of POP*n*, where *n* is the next available integer. Also, the menu is automatically available in all workspaces. Workspaces are covered in detail in Chapter 25.

Exercise 22-4

Complete the exercise on the companion website.
www.g-wlearning.com/CAD

Figure 22-23.
A—Adding a new menu. B—Commands have been added to the new menu.

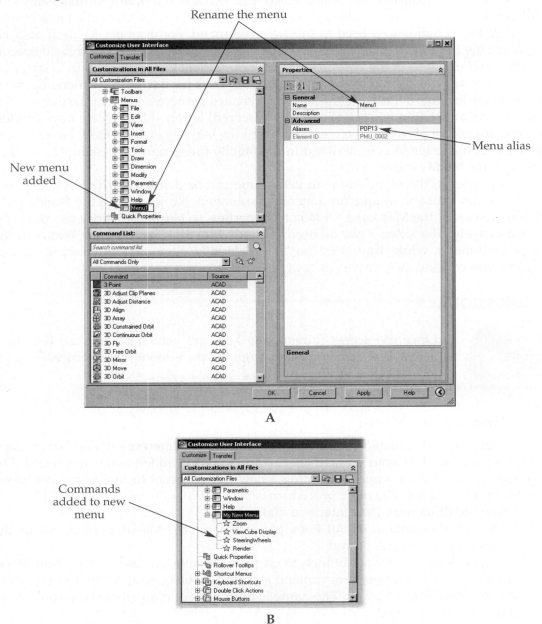

A

B

Adding a Submenu (Cascading Menu)

A *submenu* (cascading menu) is a menu contained within another menu. It can be used to help group similar commands or options. For example, when **Circle** is selected in the **Draw** menu, a submenu appears that offers the different options for drawing a circle.

Adding a submenu to a menu is similar to adding a "main" menu. First, open the **Customize User Interface** dialog box. Then, in the **Customizations in All Files** pane, expand the branch for the menu to which the submenu is to be added. Right-click on the command after which the submenu should appear. In the shortcut menu that is displayed, pick **New Sub-menu**. A new menu is added within the first menu. Notice that the icon in the tree indicates this item is a menu, not a command. Now, the submenu can be renamed to an appropriate name. Finally, drag commands from the **Command List:** pane and drop them into the new menu. See **Figure 22-24.**

Adding a Command to a Menu

To add a command to a menu, first expand the Menus branch in the **Customizations in All Files** pane in the **Customize User Interface** dialog box. Next, expand the branch for the menu to which the command will be added. Then, select a command from the **Command List:** pane and drag it into position in the tree in the **Customizations in All Files** pane. As you drag the command in the tree, an I-bar cursor is displayed. When the I-bar is below the command where you want the new command placed, release the mouse button.

Removing a Menu

If you want to permanently remove a menu, open the **Customize User Interface** dialog box. Then, expand the tree in the **Customizations in All Files** pane to display the menu to be deleted. Right-click on the menu and pick **Delete** from the shortcut menu. You can also highlight the menu in the tree and press the [Delete] key.

The above procedure is not recommended because the menu is permanently removed. To "restore" the menu in the future, it must be rebuilt. A better way to remove any unwanted menus is by deleting them from the workspace. Managing workspaces is covered in detail in Chapter 25.

Some Notes about Menus

Here are a few more things to keep in mind when developing menus.
- A menu label can be as long as needed, but should be as brief as possible for easy reading. The menu width is automatically created to fit the width of the longest item.
- Menus that are longer than the screen display are truncated to fit on the screen.
- Menus are disabled when using the **DTEXT** command after the rotation angle is entered, when creating text with **MTEXT**, when editing text created with **MTEXT**, and when using **SKETCH** after the record increment is set.

Figure 22-24.
A—A submenu has been added to the new menu. B—The menu displayed in the menu bar.

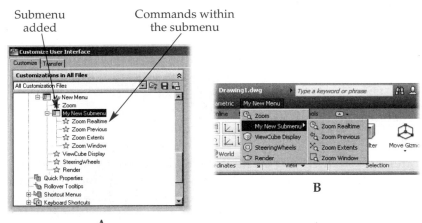

Sample Custom Commands

The following examples show how AutoCAD commands and options can be used to create commands. These custom commands can be placed on ribbon panels, tool-bars, or menus. Remember, an ampersand (&) preceding a character in a menu or item name defines the keyboard access (mnemonic) key used to enable it. The examples are listed using the following three-step process.

- Step 1. A description of the macro.
- Step 2. The keystrokes required for the macro.
- Step 3. The name and macro for the new command as entered in the **Properties** pane of the **Customize User Interface** dialog box.

Example 1

1. This **HEXAGON** command will start the **POLYGON** command and draw a six-sided polygon inscribed in a circle.
2. **POLYGON**↵
   6↵
   (select center)
   I↵
3. Name: &Hexagon
   Macro: *^C^Cpolygon;6;\i

The asterisk in front of the ^C^C repeats the command until it is canceled. The \ in front of i indicates that the macro will wait for user input, in this case the center of the polygon, before continuing.

Example 2

1. This **DOT** command draws a solid dot that is .1 unit in diameter. Use the **DONUT** command. The inside diameter is 0 (zero) and the outside diameter is .1.
2. **DONUT**↵
   0↵
   .1↵
3. Name: &Dot
   Macro: ^C^Cdonut;0;.1

Example 3

1. This **X-POINT** command sets the **PDMODE** system variable to 3 and draws an X at the pick point. The command should repeat.
2. **PDMODE**↵
   3↵
   POINT↵
   (pick the point)
3. Name: &X-Point
   Macro: *^C^Cpdmode;3;point

Example 4

1. This command, named **NOTATION**, could be used by a drawing checker or instructor. It allows the user to circle features on a drawing and then add a leader and text. It first sets the color to red, then draws a circle, snaps a leader to the nearest point that is picked on the circle, and prompts for the text. User input for text is provided, then a cancel returns the Command: prompt, and the color is set to ByLayer.

2. **-COLOR.⏎**
 RED.⏎
 CIRCLE.⏎
 (pick center point)
 (pick radius)
 LEADER.⏎
 NEA.⏎
 (pick a point on the circle)
 (pick end of leader)
 (press [Enter] *for automatic shoulder)*
 (enter text) ⏎
 (press [Enter] *to cancel)*
 -COLOR.⏎
 BYLAYER.⏎
3. Name: &Notation
 Macro: ^C^C-color;red;circle;\\leader;nea;\\;\;-color;bylayer

Example 5

1. This is a repeating command named **MULTISQUARE** that draws one-unit squares oriented at a 0° horizontal angle until the command is canceled.
2. **RECTANG.⏎**
 (pick lower-left corner)
 @1,1.⏎
3. Name: &Multisquare
 Macro: *^C^Crectang;\@1,1

PROFESSIONAL TIP

Some commands, such as the **COLOR** command, display a dialog box. Menu macros can provide input to the command line, but cannot control dialog boxes. To access the command-line version of a command, prefix the command name with a hyphen (-), as shown in Example 4 in the previous section. However, not all commands that display a dialog box have a command-line equivalent.

Exercise 22-5

Complete the exercise on the companion website.
www.g-wlearning.com/CAD

Chapter Review

Answer the following questions. Write your answers on a separate sheet of paper or complete the electronic chapter review on the companion website.
www.g-wlearning.com/CAD

1. What is an *interface element* in AutoCAD?
2. Which command is used to access the **Customize User Interface** dialog box?
3. In which pane of the **Customize User Interface** dialog box can you find all predefined commands?
4. How do you add a command to an interface element?
5. How do you remove a command from an interface element?
6. How can you copy a command to a new location on a different interface element?
7. What is a *partial CUIx file*?
8. Briefly describe how to create a custom command.
9. What is an *extended help file*?
10. Name the four drawing tools that are provided in the **Button Editor** dialog box.
11. What is the default, small size (in pixels) of the button editor drawing area?
12. Where is the **Use large buttons for Toolbars** check box located? What function does this check box perform?
13. How should you develop and test a new macro before entering it into a custom command definition?
14. Name two ways to specify an [Enter] in a macro. Which of the two methods is recommended?
15. Briefly describe the composition of the ribbon.
16. What determines which commands in a ribbon panel appear in the expanded panel?
17. How do you create a new ribbon tab and add a new panel to it?
18. Briefly describe how to customize a toolbar.
19. How do you create a new toolbar?
20. Explain how to display a toolbar.
21. Briefly describe a contextual tab.
22. What is a *drop-down list*?
23. How do you add a drop-down list to a ribbon panel?
24. How wide is a menu?
25. Interpret the following menu item.
    ^C^Crectang;\@1,1

Drawing Problems

Before customizing any toolbars, menus, or the ribbon, check with your instructor or supervisor for specific instructions or guidelines.

1. Create a new toolbar using the following information.
   A. Name the toolbar **Draw/Modify**.
   B. Copy at least three, but no more than six, commonly used drawing commands onto the new toolbar. Use only existing commands; do not create new ones.
   C. Copy at least three, but no more than six, commonly used editing commands onto the new toolbar. Use only existing commands; do not create new ones.
   D. Dock the new **Draw/Modify** toolbar at the upper-left side of the screen.

Drawing Problems - Chapter 22

2. Create a new toolbar using the following information.
   A. Name the toolbar **My 3D Tools**.
   B. Copy the following solid primitive commands onto the new toolbar.

Box	**Pyramid**
Cone	**Sphere**
Cylinder	**Torus**

   C. Copy the following view commands onto the new toolbar.

Top	**Bottom**	**Left**
Right	**Front**	**Back**

   D. Copy the following UCS commands onto the new toolbar.

3 Point	**Object**	**World**
Face UCS	**Origin**	**UCS Previous**

   E. Dock the toolbar below the toolbar created in Problem 1.

3. Create a new ribbon panel using the following information.
   A. The displayed name of the panel should be **Paper Space Viewports**.
   B. The panel should contain eight custom commands that use the **MVIEW** command to create paper space viewports:
   - **1 Viewport**—allow user to pick location
   - **1 Viewport (Fit)**
   - **2 Viewports (Horizontal)**—allow user to pick location
   - **2 Viewports (Vertical)**
   - **3 Viewports**—allow user to pick orientation and location
   - **3 Viewports (Right)**
   - **4 Viewports**—allow user to pick location
   - **4 Viewports (Fit)**
   C. Construct button graphics for the custom commands. Save the images in the default \Icons folder or create a new folder (be sure to add it to the AutoCAD support environment).
   D. Create a custom command that will switch from one viewport to another.
   E. Place a button on the panel that executes the **PLOT** command.

4. Create a new ribbon panel for inserting title block drawings. Name the panel **Title Blocks**. For the drawings to be inserted, use title block drawings that you have developed as templates or use title block drawings supplied with AutoCAD. Title block drawings with title blocks provided as dynamic blocks are available on the Autodesk website at www.autodesk.com/autocad-samples. The sample title block drawings are available for downloading by clicking on the Title Block (ansi), Title Block (arch), and Title Block (iso) links.
   A. The ribbon panel should contain five custom commands that do the following.
   - Insert an A-size title block drawing for mechanical drafting.
   - Insert a B-size title block drawing for mechanical drafting.
   - Insert a C-size title block drawing for mechanical drafting.
   - Insert a D-size title block drawing for mechanical drafting.
   - Insert an architectural title block drawing.
   B. Create button graphics for each of the custom commands. Save the images in a new folder and add the folder to the AutoCAD support environment.

5. Add a drop-down list to the **Paper Space Viewports** panel created in Problem 3. Add the five custom commands created in Problem 4 to this flyout.

6. Create a new dimensioning menu. Place as many dimensioning commands as you need in the menu. Use submenus if necessary. Include menu access (mnemonic) keys.

7. Create a menu for 3D objects. Include menu access (mnemonic) keys. The menu should include the following items.
   • At least three 3D solid objects
   • **HIDE** command
   • At least three visual style commands
   • **VPORTS** command
   • **NAVSWHEEL** command

8. Create a new menu named **Special**. The menu should include the following drawing and editing commands.

LINE	**MOVE**
ARC	**COPY**
CIRCLE	**STRETCH**
POLYLINE	**TRIM**
POLYGON	**EXTEND**
RECTANGLE	**CHAMFER**
DTEXT	**FILLET**
ERASE	

   Use submenus, if necessary. Include a separator line between the drawing and editing commands and specify appropriate menu access (mnemonic) keys.

9. Create a menu to insert a variety of blocks or symbols. These symbols can be for any drawing discipline that suits your needs. Use submenus and menu access (mnemonic) keys, if necessary.

10. Create a new ribbon panel named **My 3D Tools**. Evaluate which tools you use most often for creating and rendering 3D models. Place these commands on the new panel, even if they are already contained on another panel. The purpose of this new panel is to streamline your modeling and rendering work. Use drop-down lists as necessary. Arrange the commands on the panel so the panel does not need to be expanded to access the most frequently used commands.

11. Create a ribbon tab with the display name **My Tools**. Set the tab name to your initials. Add the panels created in Problems 3 and 10 to this tab. Review the tabs and tools to be sure they function properly.

Customizing Key and Click Actions

Learning Objectives

After completing this chapter, you will be able to:

✓ Assign shortcut keys to commands.
✓ Explain how shortcut menus function.
✓ Edit existing shortcut menus.
✓ Create custom shortcut menus.
✓ Customize an object's quick properties.
✓ Create custom rollover tooltips.
✓ Describe double-click actions.
✓ Edit double-click actions.
✓ Create custom double-click actions.

AutoCAD has many tools that can be used in "heads-up design." Heads-up design is a concept of working in which your eyes remain focused on the drawing area. For example, when dynamic input is on, you do not need to look at the command line to see the options for the current command. The options are displayed near the cursor in the drawing area. AutoCAD's shortcut menus and double-click actions also contribute to heads-up design. Shortcut menus are displayed by right-clicking. Double-click actions are initiated when an object is double-clicked. Like much of the graphic content in AutoCAD (the ribbon, toolbars, etc.), shortcut menus and double-click actions can be customized.

In order to use the shortcut menu and double-click action customization techniques discussed in this chapter, shortcut menus and double-click editing need to be enabled. To do this, open the **Options** dialog box and select the **User Preferences** tab. Then, check the **Double click editing** and **Shortcut menus in drawing area** check boxes, as shown in **Figure 23-1**.

The use of shortcut menus can be further refined by picking the **Right-click Customization...** button that appears below the check boxes. This displays the **Right-Click Customization** dialog box. See **Figure 23-2**. The settings in this dialog box allow you to define what a right-click does when in default mode, edit mode, or command mode. For this chapter, pick the **Shortcut Menu** radio buttons in the **Default Mode** and **Edit Mode** areas. Also, pick the **Shortcut Menu: always enabled** radio button in the **Command Mode** area. Then, close the **Right-Click Customization** and **Options** dialog boxes.

Figure 23-1.
The settings for
enabling shortcut
menus and double-
click editing are
found in the **Options**
dialog box.

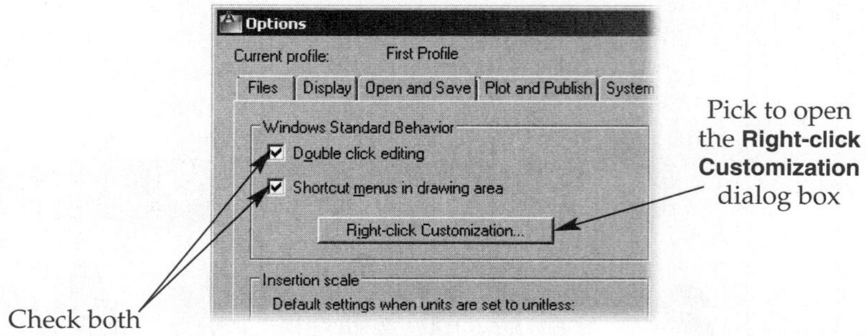

Check both

Pick to open
the **Right-click
Customization**
dialog box

Figure 23-2.
The settings in
the **Right-Click
Customization**
dialog box allow
you to define what
a right-click does
when in default
mode, edit mode, or
command mode.

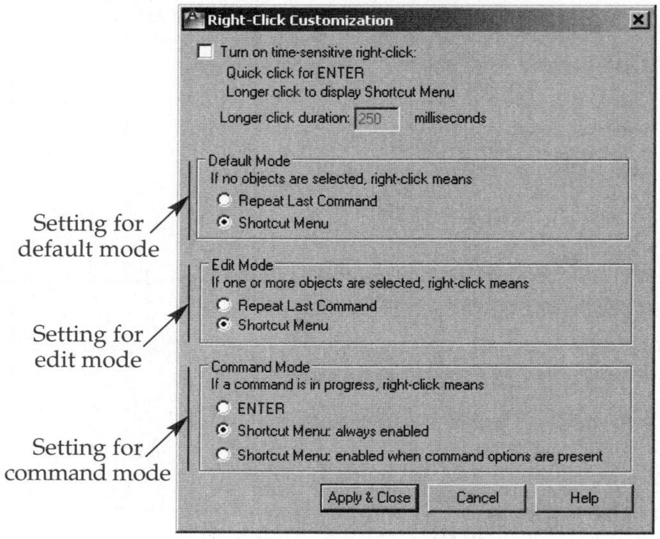

Setting for
default mode

Setting for
edit mode

Setting for
command mode

Customizing Shortcut Keys

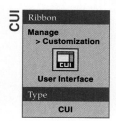

You can define your own custom shortcut keys (accelerator keys) for AutoCAD commands and custom macros. The **Customize User Interface** dialog box is used to define shortcut keys. To see the commands to which shortcut keys are assigned, expand the Keyboard Shortcuts branch in the **Customizations in All Files** pane. Then, expand the Shortcut Keys branch. All commands that have a shortcut key assigned appear in this branch. See **Figure 23-3**.

When the Shortcut Keys branch is selected, the **Shortcuts** pane is displayed in the upper-right corner of the **Customize User Interface** dialog box. A command that has a shortcut key assigned to it can be selected in this pane to display its properties in the **Information** pane in the lower-right corner of the dialog box.

Assigning a Shortcut Key

To assign a shortcut key to a command, first locate the command in the **Command List:** pane of the **Customize User Interface** dialog box. Next, drag the command into the Shortcut Keys branch in the **Customizations in All Files** pane. The command is added to the list of shortcut keys (although it may not be immediately visible) and the **Properties** pane is displayed for the command. See **Figure 23-4**. In the **Access** category of the **Properties** pane, pick in the Key(s) property text box. Next, pick the ellipsis button (...) on the right-hand end of the text box to display the **Shortcut Keys** dialog box. See **Figure 23-5**.

Figure 23-3.
Shortcut keys are added to commands in the **Customize User Interface** dialog box.

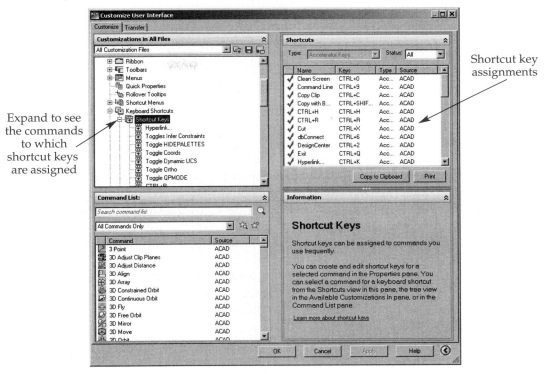

Expand to see the commands to which shortcut keys are assigned

Shortcut key assignments

Figure 23-4.
Adding a shortcut key for the **RENDER** command. Drag the **RENDER** command up from the **Command List:** pane to the Shortcut Keys branch.

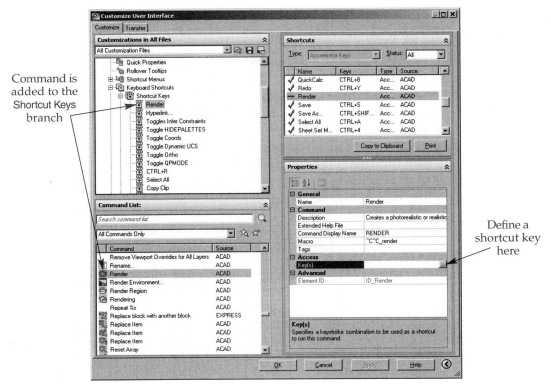

Command is added to the Shortcut Keys branch

Define a shortcut key here

Figure 23-5.
The **Shortcut Keys**
dialog box is where
a shortcut key is
specified for the
command.

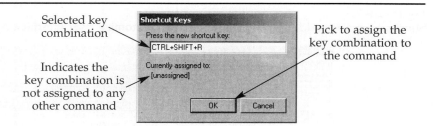

Selected key
combination

Indicates the
key combination is
not assigned to any
other command

Pick to assign the
key combination to
the command

To assign a new shortcut key to the command, pick in the text box labeled **Press the new shortcut key:** and press a combination of [Ctrl] + another key. If the shortcut key combination is currently assigned to another command, the name of the other command is displayed in the **Currently assigned to:** area. If the shortcut key combination is unassigned, pick the **OK** button to associate the shortcut key with the command. The shortcut key then appears in the Key(s) property in the **Customize User Interface** dialog box.

If you attempt to assign a shortcut key that is currently assigned to another command, an alert box appears indicating the shortcut assignment already exists and explaining the priority for using the shortcut. See **Figure 23-6**. It is not a good idea to have a shortcut key assigned to multiple commands. Be especially careful to ensure the standard Windows keyboard shortcuts are unique, such as [Ctrl]+[X] for cut, [Ctrl]+[C] for copy, and [Ctrl]+[V] for paste.

**PROFESSIONAL
TIP**

In addition to [Ctrl]+*key*, a shortcut can be [Ctrl]+[Shift]+*key*, [Ctrl]+[Alt]+*key*, or [Ctrl]+[Shift]+[Alt]+*key*. The [Caps Lock] key must be off in order to specify the [Shift] key in the **Press the new shortcut key:** text box.

Example Shortcut Key Assignment

To provide an example of customizing shortcut keys, this section shows how to assign the shortcut key [Ctrl]+[Alt]+[C] to the **CLOSE** command. Do the following:

1. Open the **Customize User Interface** dialog box.
2. Expand the Keyboard Shortcuts branch in the **Customizations in All Files** pane.
3. Expand the Shortcut Keys branch.
4. Drag the **Close** command from the **Command List:** pane into the Shortcut Keys branch. Make sure the command macro for the command is ^C^C_close.
5. In the **Properties** pane, pick in the Key(s) property text box. Then, pick the ellipsis button (...) on the right-hand side of the text box.

Figure 23-6.
This warning
appears if the
shortcut key you are
trying to assign to a
command is already
assigned to a
different command.
Avoid assigning a
shortcut key to more
than one command.

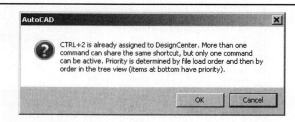

6. In the **Shortcut Keys** dialog box, pick in the **Press the new shortcut key:** text box.
7. Press the [Ctrl] key, [Alt] key, and [C] key at the same time. The message at the bottom of the dialog box should indicate that this shortcut key is unassigned.
8. Pick the **OK** button to close the **Shortcut Keys** dialog box.
9. Pick the **OK** button to close the **Customize User Interface** dialog box and apply the change.
10. Test the [Ctrl]+[Alt]+[C] shortcut key. When the shortcut key is used, either the current drawing should close or you should be prompted to save the changes to the drawing before closing.

PROFESSIONAL TIP

Shortcut keys (accelerator keys) have some specific limitations. For example, a shortcut cannot pause for user input or use repeating commands. Be aware of this when assigning shortcut keys to custom commands.

Exercise 23-1

Complete the exercise on the companion website.
www.g-wlearning.com/CAD

Examining Existing Shortcut Menus

Shortcut menus are context-sensitive menus that appear at the cursor location when using the right-hand button on the mouse (right-clicking). *Context sensitive* means that the displayed shortcut menu is dependent on what is occurring at the time of the right-click. For example, if no command is active, there is no object selection, and you right-click in the drawing area, the shortcut menu shown in **Figure 23-7A** is displayed. If

---

Figure 23-7.
A—Menu displayed when no command is active and no object is selected. B—Menu displayed when no command is active and you right-click in the command line window. C—Menu displayed when the **CIRCLE** command is active and before any point is selected.

A

B

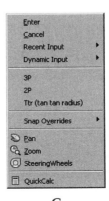

C

no command is active and you right-click in the command line window, the shortcut menu shown in **Figure 23-7B** is displayed. If the **CIRCLE** command is active and you right-click in the drawing area before any point is selected, the shortcut menu shown in **Figure 23-7C** is displayed. Other menus appear when right-clicking in other situations, too, such as when grips are being used or when an object is selected in the drawing window. In the case of a selected object, the shortcut menu is based on the type of object that is selected.

Before learning how to customize shortcut menus, examine the existing shortcut menus. Open the **Customize User Interface** dialog box and look at the **Customizations in All Files** pane. The name of this pane is based on what is selected in the drop-down list. Expand the Shortcut Menus branch in the tree. All of the existing shortcut menu names are displayed as branches. See **Figure 23-8**. There are object-specific and generic shortcut menus. Although none are included with the standard installation of AutoCAD, command-specific shortcut menus can also be created. The generic shortcut menus are:

- **Command Menu.** This menu appears when right-clicking in the drawing window while a command is active. Any command options for the active command are inserted into this menu. See **Figure 23-9**.
- **Default Menu.** This menu appears when right-clicking in the drawing window while no command is active and no objects are selected. See **Figure 23-10**.

Figure 23-8.
Existing shortcut menus are displayed as branches in the Shortcut Menus branch in the **Customize User Interface** dialog box.

Expand the Shortcut Menus branch

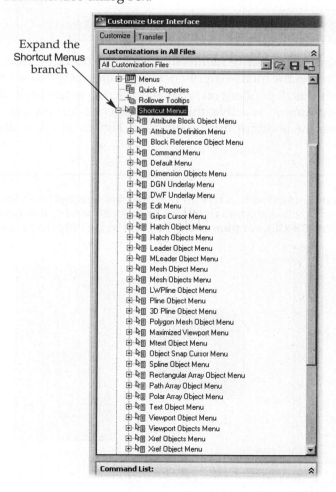

Figure 23-9.
A—The Command Menu branch in the **Customize User Interface** dialog box. B—The command shortcut menu.

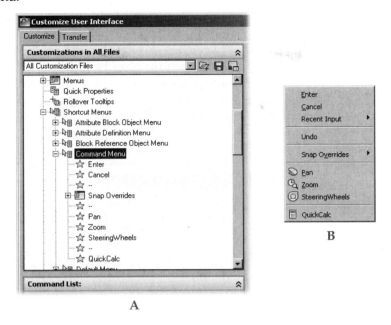

A

B

Figure 23-10.
A—The Default Menu branch in the **Customize User Interface** dialog box. B—The default shortcut menu.

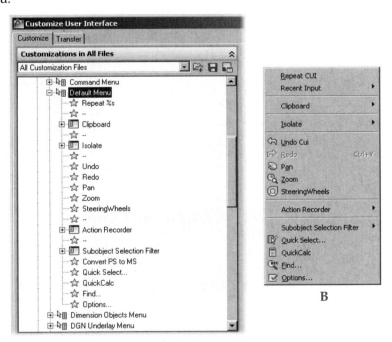

A

B

- **Edit Menu.** This menu appears when right-clicking in the drawing window when no command is active and an object is selected. In order for this menu to be displayed, the **PICKFIRST** system variable must be set to 1. If an object menu is available for the type of object selected, it is inserted into this menu. See **Figure 23-11.**
- **Grips Cursor Menu.** This menu appears when grips are being used. See **Figure 23-12.** An object must be selected and at least one grip must be hot.
- **Object Snap Cursor Menu.** This menu appears when holding down the [Shift] key and right-clicking. See **Figure 23-13.** It also appears as the Snap Overrides branch in the Command Menu branch, meaning it is displayed as a cascading menu.

Figure 23-11.
A—The Edit Menu branch in the **Customize User Interface** dialog box. B—The edit shortcut menu.

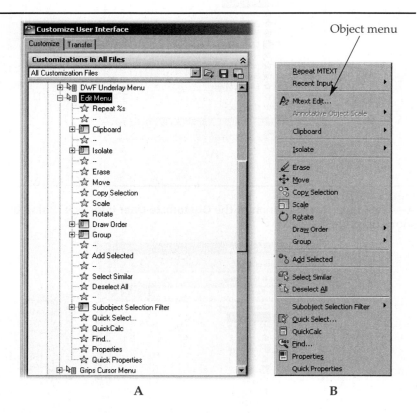

A

B

Figure 23-12.
A—The Grips Cursor Menu branch in the **Customize User Interface** dialog box. B—The grips shortcut menu.

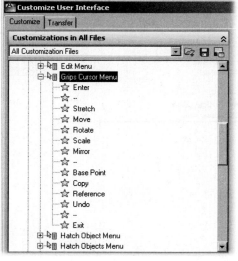

A

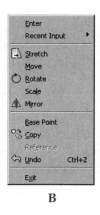

B

Figure 23-13.
A—The Object Snap
Cursor Menu branch
in the **Customize
User Interface** dialog
box. B—The object
snap shortcut menu.

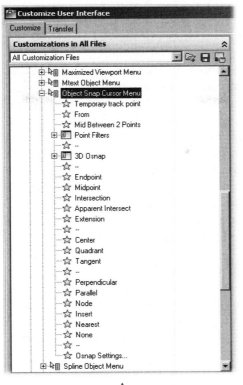

A

B

The remaining menus are object-specific menus that appear when right-clicking while a certain type of object is selected. Notice that there are menu branches named Attribute Block Object Menu, Block Reference Object Menu, Dimension Objects Menu, Hatch Object Menu, and others. These menus contain items that can be used on the type of object selected. For example, in the Shortcut Menus branch, the menu branch Dimension Objects Menu contains commands for editing the dimension style and precision.

Customizing Shortcut Menus

The existing shortcut menus can be customized by adding or removing commands. Shortcut menus can also be customized by visually grouping commands using separators. Cascading menus can be added to shortcut menus.

To add a command to a shortcut menu, open the **Customize User Interface** dialog box and, in the **Customizations in All Files** pane, expand the branch for the shortcut menu you would like to customize. Next, locate the command you wish to add in the **Command List:** pane. Then, drag the command into the desired position within the shortcut menu in the **Customizations in All Files** pane and drop it when the bar appears.

To remove a command from a shortcut menu, expand the branch for the shortcut menu in the **Customizations in All Files** pane. Highlight the command to be removed, right-click, and select **Remove** from the shortcut menu. You can also highlight the command and press the [Delete] key.

To add a separator to a shortcut menu, expand the branch for the shortcut menu in the **Customizations in All Files** pane. Highlight the command *after* which you would like the separator to be added. Right-click and select **Insert Separator** from the shortcut menu.

To rename a shortcut menu, highlight the branch in the **Customizations in All Files** pane. Then, right-click and select **Rename** from the shortcut menu. Finally, type the new name and press [Enter]. The shortcut menu can also be renamed using the **Name** property in the **Properties** pane.

To add a cascading menu to a shortcut menu, expand the branch for the shortcut menu in the **Customizations in All Files** pane. Highlight the command *after* which you would like the cascading menu to appear. Right-click and select **New Sub-menu** from the shortcut menu. A new shortcut menu branch with the default name of Menu*x* is added to the current shortcut menu. The new menu can be renamed. Now, in the **Command List:** pane, locate the commands you wish to add to the new shortcut menu. Drag the commands into the **Customizations in All Files** pane and drop them next to the name of the new shortcut menu. When the arrow appears next to the new menu name, drop the command to add it to the new shortcut menu.

Creating a new, custom shortcut menu is a two-step process. First, make a new shortcut menu and then drag commands into it. Follow these steps to make a new shortcut menu:

1. Open the **Customize User Interface** dialog box.
2. In the **Customizations in All Files** pane, right-click on the Shortcut Menus branch and select **New Shortcut Menu** in the shortcut menu that is displayed.
3. Enter a name for the shortcut menu.
4. In the **Properties** pane, add a description for the shortcut menu in the **General** category.
5. In the **Advanced** category of the **Properties** pane, add an alias. This alias is in addition to the automatic, sequential POP5*xx* alias that AutoCAD creates. Select the property, pick the ellipsis button (**...**) at the right-hand end of the text box, and type the alias in the **Aliases** dialog box that appears. Each alias must be on its own line in this dialog box. Close the **Aliases** dialog box.
6. Drag commands from the **Command List:** pane into the new shortcut menu.
7. Pick the **Apply** or **OK** button to apply the changes.

There are two types of custom shortcut menus: object specific and command oriented. The next sections describe the two types of custom shortcut menus in detail.

Creating Object-Specific Shortcut Menus

When creating an object-specific shortcut menu, there can actually be two menus available. One menu is displayed for instances when just a single object of a given type is selected. The other menu is displayed when more than one object is selected.

The name assigned to the object menu should follow the same syntax used for naming AutoCAD's default object-specific menus: *object_type* **Object Menu** or *object_type* **Objects Menu** (with an S). In this way, when looking at the shortcut menus in the **Customizations in All Files** pane in the **Customize User Interface** dialog box, you will easily recognize which object type that menu applies to and whether it is for multiple selected objects or a single selected object. The use of this syntax is optional. Menus can be named using whatever naming scheme you wish. However, it is recommended to follow the naming syntax described here.

The alias for the shortcut menu has a syntax that *must* be followed. It is this alias that AutoCAD uses in determining to which object or objects the menu applies. The syntax for the alias must take on the form of OBJECT_*type* or OBJECTS_*type* and must be exactly followed in order for AutoCAD to properly display the shortcut menu.

As an example, the following procedure creates a shortcut menu that allows access to the **LENGTHEN** and **BREAK** commands when a single line is selected.

1. Open the **Customize User Interface** dialog box.
2. Right-click on the Shortcut Menus branch in the **Customizations in All Files** pane and select **New Shortcut Menu.**

3. Name the shortcut menu **Line Object Menu**.
4. In the **Properties** pane, select the Aliases property in the **Advanced** category. Then, pick the ellipsis button to open the **Aliases** dialog box. On the second line, enter the alias OBJECT_LINE and then pick the **OK** button to close the **Aliases** dialog box. Since the **LENGTHEN** and **BREAK** commands can only be applied to a single object, be sure to use the OBJECT_*type* syntax (without the S).
5. Drag the **LENGTHEN** and **BREAK** commands from the **Command List:** pane into the Line Object Menu branch in the **Customizations in All Files** pane. See **Figure 23-14A**.
6. Pick the **OK** button to close the **Customize User Interface** dialog box and apply the changes.

Now, draw a line, select it, and right-click. Notice that **Lengthen** and **Break** entries appear in the shortcut menu. See **Figure 23-14B**. Selecting either entry executes the command. If the command accepts a preselected object, it is executed on the selected line. Neither **LENGTHEN** nor **BREAK** accepts preselected objects; you must reselect the line. Having the entries in the shortcut menu provides for quicker access to the command.

Creating Command-Oriented Shortcut Menus

When a command is being executed, any command options appear in the shortcut menu. For example, when the **CIRCLE** command prompts for a radius, you can right-click and select **Diameter** from the shortcut menu. Custom shortcut menus can be created for use when certain commands are active. This allows you to add options to the shortcut menu that is displayed when a command is active. Quicker access to object snaps and object selection methods are just a couple of applications that custom, command-oriented shortcut menus could allow for within commonly used commands.

A command-oriented shortcut menu is created in the same way as an object-oriented shortcut menu, as discussed in the previous section. However, the syntax for the alias is slightly different. The alias must be in the form of COMMAND_*command_name*, where *command_name* is the name of the command with which you want the shortcut menu associated.

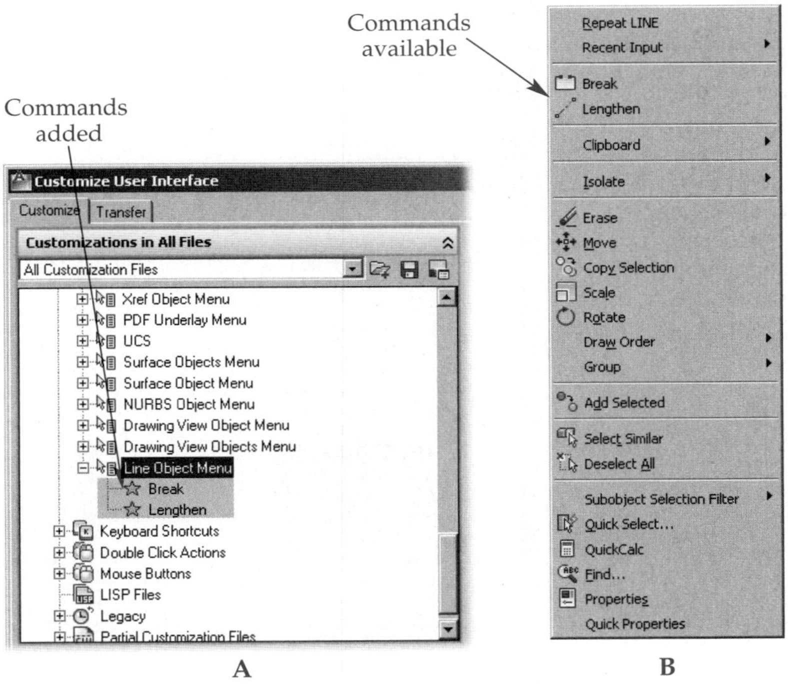

Figure 23-14.
A—The new object-specific shortcut menu is created.
B—The **BREAK** and **LENGTHEN** commands are now available in the shortcut menu.

A

B

Also, if the command step does not have any default options, such as a Select objects: prompt, right-clicking is, by default, interpreted as the [Enter] key. Therefore, in the **Right-Click Customization** dialog box, the **Shortcut Menu: always enabled** radio button must be selected in the **Command Mode** area, as described earlier in this chapter.

As an example, the following procedure creates a custom shortcut menu that displays **Select Previous**, **Select Last**, and **Select Fence** selection options at the Select objects: prompt for the **MOVE** command.

1. Open the **Customize User Interface** dialog box.
2. Right-click on the Shortcut Menus branch in the **Customizations in All Files** pane and select **New Shortcut Menu**.
3. Name the shortcut menu **Move Command Menu**.
4. In the **Properties** pane, select the Aliases property in the **Advanced** category. Then, pick the ellipsis button to open the **Aliases** dialog box. On the second line, enter the alias COMMAND_MOVE and then pick the **OK** button to close the **Aliases** dialog box. The syntax of COMMAND_*command_name* must be exactly followed in order for AutoCAD to properly display the shortcut menu.
5. Drag the **Select Previous**, **Select Last**, and **Select Fence** commands from the **Command List:** pane into the Move Command Menu branch in the **Customizations in All Files** pane. See **Figure 23-15A**.
6. Pick the **OK** button to close the **Customize User Interface** dialog box and apply the changes.

Now, initiate the **MOVE** command. At the Select objects: prompt, right-click and notice that **Select Previous**, **Select Last**, and **Select Fence** entries are available in the shortcut menu. See **Figure 23-15B**. Remember, the **Shortcut Menu: always enabled** radio button must be on in the **Right-Click Customization** dialog box for this shortcut menu to appear.

Exercise 23-2

Complete the exercise on the companion website.
www.g-wlearning.com/CAD

Figure 23-15.
A—The new command-specific shortcut menu is created. B—The **Previous, Last,** and **Fence** commands are available in the shortcut menu.

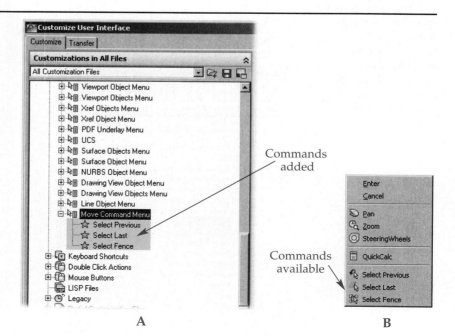

A

B

Customizing Quick Properties and Rollover Tooltips

The **Quick Properties** palette appears, by default, when you select objects in the drawing area. This is a streamlined version of the **Properties** palette that displays *quick properties*, which provide certain information about the object. In order for the **Quick Properties** palette to be displayed, the **Quick Properties** button on the status bar must be on. The palette can also be displayed by right-clicking with an object selected and selecting **Quick Properties** in the shortcut menu so it is checked.

Rollover tooltips provide quick properties in a graphic tooltip. This tooltip is displayed as the cursor is hovered over the object. A rollover tooltip is similar to the **Quick Properties** palette, but a rollover tooltip only provides information. Properties cannot be changed in the tooltip.

The information displayed in the **Quick Properties** palette and a rollover tooltip can be customized. You can specify the object types that display quick properties. Additionally, you can set which quick properties are displayed in the **Quick Properties** palette and rollover tooltip for a given object type. The steps for customizing quick properties are the same for the **Quick Properties** palette and rollover tooltips, as discussed in this section.

Open the **Customize User Interface** dialog box. In the **Customizations in All Files** pane, select the Quick Properties branch. Two columns are displayed on the right-hand side of the dialog box. See **Figure 23-16**. When you select Rollover Tooltips in the **Customizations in All Files** pane, the same two columns are displayed. However, the settings may be different between the **Quick Properties** branch and the Rollover Tooltips branch.

Figure 23-16.
When the Quick Properties branch is selected in the **Customize User Interface** dialog box, the object type list is displayed on the right-hand side of the dialog box.

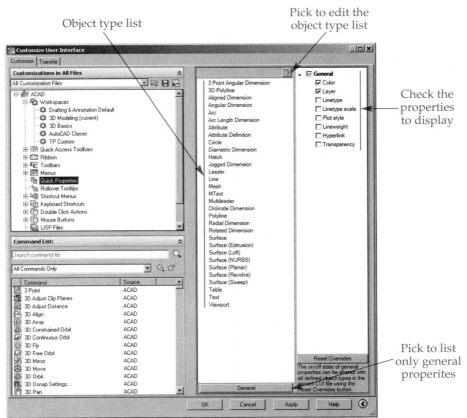

Object type list

Pick to edit the object type list

Check the properties to display

Pick to list only general properites

The left-hand column displays the *object type list*, which is a list of AutoCAD object types. You may add or remove object types by picking the **Edit Object Type List** button at the top of the column. This displays the **Edit Object Type List** dialog box, **Figure 23-17**. All available AutoCAD object types are listed in this dialog box. Those that are checked appear in the object type list in the **Customize User Interface** dialog box.

The object type list controls which objects display quick properties. The right-hand column displays a list of quick properties that can be displayed. The properties that are checked appear in the **Quick Properties** palette or rollover tooltip. Remember, the palette and tooltip can display different properties.

All objects have a General category containing similar quick properties. To display the General category properties for objects that do not have defined quick properties, pick the **General** button at the bottom of the object type list.

To customize the quick properties displayed for a specific object type, select the object in the object type list. The category list in the right-hand column displays all categories and quick properties available for that object. Check the properties that you want displayed and uncheck the properties you do not want displayed.

Figure 23-18A shows the line object type selected in the object type list. Notice that this object has 3D Visualization and Geometry categories, as well as the General category. By default, the Length property is set to display in the Geometry category, in addition to the default settings in the General category. Notice in **Figure 23-18A** that additional properties have been set to display. Once you close the **Customize User Interface** dialog box to apply the changes, the new quick properties are displayed for a selected line. See **Figure 23-18B**. You may need to pause the cursor over the **Quick Properties** palette to expand the palette. You can also right-click on the sidebar of the **Quick Properties** palette and pick **Auto-Collapse** in the shortcut menu to remove the check mark. This allows the **Quick Properties** palette to fully display each time it appears.

Figure 23-17.
Determining which object types appear in the object type list in the **Customize User Interface** dialog box.

Check the object types to display

Figure 23-18.
Customizing the quick properties for a line. A—Two properties in addition to the default properties are set to display. B—The **Quick Properties** palette contains the additional properties.

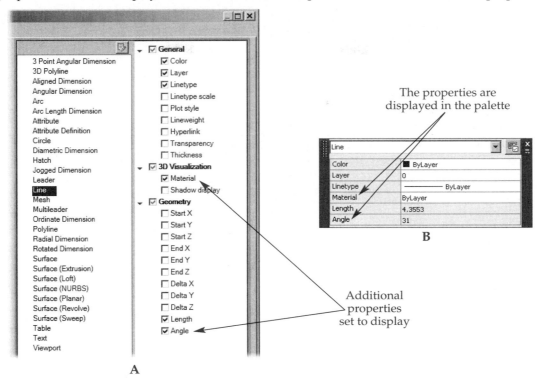

The properties are displayed in the palette

B

Additional properties set to display

A

By customizing quick properties, you may improve your AutoCAD workflow by having quick access to properties throughout the design process. For example, you may wish to have the Show History property displayed in the **Quick Properties** palette for solid primitives, sweeps, lofts, and revolutions. This will allow you to quickly show or hide the history of the solid. Additionally, you may wish to have the Material property displayed in the rollover tooltip for these objects. This will provide a quick indication of the material assigned to the object.

Customizing Double-Click Actions

By double-clicking on certain objects, an appropriate editing command is automatically executed. The command that is initiated is determined by the *double-click action* associated with the object type. Some AutoCAD objects have very specific editing tools available. For example, multiline text (mtext) objects are edited with the in-place text editor and polylines are edited with the **PEDIT** command.

A list of AutoCAD objects that have default double-click actions associated with them, other than the **QUICKPROPERTIES** command, is shown in **Figure 23-19**. If you double-click on one of the object types listed in the table, the command or macro listed in the Associated Double-Click Action column is executed. If the object type is not listed in the table, it is likely the **Quick Properties** palette is displayed, by default, when the object is double-clicked. This is the double-click action associated with most objects.

Figure 23-19.
AutoCAD objects to which a default double-click action other than **QUICKPROPERTIES** is assigned.

AutoCAD Object Type	Associated Double-Click Action
ATTDEF	DDEDIT
ATTBLOCK	EATTEDIT
ATTDYNBLOCK	EATTEDIT
ATTRIBUTE	ATTIPEDIT
BLOCK	$M=$(if,$(and,$(>,$(getvar,blockeditlock),0)),^C^C_properties,^C^C_bedit)
DYNBLOCK	$M=$(if,$(and,$(>,$(getvar,blockeditlock),0)),^C^C_properties,^C^C_bedit)
IMAGE	IMAGEADJUST
LWPOLYLINE	PEDIT
MLINE	MLEDIT
MTEXT	MTEDIT
POLYLINE	PEDIT
SECTIONOBJECT	LIVESECTION
SPLINE	SPLINEDIT
TEXT	DDEDIT
XREF	REFEDIT

Assigning Double-Click Actions

Double-click actions are assigned to specific object types in the **Customize User Interface** dialog box. In the **Customizations in All Files** pane, expand the Double Click Actions branch. All of the AutoCAD object types are listed. See **Figure 23-20.** Expand each branch and notice that many double-click actions call the **Quick Properties** palette, while the objects listed in the table in **Figure 23-19** have double-click actions that call the object-specific editing command.

Use the following procedure to change the double-click editing action associated with an object type. For this example, the **DDPTYPE** command will be associated with the point object type so the **Point Style** dialog box appears when a point object is double-clicked.

1. Open the **Customize User Interface** dialog box.
2. In the **Customizations in All Files** pane, expand the Double Click Actions branch.
3. Expand the Point branch under the Double Click Actions branch. Notice that the **QUICKPROPERTIES** command is associated with the point object type.
4. In the **Command List:** pane, select the **Point Style...** command. This is the **DDPTYPE** command, as indicated in the **Properties** pane when the command is selected.
5. Drag the **Point Style...** command from the **Command List:** pane and drop it into the Point branch in the **Customizations in All Files** pane. See **Figure 23-21.** The command replaces the existing command as there can only be one double-click action.
6. Pick the **OK** button to close the **Customize User Interface** dialog box and apply the change.

Now, draw a point using the **POINT** command. Double-click on the point and the **Point Style** dialog box appears. Select a new point style in the dialog box and pick the **OK** button. All existing points in the drawing should update to the new style. If not, use the **REGEN** command to update the display.

Figure 23-20.
All of the AutoCAD object types are displayed in the Double Click Actions branch in the **Customize User Interface** dialog box.

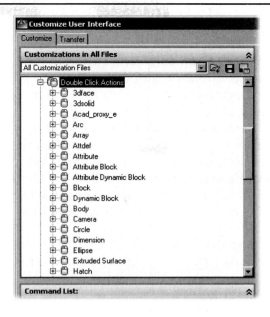

Figure 23-21.
The double-click action associated with the point object type is changed.

New double-click action assigned

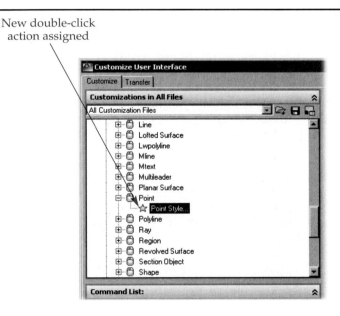

Custom Double-Click Action

You can also create a custom command and assign it to an object type as a double-click action. In this section, you will create a custom command for editing the radius of a circle to one-half of its current value when the circle is double-clicked. Here is a breakdown of what the custom command will do:

- Cancel any commands in progress. (^C^C)
- Execute the **SCALE** command with the circle that was double-clicked being the object to be scaled. (scale;)
- Set the center of the circle, which is the point last created, as the base point. (@;)
- Scale the circle to half of its original size. (0.5)

This can be a handy tool should you accidentally enter the intended diameter in response to the radius prompt for the **CIRCLE** command. On noticing you just made that mistake, double-click the circle and the problem is fixed.

Follow these steps to create a custom command and assign it as a double-click action for the circle object type:

1. Open the **Customize User Interface** dialog box.
2. Pick the **Create a new command** button in the **Command List:** pane.
3. Name the custom command **CirRadToDia**.
4. In the **Properties** pane, select the Macro property. Then, enter the macro ^C^Cscale;@;0.5 in the text box.
5. In the **Customizations in All Files** pane, expand the Double Click Actions branch and locate the Circle branch below it. Notice that the **QUICKPROPERTIES** command is currently associated with the circle object type.
6. In the **Command List:** pane, select the new **CirRadToDia** command and drag it into the Circle branch in the **Customizations in All Files** pane.
7. Pick the **OK** button to close the **Customize User Interface** dialog box and apply the change.

Draw a circle using the **CIRCLE** command. Now, double-click on the circle. The circle becomes one-half of its original size. In other words, the previous radius value is the new diameter value.

The macro you have just created is intended to be used *immediately* after incorrectly creating the circle. Using the custom double-click action assigned to the circle object type after performing other drawing operations will cause unexpected results as the center of the circle will no longer be the last point created.

Chapter Review

Answer the following questions. Write your answers on a separate sheet of paper or complete the electronic chapter review on the companion website.
www.g-wlearning.com/CAD

1. What is the key combination called that allows you to press the [Ctrl] key and an additional key to execute a command?
2. How can you disable all shortcut menus and all double-click actions?
3. In which dialog box can you define what a right-click does when in default mode, edit mode, and command mode?
4. Why are shortcut menus *context sensitive*?
5. When does the command shortcut menu appear?
6. What must the **PICKFIRST** setting be in order for the edit shortcut menu to appear?
7. Briefly describe how to create a new shortcut menu.
8. Describe the syntax for the name of an object-specific shortcut menu.
9. Why is CIRCLE_OBJECT *not* a valid alias for a shortcut menu?
10. Describe the difference between an OBJECT_*type* shortcut menu and an OBJECTS_*type* shortcut menu.
11. What is the syntax for the alias for a command-specific shortcut menu?
12. List the steps to add an alias to a shortcut menu.
13. How do you turn on the display of the **Quick Properties** palette?
14. What is a *quick property*?
15. How does a rollover tooltip differ from the **Quick Properties** palette?
16. In which dialog box do you select the objects that appear in the object type list in the **Customize User Interface** dialog box?
17. What is a *double-click action*?
18. What is the most common double-click action?
19. In which dialog box is a double-click action assigned?
20. List the basic steps for modifying the double-click action associated with an object.

Drawing Problems

Before customizing AutoCAD, check with your instructor or supervisor for specific instructions or guidelines.

1. Create a shortcut key for each of the drawing and editing commands listed below. Be sure not to use any existing shortcut keys.

LINE	**MOVE**
ARC	**COPY**
CIRCLE	**STRETCH**
POLYLINE	**TRIM**
POLYGON	**EXTEND**
RECTANGLE	**CHAMFER**
DTEXT	**FILLET**
ERASE	

2. In this problem, create an object-specific shortcut menu. The shortcut menu should be displayed when an arc object is selected. The shortcut menu should contain the **LENGTHEN** and **BREAK** commands.

3. In this problem, create a command-specific shortcut menu. The shortcut menu should be displayed when the **CIRCLE** command is active. Add the **Circle, Tan, Tan, Tan** command available in the **Customize User Interface** dialog box to the shortcut menu. Add two selection options, such as **Last** or **Window**, as described in this chapter.

4. By default, double-clicking on a circle displays the **Properties** palette. Take the steps necessary so that a **REGEN** is performed instead.

5. Create a custom command that changes the color of an object to blue. Then, assign this command as the double-click action for the hatch object type.

Chapter

Tool Palette Customization

Learning Objectives

After completing this chapter, you will be able to:

✓ Modify the appearance of the **Tool Palettes** window.
✓ Compare and contrast block insertion, hatch insertion, and command tools.
✓ Create new tool palettes from scratch and using **DesignCenter**.
✓ Add tools to existing tool palettes.
✓ Explain how tool palettes are formatted.
✓ Adjust the properties of tools.
✓ Create a flyout tool.
✓ Organize tool palettes into groups.
✓ Export and import tool palettes and tool palette groups.

Tool palettes are a user-interface method for the easy insertion of blocks and hatch patterns and for command entry. Blocks and hatch patterns can be simply dragged and dropped from a tool palette directly into a drawing. Tool palettes are contained within the **Tool Palettes** window.

Tool Palette Overview

The **TOOLPALETTES** command is used to display the **Tool Palettes** window. See **Figure 24-1**. Notice that the **Tool Palettes** window has a number of tabs on its edge. Each of these tabs corresponds to a tool palette. To make a tool palette active, pick its tab. If there are more tabs than those that can be displayed, pick on the "stack" at the bottom of the tabs to display a shortcut menu. Then, select the tool palette to display. To use a tool on any tool palette, drag it from the tool palette and drop it into the drawing.

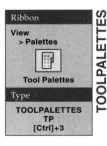

Ribbon
View
> Palettes

Tool Palettes

Type
TOOLPALETTES
TP
[Ctrl]+3

TOOLPALETTES

Figure 24-1.
The **Tool Palettes** window with the **Modeling** palette displayed.

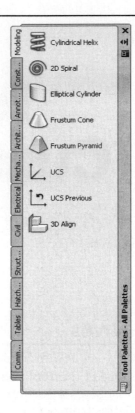

By default, the **Tool Palettes** window contains the **Modeling, Constraints, Annotation, Architectural, Mechanical, Electrical, Civil, Structural, Hatches and Fills, Tables, Command Tool Samples, Leaders, Draw, Modify, Cameras,** and **Visual Styles** palettes. There are also five light palettes. Each of the palettes is described below.

- **Modeling palette.** Contains tools for creating specific variations of some solid primitives. Also contains two UCS tools and a 3D align tool.
- **Constraints palette.** Contains tools for inserting geometric and dimensional constraints.
- **Annotation palette.** Contains blocks used for drawing annotations.
- **Architectural, Mechanical, Electrical, Civil,** and **Structural palettes.** Can be used to insert blocks that are meant to be used within the discipline for which the palette is named. Certain preset properties are already attached to the symbols, such as scale and rotation.
- **Hatches and Fills palette.** Contains some commonly used hatches and sample gradient fills that can be quickly inserted into a drawing.
- **Tables palette.** Allows you to insert sample tables in US customary (Imperial) and metric formats.
- **Command Tool Samples palette.** Allows you to execute certain commands by picking the tool.
- **Leaders palette.** Allows you to place leaders with or without text or with various types of callout balloons in both US customary (Imperial) and metric scales.
- **Draw palette.** Contains many of the same tools found in the **Draw** panel on the **Home** tab of the ribbon and some tools for inserting blocks, attaching images, and attaching xrefs.
- **Modify palette.** Contains many of the same tools found in the **Modify** panel on the **Home** tab of the ribbon.
- **Light palettes.** Contain tools for inserting many types of lights.
- **Cameras palette.** Contains tools for inserting three different cameras into the drawing. The cameras have differing lens lengths and fields of view.
- **Visual Styles palette.** Contains tools for three variations of the default visual styles.

For detailed instruction on how tool palettes can be used to insert blocks and hatch patterns, see *AutoCAD and Its Applications—Basics*. This chapter describes how to customize existing tool palettes and create your own tool palettes.

Tool Palette Appearance

There are a number of methods for altering the appearance of the **Tool Palettes** window, either for productivity or personal preference. For example, the tabs can be renamed. Right-click on the title of the tab and select **Rename Palette** from the shortcut menu. An edit box is displayed near the current name with the name highlighted. Type a new name and press [Enter] to rename the palette tab. Other methods of changing the appearance of the **Tool Palettes** window are covered in this section.

Docking

By default, the **Tool Palettes** window is floating on the right side of the screen in the AutoCAD Classic workspace and not displayed in the Drafting & Annotation and 3D Modeling workspaces. In the 3D Basics workspace, it will be displayed if displayed in the previous workspace.

When floating, the **Tool Palettes** window can be moved around the screen. Moving the window to the far side of the drawing window forces the title bar to flip to the other side of the **Tool Palettes** window so the bar is toward the outer edge of the drawing window.

By default, the **Tool Palettes** window can be docked. Moving the window outside of the drawing window to the left or right docks it. The **Tool Palettes** window cannot be docked at the top or bottom. To prevent docking, right-click on the **Tool Palettes** window title bar to display the shortcut menu. Select **Allow Docking** to remove the check mark. When a check mark appears next to **Allow Docking**, the window can be docked. If docked, it can be moved to a new, floating location by picking and holding on the title bar, dragging the window to the desired location, and releasing the pick button.

The **Tool Palettes** window can be resized just like standard windows. While floating, the top and bottom edges, the vertical area just below the tabs, and the corners can be used to resize the window. While docked, only the right or left edge can be used for resizing. As with standard windows, move the cursor to one of the edges of the window until a double arrow appears. Then, press and hold the pick button, drag the edge until the window reaches the desired size, and release the pick button.

Transparency

Using the **Tool Palettes** window while it is floating may cause occasional visibility problems because it covers up part of the drawing window. However, the window can be made partially transparent so that the part of the drawing under the window can be seen. Transparency will not be active when the window is in a docked position. Hardware acceleration may need to be turned off to set transparency, depending on your system configuration.

Right-click on the title bar of the **Tool Palettes** window and pick **Transparency...** from the shortcut menu. The window must be floating for this option to appear. The **Transparency** dialog box is then displayed, **Figure 24-2**. To set the level of transparency, move the slider in the **General** area to a lower value. The slider controls the level of transparency applied to the window. The further to the left that the slider is placed, the more transparent the **Tool Palettes** window, **Figure 24-3**. Placing the slider all of the way to the right makes the **Tool Palettes** window opaque. The slider in the **Rollover** area sets the transparency level for when the cursor is over the window. This setting must be equal to or greater than the slider setting in the **General** area. To see the transparency level when the cursor is over the window, pick the **Click to Preview** button.

Figure 24-2.
The **Transparency** dialog box is used to control the transparency of the **Tool Palettes** window.

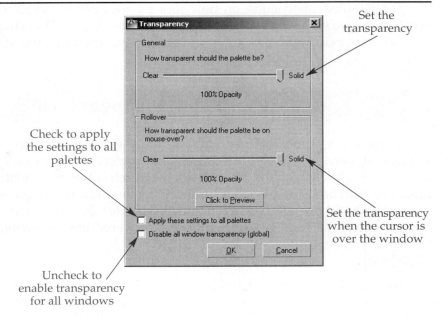

Check to apply the settings to all palettes

Set the transparency

Set the transparency when the cursor is over the window

Uncheck to enable transparency for all windows

Figure 24-3.
A—The **Tool Palettes** window has a low transparency setting.
B—The **Tool Palettes** window has a high transparency setting.

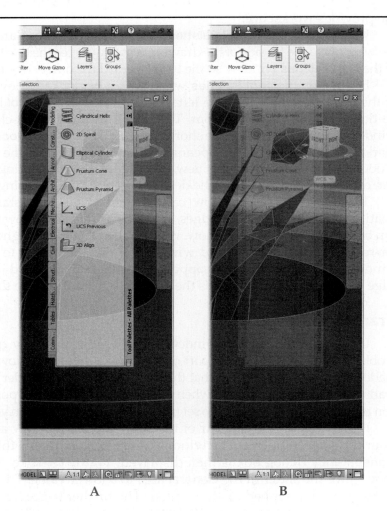

A

B

If you find a particular level of transparency that you like, but wish to make the **Tool Palettes** window opaque for a short time while you perform an operation or two, you can just use the toggle to turn off transparency. Open the **Transparency** dialog box and check the **Disable all window transparency (global)** check box. Then, when you want to go back to your previous level of transparency, simply open the **Transparency** dialog box again and uncheck the check box. You will not have to adjust the slider; just toggle the transparency back on.

Also notice in the **Transparency** dialog box the **Apply these settings to all palettes** check box. This allows you to apply the changes made in the dialog box to all AutoCAD palettes. Often, a user will prefer the same level of transparency for all palettes.

 Although transparency allows you to *see* through the **Tool Palettes** window, you cannot *work* through it. You cannot access points behind the window because the cursor is actually on the **Tool Palettes** window, not on the drawing underneath it. This is true of any AutoCAD palette.

Autohide

As useful as the **Tool Palettes** window is, it does take up a large amount of valuable drawing area. The *autohide* feature, when enabled, compresses the **Tool Palettes** window so just the title bar appears when the cursor is not over the window, **Figure 24-4**. This allows the **Tool Palettes** window to take up less room when not being used.

To turn on autohide, right-click on the title bar of the **Tool Palettes** window or pick the **Properties** button at the top of the title bar to display the shortcut menu, **Figure 24-5**. The **Properties** button is only displayed when the window is floating. Then, select **Auto-hide** from the shortcut menu. A check mark appears next to the menu item when autohide is enabled. You can also pick the **Auto-hide** button at the top of the **Tool Palettes** window title bar.

Figure 24-4.
When the autohide feature is enabled, the **Tool Palettes** window appears as only the title bar when the cursor is not over it.

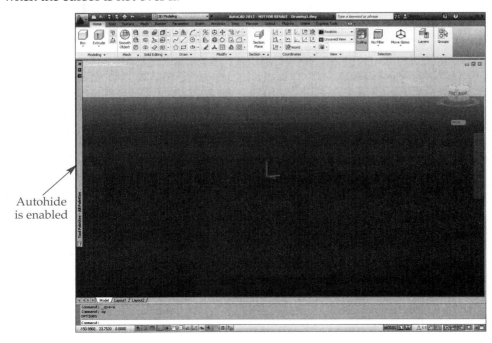

Autohide
is enabled

To use the **Tool Palettes** window when autohide is enabled, move the cursor over the title bar so that the window expands to its normal size. The tool palettes can then be used in the standard way. After using the **Tool Palettes** window, it is again hidden shortly after the cursor is no longer over the window.

Anchoring

The **Tool Palettes** window can also be *anchored*. Anchoring is a combination of docking and using autohide. When a tool palette is anchored, it is docked on the right or left side of the drawing area, but it is compressed to just a title bar. See **Figure 24-6**. To use the **Tool Palettes** window when it is anchored, move the cursor over the anchored title bar. The **Tool Palettes** window is then displayed floating next to the anchored window. It can be used just as if autohide is enabled. Once the cursor is moved off of the **Tool Palettes** window, the window is hidden.

To anchor the tool palette, docking must first be enabled. Then, right-click on the title bar of the **Tool Palettes** window or pick the **Properties** button at the top of the title bar to display the shortcut menu. Next, select either **Anchor Left <** or **Anchor Right >** in the shortcut menu. To return the window to floating mode, pick and drag the title bar back into the drawing area while the window is displayed.

PROFESSIONAL TIP

The **Properties** palette, **External References** palette, **QuickCalc** palette, **DesignCenter,** and the ribbon also have the anchoring feature. You can create a very productive drawing-window arrangement by anchoring these items along with the **Tool Palettes** window. For example, anchor the **Tool Palettes** window on one side of the drawing window and the **Properties** palette on the other side.

Figure 24-5.
Using the shortcut menu to enable autohide.

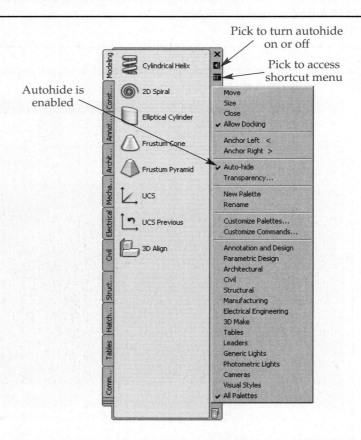

Figure 24-6.
When the **Tool Palettes** window is anchored, it is docked, but compressed to just its title bar.

The window is anchored

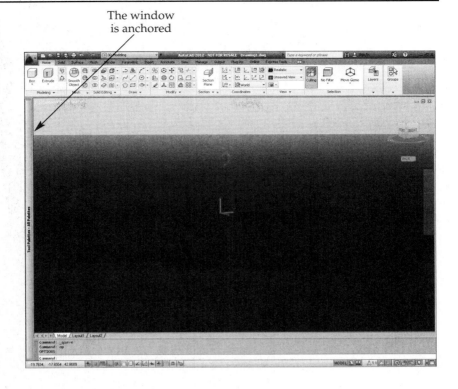

Tool Appearance

The way in which the tools are shown in the palettes can be customized. To do so, right-click in a blank area of the current tool palette (not on the title bar). Then, select **View Options...** from the shortcut menu. The **View Options** dialog box is displayed, **Figure 24-7**.

The **Image Size:** area of the dialog box is used to set the size of the tool icons in the palette. Drag the slider to the left or right to change the size. Dragging the slider to the left decreases the size of the icon. Dragging the slider to the right increases the size of the icon. To the left of the slider is a preview that represents the size of the icon.

The **View style:** area controls how the tools on the tool palettes are displayed. When the **Icon only** radio button is selected, the tools are represented as an image only, **Figure 24-8A**. Tooltips will be displayed if you hold your cursor over a tool. Selecting the **Icon with text** radio button represents the tools with an image and the tool name below the image, **Figure 24-8B**. When the **List view** radio button is selected, the tools are represented with an image and the tool name to the side of the image, **Figure 24-8C**. This is the default view style.

Figure 24-7.
The **View Options** dialog box is used to set how the tools appear in tool palettes.

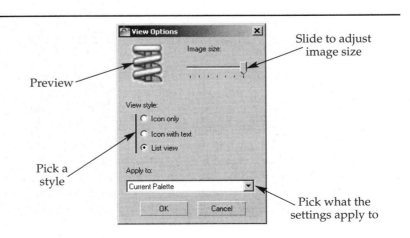

Preview

Slide to adjust image size

Pick a style

Pick what the settings apply to

Figure 24-8.
The various ways in which tools can appear in tool palettes. A—Icons only. B—Icons and text.
C—As a list.

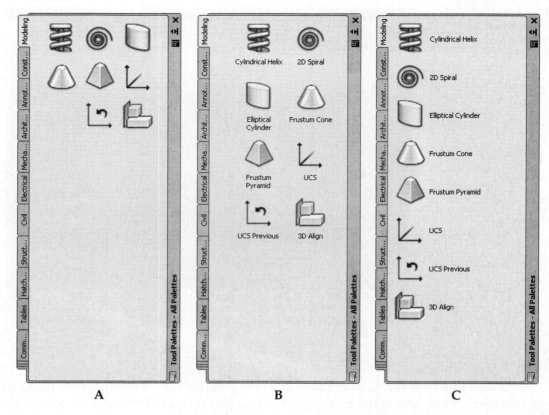

A

B

C

The **Apply to:** drop-down list at the bottom of the dialog box determines where the settings are applied. To have the settings applied to the current tool palette, select **Current Palette** from the drop-down list. To have the settings applied to all tool palettes, select **All Palettes** from the drop-down list. When finished making settings, pick the **OK** button to close the **View Options** dialog box.

Exercise 24-1

Complete the exercise on the companion website.
www.g-wlearning.com/CAD

Commands in a Tool Palette

As mentioned earlier in this chapter, the **Tool Palettes** window not only offers a means to easily insert blocks and hatch patterns, it can be used to execute commands. Make the **Tool Palettes** window active and select the **Command Tool Samples** palette, **Figure 24-9**. The tools on this palette are provided to demonstrate how tool palettes can be customized by adding commands. This tool palette is provided with the intention that it will be customized by the user. The default tools provided are:

- **Line.** Executes the **LINE** command. Notice that this tool has a small triangle, or arrow, to the right of the icon. The arrow indicates that the tool acts as a *flyout tool*, similar to flyout buttons on the ribbon. Picking the arrow displays

Figure 24-9.
Tools on the
**Command Tool
Samples** palette. A
flyout tool contains
other tools.

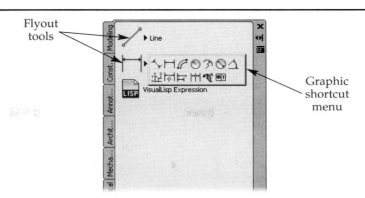

Flyout
tools

Graphic
shortcut
menu

a graphic shortcut menu containing other command tools attached to this tool, as shown in **Figure 24-9**. Selecting a command tool from the shortcut menu executes the command and causes that command image to become the default image displayed in the palette.

- **Linear Dimension**. Executes the **DIMLINEAR** command. This tool is also a flyout.
- **VisualLisp Expression**. Executes the AutoLISP expression (entget (car (entsel))). The entity data list for the selected entity is displayed on the command line. AutoLISP is discussed in Chapters 27 and 28.

While these three tools on the **Command Tools** palette can increase productivity, this palette is included to show you examples of ways in which tools can be customized to make you more productive in your own design environment. It is meant to be customized to your own needs. Customizing command tools on a tool palette is covered later in this chapter.

Adding Tool Palettes

As discussed later in this chapter, you can add new tools to tool palettes. You can also create new tool palettes and then add tools to them. To add a new tool palette, right-click on a blank area of an existing tool palette or on the title bar of the **Tool Palettes** window. Then, select **New Palette** from the shortcut menu. A new, blank palette is added and a text box appears next to the name. Type the desired name for the new palette and press [Enter].

Notice the help link at the top of the new tool palette, **Figure 24-10**. This link offers assistance in customizing tool palettes. It will disappear once you add a tool to the tool palette.

The new tool palette can be reordered within the tabs by right-clicking on the tab name to display the shortcut menu. Then, select **Move Up** or **Move Down** to reorder the palettes. You may need to do this several times in order to get the palette in the position you want. The other palettes can be reordered in this same manner. If you need to reorder multiple palettes, you can use the **Customize** dialog box. This dialog box is discussed in detail later in this chapter.

DesignCenter can be used to create a new palette fully populated with all of the blocks contained in a drawing. In the **Folders** tab of the **DesignCenter** palette, navigate to the drawing from which you are making the tool palette, right-click on the drawing name, and select **Create Tool Palette** from the shortcut menu. A new tool palette is added to the **Tool Palettes** window with the same name as the drawing file.

For example, open **DesignCenter** and navigate to the Fasteners-US drawing located in the \Sample\DesignCenter folder. Right-click on the file name and select **Create Tool Palette** from the shortcut menu, **Figure 24-11A**. A tool palette named Fasteners-US is added to the **Tool Palettes** window. All of the blocks contained in the Fasteners-US drawing are available on the tool palette, **Figure 24-11B**.

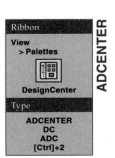

Ribbon
View
> Palettes

DesignCenter

Type

ADCENTER
DC
ADC
[Ctrl]+2

ADCENTER

Figure 24-10.
A new, blank tool palette has been created.

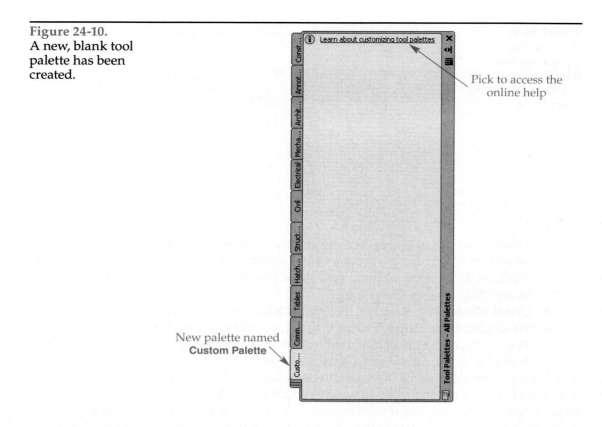

Pick to access the online help

New palette named **Custom Palette**

Figure 24-11.
A—Creating a tool palette from the blocks contained within a drawing. B—The new tool palette is added.

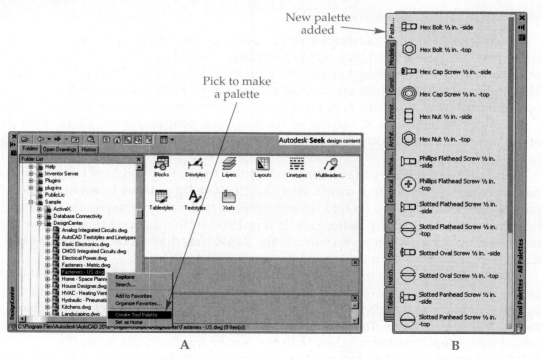

New palette added

Pick to make a palette

A

B

CAUTION

When using any "block insertion" tool from a tool palette, the block is actually being imported from the source drawing—the drawing from which the block tool on the palette was created. An error occurs if the source file has been moved or deleted. The source file must be restored to its original location to allow the tool to work or the tool must be recreated from the source drawing in the drawing's new location.

Adding Tools to a Tool Palette

Tools can be added to a tool palette in a variety of ways. Tools can be created from toolbar buttons, geometric objects in the current drawing, and objects in other drawings (via **DesignCenter**). Tools can also be copied to and pasted from the Windows Clipboard. Once tools have been added to a tool palette, they can be arranged to suit your preference. Related tools can be separated into distinct areas on the tool palette and those areas can have text labels added to them.

PROFESSIONAL TIP

Tool palettes can be thought of as an extension of ribbon tabs. You may want to add some of your favorite commands to a tool palette.

Creating a Tool from a Toolbar

Toolbar buttons can be directly dragged and dropped onto a tool palette. To do this, the **Customize** dialog box for tool palettes must be open. This is *not* the **Customize User Interface** dialog box that is used to customize the ribbon, as described in previous chapters. To display the **Customize** dialog box, use the **CUSTOMIZE** command or right-click on a tool palette and select **Customize Palettes...** from the shortcut menu.

Ribbon
Manage > Customization

Tool Palettes

Type
CUSTOMIZE

You do not actually use the **Customize** dialog box to copy a toolbar button to a tool palette, but the dialog box must be open. The **Customize** dialog box is discussed in detail later in this chapter.

Make sure the palette you want the button added to is current (on top). With the **Customize** dialog box open, move your cursor to the desired toolbar button. Flyout buttons cannot be added to a tool palette. Pick and hold on the toolbar button and drag it to the desired location in the tool palette, **Figure 24-12**. A horizontal "I-bar" appears in the tool palette to indicate where the new tool will be inserted. Drop the toolbar button when it is in the desired position. The tool is inserted in the tool palette. Close the **Customize** dialog box.

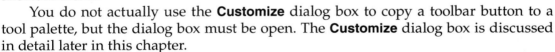

Tools cannot be dragged from the ribbon to a tool palette, only from toolbars. Toolbars are not displayed by default except in the AutoCAD Classic workspace.

Figure 24-12.
Creating a tool on a tool palette from a toolbar button.

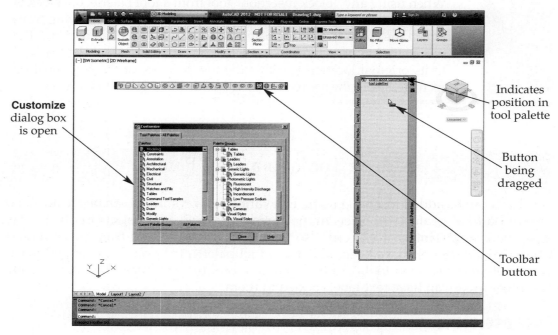

Customize dialog box is open

Indicates position in tool palette

Button being dragged

Toolbar button

Creating a Tool from an Object in the Current Drawing

Another way to add a drawing command to a tool palette is to drag an object in the current drawing, such as a line, hatch, block, dimension, camera, or light, and drop it onto the tool palette. The appropriate command to create that object is added to the tool palette. Not all objects support this method. For example, solid primitives cannot be dragged onto a tool palette to create a tool.

First, ensure that the **PICKFIRST** system variable is set to 1. Also, make sure the tool palette to which you want the tool added to is current (on top). Next, with no command active, select the desired object. Move the cursor directly onto the selected object (not a grip). Then, press and hold down either mouse button. Finally, drag the object to the desired position on the tool palette and drop it. The appropriate drawing command is inserted into the tool palette.

PROFESSIONAL TIP

You can also drag and drop blocks and hatch patterns from the current drawing onto a tool palette. To drag a block from the drawing to a tool palette, the drawing must be saved (cannot be unnamed) because the tool references the source file of the block.

Using the Windows Clipboard to Create a Tool

The copy-and-paste feature of the Windows operating system is another way to transfer objects in the drawing to a tool palette. First, ensure that the **PICKFIRST** system variable is set to 1. Then, with no command active, right-click on the selected object and select **Clipboard** from the shortcut menu and then **Copy** from the cascading menu. Next, make current the tool palette to which you want the tool added. Finally, right-click on a blank area in the tool palette and select **Paste** from the shortcut menu. The appropriate command is added as the last tool on the tool palette.

The copy-and-paste technique can also be used to transfer tools from one tool palette to another. First, make current the tool palette that contains the tool to be transferred. Right-click on the tool to transfer and select **Copy** in the shortcut menu. If you want to *move* the tool from the first tool palette to the second, select **Cut** in the shortcut menu. Next, make current the tool palette to which you want the tool transferred. Right-click and select **Paste** from the shortcut menu. The tool is added to the second tool palette as the last tool on the palette.

Adding Block and Hatch Tools from DesignCenter

Earlier, you saw how to create a tool palette consisting of all of the blocks in a single drawing by using **DesignCenter**. It is also possible to add individual blocks from a drawing to a tool palette using **DesignCenter**. To do this, open **DesignCenter**. In the **Folders** tab, navigate to the drawing that contains the desired block. Expand that drawing's branch to see the named objects within the drawing. Select the Blocks branch. The blocks that are defined in the drawing appear on the right-hand side of the **DesignCenter** palette. Make sure the tool palette you want the tool added to is current. Then, select the block in **DesignCenter** and drag it to the tool palette, **Figure 24-13**. Move the cursor to the desired position on the tool palette and drop the block. The new block tool is inserted in the tool palette.

A tool that inserts an entire drawing into the current drawing can also be added to a tool palette using **DesignCenter**. In **DesignCenter**, navigate to the drawing in the **Folders** tab. Select the drawing on the right-hand side of the **DesignCenter** palette, drag it to the tool palette, and drop it in the desired position, **Figure 24-14**. The new block tool is inserted in the tool palette. When the new tool is used, the entire drawing is inserted into the current drawing as a block.

DesignCenter can also be used to add hatch patterns to a tool palette. Hatch pattern definitions are stored in two files—acad.pat and acadiso.pat. These files are located in the user's \Support folder. In the **Folders** tab of **DesignCenter**, navigate to the acad.pat file and select it. All of the hatch patterns defined within that file are shown on the

Figure 24-13.
Adding an individual block contained within a drawing as a tool on a tool palette.

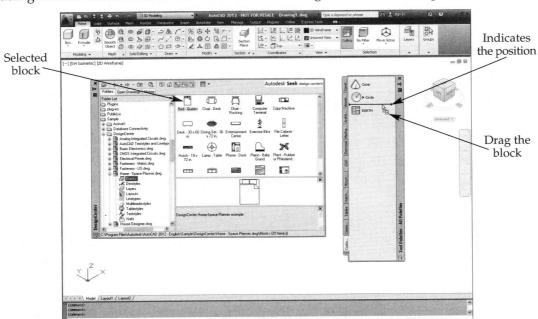

Figure 24-14.
Adding an entire drawing as a tool on a tool palette.

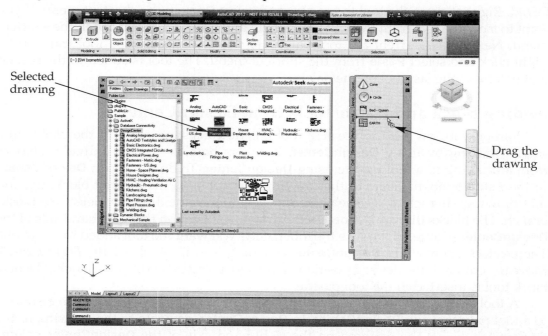

right-hand side of the **DesignCenter** palette, **Figure 24-15**. Make sure the tool palette you want the hatch pattern added to is current. Then, select the desired hatch pattern in **DesignCenter**, drag it to the tool palette, and drop it in the desired location. The new hatch tool is inserted in the tool palette.

Adding Visual Style Tools

A visual style that is available in the drawing can be added as a tool to a tool palette. To do this, first make current the tool palette to which you want the visual style added. Then, open the **Visual Styles Manager**. Select the icon for the visual style at the top of the **Visual Style Manager**, drag it to the tool palette, and drop it into position. See **Figure 24-16**.

Figure 24-15.
Hatch patterns can be selected in **DesignCenter** and dragged to a tool palette to create a new tool.

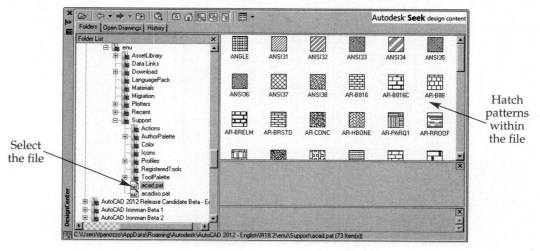

Figure 24-16.
Adding a visual style as a tool on a tool palette.

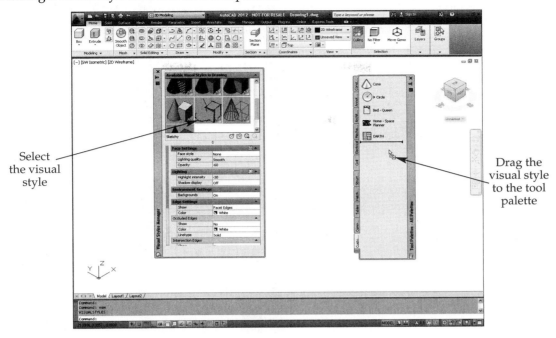

Select the visual style

Drag the visual style to the tool palette

PROFESSIONAL TIP

While they can be added to a tool palette, both visual styles and materials are more efficiently and productively managed using the **Visual Styles Manager** and **Materials Browser** palettes.

Rearranging Tools on a Tool Palette

The tools on a tool palette can be rearranged into a more productive order. To move a tool within a tool palette, simply select the tool and drag it to a new location. Remember, the horizontal I-bar indicates where the tool will be moved. In this way, you can place your drawing tools together, your block tools together, and so on.

You can further separate the tools within a tool palette by adding separator bars and text labels. Move the cursor so that it is between the two tools where you would like to add the separator. Then, right-click and select **Add Separator** from the shortcut menu. A horizontal bar is added to the tool palette. To add text to a tool palette, right-click between the two items where the label should be and select **Add Text** from the shortcut menu. A text box is displayed with the default text highlighted. Type the text that you want for the label and press [Enter]. The text label is added to the tool palette. Separators and text labels can be used together to make the visual grouping of tools even more apparent, **Figure 24-17.** You can move separators and text labels to different locations within the palette just as you can tools.

Exercise 24-2

Complete the exercise on the companion website.
www.g-wlearning.com/CAD

Figure 24-17.
Separators and text labels can be added to tool palettes to help group tools.

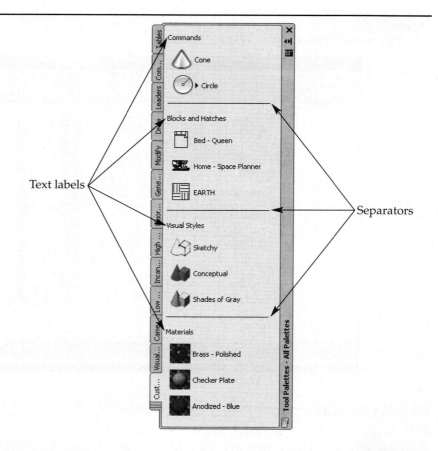

Text labels

Separators

Modifying the Properties of a Tool

A tool on a tool palette can basically do one of these operations:
- Insert a block.
- Insert a hatch pattern.
- Insert a gradient fill.
- Initiate a command.
- Insert a light.
- Insert a camera.
- Set a visual style current.

Another type of tool called a flyout is discussed later in this chapter. Each of these tools has general properties assigned to it, such as color, layer, or linetype. Each also has some tool-specific properties assigned to it, depending on the operation associated with the tool. Most of these assigned properties can be customized to your own needs.

For example, select the **Hatches and Fills** palette in the **Tool Palettes** window. Right-click on the **Curved** gradient tool and select **Properties...** in the shortcut menu. The **Tool Properties** dialog box is displayed, **Figure 24-18.** Notice that the lower half of the dialog box is divided into two sections. The **General** category contains settings for properties such as Color, Layer, and Linetype. The **Pattern** category contains settings for properties specific to this tool's operation—inserting a gradient fill. Notice that properties such as Color 1, Color 2, and Gradient angle are shown. Other types of tools will have a different category in place of the **Pattern** category. Select the **Cancel** button to close the dialog box.

Now, select the **Command Tool Samples** palette in the **Tool Palettes** window, right-click on the **VisualLisp Expression** tool in the tool palette, and select **Properties...** from

Figure 24-18.
Modifying the properties of a gradient fill tool.

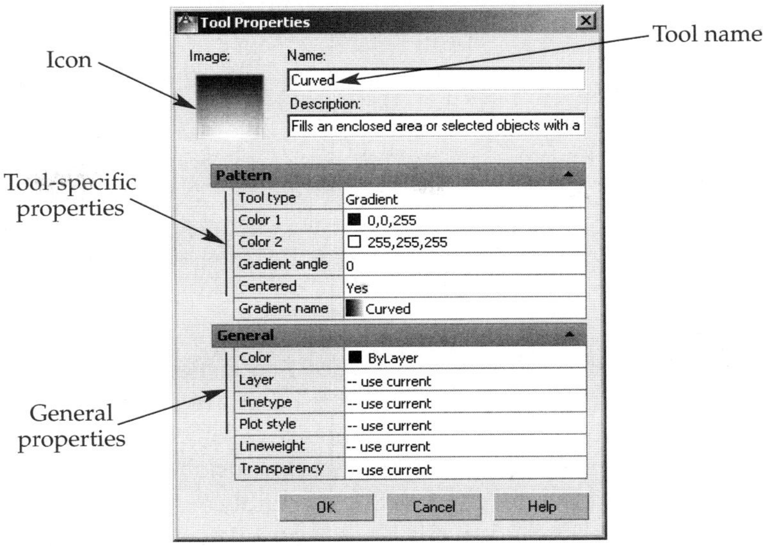

Tool name

Icon

Tool-specific properties

General properties

the shortcut menu. The **Tool Properties** dialog box is displayed, **Figure 24-19**. This is the same dialog box displayed for the **Curved** gradient tool. However, in place of the **Pattern** category is the **Command** category. The settings in the **Command** category are specific to the **VisualLisp Expression** tool. Notice that the **General** category contains the same properties available for the **Curved** gradient tool. There are some other general properties listed that are not applicable to a gradient (Linetype scale, Text style, and Dimension style). Select the **Cancel** button to close the dialog box.

Customizing General Properties

The three items at the top of the **Tool Properties** dialog box are available for all types of tools. The **Image:** area shows the image that is assigned to the tool. This can be modified by right-clicking and selecting **Specify image...** in the **Image:** area. The **Name:** text box displays the name of the tool. You can enter a new name for the tool. The **Description:** text box displays the current description of the tool. You can change the existing description or enter a new description. The name and description appear in the tooltip that is displayed when the cursor is held over the tool, **Figure 24-20**.

Figure 24-19.
Modifying the properties of a command tool.

Tool-specific properties

General properties

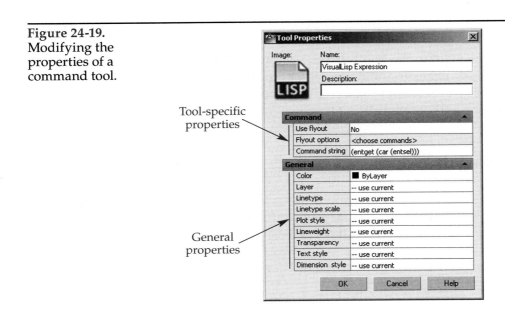

Figure 24-20.
The Name: and
Description: property
settings in the **Tool
Properties** dialog box
are displayed in the
tooltip for the tool.

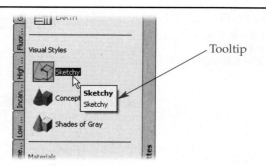

Tooltip

The **General** category in the **Tool Properties** dialog box can be used to assign specific values to the general properties that are a part of nearly all AutoCAD objects. The properties that can be customized are Color, Layer, Linetype, Linetype scale, Plot style, Lineweight, Transparency, Text style, and Dimension style. Depending on the type of tool, some of these properties may not be available.

The values assigned in the **Tool Properties** dialog box override the current property settings in the drawing when the tool is used. For instance, create three layers named Object, Hidden, and Center in the current drawing. A separate line tool can now be created for each of these layers:

1. Create a new, blank palette named **Line Tools**.
2. Add three line tools to the tool palette by copying the line tool from the **Command Tool Samples** palette.
3. Right-click on the first of these new line tools and select **Properties** from the shortcut menu to display the **Tool Properties** dialog box.
4. In the **Name:** text box, enter Line-Object as the name.
5. In the **Description:** text box, enter Draws a line on the Object layer. as the description.
6. In the **Command** category, change the Use flyout property to No. This makes the tool become a Line tool exclusively and removes the flyout that allows access to other drawing tools.
7. In the **General** category, pick the Layer property, which is set to —use current (depending on how the tool was created, this may be set to a specific layer). This means that when you draw a line using the tool, the line is drawn on the current layer.
8. In the drop-down list, select Object. Now, any lines drawn with the tool are placed on the Object layer.
9. Pick the **OK** button to close the **Tool Properties** dialog box and save the changes.
10. Repeat the above steps for the other two line tools, setting them to the Hidden and Center layers.

Now, use one of your new tools to draw a line. Notice that when you select the tool the current layer switches to the one assigned to the tool. When you finish using the tool, the current layer switches back to the previous layer. Try each of the other new tools. Experiment with some of the other general properties, such as Color, Linetype, and Lineweight.

PROFESSIONAL TIP

Creating layer-specific drawing tools as outlined in the previous section can streamline your drafting. For example, in an architectural drawing, you could have tools set to draw solid lines on a Walls layer, dashed lines on an Electrical layer, etc. This saves you from needing to set a new layer current or change properties after drawing lines.

Customizing Block Insertion Tools

When a block insertion tool is modified in the **Tool Properties** dialog box, properties specific to block insertion are displayed in a category labeled **Insert**, Figure 24-21. These properties are described as follows.

Name

The Name property contains the name of the block to be inserted. It is not usually modified.

Source File

The Source file property lists the name of and path to the source drawing containing the block. If the drawing file has been moved to a new location, specify the correct location in the text box for this property. When you pick in the text box, an ellipsis button (...) appears at the right-hand side of the box. You can pick this button to browse for the drawing file.

Scale

The Scale property value is the scale that will be applied to the block when it is inserted into the drawing. The scale is applied equally in the X, Y, and Z directions. The default value is 1.000.

Auxiliary Scale

The block is inserted at a scale calculated by multiplying the Auxiliary scale property value by the Scale property value. This drop-down list allows you to apply either the dimension scale or the plot scale to the insertion scale. The default value is None.

Rotation

The rotation angle for the block when it is inserted is set by the Rotation property. The default value is 0.

Prompt for Rotation

The Prompt for rotation property determines whether the user is prompted for a rotation angle when the block is inserted. The default value in the drop-down list is No.

Figure 24-21.
The **Tool Properties** dialog box for a block insertion tool.

Properties specific to block insertion

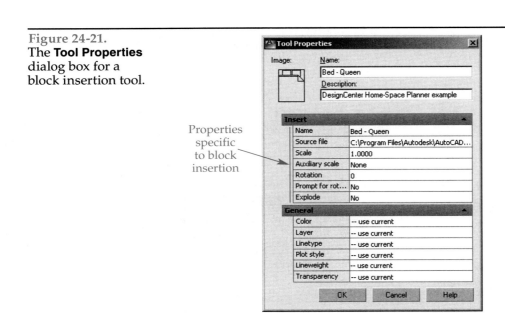

Explode

The Explode property determines whether the block is inserted as a block or as its component objects (exploded). The default value in the drop-down list is No, which means the block is inserted unexploded.

Customizing Hatch Pattern Tools

When a hatch pattern insertion tool is modified in the **Tool Properties** dialog box, properties specific to hatch patterns are displayed in the **Pattern** category, **Figure 24-22**. These properties are described as follows. The pattern type determines whether the properties are disabled or enabled.

Tool Type

The Tool type property determines if the hatch pattern is a standard hatch or a gradient fill. Choosing Gradient in the drop-down list changes the rest of the properties found in this area of the dialog box. Gradient fill properties are discussed later in this chapter.

Type

The Type property determines the type of hatch pattern. To change the type, select the property and then pick the ellipsis button (...) at the right-hand end of the entry. The **Hatch Pattern Type** dialog box is displayed, **Figure 24-23**. In the **Pattern Type:** drop-down list of this dialog box, select User-defined, Predefined, or Custom.

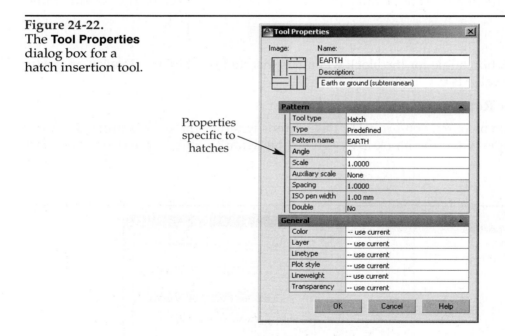

Figure 24-22.
The **Tool Properties** dialog box for a hatch insertion tool.

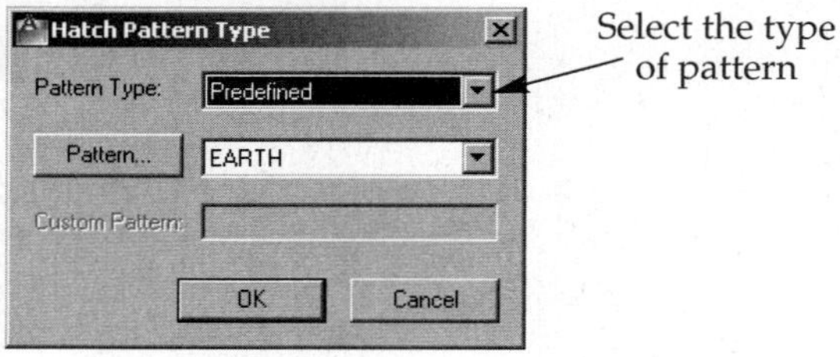

Properties specific to hatches

Figure 24-23.
The **Hatch Pattern Type** dialog box is used to determine the type of hatch. A predefined hatch pattern can also be selected in this dialog box.

Select the type of pattern

If Predefined is selected, the **Pattern...** button and drop-down list are enabled. Select a pattern from the drop-down list or pick the button to select a pattern in the **Hatch Pattern Palette** dialog box. In this dialog box, you can select which hatch pattern to use, **Figure 24-24**.

If User-defined is selected, the rest of the items in the dialog box are disabled. The properties for user-defined hatch patterns are set in the **Tool Properties** dialog box, as discussed next.

Selecting Custom disables the **Pattern...** button and drop-down list and enables the **Custom Pattern:** text box. In this text box, enter the name of the custom pattern to use.

Pattern Name

If the pattern type is set to Predefined, you can change the pattern using the Pattern name property. Select the property and pick the ellipsis button (**...**) to open the **Hatch Pattern Palette** dialog box. Refer to **Figure 24-24**. If you selected the pattern in the **Hatch Pattern Type** dialog box, you will not need to select it using this property. This property is read-only when the type is set to User-defined or Custom.

Angle

The Angle property allows you to rotate the hatch pattern. This is the same as entering a rotation when using the **HATCH** command. The default value is 0.

Scale

The Scale property determines the scale to be applied to the hatch pattern. Pick in the text box and enter a scale factor for the pattern. The default value is 1.000.

Auxiliary Scale

The hatch is inserted at a scale calculated by multiplying the Auxiliary scale property value by the Scale property value. The Auxiliary scale drop-down list allows you to apply either the dimension scale or the plot scale to the insertion scale. The default value is None.

Spacing

The Spacing property determines the spacing of lines in a user-defined hatch pattern. For other hatch pattern types, this property is read-only. To change the spacing, pick in the text box and enter the relative distance between lines. The default value is 1.000.

Figure 24-24.
Selecting a predefined hatch pattern.

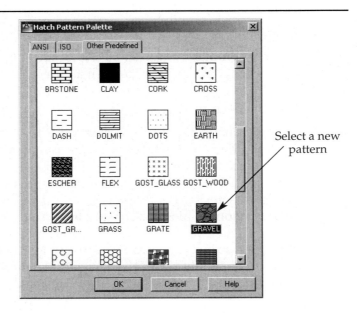

Select a new pattern

ISO Pen Width

The ISO pen width property allows you to set the pen width for user-defined ISO hatch patterns. It is read-only for other hatch patterns. The default value in the drop-down list is 1.00 mm.

Double

The Double property determines whether the user-defined hatch is a crosshatch pattern. For other hatch pattern types, this property is read-only. The default value in the drop-down list is No. To make a crosshatch pattern, select Yes.

Customizing Gradient Fill Tools

When a gradient fill hatch pattern insertion tool is modified in the **Tool Properties** dialog box, properties specific to gradient fills are displayed in the **Pattern** category. Refer to **Figure 24-18**. These properties are described as follows.

Tool Type

The Tool type property determines if the hatch pattern is a standard hatch or a gradient fill. Choosing Hatch in the drop-down list changes the rest of the properties found in this area of the dialog box, as described earlier.

Color 1 and Color 2

The Color 1 property is used to specify the first color of the gradient fill. The Color 2 property is used to specify the second color of the gradient fill. When you pick the property, a drop-down list is displayed. Pick a color from the drop-down list or choose **Select Color…** to pick a color in the **Select Color** dialog box.

Gradient Angle

The value of the Gradient angle property determines the rotation of the gradient fill. Pick in the text box and enter the number of degrees for the angle of the gradient. When the fill is created, the angle is relative to the current UCS. When you press [Enter], the **Image:** preview tile is updated to reflect the setting.

Centered

The Centered property determines whether the gradient fill is centered. The default value in the drop-down list is Yes. This creates a symmetrical fill. Selecting No shifts the gradient to the left. This is used to simulate a light source illuminating the fill from the left side.

Gradient Name

The type of gradient fill is set by the Gradient name property. AutoCAD's preset gradient fills appear in the drop-down list. Select the type of gradient fill to use.

Customizing Command Tools

When a command tool is modified in the **Tool Properties** dialog box, properties specific to commands are displayed in the **Command** category, **Figure 24-25**. These properties are described as follows.
- **Use flyout.** This property determines whether the tool is a flyout. Flyouts are discussed later in this chapter.
- **Flyout options.** This property allows you to pick which commands are associated with the flyout, as discussed later in this chapter.
- **Command string.** The command macro for the tool is entered in this text box, if the tool is not a flyout. Creating custom commands is discussed in Chapter 22. If the tool is a flyout, this text box is disabled.

Figure 24-25.
The **Tool Properties**
dialog box for a
command tool.

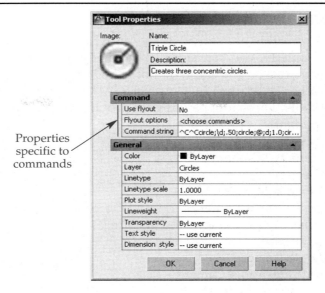

Properties
specific to
commands

For example, suppose you need a tool that will draw three concentric circles with diameters of .50, 1.00, and 1.50. This tool is shown in **Figure 24-25.** Use the following procedure.

1. In the current drawing, create a circle of any diameter at any location.
2. Drag and drop the circle onto a tool palette to create a new tool.
3. Right-click on the new **Circle** tool and select **Properties...** from the shortcut menu.
4. In the **Tool Properties** dialog box, change the following properties. The Use flyout property must be set to No to enable the Command string property.
   Name: Triple Circle
   Description: Creates three concentric circles.
   Use flyout: No
   Command string: ^C^Ccircle;\d;.50;circle;@;d;1.0;circle;@;d;1.5
5. Pick the **OK** button to close the **Tool Properties** dialog box.
6. Test the new tool.

Look closely at the command macro for the **Triple Circle** tool you just created. Can you identify each component of the macro? If not, use the tool and then display the **AutoCAD Text Window** by pressing [F2]. Using the text window, determine what function each component of the macro performs.

The **VisualLisp Expression** tool included on the **Command Tool Samples** palette is not actually a "command" tool. It is merely there to let you know that you can use a command tool to execute Visual LISP expressions. Visual LISP functions can be added in the Command string property in the **Tool Properties** dialog box. For more information on AutoLISP and Visual LISP, refer to Chapters 27 and 28 of this text.

Exercise 24-3

Complete the exercise on the companion website.
www.g-wlearning.com/CAD

Customizing Camera and Light Tools

Cameras and lights are stored in the drawing file. By placing camera and light tools on a tool palette, you can store your favorite settings for use in all drawings. Productivity is increased since you do not have to look for your favorite cameras and lights within other drawings. In fact, AutoCAD provides several tool palettes with different light tools. These tool palettes are provided for lights:

- **Generic Lights**
- **Fluorescent**
- **High Intensity Discharge**
- **Incandescent**
- **Low Pressure Sodium**

AutoCAD also has a **Cameras** tool palette containing three camera tools. These tools are provided on the **Cameras** tool palette:

- **Normal Camera**
- **Wide-Angle Camera**
- **Extreme Wide-Angle Camera**

When customizing a camera or light tool, the **Tool Properties** dialog box provides all of the properties that are required to create a camera or light. See **Figure 24-26.** For detailed information about setting up lights and cameras, see Chapters 18 and 20.

Changing a Tool Icon

You can change the icon associated with a tool. This can be done from within the **Tool Properties** dialog box or directly on the tool palette. Right-click on the icon, either in the dialog box or on the tool palette, and select **Specify image...** from the shortcut menu. The **Select Image File** dialog box is displayed. This is a standard "open" dialog box. Navigate to the image file, select it, and pick the **Open** button. The icon displays the new image.

Figure 24-26.
A—The **Tool Properties** dialog box for a light tool.
B—The **Tool Properties** dialog box for a camera tool.

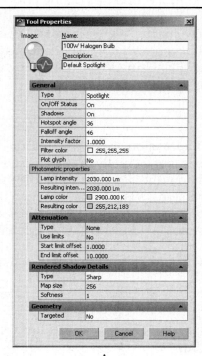

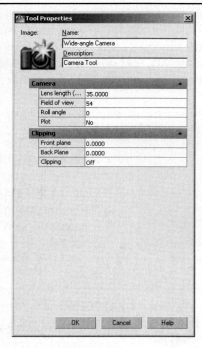

A   B

Some tools have an icon image based on the specific settings for the tool. You should not specify an image for these tools. If the tool is redefined, right-click on the tool icon in the tool palette and select **Update tool image** from the shortcut menu.

Working with Flyouts

A command tool in a tool palette can be set to function as a *flyout*. A flyout tool is similar to a flyout button found on the ribbon or a toolbar. The tool icon for a flyout displays a small arrow to the right of the tool. When the arrow is picked, a graphic shortcut menu is displayed that contains the command tools in the flyout.

To set a command tool as a flyout, open the **Tool Properties** dialog box for the tool. Then, in the **Command** category, set the Use flyout property to Yes. The tool is now a flyout. Notice that the **Command string** text box in the dialog box is disabled. You must now select the commands that will be displayed in the flyout.

The Flyout options property is used to specify which commands will be displayed in the flyout for a tool. This property is disabled unless the Use flyout property is set to Yes. To choose the commands to appear in the flyout, select the property and pick the ellipsis button (...) at the right-hand side. The **Flyout Options** dialog box appears, **Figure 24-27.**

The commands that will appear in the flyout tool are indicated with a check in the **Flyout Options** dialog box. To prevent a command from being displayed in the flyout, remove the check mark next to its name. Then, close the **Flyout Options** dialog box to return to the **Tool Properties** dialog box. Finally, close the **Tool Properties** dialog box and test the tool.

The commands displayed in the **Flyout Options** dialog box depend on which command tool is being modified. **Figure 24-27A** shows the commands available when a dimensioning command tool is modified. When a drawing command tool is modified, the commands shown in **Figure 24-27B** are available. These also act as the default tools for other commands, such as modify commands or custom commands.

Figure 24-27.
A—The commands available for a dimensioning command flyout tool.
B—The commands available for a drawing command flyout tool.

A   B

When a drawing or dimensioning command is added to a tool palette, the tool is automatically a flyout. It can then be modified to customize or remove its flyout capabilities.

Organizing Tool Palettes into Groups

Tool palettes are very useful, productive tools. You may find it beneficial to create many tool palettes to meet your design needs. However, having too many palettes visible at once can be counterproductive. The tool palette names become abbreviated in the **Tool Palettes** window or the tabs may be stacked on top of each other. This is the case if all of the default AutoCAD tool palettes are displayed. To help manage tool palettes, you can organize tool palettes into named groups and then display a single group.

By default, all of the tool palettes are shown in the **Tool Palettes** window, but the existing tool palettes are divided into named groups. Right-click on the title bar of the **Tool Palettes** window and look at the shortcut menu. At the bottom of the shortcut menu, notice the entries below the last separator, Figure 24-28. These are the palette groups. The check mark next to **All Palettes** indicates that all of the defined tool palettes are being displayed. Select **Annotation and Design** from the shortcut menu and notice that only the block-related tool palettes are visible. This is indicated in the title bar. Right-click on the title bar again and select **3D Make** from the shortcut menu. Notice that three tool palettes that may be used in an initial 3D design are displayed.

The **Customize** dialog box is used to make new palette groups or customize existing groups. To open this dialog box, right-click on the **Tool Palettes** window title bar and select **Customize Palettes...** from the shortcut menu. You can also use the **CUSTOMIZE** command. All of the currently defined tool palettes are listed in the **Palettes:** area on the left side of the dialog box, Figure 24-29. The currently defined palette groups are

Figure 24-28.
Select a palette group to display or choose to display all palettes.

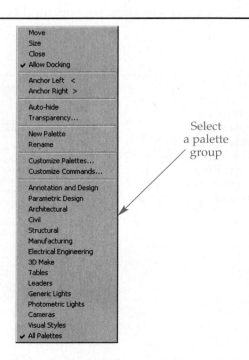

Select a palette group

Figure 24-29.
The **Customize** dialog box is used to create and manage palette groups.

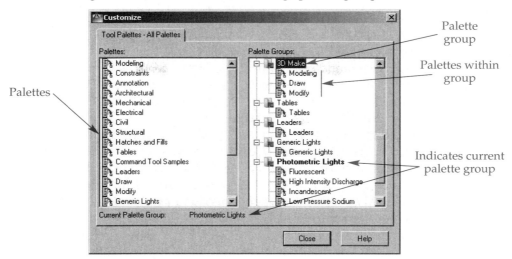

listed in the **Palette Groups:** area on the right side of the dialog box. Palette groups are shown as folders. The tool palettes contained within the group are shown in the tree below the folder. The current palette group is indicated at the bottom of the dialog box and by the bold folder name in the **Palette Groups:** area.

You can customize one of the existing palette groups by adding tool palettes to or removing tool palettes from the group. To remove a tool palette from a group, select the palette name under the group name in the **Palette Groups:** area. Then, press the [Delete] key or right-click and select **Remove** from the shortcut menu. To delete a palette group, select it in the **Palette Groups:** area and press the [Delete] key or right-click and select **Delete** from the shortcut menu. To add a palette to a group, simply select the palette in the **Palettes:** area and drag it into the group in the **Palette Groups:** area. A tool palette can be a member of more than one palette group.

To create a new palette group, right-click in a blank area of the **Palette Groups:** area and select **New Group** from the shortcut menu. A new folder appears in the tree with the default name highlighted in a text box. Type a name for the new palette group and press [Enter]. Then, drag tool palettes from the **Palettes:** area and drop them into the new palette group.

The order in which tool palettes appear in the **Tool Palettes** window is determined by their positions in the tree. To reorder the tool palettes, simply drag them to different positions within the palette group. To reorder the palette groups, drag the folders to different positions within the tree in the **Palette Groups:** area. The order in which the palette groups appear in the **Palette Groups:** area determines the order in which they appear in the shortcut menu displayed by right-clicking on the **Tool Palettes** window title bar. The order in which tool palettes appear in the **Palettes:** area is the order in which they are displayed when the **All Palettes** option is selected. By dragging a tool palette up or down in the tree, you can reorder the list.

CAUTION

When adding a palette group, be careful where you right-click. If you right-click within a group, the new palette group is nested within that group. If you do not want the palette group nested, simply drag the group to the top of the tree in the **Customize** dialog box.

Saving and Sharing Tool Palettes and Palette Groups

Tool palettes and palette groups can be exported from and imported into AutoCAD. Both operations are performed in the **Customize** dialog box. Some precautions about tool palette files are as follows:

- A tool palette file should only be imported into the same version of AutoCAD as the version from which the file was exported. Differences in the .dwg database and the .cuix file structure could cause tool palettes to not behave as intended or not work at all. CUIx files are discussed in Chapter 22.
- Tool palette files exported from AutoCAD and imported into AutoCAD LT may have tools that will not work or behave the same. For example, color property tools using a color other than an AutoCAD Color Index (ACI) color are converted to ByLayer in AutoCAD LT. Also, gradient fill tools convert to hatch tools in AutoCAD LT. Raster image tools do not work in AutoCAD LT.

To export a tool palette, right-click on the palette to be exported in the **Palettes:** area of the **Customize** dialog box and select **Export...** from the shortcut menu. To export a palette group, right-click on the name of the group in the **Palette Groups:** area and select **Export...** from the shortcut menu. The **Export Palette** or **Export Group** dialog box is displayed. These are standard Windows "save" dialog boxes. Name the file, navigate to the folder where you want to save it, and pick the **Save** button. Tool palettes are saved with a .xtp file extension. Palette groups are saved with a .xpg file extension.

To import a tool palette, right-click in the **Palettes:** area of the **Customize** dialog box and select **Import...** from the shortcut menu. To import a palette group, right-click in the **Palette Groups:** area and select **Import...** from the shortcut menu. The **Import Palette** or **Import Group** dialog box is displayed. Navigate to the folder where the file is saved, select it, and pick the **Open** button. The imported tool palette is added to the **Palettes:** area. An imported palette group is added to the **Palette Groups:** area.

Chapter Review

Answer the following questions. Write your answers on a separate sheet of paper or complete the electronic chapter review on the companion website.
www.g-wlearning.com/CAD

1. Name three ways to open the **Tool Palettes** window.
2. How do you dock the **Tool Palettes** window?
3. When enabled, what does the autohide feature do to the **Tool Palettes** window? How is the **Tool Palettes** window accessed when autohide is enabled?
4. Name two ways to toggle the autohide feature on the **Tool Palettes** window.
5. Describe how anchoring the **Tool Palettes** window differs from using the autohide feature or docking it.
6. How do you activate the transparency feature for the **Tool Palettes** window?
7. Name the three view styles in which tools can be displayed in the **Tool Palettes** window.
8. How do you create a new, blank tool palette?
9. How do you add a tool palette that contains all of the blocks in a particular drawing?
10. Which dialog box must be open to create a tool on a tool palette from a toolbar button? How is it opened?
11. What are the names of the two files that store hatch pattern definitions?
12. How do you rearrange tools on a tool palette?
13. Name four of the general properties that can be customized on individual tools on a tool palette.
14. What is the purpose of the auxiliary scale property for block and hatch insertion tools?
15. What are the two types of patterns that can be inserted using a tool in a tool palette?
16. If you are creating a command tool that performs a custom function, where is the command macro entered?
17. In which dialog box can you define the commands that will appear in a flyout tool?
18. What is the purpose of creating tool palette groups?
19. What is the extension given to an exported tool palette file?
20. How do you import a tool palette file?

Drawing Problems

Before customizing or creating any tool palettes, check with your instructor or supervisor for specific instructions or guidelines.

1. Design a complete tool palette system for your chosen discipline. Incorporate:
   - Multiple tool palettes
   - Block tools
   - Hatch tools
   - Drawing command tools (with and without flyouts)
   - Modification command tools
   - Inquiry command tools
   - Custom macro tools
   - Dimensioning command tools (with and without flyouts)
   A. On the tool palettes that use multiple types of command tools, use separators and text labels to group the types of tools.
   B. Create groups for the multiple tool palettes.

2. Export the tool palette groups created in Problem 1. Copy the files to removable media or an archive drive.

Drawing Problems - Chapter 24

User Profiles and Workspaces

Learning Objectives

After completing this chapter, you will be able to:

✓ Describe user profiles.
✓ Create user profiles.
✓ Restore a user profile.
✓ Describe workspaces.
✓ Explain the **Quick Access** toolbar.
✓ Create workspaces.
✓ Customize a workspace.
✓ Restore a workspace.
✓ Customize the **Quick Access** toolbar.

In a school or company, there is often more than one person who will use the same AutoCAD workstation. Each drafter has a unique style for creating a drawing. While there are often general rules to follow, many times the method used to arrive at the end result is not important. As you learned in previous chapters, there are many ways to customize AutoCAD. You can set screen colors and other features of the AutoCAD environment, create custom menus, and customize the ribbon. Many of these settings can be saved in a user profile or workspace. User profiles and workspaces allow you to quickly and easily restore a group of custom settings.

User Profiles

A *user profile* is a group of settings for devices and AutoCAD functions. Some of the settings and values a profile can contain are:

• Temporary drawing file location.
• Template drawing file location.
• Text display format.
• Startup dialog box display.
• Minutes between automatic saves.
• File extension for temporary files.
• Color and font settings for AutoCAD's text and graphics screens.

- Type of pointer and length of crosshairs.
- Default locations of printers, plotters (PC3 files), and plot styles (CTB/STB files).

Multiple profiles can be saved by a single user for different applications or several users can create individual profiles for their own use. A user profile should not be confused with settings found in a drawing. Template files are used to save settings relating to a drawing session, such as units, limits, object snap settings, drafting settings, grip settings, arc and circle smoothness, dimension styles, and text styles. A user profile, on the other hand, saves settings related to the performance and appearance of the software and hardware.

Creating a User Profile

A user profile is basically a collection of all of the AutoCAD environmental settings you have customized in AutoCAD *except* ribbon tabs and panels, application and pull-down menus, tool palettes, and toolbars. These customizations are usually done to make AutoCAD easier for you to use. For example, as you gain experience in AutoCAD, you realize that you may:
- Prefer the crosshairs extending to the edges of the drawing window.
- Need to have the plotters and plot styles shared on a network.
- Prefer the drawing window background color to be gray.
- Need to establish specific directories for projects.

Through the course of several drawing sessions, you have customized AutoCAD to reflect these preferences. Now, so you do not lose your preferred settings, you should create a user profile.

First, open the **Options** dialog box and pick the **Profiles** tab, **Figure 25-1**. Pick the **Add to List...** button on the right side of the tab. The **Add Profile** dialog box is opened, **Figure 25-2**. Enter a name and description. Then, pick the **Apply & Close** button to close the **Add Profile** dialog box. The current settings are saved to the user profile and the new user profile is now listed in the **Profiles** tab of the **Options** dialog box. The user profile is saved and will be available in the current drawing session and future AutoCAD drawing sessions.

Figure 25-1.
Settings can be saved in a profile.

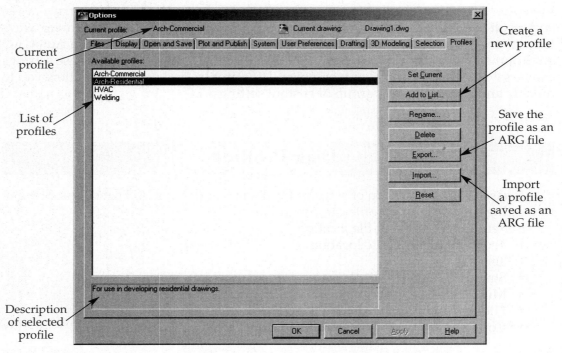

Figure 25-2.
The **Add Profile**
dialog box is used to
create a new profile.

Enter name
for new
profile

Enter
description
for new
profile

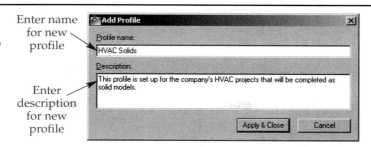

To change the name of a user profile, pick the **Rename...** button in the **Profiles** tab. In the **Change Profile** dialog box, enter a new name and description. To delete a user profile, highlight it in the **Profiles** tab and pick the **Delete** button. You cannot delete the current user profile. If you pick the **Reset** button, the highlighted user profile has all of its settings restored to AutoCAD defaults.

PROFESSIONAL TIP

Changes made to the environment (screen color, toolbar display, etc.) are automatically saved to the current profile and stored in the Windows registry.

Restoring a User Profile

Once a user profile is saved, it is available in the current drawing session and future AutoCAD drawing sessions. To set any saved user profile as the current user profile, first open the **Options** dialog box. The current profile is indicated at the top of the **Options** dialog box. Then, pick the **Profiles** tab. Highlight the name of the profile to restore in the **Available profiles:** list. Then, pick the **Set Current** button. You can also double-click on the name in the **Available profiles:** list to restore a profile. All of the settings in the user profile are applied while the **Options** dialog box is still open. Close the dialog box to return to the drawing window.

Exercise 25-1

Complete the exercise on the companion website.
www.g-wlearning.com/CAD

Importing and Exporting User Profiles

A user profile can be exported and imported. This allows you to take your user profile to a different AutoCAD workstation. A user profile is saved as an ARG file.

To export a user profile, open the **Options** dialog box and pick the **Profiles** tab. Highlight the profile to export and pick the **Export...** button. The **Export Profile** dialog box is opened, **Figure 25-3**. Then, select a folder and name the file. When you pick the **Save** button, the user profile is saved with the .arg file extension.

To import a user profile, pick the **Import...** button in the **Profiles** tab. The **Import Profile** dialog box is displayed. This is a standard "open" dialog box. Then, navigate to the appropriate folder, select the proper ARG file, and pick the **Open** button. A second dialog box named **Import Profile** is displayed. See **Figure 25-4**. You can rename the user profile, change the description, and choose to include the file path. Pick the **Apply & Close** button to complete the process. The user profile is then available in the **Profiles** tab.

Figure 25-3.
The **Export Profile** dialog box is used to save a user profile as an ARG file, which can be transferred to another AutoCAD workstation.

Select a folder

Name the user profile

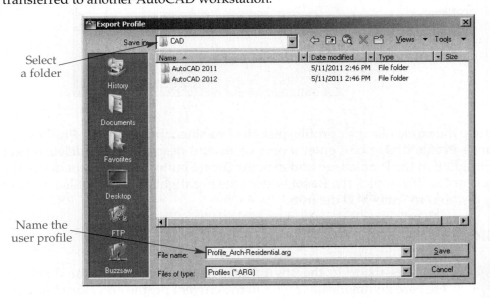

Figure 25-4.
Importing a user profile.

Profile name

Description

Check to include the path

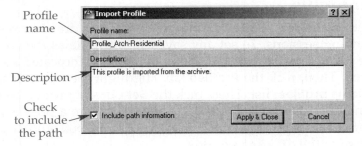

PROFESSIONAL TIP

A variety of settings can be changed in the **Options** dialog box. Settings that have the AutoCAD drawing icon located next to them can be stored in a template file (or drawing). Using the appropriate template file to begin a drawing automatically resets these settings. Settings within the **Options** dialog box that do not have the AutoCAD drawing icon next to them can typically be restored through the use of user profiles.

Exercise 25-2

Complete the exercise on the companion website.
www.g-wlearning.com/CAD

Workspaces

As you have seen, a user profile allows you to set and restore settings for screen colors, drafting settings, and file locations. On the other hand, a workspace allows you to set and restore settings for the **Quick Access** toolbar, ribbon, menus, and palettes (**DesignCenter**, **Properties** palette, **Tool Palettes** window, etc.). A *workspace* is a collection of displayed toolbars, palettes, ribbon tabs and panels, and menus and their configurations. A workspace stores not only which of these graphic tools are visible, but also their on-screen locations. A workspace does not store environmental settings, unless configured to do so by the user. This is discussed later in this chapter.

PROFESSIONAL TIP

A user profile also stores which toolbars are displayed and their location. A user profile updates when you change environmental settings or when you set a different workspace current. If a workspace is restored, the user profile updates to reflect the settings of the workspace for toolbar display and location. However, the user profile only stores which workspace is *current*, not what resides in the workspace or its settings.

Creating a Workspace

AutoCAD has four default workspaces—Drafting and Annotation, 3D Basics, 3D Modeling, and AutoCAD Classic. A workspace can be set current using the **Workspace Switching** button on the right-hand side of the status bar or the drop-down list on the **Quick Access** toolbar. See Figure 25-5. Picking this button displays a shortcut menu containing the names of all available workspaces. It also contains options for working with workspaces. When AutoCAD is launched, the workspace that was last active is restored. AutoCAD can be set up so that any changes made to the ribbon, toolbars, menus, and palettes are saved to this workspace. However, it is best to create your own workspaces.

The first step in setting up and storing your own workspace is to arrange the ribbon, toolbars, tool palettes, and menus to your liking. Refer to Chapter 22. Next, use the **WSSAVE** command to open the **Save Workspace** dialog box, Figure 25-6. You can easily access the **WSSAVE** command by selecting **Save Current As...** in the shortcut menu displayed by picking the **Workspace Switching** button. Selecting **Save Current As...** in the drop-down list on the **Quick Access** toolbar also initiates the **WSSAVE** command. In

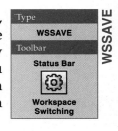

Figure 25-5.
Switching workspaces.

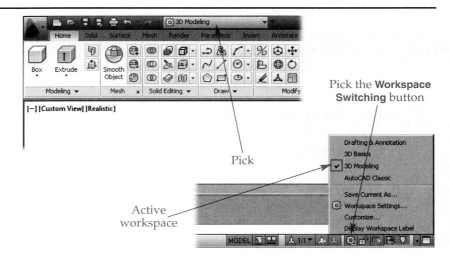

Pick

Pick the **Workspace Switching** button

Active workspace

Drafting & Annotation
3D Basics
✓ 3D Modeling
AutoCAD Classic

Save Current As...
Workspace Settings...
Customize...
Display Workspace Label

Figure 25-6.
Creating a new
workspace.

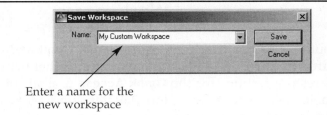

Enter a name for the
new workspace

the **Save Workspace** dialog box, enter a name for the workspace, such as Normal Design or Standard Arrangement, and then pick the **Save** button.

The current settings for toolbars, pull-down menus, and tool palettes are now stored in the new workspace, which is also made current. The workspace name also appears in the shortcut menu displayed by picking the **Workspace Switching** button and in the drop-down list on the **Quick Access** toolbar. The current workspace is indicated in the shortcut menu by a check mark and appears at the top of the drop-down list on the **Quick Access** toolbar.

PROFESSIONAL TIP

Workspaces are saved in a CUIx file. By default, they are saved to the main CUIx file (acad.cuix). See Chapter 22 for details regarding CUIx files.

Restoring a Workspace

Once a workspace has been saved, it can easily be restored. A list of available workspaces appears in the shortcut menu displayed by picking the **Workspace Switching** button and in the drop-down list on the **Quick Access** toolbar. To select a different workspace, simply pick the name of the workspace in the pop-up list. A workspace can also be restored using the command line:

Command: **WORKSPACE**↵
Enter workspace option [setCurrent/SAveas/Edit/Rename/Delete/SEttings/?] <setCurrent>: **C**↵
Enter name of workspace to make current [?] <*current*>: **MY CUSTOM WORKSPACE**↵
Command:

Notice that you can manage workspaces using this command. This command-line command can be advantageous when creating macros for custom commands. See Chapter 22 for more information on creating macros.

You can also use the **Customize User Interface** dialog box to restore a workspace. First, open the dialog box. Then, expand the Workspaces branch in the **Customizations in All Files** pane. All of the workspaces defined in the default CUIx file and any open CUIx files are displayed in this branch. The label (current) follows the name of the current workspace. To restore a workspace, right-click on its name in the Workspaces branch and select **Set Current** from the shortcut menu. Its name is now followed by (current). Pick the **OK** button to close the **Customize User Interface** dialog box and make the workspace current.

PROFESSIONAL TIP

The **WSCURRENT** system variable indicates the current workspace. You can use this system variable to restore a workspace. Simply set the system variable to the name of the workspace you want to restore.

Quick Access Toolbar

The **Quick Access** toolbar is located in the top-left corner of the screen to the right of the application menu icon, **Figure 25-7**. It contains essential tools for AutoCAD. By default, it displays the **New, Open, Save, Save As, Undo, Redo**, and **Plot** tools along with the **Workspace** drop-down list. In addition, you can display other tools on the toolbar by picking the down arrow button on the right-hand end of the **Quick Access** toolbar to display a shortcut menu. In this menu, select the tools to show and hide. The additional available tools are **Match Properties, Batch Plot, Plot Preview, Properties, Sheet Set Manager**, and **Render**. Any other defined command can be added to the toolbar by selecting **More Commands...** in the drop-down list. This displays the **Customize User Interface** dialog box.

Unlike the toolbars discussed in Chapter 22, the **Quick Access** toolbar is customized inside of the workspace. You may create more than one **Quick Access** toolbar, but only one may be active in a workspace. To create an additional **Quick Access** toolbar, open the **Customize User Interface** dialog box, right-click on the Quick Access Toolbars branch, and select **New Quick Access Toolbar**. When creating a new **Quick Access** toolbar, it is automatically populated with the tools outlined above. Customizing the **Quick Access** toolbar is discussed in the next section.

PROFESSIONAL TIP

Any tool on a ribbon panel can be quickly added to the **Quick Access** toolbar. Right-click on the tool icon and select **Add to Quick Access Toolbar** from the shortcut menu.

Customizing a Workspace

An existing workspace can be customized using the **Customize User Interface** dialog box. First, select the workspace to be customized in the Workspaces branch of the **Customizations in All Files** pane. Remember, the name of this pane will change based on the selection in the drop-down list. The **Workspace Contents** pane at the upper-right corner of the dialog box displays the contents of the selected workspace. There are Quick Access Toolbar, Toolbars, Menus, Palettes, and Ribbon Tabs branches. Refer to **Figure 25-8**. Expand a branch to see which components the workspace contains.

To customize the workspace, pick the **Customize Workspace** button at the top of the **Workspace Contents** pane. The tree in the pane turns blue to indicate you are in customization mode and the button changes to the **Done** button. Also, notice that the tree in the **Customizations in All Files** pane has changed. Several branches have disappeared and

---

Figure 25-7.
The **Quick Access** toolbar.

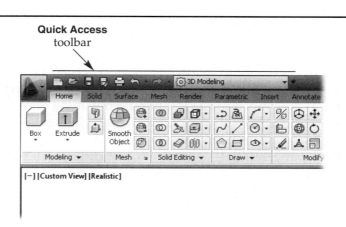

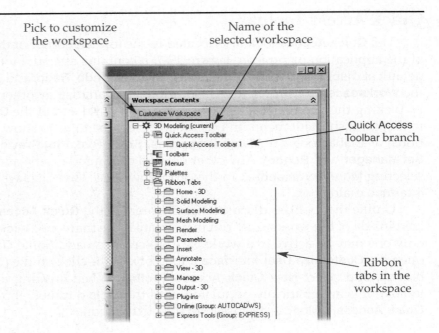

Figure 25-8.
You can customize existing workspaces.

Pick to customize the workspace

Name of the selected workspace

Quick Access Toolbar branch

Ribbon tabs in the workspace

the Menus branch has a green check mark next to it. If you expand the branches in the **Customizations in All Files** pane, you will see that a green check mark also appears next to the components that currently are in the workspace. See **Figure 25-9.**

To add a component to the workspace, pick the blank box in front of its name to place a check mark in the box. The component also appears in the **Workspace Contents** pane. To remove a component from the workspace, pick the check mark in front of its name to clear the box. The component is also removed from the **Workspace Contents** pane.

Referring to **Figure 25-9,** notice the Quick Access Toolbar branch in the **Workspace Contents** pane. By default, Quick Access Toolbar 1 appears in this branch. Additional toolbars appear if you have added them, as discussed earlier in this chapter. When in customization mode for workspaces, you may choose which **Quick Access** toolbar is displayed in the workspace. However, you cannot be in customization mode to customize a **Quick Access** toolbar. You must do this after exiting workspace customization mode.

When done customizing the workspace, pick the **Done** button in the **Workspace Contents** pane. The tree is no longer displayed in blue. Also, all branches are once again displayed in the **Customizations in All Files** pane. You can now expand a branch in the **Workspace Contents** pane, select a component name, and use the **Properties** pane to adjust the component's properties.

For example, suppose you have added the **Layers** toolbar to a workspace. In the **Properties** pane, you can set the orientation to floating, specify the location of its anchor point, and set the number of rows for the toolbar. You can also change the order in which menus appear on the menu bar by dragging them to a new location in the tree in the **Workspace Contents** pane. The top of the tree is the left-hand side of the menu bar.

In the **Customizations in All Files** pane, you can customize the **Quick Access** toolbar. The components in the Quick Access Toolbars branch are commands. To add a command to the Quick Access Toolbars branch, locate it in the **Command List:** pane. Then, drag and drop it into the branch in the **Customizations in All Files** pane. To remove a command from the Quick Access Toolbars branch, select it and press the [Delete] key.

You may have noticed that there is not a Palettes branch in the **Customizations in All Files** pane. All palettes are automatically available in all workspaces. You can, however, specify whether a palette is displayed or hidden in a workspace. You can also change other properties of a palette, such as floating/docked status, its size, and whether the autohide feature is enabled. To change the properties of a palette, first

Figure 25-9.
The appearance of the **Customize User Interface** dialog box when customizing workspace mode is active. The toolbars, menus, and ribbon tabs included in the workspace are specified by placing a check mark next to the component name.

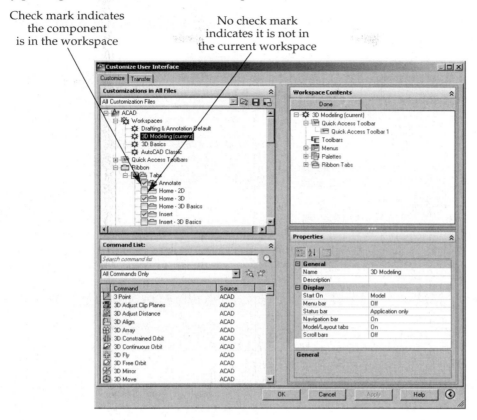

Check mark indicates the component is in the workspace

No check mark indicates it is not in the current workspace

select it in the Palettes branch in the **Workspace Contents** pane. Then, in the **Properties** pane, adjust the properties as needed. When you pick **OK** to close the **Customize User Interface** dialog box, the new default properties of the palette are set for that workspace.

For example, the **Tool Palettes** window in the AutoCAD Classic workspace is, by default, displayed floating. You can hide the **Tool Palettes** window when in the drawing area by simply picking the **Close** button (**X**). You can also drag the palette into the drawing area to a new position or dock it. However, these actions do not change the default setting for the workspace. If you restore the workspace in the future, the **Tool Palettes** window will again be floating. You must alter the default settings for the **Tool Palettes** window in the workspace. First, open the **Customize User Interface** dialog box and select the AutoCAD Classic workspace. Expand the Palettes branch in the **Workspace Contents** pane and select **Tool Palette**. Then, in the **Properties** pane, change the Show property to No. See **Figure 25-10**. Now, when you set the AutoCAD Classic workspace current, the **Tool Palettes** window will not be displayed. It can, of course, be manually displayed as needed.

You can also set up a workspace so that it displays model space or layout (paper) space when restored. By default, a workspace displays model space when it is set current. To change this, highlight the workspace name in either the **Customizations in All Files** pane or the **Workspace Contents** pane. Then, in the **Properties** pane, change the Start On property to Layout or Do not change. If Model is specified for the Start On property, model space is displayed when the workspace is restored. If Layout is specified for the Start On property, the most recently active layout tab is displayed when the workspace is restored. If Do not change is specified for the Start On property, the current tab (model space or paper space) remains active when the workspace is restored.

Figure 25-10.
Changing the default properties of a palette for a given workspace.

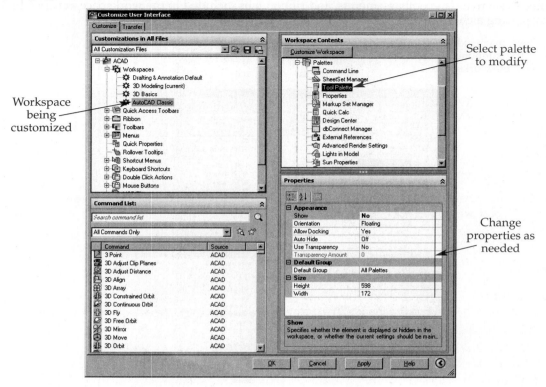

Workspace being customized

Select palette to modify

Change properties as needed

PROFESSIONAL TIP

If the **Customize User Interface** dialog box is open, but you are not in workspace customization mode, you can also remove a component from the workspace by right-clicking on it in the tree in the **Workspace Contents** pane and selecting **Remove from Workspace** in the shortcut menu. This is true for all component types except palettes, as all palettes are always available to all workspaces.

Workspace Settings

There are various settings related to workspaces. These are set in the **Workspace Settings** dialog box. See **Figure 25-11.** The **WSSETTINGS** command opens this dialog box. It can also be displayed by selecting **Workspace Settings...** from the shortcut menu displayed by picking the **Workspace Switching** button on the status bar or the drop-down list on the **Quick Access** toolbar.

At the top of the **Workspace Settings** dialog box is the **My Workspace =** drop-down list. All saved workspaces in the CUIx file appear in this list. The workspace that is selected in the list is defined as My Workspace, or known as a home workspace. The workspace that is designated as My Workspace is restored when the **My Workspace** button on the **Workspaces** toolbar is picked. This can be useful if one person primarily uses a machine, but others may temporarily use the machine with their own workspace settings. You may also find this useful if you have more than one workspace, but use one more often than all of the others. This feature can only be accessed when the **Workspaces** toolbar is displayed. This toolbar is displayed by default in the AutoCAD Classic workspace.

The **Menu Display and Order** area of the **Workspace Settings** dialog box contains a list of all workspaces saved in the current CUIx file. The order of this list determines the

Figure 25-11.
The **Workspace Settings** dialog box is used to set which workspace is My Workspace, specify which workspaces are shown in the menu and drop-down list and their order, and set whether changes are automatically saved when a different workspace is restored.

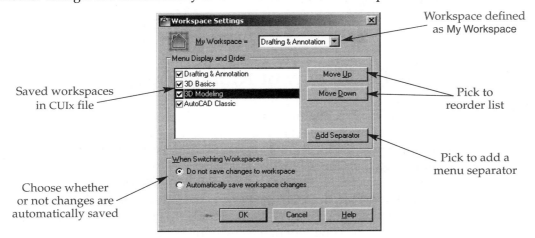

Saved workspaces in CUIx file

Choose whether or not changes are automatically saved

Workspace defined as My Workspace

Pick to reorder list

Pick to add a menu separator

order of the list that appears in the **Workspace Switching** shortcut menu, in the drop-down list on the **Quick Access** toolbar, and in the drop-down list on the **Workspaces** toolbar. The order of the list can be modified by highlighting one of the workspaces and using the **Move Up** and **Move Down** buttons. The **Add Separator** button is used to add a horizontal line, or menu separator, to the list. A separator is used to logically group workspace names within the list. The separator can be relocated within the list just like a workspace name.

You can prevent a workspace name or separator from being displayed in the menu or drop-down list by removing the check box next to its name. The check box next to the current workspace and the workspace designated as My Workspace can be cleared. However, these workspaces will always be displayed in the menu and drop-down list.

At the bottom of the **Workspace Settings** dialog box is the **When Switching Workspaces** area. The radio buttons in this area determine whether changes you have made since you last saved the workspace, such as the visibility of toolbars, are saved when you switch to a different workspace. To retain the settings as you last saved them, pick the **Do not save changes to workspace** radio button. Changes made since the workspace was last saved are discarded when a different workspace is restored. If the **Automatically save workspace changes** radio button is on, any "as you work" changes are automatically saved to the workspace when a different workspace is restored. This option is also controlled by the **WSAUTOSAVE** system variable. The default setting is 0.

At any time, you can manually save the settings to the current workspace. Pick the **Save Current As...** selection in the **Workspace Switching** shortcut menu. When the **Save Workspace** dialog box appears, select the current workspace from the drop-down list. Then, pick the **Save** button. An alert is displayed stating that a workspace with that name already exists and asking if you would like to replace it. Pick the **Yes** button to save the changes to the current workspace.

PROFESSIONAL TIP

Unlike a user profile, changes to the environment (toolbar display, menu configuration, etc.) are not necessarily automatically saved. To ensure the changes are only saved when you decide to save them, be sure the **Do not save changes to workspace** radio button is on in the **Workspace Settings** dialog box.

Exercise 25-3

Complete the exercise on the companion website.
www.g-wlearning.com/CAD

Controlling the Display of Ribbon Tabs and Panels

Right-click anywhere on the ribbon to display the shortcut menu. Notice the **Show Tabs** and **Show Panels** cascading menus. See Figure 25-12. The default ribbon tabs in the 3D Modeling workspace are **Home, Solid, Surface, Mesh, Render, Parametric, Insert, Annotate, View, Manage, Output, Plug-ins, Online,** and **Express Tools** (if installed). The **Show Panels** cascading menu in the shortcut menu displays the panels of the current tab. For example, the default panels of the **Home** tab in the 3D Modeling workspace are **Modeling, Mesh, Solid Editing, Draw, Modify, Section, Coordinates, View, Selection, Layers,** and **Groups**. In the **Show Tabs** or **Show Panels** cascading menu, the currently displayed tabs and panels have a check mark next to their name. Selecting a name toggles the visibility of the tab or panel.

While the visibility of the ribbon tabs and panels can be adjusted "on the fly" by using the **Show Tabs** or **Show Panels** menus, visibility can also be controlled by customizing the workspace. This is a more efficient method.

1. Open the **Customize User Interface** dialog box and expand the Workspaces branch in the **Customizations in All Files** pane.
2. Select the workspace for which you wish to adjust panel visibility.
3. In the **Workspace Contents** pane, pick the **Customize Workspace** button.
4. Expand the Ribbon Tabs branch in the **Workspace Contents** pane. The tabs visible in the workspace appear in the tree. The panels visible in the workspace appear as branches below the tabs.
5. In the **Customizations in All Files** pane, expand the Ribbon branch and then the Tabs branch. All available tabs appear in the branch. The currently displayed tabs have a check mark next to their name. See Figure 25-13.
6. Check the tabs to display and uncheck the tabs to hide.
7. To control which panels are displayed for a tab, expand the tab's branch in the **Workspace Contents** pane. Then, select the panel to hide and, in the **Properties** pane, change its Show property to No. To display a panel, change the property to Yes.
8. In the **Workspace Contents** pane, pick the **Done** button.
9. Exit the **Customize User Interface** dialog box.

Tabs and panels appear in the ribbon in the order in which they are displayed in the tree of the **Workspace Contents** pane. Top to bottom equals left to right when the ribbon is horizontal. As you add panels to the workspace, they are added to the bottom of the Ribbon Tabs branch. Refer to Chapter 22 for information on creating custom ribbon tabs and panels. To rearrange the tabs or panels, pick and drag them within the **Workspace Contents** pane. You do not need to be in customization mode. You can also remove a tab from the workspace by right-clicking on its name in the **Workspace Contents** pane and picking **Remove from Workspace** from the shortcut menu.

PROFESSIONAL TIP

You do not need to be in customization mode to change a panel's Show property. Also, when a panel's Show property is set to No, the panel is still available in the **Show Panels** cascading menu in the shortcut menu displayed by right-clicking on the ribbon.

Figure 25-12.
A shortcut menu is displayed when you right-click on the ribbon. A—The **Show Tabs** cascading menu. B—The **Show Panels** cascading menu. This is for the **Home** tab.

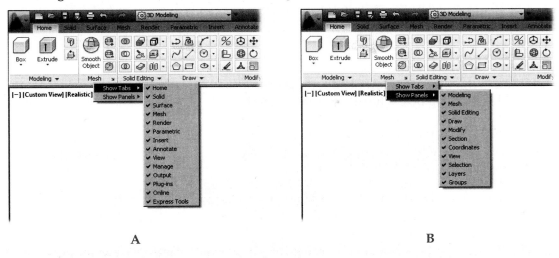

A

B

Figure 25-13.
Setting which ribbon tabs and panels are displayed for a workspace.

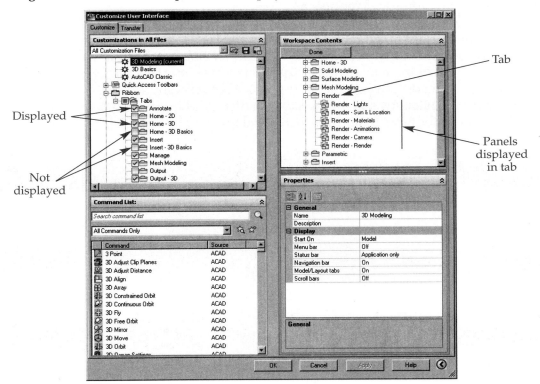

Chapter Review

Answer the following questions. Write your answers on a separate sheet of paper or complete the electronic chapter review on the companion website.
www.g-wlearning.com/CAD

1. What is a *user profile*?
2. How and why are profiles used?
3. What is the file extension used for a profile when it is exported?
4. In which dialog box is a user profile created?
5. How do you restore a user profile?
6. Why would you export a user profile?
7. Briefly describe how to import a user profile.
8. Define *workspace* as related to AutoCAD.
9. List three ways to open the **Save Workspace** dialog box.
10. How do you restore a workspace? List three methods.
11. When customizing a workspace, how do you determine which ribbon tabs, menus, and toolbars are displayed?
12. Briefly describe how to change the default settings for a palette for a given workspace.
13. List two ways to open the **Workspace Settings** dialog box.
14. How do you add a separator to the shortcut menu displayed by picking the **Workspace Switching** button?
15. How do you define My Workspace?
16. Briefly describe how to set up AutoCAD so that changes made to the environment are automatically saved to the current workspace.
17. Briefly describe how to control the visibility of ribbon tabs in a workspace when you are in customize mode in the **Customize User Interface** dialog box.
18. How can you add commands to the **Quick Access** toolbar?
19. How many **Quick Access** toolbars can be displayed in a workspace?
20. Briefly describe how to control which panels are displayed for a tab in a workspace.

Drawing Problems

Before creating any user profiles or workspaces, check with your instructor or supervisor for specific instructions or guidelines.

1. Create two user profiles, one named Model Development for modeling and one named Rendering for visual styles and rendering.
   A. Display the toolbars that contain the commands needed for each type of work.
   B. Change the color of the drawing area as needed. For example, some drafters prefer a white background when drawing. However, a black background is often desired when shading and rendering the model.

2. Export the user profiles created in Problem 1 to ARG files. Then, delete each profile from AutoCAD. Restart AutoCAD and verify that the profiles are no longer available. Next, import each profile. Restore each profile to verify the settings.

3. Create two workspaces, one named Design Development and one named Dimensioning.
   A. For the Design Development workspace, display the ribbon tabs and panels related to drawing and editing. Also, rearrange the menus to group the drawing and editing/modifying menus together. Remove any menus that are not needed.
   B. For the Dimensioning workspace, display the ribbon tabs and panels related to dimensioning the drawing. You may consider removing editing/modifying menus.
   C. Create custom ribbon tabs and panels as needed. Include flyouts when advantageous.

4. Customize the two workspaces created in Problem 3.
   A. Change default on/off status of any palettes to suit the purpose of the workspace. For example, you may want the **DesignCenter**, **Tool Palettes**, and **Properties** palettes displayed for the Design Development workspace.
   B. Enable the autohide status of all displayed palettes.
   C. Change the Dimensioning workspace so that it starts in a layout.

5. Set up the workspaces from Problem 4 so that "on the fly" changes are automatically saved when a different workspace is made current.

Actions are used to automate commonly performed tasks in AutoCAD. The steps used to create an action are shown in the action tree, which is located in the expanded **Action Recorder** panel on the **Manage** tab of the ribbon. In the example below, the steps listed in the action tree for an action titled BrickHatchEdit are shown. This action is used to create a rectangle, draw a brick hatch pattern inside the rectangle, and then edit the hatch pattern origin.

Chapter

Recording and Using Actions

Learning Objectives

After completing this chapter, you will be able to:

✓ Describe the action recorder.
✓ Create and store an action.
✓ Run an action.
✓ Modify an action.
✓ Explain the limitations of the action recorder.

In schools and workplaces, certain design elements and customizing features can be enhanced by automation. The AutoCAD *action recorder* is a tool that allows the user to automate a task or function by saving the steps needed to complete it. These steps are called an *action*. Creating a library of recorded actions allows a user to increase productivity by more efficiently working in AutoCAD.

Action Recorder Overview

The **Action Recorder** panel is located in the **Manage** tab of the ribbon. It contains seven tools: **Record/Stop**, **Insert Message**, **Insert Base Point**, **Pause for User Input**, **Play**, **Preference**, and **Manage Action Macros**. In addition, a drop-down list displays all recorded actions. See **Figure 26-1**. The expanded **Action Recorder** panel displays the action tree. The *action tree* shows the steps used to create the current action that is selected in the drop-down list. The steps are listed in order from top to bottom. The first step is located at the top of the tree.

The action recorder saves all steps in an action to an ACTM file. By saving the action to a file, it can repeatedly be used. The ACTM file name cannot contain spaces or special characters. Also, the name cannot be the same as an existing AutoCAD command name.

The folder where actions are stored is defined in the **Files** tab of the **Options** dialog box. To change the location where actions are stored, expand the Action Recorder Settings branch in the **Files** tab, **Figure 26-2**. Then, pick the Actions Recording File Location branch, pick the **Browse...** button, and locate the new default folder. The Action Recorder Settings branch also contains the Additional Actions Reading File Locations branch. Use this to define network or shared folder connections for ACTM files.

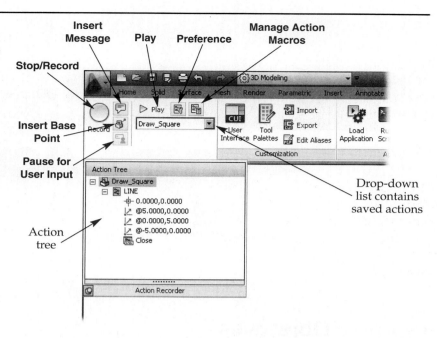

Figure 26-2.
The location where actions are stored is set in the **Files** tab of the **Options** dialog box.

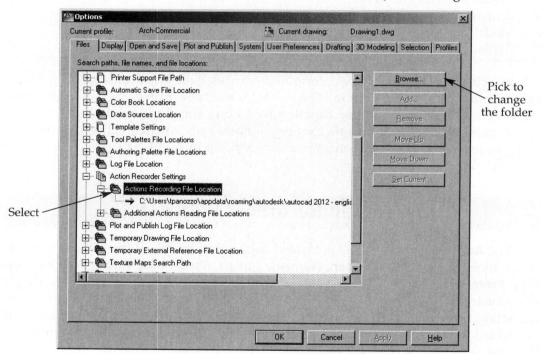

Record/Stop

The **Record** tool launches AutoCAD into record-action mode. Then, you can begin performing the steps needed to define your action. Once the **Record** tool is activated, the button changes to the **Stop** button. This is used when the steps are completed and you have finished recording your action. When recording an action, the **Action Recorder** panel will shade red.

Insert Message

When in record mode or while editing a macro, the **Insert Message** tool is enabled. This tool allows you to insert a message dialog box that will be displayed for the user during action playback. A message is especially useful for directing the user's attention to what is required. One can also be used at the beginning of the action to introduce the user to the action and how it will perform.

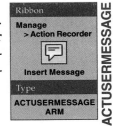

Insert Base Point

When in record mode or while editing a macro, the **Insert Base Point** tool is enabled. This tool allows you to choose a starting location for a particular command or series of commands. This is used to override the default absolute coordinates that are automatically entered while creating an object. It can be used during recording, but only when applicable. The tool is disabled when not available. Additionally, when editing a macro, you may add this option in front of a particular command.

Pause for User Input

When in record mode or while editing a macro, the **Pause for User Input** tool is enabled. Inserting a request for user input temporarily halts the action at that point during playback. The user is prompted for input, such as selecting objects or picking locations on the screen.

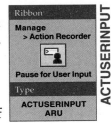

Play

The **Play** tool plays the current macro. The current macro is displayed at the top of the drop-down list in the **Action Recorder** panel. This drop-down list is discussed later in this chapter. When playing an action, the **Action Recorder** panel will shade green.

Preferences

Picking the **Preference** tool displays the **Action Recorder Preferences** dialog box, **Figure 26-3**. This dialog box contains three check boxes for setting how the **Action Recorder** panel functions. When the **Expand on playback** check box is checked, the **Action Recorder** panel is automatically expanded to show the action tree during action playback. This is unchecked by default. When the **Expand on recording** check box is checked, the **Action Recorder** panel automatically expands to display the action tree during action recording. This is checked by default. When the **Prompt for action macro name** check box is checked, the **Action Macro** dialog box is displayed after recording of the action is complete (the **Stop** button is picked). This dialog box prompts for the action name, which is the name of the ACTM file. See **Figure 26-4**.

Figure 26-3.
The **Action Recorder**
Preferences dialog
box is used to set
certain behavior of
the action recorder.

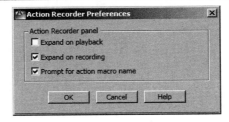

Figure 26-4.
The **Action Macro**
dialog box is used to
save an action.

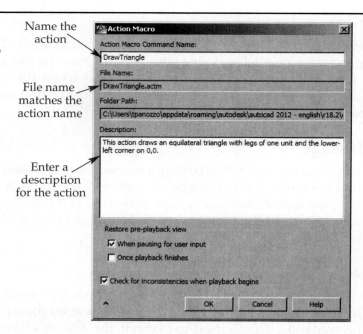

Name the
action

File name
matches the
action name

Enter a
description
for the action

Manage Action Macros

ACTMANAGER

Ribbon

Manage
> Action Recorder

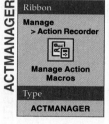

Manage Action
Macros

Type

ACTMANAGER

The **Action Macro Manager** dialog box allows you to rename saved actions or change their descriptions. Picking the **Manage Action Macros** button in the **Action Recorder** panel displays this dialog box. See **Figure 26-5**. The dialog box contains a list of actions saved in the action macro folder specified in the **Options** dialog box.

Picking the **Copy** button displays the **Action Macro** dialog box, which is also displayed by default when recording of an action is complete. Enter a name and description for the new action and set the other properties. When you pick the **OK** button, the steps in the action highlighted in the **Action Macro Manager** dialog box are saved under the new name. The original action is not altered.

Picking the **Rename** button replaces the name of the highlighted macro with an edit box. This allows you to enter a new name for the action. Renaming the action also changes the name of the ACTM file.

Picking the **Modify** button also displays the **Action Macro** dialog box. However, when you make changes to the name, description, or other properties, the action highlighted in the **Action Macro Manager** dialog box is altered.

Figure 26-5.
The **Action Macro**
Manager dialog
box is used to copy,
rename, or change
an existing macro.

List of
saved actions

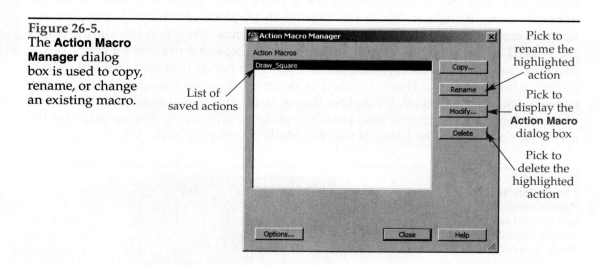

Pick to
rename the
highlighted
action

Pick to
display the
Action Macro
dialog box

Pick to
delete the
highlighted
action

Picking the **Delete** button deletes the action and associated ACTM file. Picking the **Options...** button displays the **Options** dialog box with the Action Recorder Settings branch highlighted in the **Files** tab.

Action Tree and Available Actions

The **Available Action Macro** drop-down list displays all action files (ACTM files) located in the action folder. As explained earlier, the action folder is set in the **Files** tab of the **Options** dialog box. The current action is displayed at the top of the list. To see all of the available actions, pick the drop-down list. To set a different action current, simply pick it in the drop-down list.

At the bottom of the expanded **Action Recorder** panel is the action tree. The action tree contains all of the steps in the action in graphic form. See **Figure 26-6**. The name of the action is the top of the tree. Each step is a branch in the tree. In addition, the command branches can be expanded to see the values (nodes) entered into the action.

PROFESSIONAL TIP

Inserting messages is especially useful to direct the user about the steps of an action. Inserting requests for user input in an action pauses the action and allows the user to select objects or on-screen points. These features make actions more dynamic and help ensure proper usage.

Exercise 26-1

Complete the exercise on the companion website.
www.g-wlearning.com/CAD

Figure 26-6.
The action tree contains all of the steps in the current action.

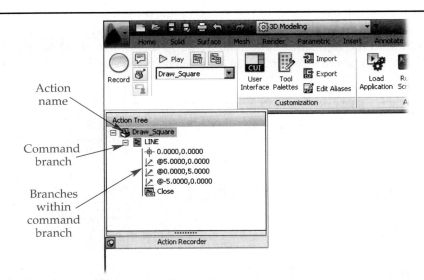

Action name

Command branch

Branches within command branch

Now that you have studied the features of the action recorder, it is time to create an action. This example action will create a layer named A-Anno-Iden, set the layer color to orange (ACI = 30), and make the layer current. The action will also set the annotation scale to 1/4″ = 1′-0″. In this section, you will create the basic action. Later, you will modify the action to make it more interactive. Follow these steps to create the basic action named AnnoQtr.

1. Start a new drawing.
2. Pick the **Record** button in the **Action Recorder** panel on the **Manage** tab of the ribbon. Notice that the panel is shaded in red and a red dot appears next to the crosshairs. These are indications that you are in record mode.
3. Display the **Layer Properties Manager** palette (it can be accessed from the **Layers** panel in the **Home** tab of the ribbon, or you can use the **LAYER** command).
4. Pick the **New Layer** button in the palette's toolbar.
5. In the edit box, name the new layer A-Anno-Iden.
6. Pick the color swatch for the layer and, in the **Select Color** dialog box, pick color 30 in the **Index Color** tab. Then, pick the **OK** button to close the **Select Color** dialog box.
7. Right-click on the newly created layer and pick **Set current** from the shortcut menu.
8. Close the **Layer Properties Manager** palette.
9. Using the **Annotation Scale** button on the status bar, set the 1/4″ = 1′-0″ scale current. See **Figure 26-7**. If the drawing is based on the acad3D.dwt template, you must type the **CANNOSCALE** command since the status bar button is unavailable.
10. Pick the **Stop** button in the **Action Recorder** panel on the **Manage** tab of the ribbon. This ends recording of the action.
11. The **Action Macro** dialog box appears when you pick the **Stop** button, unless you have turned off the "prompt for name" option.
12. In the **Action Macro** dialog box, type AnnoQtr in the **Action Macro Command Name:** text box (*Anno* is an abbreviation for annotation and *Qtr* is an abbreviation for quarter scale).
13. In the **Description:** text box, type Creates a new layer with color 30 and sets the 1/4″ = 1′-0″ annotation scale current or a similar description.
14. Pick the **OK** button in the **Action Macro** dialog box to save the action.

The AnnoQtr action is now available to run. It is automatically selected as the current action in the **Available Action Macro** drop-down list in the **Action Recorder** panel. Also, the action tree for the action is displayed in the expanded panel. See **Figure 26-8**.

Figure 26-7.
Setting the annotation scale for the action being recorded.

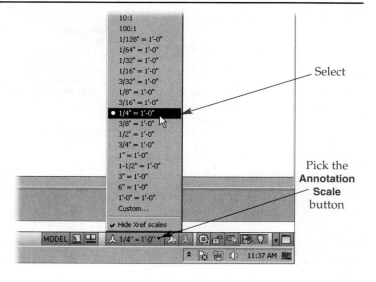

Select

Pick the **Annotation Scale** button

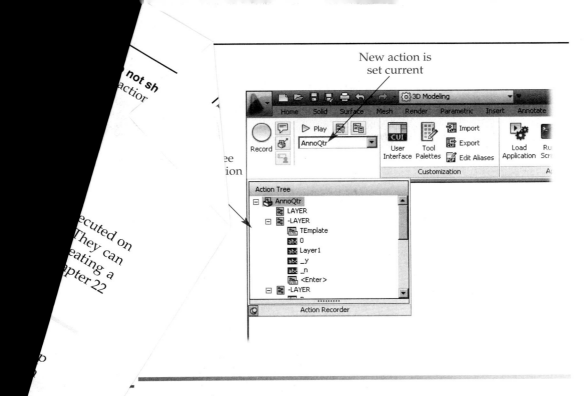

New action is
set current

In a design environment, saving actions in a shared folder on a network can streamline productivity. This creates an environment in which AutoCAD users can share their knowledge by creating actions drawn from their own design experiences. Use the Additional Actions Reading File Locations branch in the **Options** dialog box to establish the shared connection.

Exercise 26-2

Complete the exercise on the companion website.
www.g-wlearning.com/CAD

Running an Action

To run the current action, pick the **Play** button in the **Action Recorder** panel on the **Manage** tab in the ribbon. The current action is shown at the top of the **Available Action Macro** drop-down list in the **Action Recorder** panel. To run a different action, simply select it in the drop-down list and then pick the **Play** button. Another method of running an action is to type its name on the command line. This is why the action name cannot be the same as an existing AutoCAD command.

Before testing an action, be sure to delete all graphics from the screen unless needed for the action. To test the action recorded in the previous section, start a new drawing. Make sure the AnnoQtr action is set current in the **Available Action Macro** drop-down list. Then, pick the **Play** button and watch the action run (notice that the **Action Recorder** panel shades green when the **Play** button is picked). By default, when the action is complete, AutoCAD displays the **Action Macro—Playback Complete** dialog box, **Figure 26-9**. This is simply a message to the user that the action is finished. Pick the **OK** button to close the dialog box.

Figure 26-9.
This dialog box is displayed when an action is complete. If you check the **Do** **message again** check box, this dialog box will not be displayed at the end of

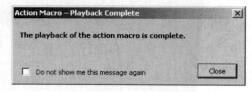

Actions run just like regular AutoCAD commands. Since they can be ex the command line, actions can be made into custom AutoCAD commands. be placed in a customized ribbon panel, toolbar, menu, or palette. When cr custom command from an action, use the syntax ^C^C_action_name. Refer to Cha for information on creating custom commands.

PROFESSIONAL TIP

Actions can streamline the customization process by creating setu procedures to minimize macro writing when creating custom commands. For example, suppose you have an annotation block that you want inserted at a set scale and on a specific layer. You can create an action similar to the example previously discussed (AnnoQtr). Then, you can insert the action in the command string prior to the block insertion. This will streamline your customization process.

Modifying an Action

The action tree can be used to edit an action. Examine the action tree for the AnnoQtr example, as shown in **Figure 26-10**. At the top of the action tree is the action name. It is always the top branch of the tree. The icon to the left of the name indicates that it is an action. The branches below the action name contain the commands and steps in the action.

Notice the dash in front of the **LAYER** command. This dash indicates the command is a command-line command. Even though you used the **Layer Properties Manager** palette when recording the action, the action recorder translated your picks into the command-line version of the steps.

Branches below the command names contain the specific information used to create each item. Each branch is automatically converted to the command-line entry. The icon next to each branch indicates the type of action or input represented by the branch. These branches are also called *nodes*. For example, reviewing the **LAYER** command branches, you can see a text icon next to the name A-Anno-Iden. This icon indicates that the branch, or node, is a text entry. Additionally, the icon for the branch where the layer color is set to 30 is a color wheel. This indicates the branch is a change in color value.

An icon with a head-and-shoulders image in the lower-right corner indicates the action will pause at this point to allow the user to enter information. This icon is called a user icon. The AnnoQtr action does not currently contain any user icons.

Now, you will modify the action recorded earlier by changing some of the steps to require user input. For example, you will have the user type the annotation scale instead of automatically setting it to 1/4″ = 1′-0″ scale. So the user knows what is required, you will add a message explaining the step. Since the scale is not automatically set, the

AutoCAD and Its Applications—Advanced

Figure 26-10.
The action tree for
the recorded action
prior to editing the
action.

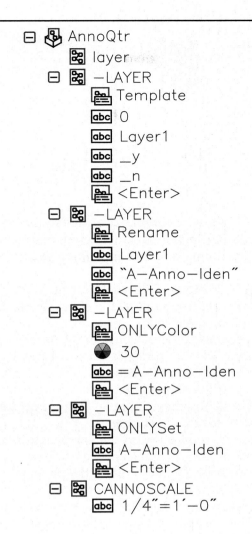

action also needs to be renamed to remove the Qtr (quarter inch scale) designation.
Complete the following steps.

1. Expand the **Action Recorder** panel to see the action tree. You may want to pin the panel in the expanded state. To pin the panel, pick the pushpin icon located in the lower-left corner of the expanded **Action Recorder** panel.

2. Right-click on the action name AnnoQtr and select **Rename** from the shortcut menu. If you want to retain the original action instead of renaming it, select **Copy...** from the shortcut menu. In either case, the **Action Macro** dialog box is displayed.

3. In the **Action Macro Command Name:** text box, type AnnoUserScale and then pick the **OK** button to rename the action. Note: the edits in Steps 2 and 3 may also be accomplished using the **Action Macro Manager** dialog box.

4. In the action tree, locate the CANNOSCALE branch. Right-click on the 1/4" = 1'-0" branch below it and select **Pause for User Input** from the shortcut menu. The value in the branch (1/4" = 1'-0") is grayed out and the icon for the branch now contains the user icon.

5. To add a message describing how to type the correct scale, right-click on the 1/4" = 1'-0" branch and select **Insert User Message...** from the shortcut menu. The **Insert User Message** dialog box appears, **Figure 26-11.**

6. In the **Insert User Message** dialog box, type the message:

   Enter the desired annotation scale on the command line. Please use the proper
      syntax. For example, type 1/8" = 1'-0" for the 1/8" Architectural scale. Include a
   space on each side of the equals symbol.

Figure 26-11.
Adding a user
message to the
action.

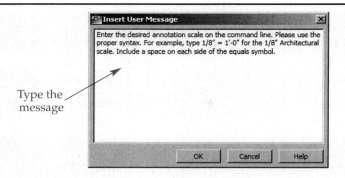

Type the
message

Notice that a User Message branch is added to the action tree above the branch that was right clicked. The action is now altered. Start a new drawing and run the altered action. Notice the dialog box that appears to guide the user through the inputs. See Figure 26-12.

To remove user input, right-click on the branch and select **Pause for User Input** from the shortcut menu to remove the check mark. The action then uses the value set when the action was recorded.

If you would like to delete a command from the action, right-click on the command branch and select **Delete** from the shortcut menu. Once deleted, the command is permanently removed from the action. This cannot be undone. If zooming or panning is done while recording the action, the action tree will display a View Change node with its appropriate icon. You may right-click and select **Delete** or **Insert User Message** from the shortcut menu to edit this node.

To edit a value, right-click on the corresponding branch and select **Edit** from the shortcut menu. The value is replaced by a text box. Type the new value and press [Enter]. Be careful; if you move the cursor off of the panel while the text box is displayed, whatever is displayed in the text box is entered as the value.

If you are creating objects such as a rectangle during recording of an action, the default coordinates are set relative to the previous point. If you wish to toggle to absolute coordinate entry, right-click on the coordinate in the action tree and uncheck **Relative to Previous** in the shortcut menu. Furthermore, if you right-click in the action tree, you may uncheck **All Points are Relative** in the shortcut menu, which will edit the entire action.

When recording the action, use the buttons on the **Action Recorder** panel to insert messages, base points, and user-input requests.

Figure 26-12.
The message
displayed to the user
during playback of
the example action.

Message to
the user

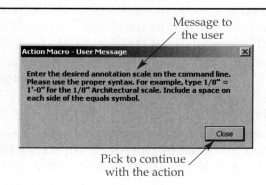

Pick to continue
with the action

Exercise 26-3

Complete the exercise on the companion website.
www.g-wlearning.com/CAD

Practical Example

This example creates a rectangle and fills it with a hatch pattern. First, you will record an action that draws a rectangle at a user-specified location and then uses the **ZOOM** command to center the rectangle in the display. The action will draw a brick hatch pattern inside of the rectangle. Then, the hatch pattern is edited to change/reset its origin. In this example, commands will be entered on the command line. Although it is more time-consuming to create an action in this way, it allows easy access to and editing of the action steps.

1. Start a new drawing.
2. Begin recording a new action.
3. Select the **Insert Base Point** button and choose any location on the screen. This allows the user to place the rectangle anywhere on the screen during future uses of the macro.
4. Type the **RECTANG** command. When prompted to specify the first corner point, pick anywhere on the screen. When prompted for the other corner point, type @120,60.
5. Type the **ZOOM** command and enter the **Extents** option. Enter the command again and type .7x to center the rectangle on the screen.
6. Type the **-HATCH** command. Use the **Select objects** option to select the rectangle.
7. Use the **Properties** option and select the hatch AR-B816. This is the architectural brick pattern. Enter a scale of 1.000 and a 0° rotation angle. Be sure to enter the values; do not press [Enter] to accept the defaults. Finish the **-HATCH** command.
8. Type the **-HATCHEDIT** command and select the brick pattern.
9. Select the **Origin** option, then select the **Set new origin** option. Pick the lower, right-hand corner of the rectangle. When prompted, store this location as the default origin.
10. Stop recording of the action and name it BrickHatchEdit.

Next, edit the action to allow user input and display messages. This will aid the user in selecting points.

11. For the <Select Objects> node of the **-HATCHEDIT** command, change it to pause for user input. This allows the user to pick the hatch to edit.
12. Insert a user message for the **-HATCHEDIT** command just above the <Select Objects> node. The message should read Select the brick hatch object. This gives instruction to the user.
13. For the point location under the **Set** option of the **-HATCHEDIT** command, change it to pause for user input. In this way, the user selects the origin point for where the hatch pattern begins.
14. Insert a user message for the point location under the **Set** option of the **-HATCHEDIT** command. The message should read Edit the hatch and select a new origin point when prompted. This gives instruction to the user.
15. Test the action. When prompted for the new origin, select the midpoint of the lower line of the rectangle. Notice the result is different from the pattern created when recording the action.

The action recorder works best with command-line versions of commands. Many commands can be run on the command line only by placing a hyphen (-) in front of the command name. For example, the **LAYER** command displays the **Layer Properties Manager** palette. However, the **-LAYER** command displays all options and settings on the command line without displaying a palette or dialog box. Not all commands have a command-line version.

Limitations of the Action Recorder

As a customizing tool, the action recorder is very powerful. However, as in all customization, there is syntax to follow and there are limitations to address. For example, the AutoCAD commands shown in **Figure 26-13** cannot be recorded in an action. Also, picking an interface element in many cases does not record the related item into the action. Zooming and panning from the navigation bar, for instance, will be recorded as a view change. In addition, picking status bar toggle items such as **Quick Properties**, **Object Snap Tracking**, and **Infer Constraints** does not record the function into the tree. However, if the **ZOOM** command is invoked from the command line, as seen in earlier practice examples, then the operation will be recorded.

The following chapters discuss AutoLISP and other programming features of AutoCAD. Actions are able to call and run AutoLISP routines during the recording process. However, the specific AutoLISP routines, Object ARX files, .NET assemblies, and associated files must all be loaded into AutoCAD in order for the programming feature to function during playback of the action. See Chapters 27 through 29 for further details on programming AutoCAD. These chapters are located on the companion website.

When creating an action for use in a professional or school environment, you must always consider the source files. For example, if you create an action that reads from a template file or loads particular blocks, those external files must always be either loaded into AutoCAD or accessible to the user.

Figure 26-13.
These commands cannot be used in an action.

ACTBASEPOINT
ACTMANAGER
ACTSTOP
ACTUSERINPUT
ACTUSERMESSAGE
-ACTUSERMESSAGE
DXFIN
EXPORTLAYOUT
FILEOPEN
NEW
OPEN
PARTIALOPEN
PRESSPULL
QNEW
RECOVER
TABLEDIT
XOPEN

Finally, you cannot append an existing action. If you need to add steps to the action, you must re-record the existing steps along with the new steps. For this reason, it is very important to plan an action before recording it.

PROFESSIONAL TIP

You can create actions with other Autodesk applications, such as AutoCAD Architecture. If you create an action using AEC objects, for example, the action can only properly function if the action is played back in AutoCAD Architecture.

Chapter Review

Answer the following questions. Write your answers on a separate sheet of paper or complete the electronic chapter review on the companion website.
www.g-wlearning.com/CAD

1. What is an *action*?
2. What is used to record an action?
3. In which file format is an action saved?
4. By default, where are action files stored?
5. What is the *action tree*?
6. How do you set an action current?
7. Name two ways to play an action.
8. How do you pause an action for user input?
9. How can you display a custom message in an action?
10. Briefly describe how to change the value associated with a branch in an action.

Drawing Problems

1. In this problem, you will create an action that sets up a drawing for architectural work. The action will create layers and set the units. The action should:
   A. Display a message to the user indicating what the action will do.
   B. Create a layer named A-Wall and set its color to ACI 113 (or the color of your choice).
   C. Create a layer named A-Door and set its color to ACI 31 (or the color of your choice).
   D. Create a layer named A-Demo and set its color to ACI 1 (or the color of your choice).
   E. Set the units to architectural with a precision of 1/4".
   F. Save the action as ArchSetup.actm. If the file is saved in a folder other than the default, make sure that folder is added to the AutoCAD search path.

2. In this problem, you will create an action that allows the user to select objects and then places those objects at an elevation of 12 units. The action should:
   A. Display a message to the user indicating what the action will do.
   B. Prompt the user to select objects.
   C. Change the elevation property of all objects to 12 units.
   D. Save the action as 1FtElevChange.actm. If the file is saved in a folder other than the default, make sure that folder is added to the AutoCAD search path.

3. In this problem, you will modify the action created in Problem 2. Edit the action to allow the user to enter the new elevation. Be sure to display a message indicating what is being asked of the user. This is a good tool for flattening linework or resetting solids. Save the action as ElevChange.actm. If the file is saved in a folder other than the default, make sure that folder is added to the AutoCAD search path.

Drawing Problems - Chapter 26

The following chapters are available on the companion website. Chapters 27, 28, and 29 discuss various aspects of programming AutoCAD, including working with AutoLISP and creating dialog boxes using DCL. Chapter 30 covers image management, working with different graphics files, and creating DWF and PDF files from AutoCAD model drawings.
www.g-wlearning.com/CAD

Programming AutoCAD

Image Management and Drawing Distribution

Index

The index entries shown in black refer to chapters located in this printed book. The index entries shown in color or page number listings in color refer to chapters that are located on the companion website.

 www.g-wlearning.com/CAD

The index entries shown in color refer to chapters that are located on the companion website.

The index entries shown in color refer to chapters that are located on the companion website.

The index entries shown in color refer to chapters that are located on the companion website.

Index–Advanced

The index entries shown in color refer to chapters that are located on the companion website.

Index–Advanced

imprints, 355, 358
    extruding, 359
inactive state, 96
index of refraction (IOR), 460
indirect illumination, 523
InfoCenter Settings dialog box, 571
information grouping, 371
Insert Base Point tool, 683
insertion scale, 575
Insert Message tool, 683
INSUNITS system variable, 575
integers, 697
integrated development
    environment (IDE), 703
interface elements, 560, 581
interface, overview, 21–24
INTERFERE command, 55–57
INTERSECT command, 55, 208
intersection, 54
intersection boundary, 399
INTERSECTIONCOLOR system
    variable, 429
INTERSECTIONDISPLAY system
    variable, 428
intersection fill, 399
ISOLATEOBJECTS command, 118
isolines, 52–53, 282, 378
ISOLINES system variable,
    377–378, 428, 433, 563
isometric viewpoint presets, 25–26

J

jogs, 396–397
justification, 385

K

key position, 132
KML, 488–489

L

label, 736
Lamp Color dialog box, 503
Lamp Intensity dialog box, 503
LAYER command, 688, 692
-LAYER command, 688, 692
leaders, in 3D, 372–373
lens length, 545
LIGHT command, 494, 497, 499–500
LIGHTGLYPHDISPLAY command,
    487
light glyphs, 487
lighting, 432–433
lighting units, 488
LIGHTINGUNITS system variable,
    488, 494, 496, 503, 506,
    521, 529

LIGHTLIST command, 501
lights, 439
    determining proper
        intensity, 504
    properties, 484–486
    types, 483–484
Lights in Model palette, 501
LINE command, 60
lineweight settings, 576–577
LISP, 695
list, 696
list function, 719, 722
list processing, 695
LIVESECTION command, 398
live sectioning, 125, 393, 397–400
load function, 709
Load/Unload Applications dialog
    box, 708
LOFT command, 198, 222–223,
    227–228, 276, 278, 287
LOFTNORMALS system variable,
    228
lofts, 217
    cross sections, 223–227
    guide curves, 227–228
    path, 228
loft surface, 278
LOGFILEOFF command, 567
LOGFILEON command, 567
log files, 567
Long String Editor dialog box, 594
Look tool, 111, 116
loop, 61
luminance, 461

M

macro, 592–594
Manual Performance Tuning
    dialog box, 128
mapped material, 464
maps, 464–475
    checker, 466
    gradient, 466, 468–469
    image, 464–466
    marble, 471–472
    noise, 472–473
    speckle, 473–474
    texture editor, 464–465
    tiles, 469, 471
    waves, 474–475
    wood, 475
marble map, 471–472
masonry, 452
MASSPROP command, 411
MATBROWSEROPEN command,
    446
MATEDITOROPEN command, 450
material, 445

MATERIALATTACH command, 448
material mapping, 476
material maps, adjusting, 476–477
material-level adjustments, 476
MATERIALMAP command, 476
materials
    applying and removing,
        448–449
    creating and modifying,
        451–463
    creating by using an existing
        material type, 451–454
    creating from scratch, 454–462
    display options, 450
    types, 452–454
Materials Browser, 446–448
Materials Editor, 450
materials library, 445
matte, 484
MAXACTVP system variable, 179
MEASUREGEOM command, 61–62
MENUBAR command, 606
menus, 606
    adding commands to, 609
    creating new, 607
mesh, 275
    closing gaps, 82–83
    collapsing face or edge, 83
    creating gaps, 82
    editing, 77–84
    merging faces, 84
    spinning faces, 85
MESHCAP command, 82–83
MESHCOLLAPSE command, 83
MESH command, 68–69
MESHCREASE command, 80
MESHEXTRUDE command, 79
mesh forms, 85–87
MESHMERGE command, 84
mesh modeling, 67–88
    applying crease, 80–81
    converting, 71–74
    drawing mesh forms, 85–87
    drawing mesh primitives,
        69–70
    editing meshes, 77–85
    removing crease, 82
    smoothing and refining, 75–76
    tessellation division values,
        68–69
mesh object, 32
MESHOPTIONS command, 71
MESHPRIMITIVEOPTIONS
    command, 70
mesh primitives, 69–70
MESHREFINE command, 76
MESHSMOOTH command, 68, 72
MESHSMOOTHLESS command,
    75–76

The index entries shown in color refer to chapters that are located on the companion website.

projection, 97–98
PROJECTGEOMETRY command, 293–294
PROJECTNAME system variable, 749
prompt function, 712, 714
PROPERTIES command, 237
Properties palette, 36, 237–238, 282, 313–314, 329, 367, 438
PROXYWEBSEARCH system variable, 570
PSOLHEIGHT system variable, 47
PSOLWIDTH system variable, 47
publish, 763
PUBLISH command, 759, 763
PYRAMID command, 49

Q

QSAVE command, 565
QTEXTMODE system variable, 563
Quick Access toolbar, 21, 671
QUICKPROPERTIES command, 629–630, 632
Quick Text mode, 563

R

radial dimensions, in 3D, 372–373
radian angle measurement, 722–724
rapid prototyping, 416
raster files, 745–746
    controlling image file displays, 750–752
    exporting drawings to, 754
    inserting, 747, 749
    managing attached images, 749–750
    uses in AutoCAD, 753
RASTERPREVIEW system variable, 565
raytrace shadow, 506
raytracing, 522
real numbers, 697–698
Record tool, 683
RECTANGLE command, 60
rectangular 3D coordinates, 18–19
REDO command, 576
REDRAW command, 32
reflectivity, 457, 484–485
REGENALL command, 191, 563
REGEN command, 30, 32, 191, 563, 630
REGION command, 34, 60, 207
region model, 60
regions, 59–60
    area calculation, 61–62
    converting to, 34

creating with **Boundary** command, 61
extruding, 199
relative coordinates, 144
RENDER command, 28, 31, 93, 438
RENDERENVIRONMENT command, 530
RENDEREXPOSURE command, 529
rendering, 31
    advanced render settings, 519–528
    introduction to, 438–441
    lights, 439
    scene, 440–441
RENDERPRESETS command, 528
Render window, 441, 517–519
REVOLVE command, 198, 276, 322, 332
    creating surfaces, 202–203
revolved-surface mesh, 86–87
REVSURF command, 68, 85–86
Rewind tool, 109, 112
ribbon
    adding drop-down lists, 600–601
    associating tool palette groups with tabs, 602
    button properties, 598–599
    customizing panels, 599–600
    overview, 595–602
    tabs and panels, 596, 598
RIBBONICONRESIZE system variable, 592
Right-Click Customization dialog box, 574, 615–616
right-hand rule, 19–20
roll arrows, 98
rollover tooltips, customizing, 627–629
ROTATE command, 246, 313, 315, 318–319, 323, 326, 331, 368
rotate gizmo, 246
roughness, 485
RTDISPLAY system variable, 563
RTPAN command, 106
RTZOOM command, 107
ruled-surface mesh, 86
RULESURF command, 68, 85–86

S

sampling, 521–522
Save As DWFx dialog box, 760–761
SAVE command, 565
Save Drawing As dialog box, 757
SAVEFIDELITY system variable, 565
SAVEIMG command, 754

SAVETIME system variable, 566–567
Save Workspace dialog box, 669–670
SCALE command, 313, 315, 323, 327, 631
section boundary state, 392
section object, 389–390
    editing, 394–397
    properties, 393–394
    states, 392–393
SECTIONPLANE command, 380, 389, 392, 398, 400
SECTIONPLANEJOG command, 396
section planes
    creating, 389–391
    editing and using, 392–403
section plane state, 392
SECTIONPLANETOBLOCK command, 401
section view, updating, 403
section volume state, 393
security settings, 568
Select Color dialog box, 455
selection cycling, 239–241, 313
self illumination, 461–462
self intersecting, 51
separators, 585–586
SEQUENCEPLAY command, 140
setq function, 700–701, 704, 712–713
setvar function, 718
shaded display, 28
shadow-mapped shadow, 505–506
shadows, 504–506, 522
shell, 356–358, 362
shortcut keys, 607
    customizing, 616–619
shortcut menus, 574, 619–623
    command-oriented, 625–626
    customizing, 623–626
    object-specific, 624–625
shot, 121
    displaying or replaying, 137
SHOWHIST system variable, 238–239
show motion, 121–123
ShowMotion toolbar, 122–123, 129, 138–140
significant digits, 699
silhouette, 378
SKETCH command, 609
sky illumination, 487
SLICE command, 261–265
slicing solids, 261–265
slideout, 586
SMOOTHMESHCONVERT system variable, 73

The index entries shown in color refer to chapters that are located on the companion website.

The index entries shown in color refer to chapters that are located on the companion website.

Index-Advanced

The index entries shown in color refer to chapters that are located on the companion website.